Library of Congress Cataloging-in-Publication Data

Peyghambarian, Nasser
Introduction to semiconductor optics / Nasser Peyghambarian, Stephan W. Koch, Andre Mysyrowicz.
p. cm.
Includes bibliographical references and index.
ISBN 0-13-638990-2
1. Semiconductors--Optical properties. I. Koch, S. W. (Stephan W.) II. Mysyrowicz, Andre. III. Title.
QC611.6.06P45 1993
621.3815'2--dc20 92-45179
CIP

Publisher: Alan Apt
Production Editor: Mona Pompili
Copy Editor: Shirley Michaels
Cover Designer: Bruce Kenselaar
Prepress Buyer: Linda Behrens
Manufacturing Buyer: Dave Dickey
Editorial Assistant: Shirley McGuire
Supplements Editor: Alice Dworkin

A Simon & Schuster Company
Englewood Cliffs, New Jersey 07632

Printed in the United States of America

10 9 8 7 6 5 4 3 2 1

ISBN 0-13-638990-2

Prentice-Hall International (UK) Limited, *London*
Prentice-Hall of Australia Pty. Limited, *Sydney*
Prentice-Hall Canada, Inc., *Toronto*
Prentice-Hall Hispanoamericana, S.A., *Mexico*
Prentice-Hall of India Private Limited, *New Delhi*
Prentice-Hall of Japan, Inc., *Tokyo*
Simon & Schuster Asia Pte. Ltd, *Singapore*
Editora Prentice-Hall do Brasil, Ltda., *Rio de Janeiro*

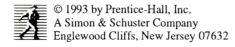

CONTENTS

Preface

This is an introductory-level book in the rapidly evolving field of solid-state optoelectronics. The purpose of this book is to provide a text for teaching graduate as well as undergraduate students the various aspects of optical and optoelectronic phenomena and devices associated with them. An effort is made to lay out the very basic principles and yet to describe the most practical applications. Thus, this book is intended to be useful not only for physicists, but also for electrical and computer engineering and material and optical sciences majors. Even though the major part of this book is devoted to optoelectronic properties of semiconductors, metals and insulators are also included.

Some of the chapters are introductory and discuss the basic phenomena of optics of solids. The more advanced chapters are focused on the most recent developments in the field and may prove to be useful for those entering this field. The entire book can be taught in two semesters. For a one-semester course, however, we suggest teaching parts of Chaps. I-VI, VIII, X, XIII, XIV, and XVI-XVIII. The more difficult sections and chapters are marked by an asterisk and may be omitted for teaching a one-semester course.

We are grateful to critical reading of the earlier versions of the book by G. Kojoian, G. Moddel, D. Fröhlich, J. P. Foing, J. Paye, M. Jackson, and students of the "Introduction to Solid State Optics" class at the Optical Sciences Center of the University of Arizona, to J. Sokoloff for writing a paragraph on the point group discussion of the Raman scattering section of Chap. IV, and to H. M. Gibbs and H. Haug for their support and inspiration.

Our special thanks goes to Linda Schadler for her patience in the preparation of this book and her ability to follow our numerous revisions and changes. Her help was invaluable to the completion of this project. We also thank J. Abbott for preparing the artwork. The camera-ready manuscript was produced using Scroll Systems PS™ Technical Word Processor.

We greatly appreciate financial research support from The National Science Foundation (Light Wave Technology and Quantum Electronics Waves and Beams Division), The Army Research Office, The Air Force Office of Scientific Research, Rome Laboratories, Strategic Defense Initiative, the Optical Circuitry Cooperative of the University of Arizona, NATO, and NEDO (Japan). Support from the OCC member companies is also appreciated. Without the support of these agencies, we would not have been able to complete this work.

Chapter I
INTRODUCTION

Stimulated by the invention of the transistor in the early 1950s, the electronic properties of semiconductors have become research and engineering topics of high interest. The results of these investigations have led to an electronic revolution with numerous applications of all solid electronic devices, replacing the old technology based on vacuum tubes. An extraordinarily large number of scientific papers, review articles, research books, and textbooks have been published which deal with the semiconductor electronic properties, all the way from basic physics to device application topics.

In comparison, the optical properties of semiconductors have not been studied with the same intensity, primarily because they have not yet led to device applications comparable to their electronic counterparts. The main exception to this statement is clearly the semiconductor laser, which is currently mass-produced and is finding increasing applications in industrial and household appliances.

The semiconductor laser is a good example that shows how the optical and electronic properties of semiconductors are in fact intimately related. It was realized in the early sixties that a semiconductor laser is possible when it was found that stimulated emission of radiation could occur in semiconductors as a consequence of the recombination of carriers, which were injected across a p-n junction. If one wants to understand the light-emission process, it is essential to first understand the electronic properties of the mobile carriers, i.e., the electrons in the semiconductor. However, it is clear that the optical response of the semiconductor must also be fully understood. Other examples illustrating the intimate connection between electronic and optical properties of semiconductors are all-solid light detectors and electro-optical and all-optical switching devices.

To provide the basis for the analysis of optical semiconductor devices, it is necessary to study the basic optical properties of the material itself. When we look at semiconductor materials we notice that they usually have a color; e.g., bright yellow for CdS, orange for ZnSe, red for Cu_2O, or metallic black for GaAs. In this book we will relate this appearance to the electronic states of the material. More specifically, we will show that light absorption and emission characteristics are determined in a very fundamental way by the material band structure, and specifically by the bandgap; i.e., the energy difference between the *valence band*, the energetically highest band which is fully occupied by electrons, and the *conduction band*, the first partially or completely empty band above the valence band.

To be more specific, let us consider an experiment where we use a light source (typically a laser beam) which shines on a semiconductor material. The schematics of such an experiment are shown in Fig. 1.1(a). We measure the transmission of the laser beam as a function of laser frequency and may obtain a result similar to the one shown in Fig. 1.1(b).

The light transmission which can be detected (e.g., by a photomultiplier tube) is related to the absorption coefficient of the semiconductor. The absorption in Fig. 1.1(b) is typical of a high-quality semiconductor material such as GaAs. Its spectrum consists of a sharp line on the low-energy side and a more-or-less structureless absorption on the high-energy side.

This book explains that the basic absorption process in a semiconductor is related to the destruction of a light quantum (a photon) and the simultaneous promotion of an electron from the originally filled valence band into the originally empty conduction band. Hence, the photon, by removing an electron from the valence band, creates a hole in the valence band and creates an electron in the previously empty conduction band. We denote this process as photon absorption through electron-hole-pair

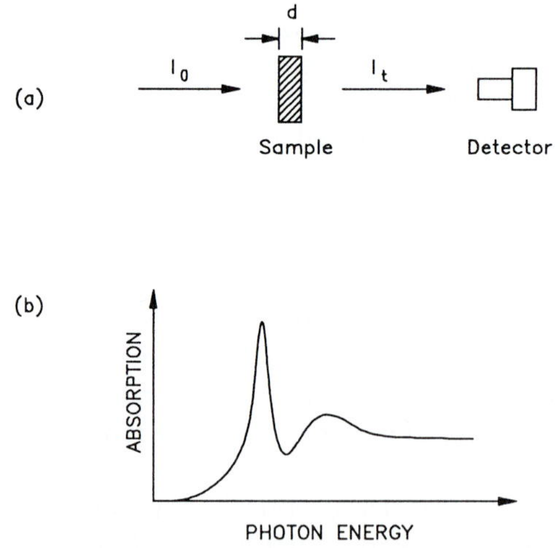

Figure 1.1. (a) Schematics of an experiment in which transmission of a laser beam through a semiconductor is detected. I_0 is the input light intensity and I_t is the intensity of the transmitted light. (b) A typical absorption spectrum of the semiconductor as a function of photon energy.

creation. Since the electrons carry a charge, Coulomb interaction processes are very important in the understanding of this process. For example, the sharp line in the absorption spectrum in Fig. 1.1(b) is a direct manifestation of the Coulomb interaction, since this line originates from the Coulomb attraction between the electron and the hole, which leads to the formation of a bound electron-hole-pair state, the *exciton*. These excitons, which are in many respects similar to hydrogen atoms, play an important role in semiconductor optics, and we will discuss their properties in some detail in the linear regime, when the light intensity is low and the optical properties of the semiconductor are independent of the light intensity.

The optical properties of semiconductors in the presence of an external force will also be discussed. For example, when a dc electric field is applied across the material, the band edge absorption changes significantly. The exciton absorption resonance shifts and broadens in response to a dc field, as schematically shown in Fig. 1.2.

Semiconductors also exhibit nonlinear behavior, i.e., the absorption coefficient and refractive index are not constants of the medium but change as a function of the intensity of the exciting light. To show such nonlinear characteristics, let us assume that we do the same experiment as in Fig. 1.1, but now an additional, second intense laser beam is applied, as shown schematically in Fig. 1.3(a). The resulting absorption spectrum of the first beam without the second beam is shown by the solid curve in Fig.

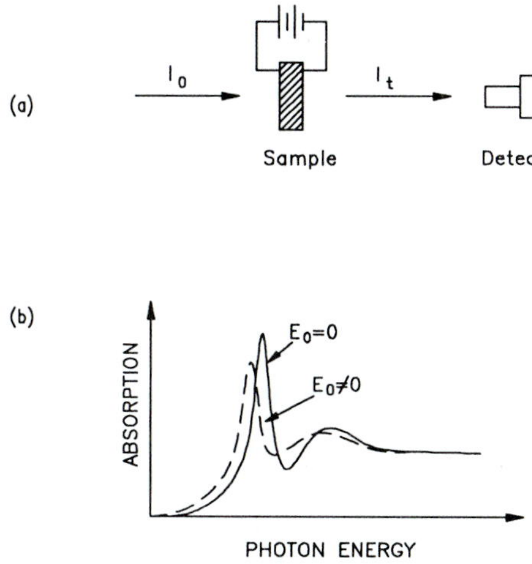

Figure 1.2. (a) Experimental setup to measure the electroabsorption of a semiconductor. A dc electric field is applied across the sample. I_0 and I_t are the incident and transmitted laser intensities, respectively. (b) Typical absorption spectra of a semiconductor with no field applied (solid curve) and in the presence of the field (dashed curve).

1.3(b). With the second intense beam present, the absorption spectrum changes as displayed by the dashed curve in Fig. 1.3(b). As we can see from this figure, strong laser excitation leads to significant changes of the linear absorption spectrum. In the presence of the strong beam, the exciton resonances disappear, the absorption gradually decreases, and a region of negative absorption eventually appears. Negative absorption means light amplification rather than absorption, and the spectral region of negative absorption is called the region of *optical gain*. This gain is responsible for the operation of the semiconductor laser mentioned at the beginning of this chapter.

Generally, we denote the changes of the semiconductor absorption spectrum, which occur as a consequence of the optical excitation with a strong beam, as the *absorptive optical nonlinearities*. These absorptive optical nonlinearities are meant to be distinguished from *dispersive optical nonlinearities*, which are based on the excitation-dependent changes of the index of refraction of the material. We will discuss in this book the fundamental relations which exist between semiconductor absorption and dispersion, and we will show how the absorptive and dispersive nonlinearities are related.

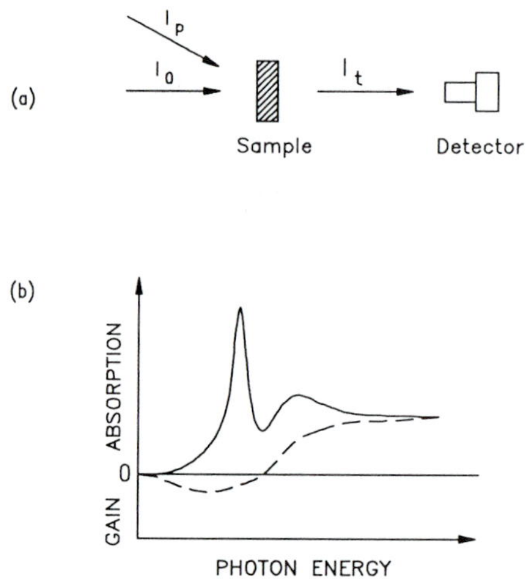

Figure 1.3. (a) Experimental setup to measure the transmission of a laser beam with low intensity, I_0, in the presence of a strong beam with intensity I_p. (b) A typical absorption spectrum of the semiconductor in the absence of the strong beam (solid curve) and in the presence of the strong beam (dashed curve). We see absorption decrease (bleaching) and the development of optical gain (negative absorption) when the pump beam is present.

We mentioned before that optical absorption is equivalent to electron-hole-pair creation. Hence, the absorption of intense light leads to the creation of many electron-hole pairs, and it is the interaction among these generated carriers which leads to the nonlinearities shown in Fig. 1.3(b). The electrons and holes in a semiconductor obey the laws of quantum mechanics, and we have to use this formalism to analyze their properties. We will argue that the various interactions in the quantum mechanical system of many electron-hole pairs are responsible for the significant modifications of the optical material properties, such as bleaching of the exciton resonance through screening of the Coulomb interaction, density-dependent changes of the semiconductor bandgap (bandgap renormalization), as well as the bleaching of the absorption through state filling (blocking of the states available for absorption).

The physical processes responsible for the optical nonlinearities are very interesting in their own right, since they usually involve fundamental interaction mechanisms. For example, the screening of the Coulomb interaction through the charge carriers is also studied widely in metals, plasma physics, and many other fields dealing with Coulomb effects. The phenomenon of state filling is a manifestation of the quantum mechanical nature of electrons and holes. As *Fermi particles* they can occupy each quantum state only once; hence, if an absorbing state is already filled, it is not available for further absorption and we say that this absorption channel is "blocked," or the absorption is "bleached."

To understand the basic physics behind these processes, we devote several chapters of this book to detailed discussions of the concepts of optical response, excitons in semiconductors, screening and plasmons, and linear, nonlinear, and electro-optical properties of semiconductors. We describe a variety of real semiconductor materials and analyze the dominant features of their linear and nonlinear optical properties as well as their device application potential. Furthermore, to gain a broader overview we discuss basic features of the optical properties of simple metals and insulators and optical properties associated with lattice vibrations (phonons).

In addition to the steady-state optical properties, it is very interesting to investigate the semiconductor response under short-pulse excitation. Significant progress has been made in the last decade in the production of optical pulses as short as a few femtoseconds (1 fs = 10^{-15} s). Some of the physics behind these femtosecond lasers will be discussed in this book. The availability of these short pulses makes it possible to study semiconductor optical properties, not just in equilibrium, but also under strong nonequilibrium conditions. Furthermore, it opens the possibility to develop compact all-solid devices with ultrafast optical switching.

If we shine light on a semiconductor with a short-pulse laser emitting photons having energy substantially larger than the bandgap energy, the light generates electron-hole pairs around the excitation energy. This process is shown schematically in Fig. 1.4(a). Following the excitation process, the carriers gradually relax and assume equilibrium distributions within their band. This evolution is schematically depicted in Fig. 1.4(b). Generally, after the excitation with femtosecond pulses, it is possible to classify the dynamics of optically generated electron-hole-pair dynamics into three dynamic regimes. First, at very early times, the light field introduces a coherent coupling between valence- and conduction-band states, thus driving an oscillating polarization into the matter. This is called the *collision-free* or *coherent* regime. The semiconductor electrons undergo rapid transition oscillations (Rabi flopping) between the coupled band states, not unlike the coherent electronic excitations in an atomic two-level atom. In the second regime, phase-relaxation processes such as carrier-carrier and carrier-phonon scattering become dominant, causing "real absorption," i.e., the electrons and holes, which are in nonthermal distributions in their respective bands, are no longer coherently coupled to

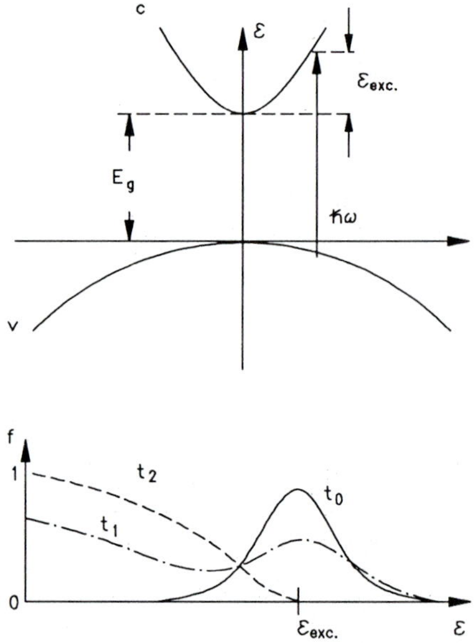

Figure 1.4. Schematics of carrier excitation high into the band, carrier relaxation, and quasi-equilibrium distributions. (a) Excitation of a semiconductor with an energy $\hbar\omega > E_g$ leads to the generation of nonequilibrium conduction-band carrier distributions, $f_c(\mathcal{E})$, in the valence (v) and conduction (c) bands. (b) The initial nonequilibrium carrier distribution $f(\mathcal{E})$, $t_0 = 0$ (solid curve), relaxes due to carrier-carrier and carrier-phonon collisions in time t_1 (dashed curve) and eventually, at time t_2, becomes a thermal Fermi-Dirac distribution (dashed-dotted curve).

the light field. Further collisions and scattering processes lead to the evolution of the nonthermal distributions toward a quasi-thermal equilibrium state. This state is said to be in a quasi equilibrium, because true equilibrium implies that the electrons fall back into the valence band. In this third *quasi-thermal* or *hydrodynamic* regime, the origin of the optical nonlinearities are the above-mentioned many-body effects of Coulomb screening, state filling, etc.

Some of the very exciting application potentials of semiconductors for optical and electronics purposes result from the fact that the growth of these materials can be manipulated to produce new material combinations of extremely small dimensions with predictable characteristics. The recent advances in crystal-growth techniques have made it possible to realize *semiconductor nanostructures*, which are so small that their electronic and optical properties deviate substantially from those of bulk materials. In these nanostructures the quantum-mechanical wave functions of electrons and holes are confined inside the material, giving rise to the so-called *quantum-confinement effects*.

The most common of these semiconductor structures are *quantum wells*, where the electrons are confined in one space dimension. Semiconductor quantum wells using, for example, GaAs-based technology, can be realized by various techniques such as *molecular beam epitaxy* to deposit several GaAs layers between layers of a material with a wider bandgap, such as $Ga_xAl_{1-x}As$, with an aluminum concentration typically in the range of $0 < x < 0.4$. If the arrangement of GaAs and such a $GaAs/Ga_xAl_{1-x}As$ is repeated several times, we speak of a *multiple quantum well* (MQW). Such a single quantum well is shown schematically in Fig. 1.5.

Besides the carrier confinement along one space dimension as in quantum-well structures, even stronger confinement is possible. If we introduce two-dimensional confinement, the structure is called a *quantum wire*, and if the confinement exists in all three space dimensions, we speak of a *quantum dot*. These structures are shown schematically in Fig. 1.6 and compared to bulk and quantum-well structures. The basic physical and optical properties of these semiconductor structures will be discussed in two chapters of this book. We analyze the basis of quantum-confinement effects and discuss the confinement-induced modifications of the optical properties.

The final three chapters of this book are entirely devoted to devices based on the linear, nonlinear, and electro-optical properties of semiconductors. We discuss the phenomena of optical bistability, optical switching, the operational principle of optical logic gates, and waveguide devices. In the final two chapters we discuss the fundamentals of the semiconductor laser and other optical devices such as detectors and modulators.

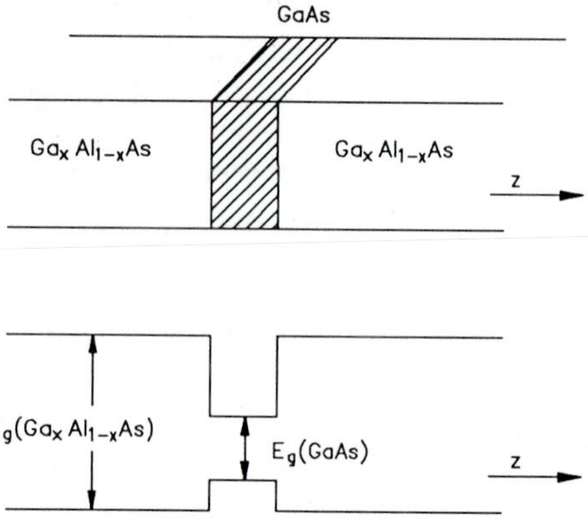

Figure 1.5. Schematic drawing of a quantum well showing well and barrier materials as well as the variation of the bandgap energy.

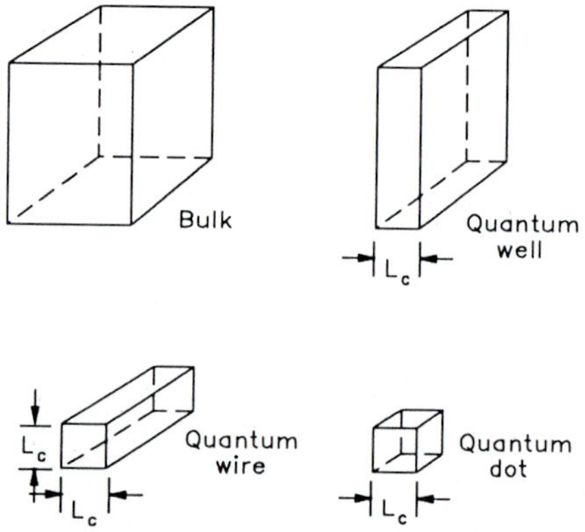

Figure 1.6. Schematic drawing of bulk material, a quantum well, a quantum wire, and a quantum dot. The confinement length L_c is larger than the atomic lattice constant, but sufficiently small that quantum-confinement effects are important.

Generally, the material in this book is presented at a level which should make it understandable to senior undergraduate or junior graduate students in physics, applied physics, optical sciences, or electrical engineering, with an interest in the optics of solids. The considerable interest in the area of semiconductor optics during the last decade has been stimulated by the numerous applications in optoelectronic devices and systems. In particular, the nonlinear optical properties of semiconductor materials and devices have attracted much attention, and this area is currently the topic of many research programs.

Chapter II
BASIC CONCEPTS IN CRYSTALS

Solid materials typically consist of about 10^{23} atoms/cm^3. They can quite generally be divided into the two categories of crystalline and amorphous structures. Crystalline structures are periodic arrangements of atoms or molecules [Fig. 2.1(a)], whereas amorphous solids, such as glasses, have practically no periodicity [Fig. 2.1(b)]. Since this book deals mainly with crystalline materials, we use the following sections to review some of the basic concepts that are helpful to characterize the optical properties of such systems.

(a) (b)

Figure 2.1. A two-dimensional schematic representation of the arrangement of atoms in (a) a crystalline solid and (b) an amorphous solid. The open circles and dots represent two different kinds of lattice ions.

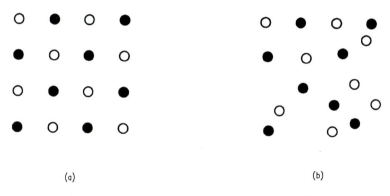

2-1. Direct and Reciprocal Lattices

The direct or real lattice of a crystal represents the periodic arrangement of the constituent atoms (or molecules). Associated with each lattice point is a *basis* of atoms. The number of atoms in the basis is at least one, as in simple metals, but can reach several thousand in certain organic crystals. The periodic repetition of the basis generates the crystal. In Fig. 2.2 we show an example where six atoms for each lattice point constitute the basis. The lattice points of a crystal can be described by a primitive lattice vector,

$$\mathbf{R} = n_1\mathbf{a} + n_2\mathbf{b} + n_3\mathbf{c} \;, \tag{2.1}$$

where n_1, n_2, and n_3 assume all integer values, 0, ±1, ±2, ... The vectors **a**, **b**, and **c** are the three fundamental translation (or primitive) vectors. The smallest region in the crystal that can be translated by the primitive lattice vector to reproduce a similar region of the crystal is called the *primitive unit cell.* In a given crystal there are many possible primitive unit cells. An example of such a unit cell is the area ABCD in Fig. 2.2. Successive translations of the primitive cell fill the entire crystal volume

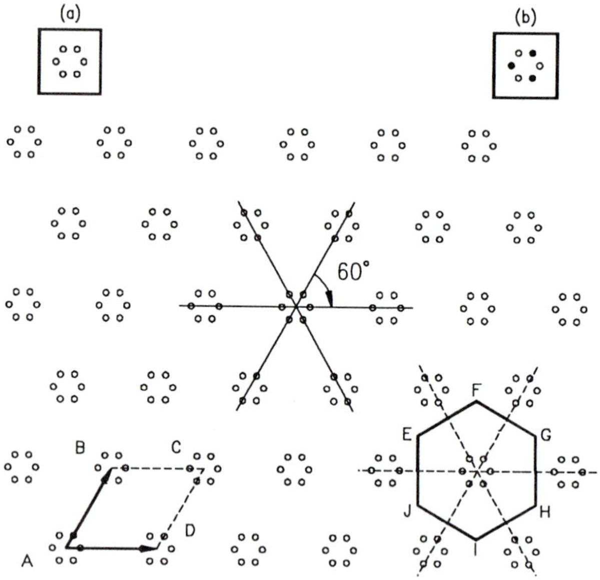

Figure 2.2. A two-dimensional lattice with six atoms in the basis. The area ABCD is a primitive unit cell, and the area EFGHIJ is a Wigner-Seitz primitive cell. The lattice has an inversion symmetry when the basis consists of identical atoms as in inset (a). There is no inversion symmetry when the atoms of the basis are not identical, as in inset (b).

without voids or overlap. The volume of the primitive cell is $V = |\mathbf{a}\cdot(\mathbf{b}\times\mathbf{c})|$, as shown in Fig. 2.3. Note that the vectors **a**, **b**, and **c**, which span the unit cells, are generally not parallel to the rectangular cartesian unit vectors **x**, **y**, and **z**. Often a special type of unit cell called the Wigner-Seitz cell is used, which can be constructed in the following way: Draw lines connecting a given lattice point to all nearby equivalent lattice points. Intersect the midpoint of each connecting line with a perpendicular line (or plane). The perpendicular bisectors, when connected to each other, enclose an area in two dimensions (or a volume in three dimensions). The smallest area (or volume) constructed in the direct lattice, with this procedure, is the Wigner-Seitz primitive cell. An example of a two-dimensional Wigner-Seitz primitive cell is the area EFGHIJ in Fig. 2.2.

It is also useful to introduce the *reciprocal lattice* in addition to the real lattice. The reciprocal lattice, which describes the crystal in the wave vector (or momentum) space, is related to the real lattice via Fourier transformation. The importance of the reciprocal lattice stems from the fact that it is often convenient to describe electronic interaction processes in reciprocal or momentum space. Consider two points **r** and **r′** in the direct lattice which are related by $\mathbf{r}' = \mathbf{r} + \mathbf{R}$, where **R** is given by Eq. (2.1). The environment around the point **r′** is completely equivalent to that associated with the lattice point **r**. Thus, if an arbitrary function $\rho(\mathbf{r})$ in the crystal [$\rho(\mathbf{r})$ may be, for example, the charge distribution in the crystal] has the translational periodicity of the direct lattice, then

$$\rho(\mathbf{r}) = \rho(\mathbf{r} + \mathbf{R}) \; . \tag{2.2}$$

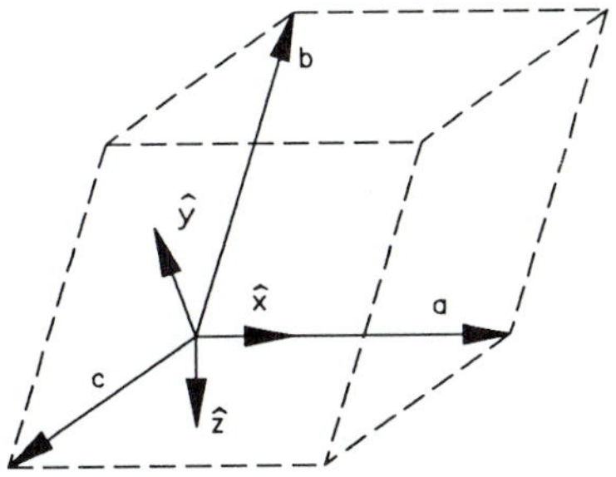

Figure 2.3. A cell of minimum volume, or the primitive cell.

This function may be expanded in terms of a Fourier series,

$$\rho(\mathbf{r}) = \sum_{\mathbf{G}} S_{\mathbf{G}}\, e^{i\mathbf{G}\cdot\mathbf{r}} \,. \tag{2.3}$$

The expansion coefficients $S_{\mathbf{G}}$ are given by

$$S_{\mathbf{G}} = \frac{1}{\Omega_c} \int_{\Omega_c} d\mathbf{r}\, \rho(\mathbf{r})\, e^{-i\mathbf{G}\cdot\mathbf{r}} \,, \tag{2.4}$$

where Ω_c is the volume of the unit cell in the direct lattice. The vector $\mathbf{G}$ has the dimensions of reciprocal length. Substituting Eq. (2.4) into Eq. (2.2) gives

$$\rho(\mathbf{r} + \mathbf{R}) = \sum_{\mathbf{G}} S_{\mathbf{G}}\, e^{i\mathbf{G}\cdot(\mathbf{r} + \mathbf{R})} = \rho(\mathbf{r}) = \sum_{\mathbf{G}} S_{\mathbf{G}}\, e^{i\mathbf{G}\cdot\mathbf{r}} \,, \tag{2.5}$$

which can be satisfied only if

$$e^{i\mathbf{G}\cdot\mathbf{R}} = 1 \,, \tag{2.6}$$

which yields

$$\mathbf{G}\cdot\mathbf{R} = n\, 2\pi \,,\ n = 0, \pm 1, \pm 2, \ldots \,. \tag{2.7}$$

Writing the vector $\mathbf{G}$ as

$$\mathbf{G} = h_1\mathbf{A} + h_2\mathbf{B} + h_3\mathbf{C}, \quad \text{where } h_1, h_2, h_3 = 0, \pm 1, \pm 2, \ldots \tag{2.8}$$

and with the help of Eq. (2.1), we find that Eq. (2.7) can be satisfied if

$$\mathbf{A}\cdot\mathbf{a} = \mathbf{B}\cdot\mathbf{b} = \mathbf{C}\cdot\mathbf{c} = 2\pi \tag{2.9a}$$

and

$$\mathbf{A}\cdot\mathbf{b} = \mathbf{A}\cdot\mathbf{c} = \mathbf{B}\cdot\mathbf{a} = \mathbf{B}\cdot\mathbf{c} = \mathbf{C}\cdot\mathbf{a} = \mathbf{C}\cdot\mathbf{b} = 0 \,. \tag{2.9b}$$

Consequently, **A** is perpendicular to both **b** and **c** and, thus, is proportional to $\mathbf{b} \times \mathbf{c}$. To determine the proportionality constant, we use Eq. (2.9a) and find

$$\mathbf{A} = 2\pi \frac{\mathbf{b} \times \mathbf{c}}{\mathbf{a} \cdot (\mathbf{b} \times \mathbf{c})} . \tag{2.10a}$$

Correspondingly, one can verify that

$$\mathbf{B} = 2\pi \frac{\mathbf{c} \times \mathbf{a}}{\mathbf{b} \cdot (\mathbf{c} \times \mathbf{a})} = 2\pi \frac{\mathbf{c} \times \mathbf{a}}{\mathbf{a} \cdot (\mathbf{b} \times \mathbf{c})} \tag{2.10b}$$

and

$$\mathbf{C} = 2\pi \frac{\mathbf{a} \times \mathbf{b}}{\mathbf{c} \cdot (\mathbf{a} \times \mathbf{b})} = 2\pi \frac{\mathbf{a} \times \mathbf{b}}{\mathbf{a} \cdot (\mathbf{b} \times \mathbf{c})} . \tag{2.10c}$$

The vector **G**, with components given by Eqs. (2.10a-c), is referred to as the reciprocal lattice vector. The equivalent of the Wigner-Seitz unit cell in the direct lattice is called the first *Brillouin zone* in the reciprocal lattice. As an example, consider the first Brillouin zone of the simple cubic lattice with the unit cell shown in Fig. 2.4. The lattice vectors in this case can be written as

$$\mathbf{a} = a\mathbf{x} \quad , \quad \mathbf{b} = a\mathbf{y} \quad , \quad \mathbf{c} = a\mathbf{z} \; , \tag{2.11}$$

and we use Eqs. (2.10a-c) to calculate the reciprocal lattice vectors and the boundaries of the first Brillouin zone. Inserting Eq. (2.11) into Eq. (2.10) yields the reciprocal lattice vectors

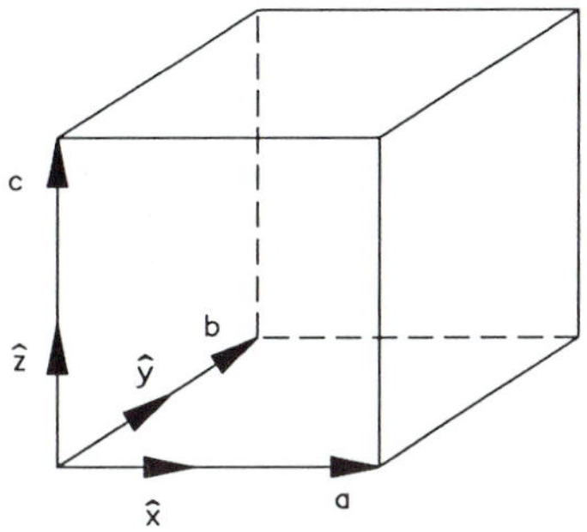

Figure 2.4. The primitive cell of a simple cubic lattice.

$$\mathbf{A} = 2\pi \frac{\mathbf{a}^2\mathbf{x}}{\mathbf{a}\mathbf{x}\cdot\mathbf{a}^2\mathbf{x}} = \frac{2\pi}{a}\mathbf{x} \quad \mathbf{B} = \frac{2\pi}{a}\mathbf{y}\ , \quad \mathbf{C} = \frac{2\pi}{a}\mathbf{z}\ . \qquad (2.12)$$

The boundaries of the first Brillouin zone of this cubic lattice are the planes normal to the six reciprocal lattice vectors, ±**A**, ±**B**, and ±**C**, intersecting the lattice vectors at their midpoints,

$$\pm\frac{1}{2}\mathbf{A} = \pm\frac{\pi}{a}\mathbf{x} \qquad (2.13a)$$

$$\pm\frac{1}{2}\mathbf{B} = \pm\frac{\pi}{a}\mathbf{y} \qquad (2.13b)$$

$$\pm\frac{1}{2}\mathbf{C} = \pm\frac{\pi}{a}\mathbf{z}\ . \qquad (2.13c)$$

These six planes form a cube of edge length $2\pi/a$ and volume $(2\pi/a)^3$. Electrons in crystals are characterized by the wave vector **k** of their quantum-mechanical wave function. The crystal electron momentum given by $\hbar\mathbf{k}$ represents the momentum of the electron inside the crystal. It is important to note that the crystal electron momentum is generally different from the momentum of a free electron. In a cubic lattice the electron momentum can take values in the range $-\pi/a \leq k \leq \pi/a$ for each of the components, k_x, k_y, or k_z. We note that an electron with zero momentum, $k = 0$, is represented by a point located at the center of the Brillouin zone. A point at the zone edge corresponds to a standing electron wave with maximum **k** value, $k = \pi/a$.

2-2. Crystal Planes and Directions

The orientation of a plane in a crystal may be specified by a set of three numbers called *Miller indices*. The Miller indices may be determined as follows:

1. Find the intercepts of the plane with the axes **a**, **b**, and **c** and express them in terms of lattice constants.

2. Take the reciprocals of these three numbers. In general, the reciprocals will be fractions. Rewrite these fractions such that they have the smallest possible common denominator. The numerators h, k, and ℓ of these fractions are the Miller indices. For a plane parallel to one of the axes, the intercept occurs at infinity, and the corresponding Miller index is zero.

As an example, consider the plane shown in Fig. 2.5. The intercepts of the plane occur at 3, 4, 2. After taking reciprocals this becomes 1/3, 1/4, 1/2, which is equivalent to 4/12, 3/12, 6/12. Thus, the Miller indices are 4, 3, and 6, and the corresponding plane is labeled the (436) plane. Some examples of planes in a cubic lattice are shown in Fig. 2.6. Lattice planes are denoted as "equivalent planes" if they have the same density of atoms and interplanar spacing. Examples of equivalent planes in a cubic lattice are $(h\ k\ \ell)$, $(\bar{h}\ k\ \ell)$, $(h\ \bar{k}\ \ell)$, $(h\ k\ \bar{\ell})$, $(\bar{h}\ \bar{k}\ \ell)$, $(k\ h\ \ell)$, $(k\ \ell\ h)$, etc. The "bar" above the Miller indices refers to a negative number.

Directions in a crystal are often specified by triplets of indices such that the direction of a vector $h\mathbf{a} + k\mathbf{b} + \ell\mathbf{c}$ is denoted as $[h\ k\ \ell]$. For example, the x axis is the [100] direction and the z axis is the [001] direction. In cubic crystals, the direction $[h\ k\ \ell]$ is perpendicular to the plane $(h\ k\ \ell)$.

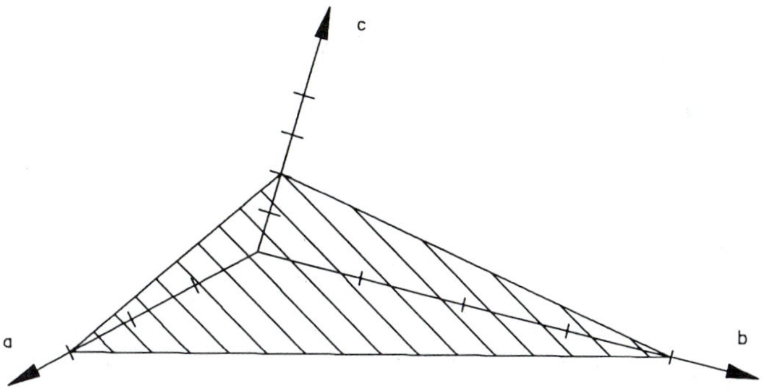

Figure 2.5. The (436) lattice plane in a crystal lattice.

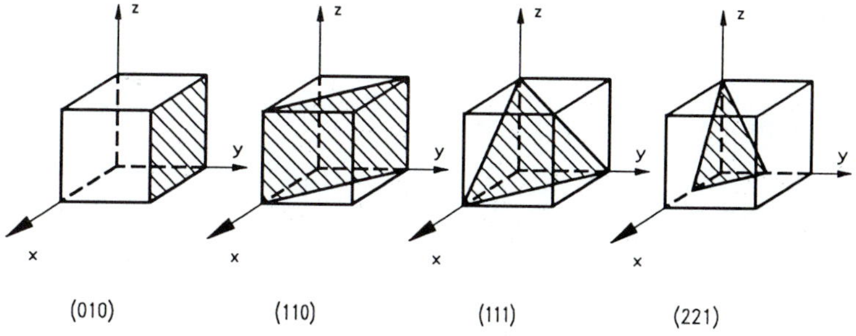

Figure 2.6. Indices of some important planes in a cubic crystal.

2-3. Crystal Symmetries

A crystal is characterized by a high degree of symmetry. These symmetry properties have a direct bearing on whether a particular optical transition is allowed. The first obvious symmetry of a crystal is translational symmetry. As we will discuss in more detail in the next section, the lattice periodicity leads to an electron wave function that has a plane wave factor given by $\exp(i\mathbf{k}\cdot\mathbf{r})$, where $\mathbf{k}$ is the wave vector that determines the momentum of the electron in the crystal. Optical transitions in crystals must obey conservation of momentum. This brings severe restrictions to the number of transitions that are allowed.

Space-inversion symmetry is another crystal symmetry. It is defined with respect to a particular point, for example, a lattice point which is then taken as the origin. For a system in which the potential $W(\mathbf{r})$ does not change when $\mathbf{r}$ is replaced by $-\mathbf{r}$, such that $W(\mathbf{r}) = W(-\mathbf{r})$, then the electron wave function $\psi(\mathbf{r})$ has the property,

$$\psi(\mathbf{r}) = \psi(-\mathbf{r}) \tag{2.14}$$

or

$$\psi(\mathbf{r}) = -\psi(-\mathbf{r}) \; . \tag{2.15}$$

These eigenfunctions in Eqs. (2.14) and (2.15) are said to have even or odd parity, respectively. Figure 2.7 displays examples of an even and odd wave function. Replacing $\mathbf{r}$ by $-\mathbf{r}$ does not alter the kinetic energy, $\mathscr{E}_{kin}$, of a physical system. Therefore, the Hamiltonian, $H = H_{kin} + W(\mathbf{r})$, of the system remains unchanged. To see the above arguments, consider a nondegenerate eigenfunction $\psi_\alpha(\mathbf{r})$ associated with a Hamiltonian such that

$$H\psi_\alpha(\mathbf{r}) = \mathscr{E}_\alpha \psi_\alpha(\mathbf{r}) \; . \tag{2.16}$$

Replacing $\mathbf{r}$ by $-\mathbf{r}$ in Eq. (2.16) yields

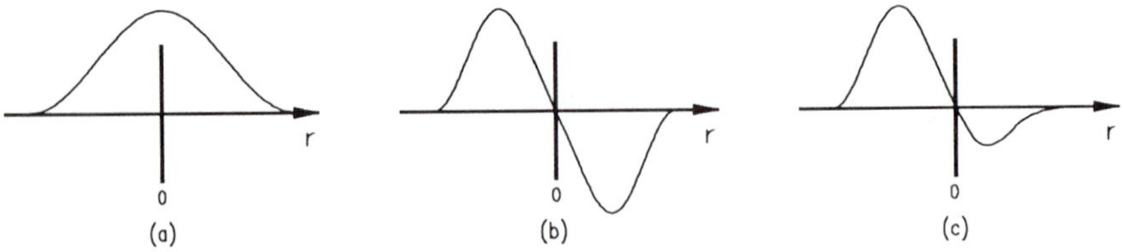

Figure 2.7. Examples of (a) an even wave function, (b) an odd wave function, and (c) a wave function with no well-defined parity.

$$H\psi_\alpha(-\mathbf{r}) = \mathcal{E}_\alpha \psi_\alpha(-\mathbf{r}) \,. \tag{2.17}$$

Note that in arriving at Eq. (2.17) we used $H(\mathbf{r}) = H(-\mathbf{r})$, since $H_{kin}(\mathbf{r}) = H_{kin}(-\mathbf{r})$, and by our assumption that $W(\mathbf{r}) = W(-\mathbf{r})$. Both $\psi_\alpha(\mathbf{r})$ and $\psi_\alpha(-\mathbf{r})$ are eigenfunctions corresponding to the same energy eigenvalue, $\mathcal{E}_\alpha$. However, for a nondegenerate system there can be only one eigenfunction for a given eigenvalue. Therefore,

$$\psi_\alpha(\mathbf{r}) = C\psi_\alpha(-\mathbf{r}) \,, \tag{2.18}$$

where C is some constant to be evaluated. Again, replacing $\mathbf{r}$ by $-\mathbf{r}$ in Eq. (2.18) we obtain

$$\psi_\alpha(-\mathbf{r}) = C\psi_\alpha(\mathbf{r}) = C[C\psi_\alpha(-\mathbf{r})] = C^2\psi_\alpha(-\mathbf{r}) \,. \tag{2.19}$$

Thus, $C^2 = 1$ and

$$C = \pm 1 \,. \tag{2.20}$$

Consequently, whenever $W(\mathbf{r}) = W(-\mathbf{r})$ the nondegenerate eigenfunctions of the Hamiltonian have the following property, $\psi_\alpha(\mathbf{r}) = \psi_\alpha(-\mathbf{r})$, and $\psi_\alpha(\mathbf{r}) = -\psi_\alpha(-\mathbf{r})$, corresponding to even and odd parity eigenfunctions. Parity conservation is obeyed in optical transitions.

Let us now discuss a general perturbing potential, $V(\mathbf{r})$, which may be due to the presence of applied field. The matrix element related to the perturbing potential term $V(\mathbf{r})$ is

$$V_{12} = \langle\psi_2|V(\mathbf{r})|\psi_1\rangle \,, \tag{2.21}$$

where ψ_1 is the initial state and ψ_2 is the final state. Equation (2.21) may be also written in integral form as

$$V_{12} = \int d\mathbf{r}\; \psi_2^*(\mathbf{r})\; V(\mathbf{r})\; \psi_1(\mathbf{r}) \,. \tag{2.22}$$

The integration is over the entire space; therefore, the integration results do not change when $\mathbf{r}$ is replaced by $-\mathbf{r}$; i.e.,

$$V_{12} = \int d\mathbf{r}\; \psi_2^*(-\mathbf{r})\; V(-\mathbf{r})\; \psi_1(-\mathbf{r}) = \int d\mathbf{r}\; \psi_2^*(\mathbf{r})\; V(\mathbf{r})\; \psi_1(\mathbf{r}) \,. \tag{2.23}$$

Assume that the perturbing term $V(\mathbf{r})$ is an even function [i.e., $V(\mathbf{r}) = V(-\mathbf{r})$] and that $\psi_1(\mathbf{r})$ and $\psi_2(\mathbf{r})$ have definite parity. Equation (2.23) becomes an identity if $\psi_1(\mathbf{r})$ and $\psi_2(\mathbf{r})$ have the same parity, either both odd or both even. If $\psi_1(\mathbf{r})$ and $\psi_2(\mathbf{r})$ have opposite parity, then Eq. (2.23) becomes

$$V_{12} = \int d\mathbf{r}\, \psi_2^*(\mathbf{r})\, V(\mathbf{r})\, \psi_1(\mathbf{r})$$

$$= \int d\mathbf{r}\, \psi_2^*(-\mathbf{r})\, V(-\mathbf{r})\, \psi_1(-\mathbf{r})$$

$$= -\int d\mathbf{r}\, \psi_2^*(\mathbf{r})\, V(\mathbf{r})\, \psi_1(\mathbf{r}) = -V_{12}\,, \tag{2.24}$$

which can be correct only if $V_{12} \equiv 0$. Therefore, when $V(\mathbf{r})$ is an even function, a transition can occur only between wave functions of equal parity; when $V(\mathbf{r})$ is an odd function, a transition can result only between states of different parity. Strong optical absorption is associated with a perturbation which is proportional to $e\mathbf{r}$, where e is the electric charge and $\mathbf{r}$ is the electron displacement; hence, it has an odd symmetry. Therefore, strong optical absorption in crystals with inversion symmetry occurs between states of opposite parity. Note that crystals without a center of inversion have no well-defined parity; i.e., parity is not a good quantum number in this case. For example, there is no inversion symmetry when the basis is given by the inset (b) in Fig. 2.2, while the crystal has an inversion symmetry for inset (a) of that figure. Therefore, the basis of inset (b) has no well-defined parity. In such a case, the electronic wave function generally consists of a symmetric and an antisymmetric part, and parity selection rules do not apply.

Rotation around a specified axis is another symmetry operation. Again, it is possible to reproduce the crystal by rotation around an axis going through a lattice point. In the example of Fig. 2.2, the pattern is identical after rotation around an axis perpendicular to the shown plane by any angle which is a multiple of 60^0. This crystal structure is said to have sixfold rotation symmetry. Note that the degree of rotation symmetry is reduced if all atoms forming a basis are not identical. Note also that the same rotation symmetries apply both in real and reciprocal space.

In general, a crystal is said to have an n-fold rotation axis when the rotation angle is $2\pi/n$. Here n is an integer that can only take the values of n = 1, 2, 3, 4, 6. With any of these values of the rotation angles, one can fill all lattice space with unit cells without any voids. For example, n = 5 is not possible because a lattice with pentagon unit cells cannot be constructed since it is not possible to fill all space with a connected array of pentagons (see Prob. 2.2).

Reflection symmetry is yet another possible crystal symmetry. The symmetry element of a reflection operation is the plane of reflection (mirror plane). For example, in Fig. 2.2 a plane through any line of lattice points is a mirror plane.

All the symmetry operations around a lattice point that leave the lattice invariant constitute the lattice point group. There are only 32 crystal point groups (i.e., rotation, inversion, and mirror symmetry operations about a lattice point) which leave the crystal unchanged. The cubic crystal group has the highest symmetry. If the cubic crystal has an inversion center, its point group is denoted by O_h. Otherwise, it belongs to the T_d group. There are several books devoted to the classification of crystal structures, and the interested reader may consult the references of this chapter for additional information. The point groups in Fig. 2.8 are some examples.

If the crystal belongs to a particular point group defined by symmetry operations about a lattice point which leaves the crystal invariant, such as rotation, inversion, and mirror symmetry (but not a translational symmetry where no lattice point is kept fixed), the local microscopic environment (i.e., the local potential acting on the electron) must also satisfy certain symmetries. For example, in Fig. 2.2 we see that the crystal belongs to different point groups depending on whether the six atoms of the basis are identical or not. Therefore, it becomes meaningful to identify the electron states according to the point-group symmetries and label electron wave

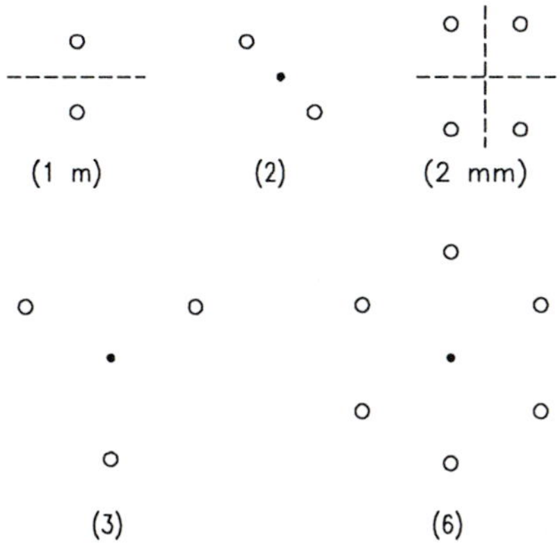

Figure 2.8. Examples of point groups. (1 m) describes the presence of one minor plane, (2 mm) refers to two minor planes, and the numbers (2), (3), and (6) describe two, three, and sixfold rotation symmetry axes (perpendicular to the plane of the atoms).

functions according to point symmetry representations. For instance, the center of the Brillouin zone, which is the most important part of the Brillouin zone as far as the onset of optical transitions in most crystals is concerned, is called the Γ point. The symmetry representation Γ_1 corresponds to a totally symmetric electron wave function, the equivalent of an s orbital in atomic physics. Representation Γ_5^- in the 0_h group refers to the symmetry of the dipole operator. The - sign indicates the negative parity associated with the dipole operator with odd symmetry. From arguments derived using group theory, it is possible to determine which optical transitions are forbidden, because they would violate crystal invariance with respect to symmetry operations of the crystal point group. Group symmetry selection rules are the analog of the angular selection rules in atomic spectroscopy.

An important material parameter, as far as optical transitions are concerned, is the dielectric constant ϵ, defined as

$$\mathbf{D} = \epsilon \mathbf{E} \,, \tag{2.25}$$

where **D** is the displacement vector and **E** is the electric field. The dielectric constant is influenced by the crystal structures and symmetries. In general, ϵ is not a scalar but a tensor of rank 2, so that Eq. (2.25) becomes

$$\begin{bmatrix} D_x \\ D_y \\ D_z \end{bmatrix} = \begin{bmatrix} \epsilon_{11} & \epsilon_{12} & \epsilon_{13} \\ \epsilon_{21} & \epsilon_{22} & \epsilon_{23} \\ \epsilon_{31} & \epsilon_{32} & \epsilon_{33} \end{bmatrix} \begin{bmatrix} E_x \\ E_y \\ E_z \end{bmatrix} . \tag{2.26}$$

In order to know how many independent elements of the dielectric tensor exist, we can use the symmetry properties of the crystal. Only the rotational symmetry considerations are needed for characterizing the dielectric tensor. However, for other material parameters (such as the electro-optic parameter), reflection and inversion symmetry properties are needed in addition to the rotational symmetry. As an example, let us consider the ϵ tensor for a hexagonal crystal, shown in Fig. 2.9(a), which has a sixfold rotation axis, i.e., where the rotation angle is $\theta = 2\pi/6$. It can be proven that the ϵ tensor is a symmetric tensor; that is, $\epsilon_{ij} = \epsilon_{ji}$ for $i \neq j$. In other words, $\epsilon_{12} = \epsilon_{21}$, $\epsilon_{13} = \epsilon_{31}$, and $\epsilon_{23} = \epsilon_{32}$ (see Prob. 2.3), and

$$\epsilon = \begin{bmatrix} \epsilon_{11} & \epsilon_{12} & \epsilon_{13} \\ \epsilon_{12} & \epsilon_{22} & \epsilon_{23} \\ \epsilon_{13} & \epsilon_{23} & \epsilon_{33} \end{bmatrix} . \tag{2.27}$$

In order to obtain additional reduction of the dielectric tensor, we now rotate the coordinate system by $\theta = 2\pi/6$ about the sixfold axis, which corresponds to an invariant operation, as shown in Fig. 2.9(b). We can see that the new coordinates (x′, y′) are related to the old coordinates (x, y) through

$$\begin{cases} x' = x\cos(\pi/3) + y\sin(\pi/3) = x/2 + (3/2)^{1/2}y \\ y' = -x\sin(\pi/3) + y\cos(\pi/3) = -(3/2)^{1/2}x + y/2\ . \end{cases} \tag{2.28}$$

Equation (2.26) in the new coordinate system is

$$\begin{bmatrix} D'_x \\ D'_y \\ D'_z \end{bmatrix} = \begin{bmatrix} \epsilon_{11} & \epsilon_{12} & \epsilon_{13} \\ \epsilon_{12} & \epsilon_{22} & \epsilon_{23} \\ \epsilon_{13} & \epsilon_{23} & \epsilon_{33} \end{bmatrix} \begin{bmatrix} E'_x \\ E'_y \\ E'_z \end{bmatrix}, \tag{2.29}$$

where the dielectric tensor is the same in the old and new coordinate systems. Using Eq. (2.28) to express the quantities in the rotated frame, Eq. (2.29) becomes

$$\begin{bmatrix} D_x/2 + \sqrt{3}D_y/2 \\ -\sqrt{3}D_x/2 + D_y/2 \\ D_z \end{bmatrix} = \begin{bmatrix} \epsilon_{11} & \epsilon_{12} & \epsilon_{13} \\ \epsilon_{12} & \epsilon_{22} & \epsilon_{23} \\ \epsilon_{13} & \epsilon_{23} & \epsilon_{33} \end{bmatrix} \begin{bmatrix} E_x/2 + \sqrt{3}E_y/2 \\ -\sqrt{3}E_x/2 + E_y/2 \\ E_z \end{bmatrix}. \tag{2.30}$$

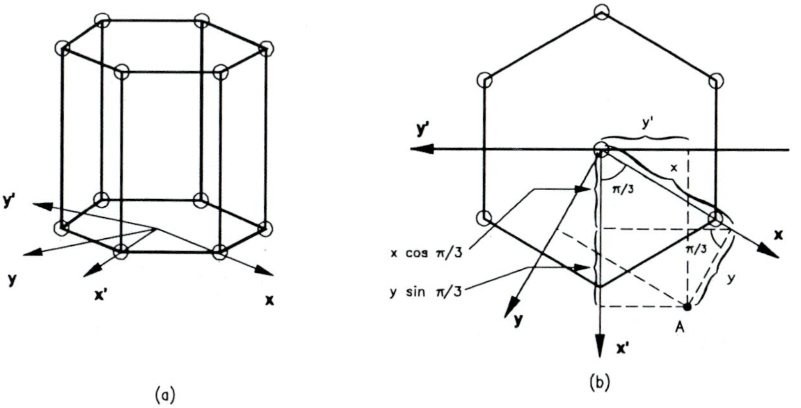

Figure 2.9. (a) Schematic representation of a hexagonal crystal. (b) Rotation of a hexagonal crystal about the sixfold axis by 2π/6. The sixfold axis goes through the center of the hexagon.

Expressing **D** in terms of **E** using Eq. (2.25) with Eq. (2.27), we obtain the following three equations from Eq. (2.30):

$$(\epsilon_{11}E_x + \epsilon_{12}E_y + \epsilon_{13}E_z)/2 + \sqrt{3}(\epsilon_{12}E_x + \epsilon_{22}E_y + \epsilon_{23}E_z)/2$$

$$= \epsilon_{11}(E_x/2 + \sqrt{3}E_y/2) + \epsilon_{12}(-\sqrt{3}E_x/2 + E_y/2) + \epsilon_{13}E_z \ , \qquad (2.31a)$$

$$-\sqrt{3}(\epsilon_{11}E_x + \epsilon_{12}E_y + \epsilon_{13}E_z)/2 + (\epsilon_{12}E_x + \epsilon_{22}E_y + \epsilon_{23}E_z)/2$$

$$= \epsilon_{12}(E_x/2 + \sqrt{3}E_y/2) + \epsilon_{22}(-\sqrt{3}E_x/2 + E_y/2) + \epsilon_{23}E_z \ , \qquad (2.31b)$$

and

$$\epsilon_{13}E_x + \epsilon_{23}E_y + \epsilon_{33}E_z$$

$$= \epsilon_{13}(E_x/2 + \sqrt{3}E_y/2) + \epsilon_{23}(-\sqrt{3}E_x/2 + E_y/2) + \epsilon_{33}E_z \ . \qquad (2.31c)$$

Since the components of **E** are arbitrary, Eq. (2.31a) can be satisfied only if

$$\epsilon_{13} = \epsilon_{23} = 0 \ . \qquad (2.32)$$

Similarly, Eq. (2.31b) implies

$$\epsilon_{12} = 0, \qquad \epsilon_{11} = \epsilon_{22} \ . \qquad (2.33)$$

Thus, the dielectric tensor for a hexagonal crystal takes the form

$$\epsilon = \begin{bmatrix} \epsilon_{11} & 0 & 0 \\ 0 & \epsilon_{11} & 0 \\ 0 & 0 & \epsilon_{33} \end{bmatrix} . \qquad (2.34)$$

Only the diagonal matrix elements are nonzero with two principal dielectric constants: ϵ_{33} along the z axis and ϵ_{11} in the xy plane. The behavior of the dielectric tensor in tetragonal and trigonal crystals is similar to that of hexagonal crystals. These three crystal systems are referred to as *optically uniaxial crystals*. The z axis in such crystals is called the optical c axis. The dielectric tensor has values that are different parallel to and perpendicular to the c axis. An example of such a uniaxial crystal is cadmium sulfide, CdS, which will be discussed in more detail in Sec. 7-4.

In a similar way it can be shown that for cubic crystals the off-diagonal matrix elements of the dielectric tensor are zero and the diagonal elements are identical. Thus, the dielectric constant is a scalar in cubic crystals (see Prob. 2.4).

2-4. Energy Bands and Bloch Wave Functions

A crystal contains a large number of electrons per unit volume, generally of the order $n \sim 10^{24}\ cm^{-3}$. Some of these electrons fill the inner shells of the atoms and, therefore, are tightly bound to a particular nucleus. These electrons are shielded from the influence of neighboring atoms by the outer electronic shells, and their properties are only weakly modified in comparison to those of free atoms. The corresponding energy levels are similar to the sharp energy levels of isolated atoms. The outer electronic shells, on the other hand, feel a weaker attractive potential from the nuclei because the Coulomb interaction is screened by the electrons in the inner shells. The properties of the outer electrons differ considerably from those of their counterparts in isolated atoms. These electrons are delocalized, meaning that their wave function is spread over the entire crystal. Furthermore, their energy spectrum consists of sets of very closely spaced energy levels forming what are called *energy bands.* These energy bands are separated from other bands by *energy gaps* which are energy intervals in which no allowed electron energy values exist .

To understand the behavior of electrons in a crystal lattice, we have to explicitly take into account the crystal potential $W(\mathbf{r})$. The crystal potential arises from the periodic charge distribution associated with the ion cores in the neighborhood of each lattice site. Thus, the crystal potential has the full periodicity of the lattice, i.e.,

$$W(\mathbf{r}) = W(\mathbf{r} + \mathbf{R}) \ , \tag{2.35}$$

for any vector $\mathbf{R}$ in the direct lattice. Figure 2.10 schematically shows this periodic potential.

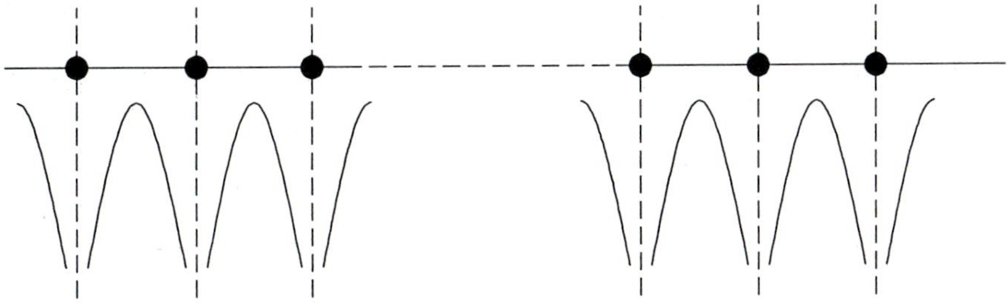

Figure 2.10. Schematics of the periodic lattice potential.

We first treat the problem qualitatively by presenting a simplified physical picture and then develop a more rigorous mathematical treatment. We start with the energy levels of an isolated atom and follow the evolution of its energy levels to energy bands as additional atoms are brought together to form a solid. As an example, consider the sodium atom consisting of a nucleus and eleven electrons, with the electrons experiencing the Coulomb potential of the nucleus. The electronic configuration of Na is $1s^2 2s^2 2p^6 3s^1$, indicating that the 1s and 2s levels are each fully occupied by two electrons, the 2p level is also full with six electrons, and the 3s level is only half full with one electron. These levels, together with the atomic Coulomb potential, are schematically shown in Fig. 2.11(a). Now consider a complex consisting of two sodium atoms. When the two atoms are far apart, the electrons in each atom do not feel the Coulomb potential of the other atom, and the energy levels of the system of two atoms, 1s, 2s, 2p, and 3s, are each doubly degenerate in the sense that each electron in one atom has its counterpart with exactly the same energy level structure in the other atom. However, when the atoms are brought close together, this degeneracy is removed as a result of the mutual interaction between the two atoms. Each of the energy levels becomes a "doublet," as shown in Fig. 2.11(b). The closer the atoms, the larger the interaction potential and, consequently, the larger the splitting between the levels in each "doublet." Furthermore, since the inner shell electrons, such as the electrons in the 1s level, are tightly bound to their nuclei, they do not experience the interatomic interaction potential as strongly as the outer shell electrons with weak binding. Therefore, the splitting of the outer shell levels is larger than that of the inner shell levels.

Now consider a complex consisting of three sodium atoms. In this case each atomic level is split into three sublevels. Similarly, for four atoms in close vicinity of each other, each atomic level is split into four sublevels, and so on. When the number of atoms increases to a large number N, each atomic energy level is split into N sublevels [see Fig. 2.11(c)]. In a cubic centimeter ($1\ cm^3$) of a crystal, the number N reaches approximately 10^{23}.

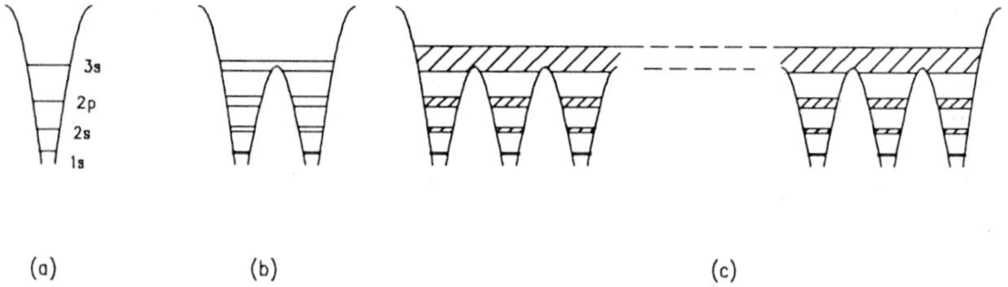

Figure 2.11. Schematics of energy levels in (a) free Na atom, (b) a complex of two Na atoms, and (c) a complex of many sodium atoms, representing solid sodium.

Therefore, the spacing between the sublevels is so small, on the order of 10^{-23} to 10^{-22} eV, that the combination of sublevels may be viewed as a band of energy. The energy regions between the bands are the energy gaps (see Fig. 2.12). There are no allowed energy levels in the energy gaps, as there are no allowed energy levels between the discrete energy levels of isolated atoms. This qualitative description of band formation in crystals is highly simplified. In reality, it is not correct that a particular band originates entirely from a specific atomic level. Energy bands are derived from different atomic energy levels, as shown schematically in Fig. 2.12. Nevertheless, the extremum of a given band often has a predominant character that can be associated with a particular atomic state, e.g., s or p.

To treat the problem more quantitatively, we consider the Schrödinger equation for the electron of mass m_0 in the periodic potential $W(\mathbf{r})$,

$$\left[-\frac{\hbar^2}{2m_0}\nabla^2 + W(\mathbf{r})\right]\psi(\mathbf{r}) = \mathcal{E}\psi(\mathbf{r}) \ . \tag{2.36}$$

Bloch's theorem states that the solution of this Schrödinger equation is of the form

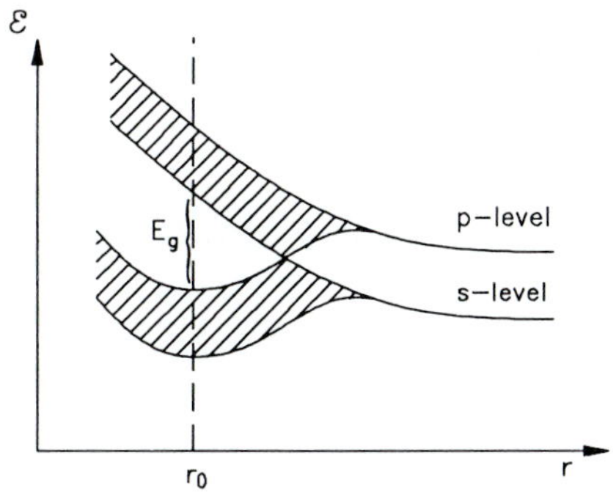

Figure 2.12. Energy band formation in a crystal. The energy levels s and p of isolated atoms merge into energy bands separated by a forbidden energy gap E_g. r_0 is the lattice constant, which is the length of the unit cell of the crystal at equilibrium.

$$\boxed{\psi_{n\mathbf{k}}(\mathbf{r}) = \frac{e^{i\mathbf{k}\cdot\mathbf{r}}}{V^{1/2}}\, u_{n\mathbf{k}}(\mathbf{r}) \quad \text{or} \quad \psi_{n\mathbf{k}}(\mathbf{r}+\mathbf{R}) = e^{i\mathbf{k}\cdot\mathbf{R}}\, \psi_{n\mathbf{k}}(\mathbf{r})\,,}$$

Bloch theorem (2.37)

where $u_{n\mathbf{k}}(\mathbf{r})$ is a function that has the full translational symmetry of the lattice so that

$$u_{n\mathbf{k}}(\mathbf{r}+\mathbf{R}) = u_{n\mathbf{k}}(\mathbf{r})\,. \tag{2.38}$$

In Eq. (2.37) V is the volume of the crystal and the index pair **nk** labels the energy eigenstates in a given band n. The wave function of Eq. (2.37) has the character of a free running wave with an amplitude modulated by the periodic lattice. The memory of the original atomic function is kept in the modulation of the Bloch function $u_{n\mathbf{k}}(\mathbf{r})$. Figure 2.13 schematically shows the behavior of a Bloch wave function.

In the above equations **k** is a vector in the reciprocal lattice. **k** can always be confined to the first Brillouin zone. If **k′** is not in the first Brillouin zone, it can be reduced into it. We can write $\mathbf{k}' = \mathbf{k} + \mathbf{G}$, where **k** is in the first Brillouin zone and **G** is the reciprocal lattice vector. Wave vectors **k** and **k′** are equivalent because $\exp(i\mathbf{k}'\cdot\mathbf{R}) = \exp(i(\mathbf{k}+\mathbf{G})\cdot\mathbf{R}) = \exp(i\mathbf{k}\cdot\mathbf{R})$ [where we used Eq. (2.6)], which makes the Bloch wave

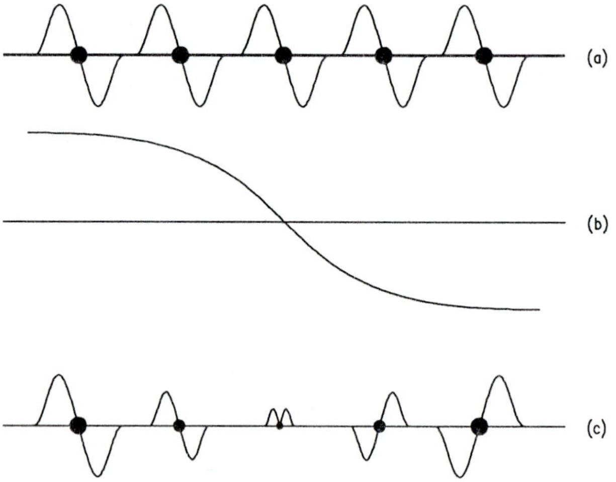

Figure 2.13. Schematic representation of a Bloch function with p-like character; (a) p-like orbitals at each atomic site, $u(\mathbf{r})$; (b) free-running part of the wave function, $\exp(i\mathbf{k}\cdot\mathbf{r})$; and (c) the total Bloch wave function, $u(\mathbf{r})\exp(i\mathbf{k}\cdot\mathbf{r})$.

functions given by Eq. (2.37) equivalent for **k** and **k′**. Thus, the allowed **k** values in a simple cubic lattice may be restricted to the range $-\pi/a \leq |\mathbf{k}| \leq \pi/a$, with a being the length of the unit cell.

The energy eigenvalues $\mathcal{E}_n(\mathbf{k})$ are the energy bands. The information contained in the functions $\mathcal{E}_n(\mathbf{k})$ is referred to as the *band structure* of the crystal. For a given band n, $\mathcal{E}_n(\mathbf{k})$ usually does not have a simple analytical form, except in the vicinity of special regions of the Brillouin zone with high symmetry, such as the Γ point at the center of the Brillouin zone. Furthermore, it can be shown that an electron in a band n with wave vector **k** has the velocity

$$v_n(\mathbf{k}) = \frac{1}{\hbar} \nabla_k \mathcal{E}_n(\mathbf{k}) \ . \tag{2.39}$$

2-5. Tight-Binding Model for Energy Bands*

In the previous section we qualitatively discussed the formation of electron energy bands resulting from the periodic potential of the crystal. In actual crystals, a calculation of electron band energies is a formidable task, which always requires a certain degree of approximation on the form of the potential W(**r**). Here we give a simplified treatment of the band structure in the *tight-binding model.* In this model the atomic potential is assumed to be so strong that the electron is essentially localized at a single atom, and the wave functions for neighboring sites have little overlap. Consequently, there is basically no overlap between electron wave functions that are separated by two or more lattice sites. This approximation is a reasonable approximation for the narrow, inner bands in solids, such as the three-dimensional bands in transition metals like Fe.

Let us assume the crystal has N atoms with single electron wave function ϕ_k and energy $\mathcal{E}_k$. The Schrödinger equation for electrons in the atom located at the lattice point $\mathbf{r}_\ell$ is then [see Fig. 2.14(a)]

$$H_0\phi_k(\mathbf{r} - \mathbf{r}_\ell) = \mathcal{E}_k \phi_k(\mathbf{r} - \mathbf{r}_\ell) \ , \tag{2.40}$$

with the Hamiltonian for electrons in the single atom,

$$H_0 = -\frac{\hbar^2}{2m_0} \nabla^2 + W(\mathbf{r} - \mathbf{r}_\ell) \ , \tag{2.41}$$

where $W(\mathbf{r} - \mathbf{r}_\ell)$ is the potential of the ℓth atom. The total potential is the sum of the single atom potentials,

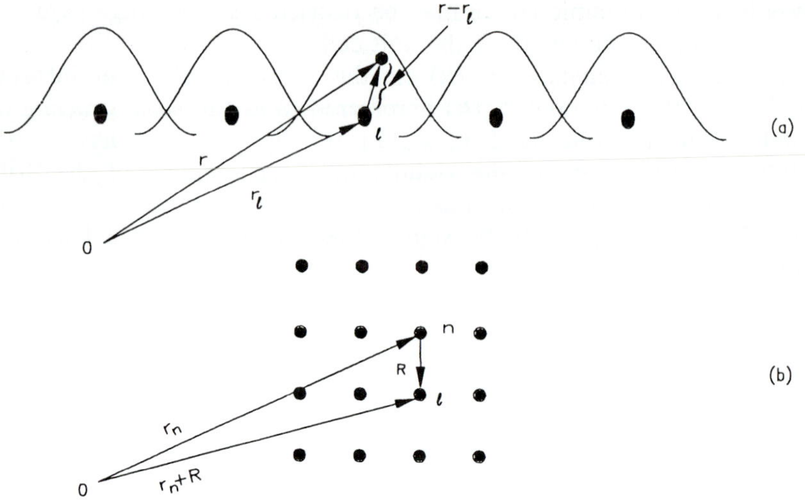

Figure 2.14. Schematic representation of (a) the electron wave functions for the tight-binding approximation and (b) the neighboring lattice sites $\mathbf{r}_n$ and $\mathbf{r}_\ell$ that are separated from each other by the direct lattice vector $\mathbf{R}$.

$$W(\mathbf{r}) = \sum_{\ell} W(\mathbf{r} - \mathbf{r}_\ell) , \tag{2.42}$$

where the summation is over all of the atoms in the crystal. If we let the total wave function and energy for the system be $\psi_\mathbf{k}(\mathbf{r})$ and $\mathcal{E}$, respectively, the Schrödinger equation for the system of N atoms becomes

$$H\psi_\mathbf{k}(\mathbf{r}) = \left[-\frac{\hbar^2}{2m_0} \nabla^2 + \sum_{\ell} W(\mathbf{r} - \mathbf{r}_\ell) \right] \psi_\mathbf{k}(\mathbf{r}) = \mathcal{E}\psi_\mathbf{k}(\mathbf{r}) . \tag{2.43}$$

To solve Eq. (2.43) we write $\psi_\mathbf{k}(\mathbf{r})$ as a linear combination of atomic wave functions,

$$\psi_\mathbf{k}(\mathbf{r}) = \sum_{n} e^{i\mathbf{k}\cdot\mathbf{r}_n} \phi_\mathbf{k}(\mathbf{r} - \mathbf{r}_n) . \tag{2.44}$$

This technique is also referred to as LCAO, an acronym for the linear combination of atomic orbitals. The wave function (2.44) obviously satisfies the Bloch theorem given by Eq. (2.37) as it should (see Prob. 2.6). The ground-state energy can be computed using the wave functions of Eq. (2.44) as

$$\mathcal{E} = \frac{\int d\mathbf{r}\ \psi_{\mathbf{k}}^{*}(\mathbf{r})\ H\ \psi_{\mathbf{k}}(\mathbf{r})}{\int d\mathbf{r}\ \psi_{\mathbf{k}}^{*}(\mathbf{r})\ \psi_{\mathbf{k}}(\mathbf{r})} . \tag{2.45}$$

The denominator of Eq. (2.45) may be written using Eq. (2.44) in the form of

$$\int d\mathbf{r}\ \psi_{\mathbf{k}}^{*}(\mathbf{r})\ \psi_{\mathbf{k}}(\mathbf{r}) = \int d\mathbf{r} \sum_{m} e^{-i\mathbf{k}\cdot\mathbf{r}_m}\ \phi_{\mathbf{k}}^{*}(\mathbf{r} - \mathbf{r}_m) \sum_{n} e^{i\mathbf{k}\cdot\mathbf{r}_n}\ \phi_{\mathbf{k}}(\mathbf{r} - \mathbf{r}_n)$$

$$= \sum_{n,m} e^{i\mathbf{k}\cdot(\mathbf{r}_n - \mathbf{r}_m)} \int d\mathbf{r}\ \phi_{\mathbf{k}}^{*}(\mathbf{r} - \mathbf{r}_m)\ \phi_{\mathbf{k}}(\mathbf{r} - \mathbf{r}_n) , \tag{2.46}$$

where n and m refer to different lattice sites. The integral in Eq. (2.46) is the overlap integral involving atomic sites m and n. We assume small wave function overlap between electrons of different lattice sites and classify the contributions of the neighboring atoms to the total energy by the amount of the wave function overlap. Thus, we only keep the leading contribution of the overlap integral. Hence, we take only the contribution for m = n from the integral in Eq. (2.46); i.e.,

$$\int d\mathbf{r}\ \phi_{\mathbf{k}}^{*}(\mathbf{r} - \mathbf{r}_m)\ \phi_{\mathbf{k}}(\mathbf{r} - \mathbf{r}_n) \simeq \delta_{n,m} . \tag{2.47}$$

Then the denominator of Eq. (2.45) becomes

$$\int d\mathbf{r}\ \psi_{\mathbf{k}}^{*}(\mathbf{r})\ \psi_{\mathbf{k}}(\mathbf{r}) = \sum_{n,m} e^{i\mathbf{k}\cdot(\mathbf{r}_n - \mathbf{r}_m)}\ \delta_{n,m} = \sum_{n=1}^{N} 1 = N\ , \tag{2.48}$$

where N is the total number of atoms in the crystal. The integral in the numerator of Eq. (2.45) may be written as

$$\int d\mathbf{r}\ \psi_{\mathbf{k}}^{*}(\mathbf{r})\ H\ \psi_{\mathbf{k}}(\mathbf{r}) =$$

$$\sum_{n,m} e^{i\mathbf{k}\cdot(\mathbf{r}_n - \mathbf{r}_m)} \int d\mathbf{r}\ \phi_{\mathbf{k}}^{*}(\mathbf{r} - \mathbf{r}_m) \left[-\frac{\hbar^2}{2m_0} \nabla^2 + \sum_{\ell} W(\mathbf{r} - \mathbf{r}_\ell) \right] \phi_{\mathbf{k}}(\mathbf{r} - \mathbf{r}_n)\ . \tag{2.49}$$

Here we only take the direct contribution, $\mathbf{r}_m = \mathbf{r}_n$, and the nearest-neighbor contribution, $\mathbf{r}_m = \mathbf{r}_n + \mathbf{R}$, into account where $\mathbf{R}$ is the direct lattice vector [see Fig. 2.14(b)]. Equation (2.49) becomes

$$\int d\mathbf{r}\ \psi_{\mathbf{k}}^{*}(\mathbf{r})\ H\ \psi_{\mathbf{k}}(\mathbf{r}) =$$

$$\sum_{n} \int d\mathbf{r}\ \phi_{\mathbf{k}}^{*}(\mathbf{r} - \mathbf{r}_n) \left[-\frac{\hbar^2}{2m_0} \nabla^2 + \sum_{\ell} W(\mathbf{r} - \mathbf{r}_\ell) \right] \phi_{\mathbf{k}}(\mathbf{r} - \mathbf{r}_n) +$$

$$\sum_{R \neq 0} \sum_{n} e^{-i\mathbf{k}\cdot\mathbf{R}} \int d\mathbf{r}\ \phi_{\mathbf{k}}^{*}(\mathbf{r} - \mathbf{r}_n - \mathbf{R}) \left[-\frac{\hbar^2}{2m_0} \nabla^2 + \sum_{\ell} W(\mathbf{r} - \mathbf{r}_\ell) \right] \phi_{\mathbf{k}}(\mathbf{r} - \mathbf{r}_n), \tag{2.50}$$

where the first term in Eq. (2.50) was obtained by setting $\mathbf{r}_m = \mathbf{r}_n$, and the second term was obtained by using $\mathbf{r}_m = \mathbf{r}_n + \mathbf{R}$. The summation over R is included in the second term of Eq. (2.50) to take into account all of the nearest neighbors. The potential energy may be divided into a component for atomic site $\ell = n$ and the contribution from the remaining atoms. That is,

$$\sum_{\ell} W(\mathbf{r} - \mathbf{r}_\ell) = W(\mathbf{r} - \mathbf{r}_n) + \sum_{\ell \neq n} W(\mathbf{r} - \mathbf{r}_\ell) . \tag{2.51}$$

Substituting Eq. (2.51) into Eq. (2.50) we get

$$\int d\mathbf{r}\, \psi_k^*(\mathbf{r})\, H\, \psi_k(\mathbf{r}) = \sum_n \int d\mathbf{r}\, \phi_k^*(\mathbf{r} - \mathbf{r}_n) \left[\left(-\frac{\hbar^2}{2m_0} \nabla^2 + W(\mathbf{r} - \mathbf{r}_n) \right) \right.$$

$$\left. + \sum_{\ell \neq n} W(\mathbf{r} - \mathbf{r}_\ell) \right] \phi_k(\mathbf{r} - \mathbf{r}_n) + \sum_{R \neq 0} \sum_n e^{-i\mathbf{k}\cdot\mathbf{R}} \int d\mathbf{r}\, \phi_k^*(\mathbf{r} - \mathbf{r}_n - \mathbf{R})$$

$$\times \left[\left(-\frac{\hbar^2}{2m_0} \nabla^2 + W(\mathbf{r} - \mathbf{r}_n) \right) + \sum_{\ell \neq n} W(\mathbf{r} - \mathbf{r}_\ell) \right] \phi_k(\mathbf{r} - \mathbf{r}_n) . \tag{2.52}$$

But, since $(-\hbar^2/2m_0)\, \nabla^2 + W(\mathbf{r} - \mathbf{r}_n) = H_0$, the first term on the RHS of Eq. (2.52) gives

$$\int d\mathbf{r}\, \phi_k^*(\mathbf{r} - \mathbf{r}_n) \left[-\frac{\hbar^2}{2m_0} \nabla^2 + W(\mathbf{r} - \mathbf{r}_n) \right] \phi_k(\mathbf{r} - \mathbf{r}_n) = \mathscr{E}_k , \tag{2.53}$$

where we used Eq. (2.40). Also,

$$\int d\mathbf{r}\, \phi_k^*(\mathbf{r} - \mathbf{r}_n - \mathbf{R}) \left[-\frac{\hbar^2}{2m_0} \nabla^2 + W(\mathbf{r} - \mathbf{r}_n) \right] \phi_k(\mathbf{r} - \mathbf{r}_n)$$

$$= \mathscr{E}_k \int d\mathbf{r}\, \phi_k^*(\mathbf{r} - \mathbf{r}_n - \mathbf{R})\, \phi_k(\mathbf{r} - \mathbf{r}_n) = 0 , \tag{2.54}$$

since $\mathbf{R} \neq 0$. Equation (2.52) then becomes

$$\int d\mathbf{r}\, \psi_{\mathbf{k}}^{*}(\mathbf{r})\, H\, \psi_{\mathbf{k}}(\mathbf{r}) = \sum_{n=1}^{N} \mathcal{E}_{\mathbf{k}}$$

$$+ \sum_{n} \sum_{\ell \neq n} \int d\mathbf{r}\, \phi_{\mathbf{k}}^{*}(\mathbf{r} - \mathbf{r}_n)\, W(\mathbf{r} - \mathbf{r}_\ell)\, \phi_{\mathbf{k}}(\mathbf{r} - \mathbf{r}_n)$$

$$+ \sum_{\mathbf{R} \neq 0} \sum_{n} e^{-i\mathbf{k}\cdot\mathbf{R}} \sum_{\ell \neq n} \int d\mathbf{r}\, \phi_{\mathbf{k}}^{*}(\mathbf{r} - \mathbf{r}_n - \mathbf{R})\, W(\mathbf{r} - \mathbf{r}_\ell)\, \phi_{\mathbf{k}}(\mathbf{r} - \mathbf{r}_n) \tag{2.55}$$

or

$$\int d\mathbf{r}\, \psi_{\mathbf{k}}^{*}(\mathbf{r})\, H\, \psi_{\mathbf{k}}(\mathbf{r}) = \mathcal{E}_{\mathbf{k}} N - A\, N - N \sum_{\mathbf{R} \neq 0} B\, e^{-i\mathbf{k}\cdot\mathbf{R}}\,, \tag{2.56}$$

where we have defined

$$A = -\sum_{\ell \neq n} \int d\mathbf{r}\, \phi_{\mathbf{k}}^{*}(\mathbf{r} - \mathbf{r}_n)\, W(\mathbf{r} - \mathbf{r}_\ell)\, \phi_{\mathbf{k}}(\mathbf{r} - \mathbf{r}_n) \tag{2.57}$$

and

$$B = -\sum_{\ell \neq n} \int d\mathbf{r}\, \phi_{\mathbf{k}}^{*}(\mathbf{r} - \mathbf{r}_n - \mathbf{R})\, W(\mathbf{r} - \mathbf{r}_\ell)\, \phi_{\mathbf{k}}(\mathbf{r} - \mathbf{r}_n)\,. \tag{2.58}$$

Substituting for the numerator and denominator of Eq. (2.45) from Eqs. (2.56) and (2.48), respectively, we get

$$\mathcal{E} = \mathcal{E}_k - A - \sum_{R \neq 0} B\, e^{-i\mathbf{k}\cdot\mathbf{R}} . \tag{2.59}$$

Equation (2.59) shows that the electron energy is wave vector dependent, implying a formation of energy bands. The coefficient A in Eq. (2.59) gives the shift of the atomic energy level $\mathcal{E}_k$ (renormalization of the atomic energy), and B determines the width of the energy band. We can now use this result to calculate the tight-binding energy bands of a cubic crystal with lattice spacing a. For such a solid there are six nearest neighbors at **R** = ±ax, ±ay, ±az. Equation (2.59) gives

$$\mathcal{E} = \mathcal{E}_k - A - B\,[e^{ik_x a} + e^{-ik_x a} + e^{ik_y a} + e^{-ik_y a} + e^{ik_z a} + e^{-ik_z a}] \tag{2.60}$$

or

$$\mathcal{E} = \mathcal{E}_k - A - 2B\,[\cos(k_x a) + \cos(k_y a) + \cos(k_z a)] . \tag{2.61}$$

Thus, the energy bands for a cubic lattice in the tight-binding approximation have cosine shape. Figure 2.15(a) displays one such band schematically in the first Brillouin zone in one-dimension (ignoring k_y and k_z components). In this case, the lowest energy in the band (i.e., the band minimum) is at

$$\mathcal{E}(k = 0) = \mathcal{E}_k - A - 2B , \tag{2.62}$$

which is below the atomic energy value $\mathcal{E}_k$. The bandwidth is 4B, indicating that B controls the width of the energy band as mentioned earlier. The bandwidth becomes narrower as the lattice spacing increases, due to a decrease in interaction among atoms. In Fig. 2.15(a) only one of the energy bands is plotted. Generally, it can be shown that for an attractive crystal potential $B < 0$ for p bands, while $B > 0$ for s bands. Thus, the s bands are concave up and p bands are concave down around $k = 0$. Figure 2.15(b) displays two bands; the upper band is of s character, while the lower band has p character. This situation is realized in some semiconductors.

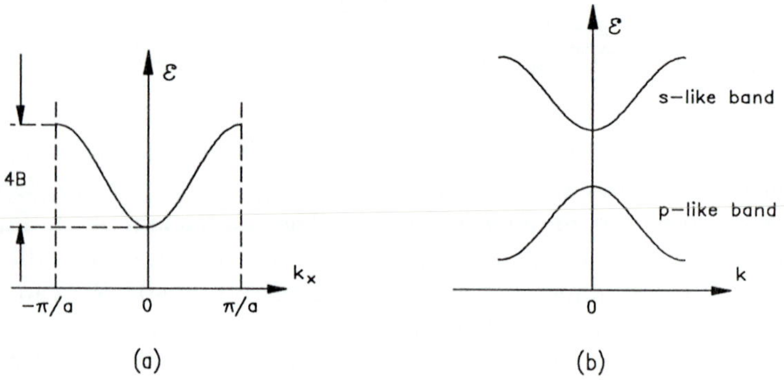

Figure 2.15. (a) One energy band in the tight-binding approximation for a cubic crystal of side a along a direction such as x axis. (b) Schematics of an s-like and a p-like band in a solid.

2-6. Effective Mass

In the preceding sections we have seen that the discrete electronic energy levels of isolated atoms are replaced in crystals by broad energy bands separated by energy gaps forbidden to electrons. We consider now, in more detail, the dispersion of an energy band, i.e., its dependence upon wave number, $\mathcal{E}_n(k)$. We first examine $\mathcal{E}_n(k)$ near an extremum, say at $k = k_0$. Very often, such an extremum occurs for $k = 0$ at the center of the Brillouin zone, as shown in Fig. 2.15.

A Taylor expansion of the energy dispersion $\mathcal{E}(k)$ around k_0 yields

$$\mathcal{E}(k) = \mathcal{E}_0 + \left.\frac{d\mathcal{E}}{dk}\right|_{k = k_0} (k - k_0) + \frac{1}{2} \left.\frac{d^2\mathcal{E}}{dk^2}\right|_{k = k_0} (k - k_0)^2 + \ldots \tag{2.63}$$

For sufficiently small values of $(k - k_0)$, we can ignore the higher-order terms not written explicitly in Eq. (2.63). Assuming that the band $\mathcal{E}(k)$ has an extremum for $k = k_0$, we have

$$\left.\frac{d\mathcal{E}}{dk}\right|_{k = k_0} = 0 \; , \tag{2.64}$$

and using Eq. (2.64), Eq. (2.63) becomes

$$\mathcal{E}(k) = \mathcal{E}_0 + \frac{\hbar^2(k - k_0)^2}{2m_e}, \tag{2.65}$$

with

$$\boxed{m_e = \frac{\hbar^2}{d^2\mathcal{E}/dk^2\big|_{k = k_0}}.}$$

effective mass (2.66)

Equation (2.65) describes a parabolic energy band with a constant mass m_e, referred to as the effective mass. Equation (2.66) shows that this effective mass is inversely proportional to the curvature of $\mathcal{E}(k)$. The value of m_e may be quite different from m_0, the mass of the electron in vacuum. The treatment we adopted here is referred to as the *effective-mass approximation.* In the framework of this approximation, the dynamical behavior of an electron in a crystal is treated as being the same as that of a free particle except for the different mass, m_e. To a first approximation, this change of the electronic mass represents the effect of the periodic crystal potential acting on the electron. It is possible to generalize the concept of effective mass to regions where $\mathcal{E}(k)$ does not have a parabolic shape, for instance, where higher-order terms in Eq. (2.63) are not negligible. In that case, m_e is no longer constant and the dynamical behavior of the electron in the crystal is that of a particle with variable (k-dependent) mass.

Since m_e is inversely proportional to the curvature of $\mathcal{E}$ versus k, the steeper the parabola, the smaller the effective mass. Unlike the gravitational mass, the effective mass can be positive, negative, zero, or even infinite. An infinite effective mass corresponds to an electron localized at a lattice site. In that case, the electron bound to the nucleus assumes the total mass of the crystal. A negative effective mass corresponds to a concave downward band. Negative m_e simply means that the electron moves in the opposite direction to the applied force. Figure 2.16 schematically displays the shape of the energy bands for different effective masses.

If the band structure of a given medium is not isotropic (i.e., if the energy surface is not spherical) the effective mass becomes a tensor, and the definition (2.66) has to be generalized as

$$\frac{1}{m_{e,ij}} = \frac{1}{\hbar^2}\frac{\partial^2\mathcal{E}}{\partial k_i \partial k_j}. \tag{2.67}$$

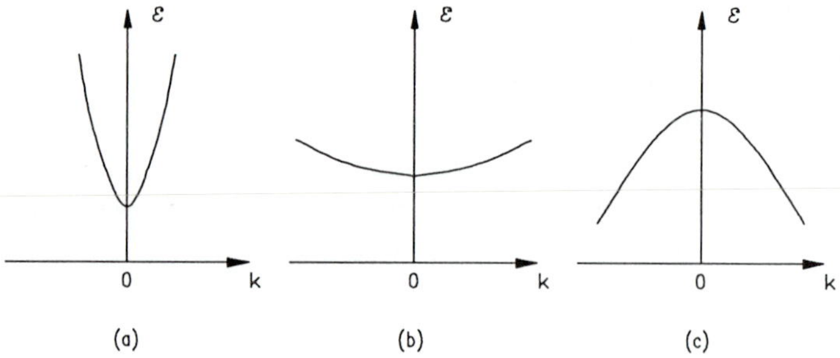

Figure 2.16. Representation of bands with different effective masses; (a) the band has a small positive effective mass; (b) the band has a large positive effective mass; and (c) the band has a negative effective mass.

Thus, in three dimensions the effective mass may have nine components. The different components describe the fact that a force in one direction produces a different acceleration of the electron than a force in another direction.

The *average* (or *density of states*) *effective mass* is obtained by suitably averaging the components, m_{ij}. For example, an ellipsoidal energy surface has three mass components, m_1, m_2, and m_3, resulting from the energy expression

$$\mathcal{E} = \frac{\hbar^2 k_1^2}{2m_1} + \frac{\hbar^2 k_2^2}{2m_2} + \frac{\hbar^2 k_3^2}{2m_3}, \tag{2.68}$$

and corresponding to the principal axes of the ellipsoid. In this case the average mass is

$$m_e = (m_1\, m_2\, m_3)^{1/3}. \tag{2.69}$$

The average mass may be used for the calculation of the density of states (see next section) and the mass we need in optical transitions (such as the exciton mass, as explained in Chap. VI).

2-7. Classification of Solids

In the previous sections we introduced a simple model showing the formation of energy bands in crystals. We have also seen how the symmetry properties of the periodic crystal structure lead to severe restrictions concerning the nature of the electron wave functions. Now we examine how the available electrons inside a crystal fill up different allowed energy bands, and how the optical and electrical properties of a crystal are crucially determined by the degree of filling of the last occupied band.

We first recall the Pauli principle, which is obeyed by all particles with a half-integer spin, such as electrons. Particles with half-integer spin are called Fermions. The Pauli principle states that each particular quantum state can be occupied by only one Fermion. A second identical Fermion is excluded from occupying that particular quantum state. This is the reason why the Pauli principle is also called the *exclusion principle.* We start filling the available energy states of the crystal from the lowest energy states up, until we exhaust the number of available electrons. One of the following situations will result: The last occupied band is either completely filled, or it is only partially filled. Here we assume the crystal temperature is very low so that the crystal is in its ground state, without any thermal excitations. The energetically highest, *completely filled band* is called the *valence band.* The next highest band, which may be empty or partially filled is called the *conduction band.* The crystal is a *metal* if the conduction band is partially filled [see Fig. 2.17(a)], while it is an *insulator* if the conduction band is empty [see Fig. 2.17(b)]. Metals are good conductors of electricity because an applied electric field easily scatters electrons from occupied to empty states within the same partially filled conduction band. Such electronic transitions within the same band are referred to as the intraband scattering processes.

By contrast, electronic intraband scattering is impossible in an insulator since there are no free valence-band states into which the electrons can scatter in the fully occupied valence band. Electrons must acquire at least the forbidden gap energy in order to contribute to an electric current. Electronic transitions between separate bands are called interband transitions.

An insulator with a small energy gap between valence and conduction bands is a semiconductor [see Fig. 2.17(c)]. The distinction between semiconductors and insulators is not a very sharp one. Wide-gap semiconductors (i.e., semiconductors with a bandgap of a few electron volts) may also be called insulators, and vice versa. Typically, the energy gap in a semiconductor is of the order of the photon energy of visible light, or less. For example, the elemental semiconductors, silicon (Si) and germanium (Ge), have bandgaps of approximately 1 eV and 0.7 eV,

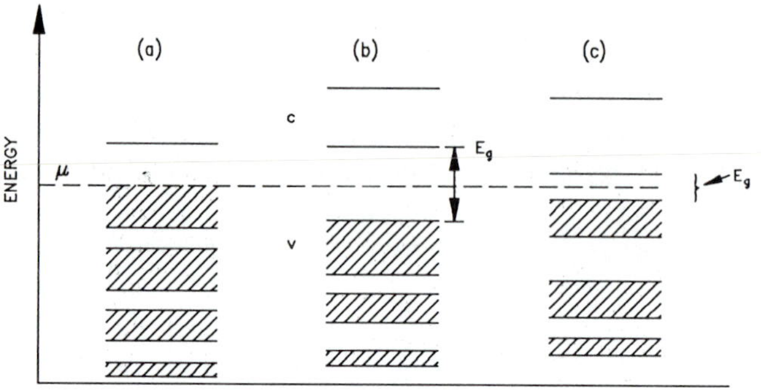

Figure 2.17. Simple band schematic for (a) a metal such as Na, (b) an insulator such as NaCl, and (c) a semiconductor such as GaAs. Note that the energetically highest occupied band in the metal is only partially filled, whereas it is completely filled in the ground state of an insulator or semiconductor. The dashed line that goes through the middle of the conduction band in the metal but goes in the middle of the forbidden gap in the insulator or semiconductor is called the Fermi level, or the chemical potential (see next section for more details).

respectively. A compound semiconductor like cadmium sulfide (CdS) has a gap, $E_g \simeq 2.5$ eV, in contrast to diamond, which is a good insulator with $E_g \simeq 6$ eV.

Metals, insulators, and semiconductors may also be distinguished by their unique optical characteristics. Insulators and semiconductors are optically transparent for energies below the bandgap energy, E_g, and absorbant at higher energies. A photon with an energy less than E_g cannot promote an electron from the valence band to an empty conduction band state and, therefore, the crystal is transparent. (We will examine this aspect in more detail in Chap. V, where we will also show that absorption may occur for energies slightly less than E_g due to excitonic effects.) On the other hand, photons with energy larger than the bandgap can be absorbed and promote electrons from filled valence bands to empty conduction bands. The insulator then becomes a better conductor of electricity. This phenomenon is referred to as photoconduction. In contrast, metals, which are good conductors of electricity, do not transmit light (except at very high frequencies in the ultraviolet).

2-8. Occupation Probability and Density of States

In this section we review some of the important properties of an ideal gas of Fermions in thermodynamic equilibrium. Electrons with spin s and wave vector **k** may occupy possible energy states denoted by $\{\mathbf{k},s\}$. The probability that the state $\{\mathbf{k},s\}$ is occupied can only vary between zero and one, since the Pauli exclusion principle states that each quantum state can be occupied by one Fermion at most. The corresponding probability distribution function is the well-known Fermi-Dirac distribution,

$$\boxed{f(\mathbf{k},s) = \frac{1}{1 + \exp[(\mathcal{E}(k) - \mu)/k_B T]} \equiv f(k) \equiv f(\mathcal{E})\ .}$$

Fermi-Dirac distribution (2.70)

Since the distribution function (2.70) does not depend on spin s, and since the energy $\mathcal{E}(k)$ is a function of the magnitude of the wave vector **k**, we have let $f(\mathbf{k},s) \equiv f(k)$. Also, because the distribution function is an explicit function of energy, we have set $f(k) \equiv f(\mathcal{E})$. The parameter μ in Eq. (2.70) is called the *chemical potential*. It is a measure of the change in the energy of a system of electrons if an additional electron is added. The chemical potential is a function of temperature and electron density. When $\mathcal{E} = \mu$, the Fermi function has the value 1/2, since $\exp(\mathcal{E}-\mu)/k_B T \rightarrow 1$ for $\mathcal{E} \rightarrow \mu$. For $T = 0$, the chemical potential is often denoted as "Fermi energy" $\mathcal{E}_F$. In the low-temperature limit, when $T \rightarrow 0$, the probability distribution approaches 1 for $\mathcal{E} < \mu$ because then $(\mathcal{E} - \mu) < 0$, making $(\mathcal{E} - \mu)/k_B T \rightarrow -\infty$, $\exp(\mathcal{E} - \mu)/k_B T \rightarrow 0$, and, therefore, $f(\mathcal{E}) \rightarrow 1$. On the other hand, for $\mathcal{E} > \mu$, we have $(\mathcal{E} - \mu) > 0$ and $(\mathcal{E} - \mu)/k_B T \rightarrow \infty$, causing $\exp(\mathcal{E} - \mu)/k_B T \rightarrow \infty$ and, consequently, forcing $f(\mathcal{E}) \rightarrow 0$. Thus, in the zero temperature limit, the Fermi function becomes

$$f(\mathcal{E}) = \begin{cases} 1 & \text{for } \mathcal{E} < \mu \\ 0 & \text{for } \mathcal{E} > \mu \end{cases}, \qquad (2.71)$$

indicating simply that all states with $\mathcal{E} < \mu$ are occupied, while all states with $\mathcal{E} > \mu$ are unoccupied. In Fig. 2.18 we plot the Fermi function for three different temperatures. The solid curve for $T = 10$ K approximates the $T = 0$ K result of Eq. (2.71) quite well.

To determine μ, we sum the occupation probability over all possible states to obtain the total number of particles in the system, N; i.e.,

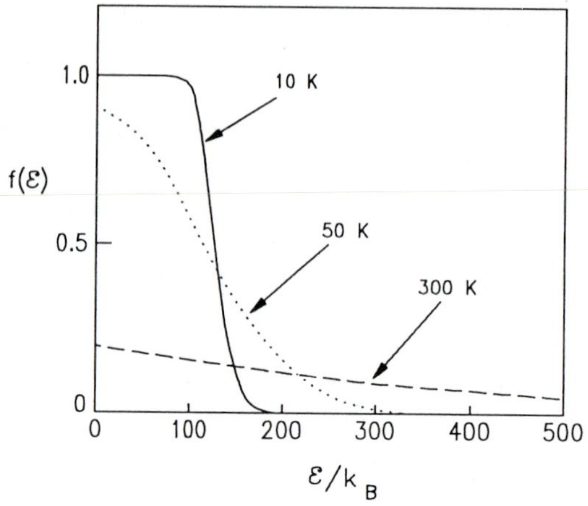

Figure 2.18. Fermi-Dirac distribution function, $f(\mathcal{E})$, as a function of $\mathcal{E}/k_B$ for the particle density $n = 10^{18}\ cm^{-3}$ and three different temperatures. Note that the specified carrier density determines the chemical potential μ in Eq. (2.70).

$$\sum_{\mathbf{k},s} f(k) = N . \tag{2.72}$$

The chemical potential has to be adjusted such that for any temperature and energy $\mathcal{E}(k)$, Eq. (2.72) is satisfied. In order to evaluate the sum in Eq. (2.72), it is convenient to convert the sum over **k** into an integral. This conversion is possible because there is a large number of electronic states with infinitesimally small spacings between them. In solids there are 6.022×10^{23} atoms per mole (Avogadro's number) and ρ_m/M' moles per unit volume, where ρ_m is mass density and M' is the atomic weight of the elements. For atoms with Z valence electrons (Z = 1 for Na and Z = 2 for Mg, etc.), the number of conduction electrons per unit volume V becomes $n = N/V = 6.022 \times 10^{23} \times \rho_m Z/M'$, which is on the order of 10^{22} to 10^{23} cm^{-3}. Thus, it is useful to introduce the concept of the density of states in order to facilitate the conversion of the sum into an integral. The density of states may be obtained by re-examining the Schrödinger equation for noninteracting electrons,

$$-\frac{\hbar^2}{2m_0} \nabla^2\psi(\mathbf{r}) = \mathcal{E}\psi(\mathbf{r}) . \tag{2.73}$$

The solution of this equation for electrons in a periodic lattice with the volume $V = L_x L_y L_z$ is

$$\psi(\mathbf{r}) = \frac{1}{V^{1/2}} e^{i\mathbf{k}\cdot\mathbf{r}} = \frac{1}{V^{1/2}} e^{i(k_x x + k_y y + k_z z)} . \tag{2.74}$$

The periodic boundary condition gives

$$\psi(x + L_x, y, z) = \psi(x, y + L_y, z) = \ldots = \psi(x, y, z) , \tag{2.75}$$

or explicitly,

$$e^{ik_x(x + L_x)} = e^{ik_x x} ,$$
$$e^{ik_y(y + L_y)} = e^{ik_y y} \ldots , \tag{2.76}$$

giving

$$k_x = \frac{2\pi n_x}{L_x} , \; k_y = \frac{2\pi n_y}{L_y} , \; k_z = \frac{2\pi n_z}{L_z} , \tag{2.77}$$

where n_x, n_y, and n_z can assume the values 0, ±1, ±2, ... The energy eigenvalue is then

$$\mathscr{E} = \frac{\hbar^2 k^2}{2m_0} = \frac{2\hbar^2\pi^2}{m_0} \left(\frac{n_x^2}{L_x^2} + \frac{n_y^2}{L_y^2} + \frac{n_z^2}{L_z^2} \right) . \tag{2.78}$$

Equation (2.78) indicates that the electron energy is discrete. In order to calculate the density of these discrete electronic states, we note that the separation, Δk_i, between two allowed k_i values is $2\pi/L_i$, i = x, y, z (see Fig. 2.19). Therefore, there is one allowed value of **k** per unit k-space volume

$$\Delta k = \Delta k_x \, \Delta k_y \, \Delta k_z = \frac{2\pi}{L_x} \frac{2\pi}{L_y} \frac{2\pi}{L_z} = \frac{8\pi^3}{V} . \tag{2.79}$$

The sum in Eq. (2.72) may be written slightly differently by noting that Δk is a constant, independent of k,

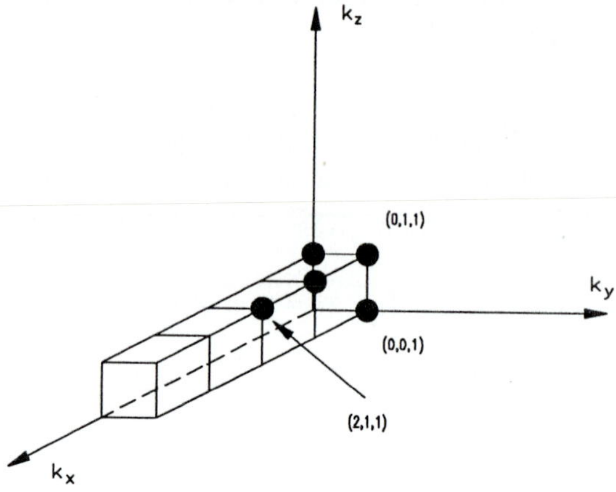

Figure 2.19. Allowed values of k. Each division has a length of $2\pi/L$, where $L = L_x, L_y$, or L_z in corresponding directions.

$$\sum_{\mathbf{k}} = \frac{1}{\Delta k} \sum_{\mathbf{k}} (\Delta k) \,. \tag{2.80}$$

Substituting for Δk from Eq. (2.79) for the prefactor of Eq. (2.80) we have

$$\sum_{\mathbf{k}} = \frac{V}{8\pi^3} \sum_{\mathbf{k}} (\Delta k) \,. \tag{2.81}$$

The sum may now be replaced by an integral as

$$\sum_{\mathbf{k}} \rightarrow \frac{V}{8\pi^3} \int d\mathbf{k} \,, \tag{2.82}$$

where $d\mathbf{k} = k^2 \, dk \, \sin\theta \, d\theta \, d\phi$ is the volume element in the three-dimensional k space. If the function under the integral depends only on the magnitude of $\mathbf{k}$, the angle integrations simply yield a prefactor 4π. Equation (2.72) may now be evaluated using Eq. (2.82),

$$N = 2\,\frac{V}{8\pi^3}\,4\pi \int dk\; k^2\, f(k)\ , \tag{2.83}$$

where the first factor of 2 is the result of the summation over the two possible spin values (±1/2) of the electrons. Combining the various factors, Eq. (2.83) can be written as

$$n = \frac{N}{V} = \frac{1}{\pi^2} \int dk\; k^2\, f(k)\ . \tag{2.84}$$

The k integration in Eq. (2.84) may be replaced by an integration over the energy; i.e.,

$$n = \int d\mathcal{E}\; g(\mathcal{E})\, f(\mathcal{E})\ , \tag{2.85}$$

where we have introduced the energy density of states $g(\mathcal{E})$, expressing the number of states per energy interval $d\mathcal{E}$ per spin orientation per unit volume of real space. Since Eqs. (2.84) and (2.85) both yield the density of electrons n, we must have

$$\frac{1}{\pi^2} \int dk\; k^2\, f(k) = \int d\mathcal{E}\; g(\mathcal{E})\, f(\mathcal{E})\ . \tag{2.86}$$

According to Eq. (2.70), $f(\mathcal{E}) = f(k)$, allowing evaluation of $g(\mathcal{E})$ from Eq. (2.86):

$$g(\mathcal{E})d\mathcal{E} = \frac{k^2}{\pi^2}\,dk\ ,\ \text{or}\ g(\mathcal{E}) = \frac{k^2}{\pi^2}\left(\frac{d\mathcal{E}}{dk}\right)^{-1}\ . \tag{2.87}$$

Using Eq. (2.78) we obtain

$$g(\mathcal{E}) = \frac{k}{\pi^2}\frac{m_0}{\hbar^2} = \frac{m_0}{\pi^2\hbar^2}\left(\sqrt{\frac{2m_0\mathcal{E}}{\hbar^2}}\right) \tag{2.88}$$

or

$$\boxed{g(\mathcal{E}) = \frac{1}{2\pi^2}\left(\frac{2m_0}{\hbar^2}\right)^{3/2} \mathcal{E}^{1/2} ,}$$

density of states (2.89)

which is called the density of states. The concept of density of states is important for the understanding of the optical response of semiconductors. It expresses how many states are available in the system in an energy interval comprised between $\mathcal{E}$ and $\mathcal{E} + d\mathcal{E}$. The density of states is plotted in Fig. 2.20. The electron density may now be calculated from Eq. (2.85) using Eqs. (2.89) and (2.70),

$$n = \frac{1}{2\pi^2}\left(\frac{2m_0}{\hbar^2}\right)^{3/2} \int_0^\infty d\mathcal{E}\, \frac{\sqrt{\mathcal{E}}}{1 + \exp[(\mathcal{E} - \mu)/k_B T]} . \tag{2.90}$$

Unfortunately, this integral does not have a closed form analytic solution, and μ cannot be expressed as a simple function of density and temperature. However, to a very good approximation, $\mu(n, T)$ can be written as

$$\frac{\mu}{k_B T} \simeq \ln(\nu) + K_1 \ln(K_2\nu + 1) + K_3\nu , \tag{2.91}$$

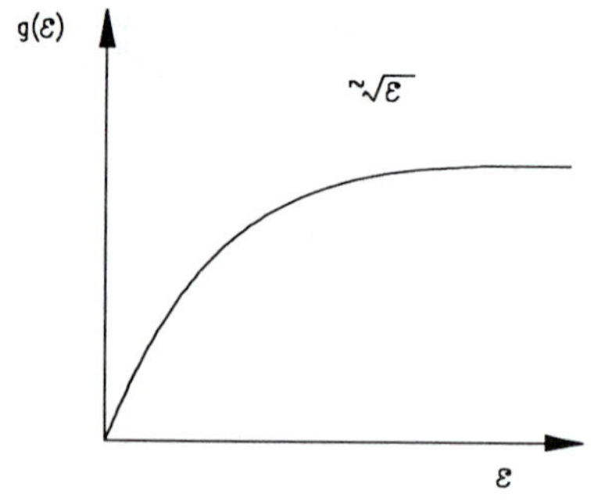

Figure 2.20. Density of energy states for electrons.

where $\nu = n/n'_0$, $4n'_0 = (m_0 k_B T/\hbar^2\pi)^{3/2}$, $K_1 = 4.8966851$, $K_2 = 0.04496457$, and $K_3 = 0.1333760$ (see Ref. 2.5 for more details).

To calculate the Fermi energy at T = 0, we use the degenerate Fermi function of Eq. (2.71) and obtain from Eq. (2.85)

$$n = \frac{1}{2\pi^2}\left(\frac{2m_0}{\hbar^2}\right)^{3/2}\int_0^{\mathcal{E}_F} d\mathcal{E}\,\mathcal{E}^{1/2}$$

$$= \frac{1}{2\pi^2}\left(\frac{2m_0}{\hbar^2}\right)^{3/2}\frac{2}{3}\mathcal{E}_F^{3/2}, \tag{2.92}$$

with the result

$$\mathcal{E}_F = \frac{\hbar^2}{2m_0}(3\pi^2 n)^{2/3}. \tag{2.93}$$

For typical metals, the value of the Fermi energy is 2.5 eV - 5 eV. To use the derived equations for the electrons in a crystal, one has to replace the free electron mass m_0 by the appropriate effective electron mass m_e.

It is customary to speak of the "classical regime," if the quantity $\exp(\mathcal{E} - \mu)/k_B T >> 1$. Then the "1" in the denominator of the Fermi function, Eq. (2.70), may be ignored in comparison with $\exp(\mathcal{E} - \mu)/k_B T$, and the Fermi-Dirac distribution reduces to

$$f(\mathcal{E}) \simeq e^{\mu/k_B T}\exp(-\mathcal{E}/k_B T). \tag{2.94}$$

Inserting Eq. (2.94) into Eq. (2.85) allows analytical evaluation of the integral and determination of the chemical potential. We find

$$n = \frac{e^{\mu/k_B T}}{2\pi^2}\left(\frac{2m_0}{\hbar^2}\right)^{3/2}\int_0^{\infty} d\mathcal{E}\,\mathcal{E}^{1/2}\exp(-\mathcal{E}/k_B T)$$

$$= \frac{e^{\mu/k_B T}}{2\pi^2}\left(\frac{2m_0 k_B T}{\hbar^2}\right)^{3/2}\int_0^{\infty} dx\,x^{1/2}e^{-x}$$

$$= \frac{1}{4}e^{\mu/k_B T}\left(\frac{2m_0 k_B T}{\hbar^2\pi}\right)^{3/2}, \tag{2.95}$$

where the value of the integral, $\pi^{1/2}/2$, was inserted. Equation (2.95) gives the chemical potential

$$e^{\mu/k_B T} = 4n \left(\frac{\hbar^2\pi}{2m_0 k_B T}\right)^{3/2} , \tag{2.96}$$

and the distribution of Eq. (2.94) becomes

$$f(\mathcal{E}) = 4n \left(\frac{\hbar^2\pi}{2m_0 k_B T}\right)^{3/2} \exp(-\mathcal{E}/k_B T) , \tag{2.97}$$

which is the well-known Maxwell-Boltzmann distribution function. The approximation leading to Eq. (2.94) is also known as the nondegenerate, or classical limit of Fermi-Dirac statistics. The dashed curve in Fig. 2.18, which represents the Fermi-Dirac distribution function for T = 300 K, is indistinguishable from the Maxwell-Boltzmann distribution for the parameters chosen, consistent with Eq. (2.97).

For completeness, it should be mentioned that particles with integer spins, which are called Bosons, do not follow Fermi-Dirac statistics, but obey the Bose-Einstein (BE) distribution law. The BE distribution function is similar to Eq. (2.70) except that the +1 in the denominator is replaced by a -1. We will see later that photons and phonons follow BE statistics with zero chemical potential, $\mu = 0$. Like the Fermi-Dirac statistics, the BE distribution also converges toward the Maxwell-Boltzmann distribution in the classical limit.

2-9. Electrons and Holes

As we have seen, the energy bands in a given crystal may be empty or completely or partially filled with electrons. When an electric field is applied, no current can arise from an entirely empty band, since no charge carriers are available. Consider now a completely filled band. The current density arising from this band is (following Ref. 2.10)

$$J = -ne\bar{v} , \tag{2.98}$$

where n is the electron density in that band and $\bar{v}$ is the average electron velocity given by

$$\bar{v} = \frac{1}{nV} \sum_i v_i \,. \tag{2.99}$$

Equation (2.98) may be written as

$$J = -\frac{e}{V} \sum_i v_i \,. \tag{2.100}$$

But if the band is entirely filled, the summation is zero. For every state of positive velocity, $1/\hbar\ (\partial\mathcal{E}/\partial k)$, with positive slope (such as point B in Fig. 2.21), there is a corresponding state of negative velocity and equal magnitude with negative slope at $k' = -k$ (such as point A in Fig. 2.21).

Let us now consider a band that is completely filled by electrons except for one state with velocity v_s. Then

$$J = -\frac{e}{V} \sum_{i \neq s} v_i = -\frac{e}{V}\left[\sum_i v_i - v_s\right] = \frac{e}{V} v_s \,. \tag{2.101}$$

Equation (2.101) shows that the total current of all the electrons in the band with one vacant state is equivalent to a current, due to the motion of one particle with a positive charge, +e, occupying the respective state. Such a fictitious particle is a *hole*. (Note that under no applied field the current is still zero, since v_s, which is related to the slope of the energy band at k = 0, is zero. However, with the applied field there is a current

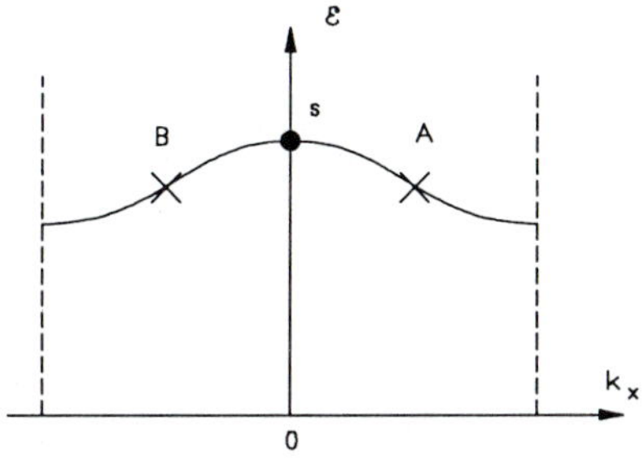

Figure 2.21. A vacancy exists in state s. All the other levels on this band are assumed to be filled.

since v_s is no longer zero.) For clarity we will use the term holes only for missing electrons in otherwise full valence bands, which are concave downward as shown in Fig. 2.16(c). The effective electron mass in such a band is negative because $d^2\mathcal{E}/dk^2$ is negative. Therefore, the holes in this band have positive effective masses. In order to see this, we note that the total energy of the band does not change when we go from the electron to the *electron-hole representation* where the absence of an electron is described by the presence of a hole. For a band with nV = N′, electron states of which N′ - 1 states are occupied and one state is empty, the total energy can be written as

$$\mathcal{E}_{total} = (N' - 1)\mathcal{E}_e , \tag{2.102}$$

where we assume, for simplicity, that each electron has the same energy, $\mathcal{E}_e$. This energy should be the same as the energy of a band with N′ electron states and one hole state in the electron-hole representation. Equation (2.102) may be written as

$$\mathcal{E}_{total} = (N' - 1)\mathcal{E}_e = N'\mathcal{E}_e + \mathcal{E}_h , \tag{2.103}$$

where $\mathcal{E}_h$ is the energy of the hole state, which obviously must satisfy

$$\mathcal{E}_h = -\mathcal{E}_e . \tag{2.104}$$

Introducing the effective hole mass through the relation

$$\mathcal{E}_h = \frac{\hbar^2 \mathbf{k}_h^2}{2m_h} , \tag{2.105}$$

and using

$$\mathcal{E}_e = \frac{\hbar^2 \mathbf{k}_e^2}{2m_e} , \tag{2.106}$$

we see immediately that Eq. (2.104) can only be fulfilled if

$$m_h = -m_e . \tag{2.107}$$

Therefore, the negative electron mass of Fig. 2.16(c) translates into a positive hole mass. Furthermore, in the presence of an electric field **E** one may write

$$\hbar \frac{d\mathbf{k}_e}{dt} = -e\mathbf{E} \tag{2.108}$$

for electrons and

$$\hbar \frac{d\mathbf{k}_h}{dt} = +e\mathbf{E} \tag{2.109}$$

for holes, which leads to

$$\mathbf{k}_h = -\mathbf{k}_e \ . \tag{2.110}$$

The velocities of the electron and hole are related by

$$\hbar\mathbf{k}_e = m_e \boldsymbol{v}_e \quad \text{and} \quad \hbar\mathbf{k}_h = m_h \boldsymbol{v}_h \ , \tag{2.111}$$

and

$$\boldsymbol{v}_h = \frac{\hbar\mathbf{k}_h}{m_h} = \frac{-\hbar\mathbf{k}_e}{-m_e} = \boldsymbol{v}_e \ , \tag{2.112}$$

showing that the velocities of the electron and the hole are the same.

2-10. Problems

2.1. (a) Prove that the reciprocal lattice vector $\mathbf{G} = h\mathbf{A} + k\mathbf{B} + \ell\mathbf{C}$ is perpendicular to the plane $(hk\ell)$ in the direct lattice.

(b) Let the distance between two adjacent planes parallel to the plane $(hk\ell)$ in the direct lattice be d. Show that $d = 2\pi/|\mathbf{G}|$, where $\mathbf{G} = h\mathbf{A} + k\mathbf{B} + \ell\mathbf{C}$.

(c) For a simple cubic lattice of side a, show that $d^2 = a^2/(h^2 + k^2 + \ell^2)$.

2.2. Prove that crystal lattices do not have fivefold rotation axes.

2.3. Show that the dielectric tensor for a homogeneous nonabsorbing medium is a symmetric tensor with $\epsilon_{ij} = \epsilon_{ji}$. (Hint: use the fact that the energy density arising from the electric field in the medium is 1/2 $\mathbf{E} \cdot \mathbf{D}$.)

2.4. Using the rotation symmetry properties of simple cubic crystals, show that the off-diagonal matrix elements for the dielectric tensor are zero and the diagonal elements are identical.

2.5. Calculate the degeneracy of the lowest three energy bands at the Γ point for a face-centered cubic lattice in the free-electron approximation. The Γ point is the point k = 0 in momentum space. (Degeneracy is defined as the independent number of wave functions for the same energy. For simplicity ignore the spin degeneracy.)

2.6. Show that the wave function for the tight-binding model

$$\psi_k(\mathbf{r}) = \sum_n e^{i\mathbf{k}\cdot\mathbf{r}_n}\, \phi_k(\mathbf{r} - \mathbf{r}_n)\,, \tag{2.113}$$

satisfies the Bloch theorem, Eq. (2.37).

2.7. For a one-dimensional solid in the tight-binding approximation, one finds

$$\mathcal{E} = \mathcal{E}_k - A - 2B \cos ka\,. \tag{2.114}$$

(a) By expanding the function, calculate the effective mass.

(b) Show that Eq. (2.66) gives the same value for the effective mass.

2.8. In solid state physics one often wants to compute the bulk properties without dealing with surface effects. To do this formally, it is common to introduce periodic boundary conditions where the wave function stays the same when the translation $\mathbf{r} \rightarrow \mathbf{r} + N\mathbf{a}_i$ is performed, where N is the total number of primitive cells in the crystal, and $\mathbf{a}_i$ are the three primitive lattice vectors. Consider a one-dimensional solid with N unit cells each of length a. Use the periodic boundary condition and show that the number of allowed states is equal to the number of unit cells.

2-11. References

1. A. O. E. Animalu, *Intermediate Quantum Theory of Crystalline Solids* (Prentice Hall, Englewood Cliffs, New Jersey, 1977).

2. N. W. Ashcroft and N. D. Mermin, *Solid State Physics* (Holt, Rinehart, and Winston, New York, 1976).

3. M. J. Buerger, *X-Ray Crystallography* (Wiley, New York, 1942).

4. N. Hamermesh, *Group Theory* (Addison-Wesley, Reading, Massachusetts, 1964).

5. H. Haug and S. W. Koch, *Quantum Theory of the Optical and Electronic Properties of Semiconductors*, Second Edition (World Scientific, Singapore, 1993).

6. C. Kittel, *Introduction to Solid State Physics* (John Wiley & Sons, New York, 1976).

7. G. F. Koster, J. O. Dimmock, R. G. Wheeler, and H. Statz, *Properties of the Thirty-Two Point Groups* (Massachusetts Institute of Technology Press, Cambridge, Massachusetts, 1963).

8. R. A. Levy, *Principles of Solid State Physics* (Academic Press, New York, 1968).

9. G. D. Mahan, *Many-Particle Physics* (Plenum Press, New York, 1981).

10. J. P. McKelvey, *Solid State and Semiconductor Physics* (Harper & Row, New York, 1966).

11. M. A. Omar, *Elementary Solid State Physics: Principles and Applications* (Addison-Wesley, Massachusetts, 1975).

12. M. Tinkham, *Group Theory and Quantum Mechanics* (McGraw-Hill, New York, 1964).

13. S. Wang, *Fundamentals of Semiconductor Theory and Device Physics* (Prentice Hall, Englewood Cliffs, New Jersey, 1989).

14. C. M. Wolfe, N. Holonyak, Jr., G. E. Stillman, *Physical Properties of Semiconductors* (Prentice Hall, Englewood Cliffs, New Jersey, 1989).

Chapter III
BASIC CONCEPTS OF OPTICAL RESPONSE

In this chapter we discuss the basic concepts needed to analyze the optical response of crystals. First, we look for solutions of Maxwell's equation describing waves which can propagate inside the medium. The variation of the frequency of the waves as a function of their wave vectors, called the dispersion relation, is derived both for longitudinal and transverse waves. Then a simple oscillator model is applied to describe the response of the crystal. Several features of this oscillator model are discussed, since it has some experimental success in describing the optical properties of metals. We introduce the concept of plasma frequency and plasma oscillations, describing collective excitations of the entire electron gas. The complex dielectric function for the oscillator model is derived, and we establish the Kramers-Kronig relations between the real and imaginary part of this function. The connection between the dielectric function and optical absorption and refraction is discussed. As we will see in later parts of this book, most of the concepts outlined in this chapter have validity well beyond the framework in which we give the explicit derivations.

3-1. Dispersion Relation

We start from Maxwell's equations in the absence of external charges (given in CGS - Gaussian units; see appendix for MKS units):

$$\nabla \cdot \mathbf{D} = 0 \qquad (3.1)$$

$$\nabla \cdot \mathbf{H} = 0 \qquad (3.2)$$

$$\nabla \times \mathbf{E} = -\frac{1}{c}\frac{\partial \mathbf{H}}{\partial t} \qquad (3.3)$$

$$\nabla \times \mathbf{H} = \frac{1}{c}\frac{\partial \mathbf{D}}{\partial t} \qquad (3.4)$$

Maxwell's equations without external charges

Here **D** is the displacement field and **H** is the magnetic field for which **H** = **B** at optical frequencies. The Fourier-transform of the displacement field, $\mathbf{D}(\omega,\mathbf{q})$, can be expressed in terms of the Fourier-transform of the electric field, $\mathbf{E}(\omega,\mathbf{q})$, through the dielectric function, $\epsilon(\omega,\mathbf{q})$

$$\mathbf{D}(\omega,\mathbf{q}) = \epsilon(\omega,\mathbf{q})\ \mathbf{E}(\omega,\mathbf{q})\ . \tag{3.5}$$

In general, the dielectric function is a tensor, relating the different components of **D** and **E** [see Chap. II, Eqs. (2.25)-(2.27)].

We now examine the dispersion relation between the frequency, $\omega(\mathbf{q})$, and the photon wave vector, **q**. Consider an electromagnetic wave with electric field given by

$$\mathbf{E} = \mathbf{E}_0\, e^{i(\mathbf{q}\cdot\mathbf{r}\ -\ \omega t)}\ . \tag{3.6}$$

Inserting Eq. (3.6) into the first of Maxwell's equations, Eq. (3.1), and using Eq. (3.5), we obtain

$$\epsilon\ \nabla\cdot\mathbf{E} = \epsilon\ i\mathbf{q}\cdot\mathbf{E} = 0\ . \tag{3.7}$$

This relation has to be satisfied for both longitudinal and transverse waves. For a longitudinal wave, which is a wave whose propagation direction is parallel to the direction of its amplitude, $\mathbf{q} \parallel \mathbf{E}_0$, we have

$$\mathbf{q}\cdot\mathbf{E} \neq 0\ . \tag{3.8}$$

Hence, Eq. (3.7) can only be satisfied if

$$\boxed{\epsilon = \epsilon_\ell(\omega,\mathbf{q}) = 0\ .} \tag{3.9}$$

dispersion relation for longitudinal waves

This important relation indicates that every time the dielectric function of the medium vanishes, a propagating longitudinal wave can be sustained by the medium. For transverse waves, on the other hand, **q** is perpendicular to $\mathbf{E}_0$ so that

$$\mathbf{q}\cdot\mathbf{E} = 0\ , \tag{3.10}$$

and Eq. (3.7) is always satisfied. Thus, the transverse dielectric function ϵ cannot be determined from this equation. To determine the dielectric function for transverse waves, we use the third and fourth of Maxwell's equations. Operating $\nabla \times$ on both sides of Eq. (3.3) and substituting for $\nabla \times \mathbf{H}$ from Eq. (3.4) gives

$$\nabla \times (\nabla \times \mathbf{E}) = -\frac{1}{c}\frac{\partial}{\partial t}(\nabla \times \mathbf{H}) = -\frac{1}{c^2}\frac{\partial^2 \mathbf{D}}{\partial t^2} . \tag{3.11}$$

Using the equality $\nabla \times (\nabla \times \mathbf{E}) = \nabla(\nabla \cdot \mathbf{E}) - \nabla^2 \mathbf{E}$ and $\mathbf{D} = \epsilon \mathbf{E}$, and noting again that for a transverse wave, $\nabla \cdot \mathbf{E} = 0$, we may write Eq. (3.11) as

$$\boxed{\nabla^2 \mathbf{E} = \frac{\epsilon}{c^2}\frac{\partial^2}{\partial t^2}\mathbf{E} .}$$

transverse wave equation (3.12)

Inserting Eq. (3.6) into Eq. (3.12) and evaluating the derivatives yields

$$\left(q^2 - \frac{\omega^2}{c^2}\epsilon\right)\mathbf{E} = 0 , \tag{3.13}$$

which requires

$$\boxed{\epsilon = \epsilon_t(\omega,\mathbf{q}) = \frac{q^2c^2}{\omega^2} .}$$

dispersion relation for transverse waves (3.14)

This equation relates the frequency of a propagating transverse wave inside the medium to its wave vector. A transverse wave can propagate as a stable solution only if this relation is satisfied.

It is usually the case that the dielectric function is a complex quantity,

$$\epsilon(\omega,\mathbf{q}) = \mathrm{Re}[\epsilon(\omega,\mathbf{q})] + i\,\mathrm{Im}[\epsilon(\omega,\mathbf{q})] \equiv \epsilon'(\omega,\mathbf{q}) + i\epsilon''(\omega,\mathbf{q}) , \tag{3.15}$$

where the index t is dropped for simplicity of notation. Inserting this into the wave equation (3.12) yields

$$\nabla^2 \mathbf{E} = \frac{\epsilon'}{c^2}\frac{\partial^2}{\partial t^2}\mathbf{E} + i\,\frac{\epsilon''}{c^2}\frac{\partial^2}{\partial t^2}\mathbf{E} . \tag{3.16}$$

To solve this equation, we make the ansatz,

$$\mathbf{E} = \mathbf{E}_0 \, e^{-i\omega t} \, e^{i(q + i\kappa)z} , \tag{3.17}$$

where we used Eq. (3.6) and let $q \rightarrow q + i\kappa$, and for simplicity we assumed one-dimensional propagation in z direction. The quantity κ is the *extinction coefficient* which is related to the absorption coefficient α, as shown in the following. Inserting Eq. (3.17) into Eq. (3.16) and evaluating the derivative yields

$$(q + i\kappa)^2 = \frac{\omega^2}{c^2} (\epsilon' + i\,\epsilon'') . \tag{3.18}$$

The complex Eq. (3.18) can be separated into two real equations for the real and imaginary parts. These equations are

$$q^2 - \kappa^2 = \frac{\omega^2}{c^2} \epsilon' \tag{3.19}$$

and

$$2\,\kappa\,q = \frac{\omega^2}{c^2} \epsilon'' . \tag{3.20}$$

The *refractive index* $n(\omega)$ is introduced as the ratio between the wave number q in the medium and the wave number $q_0 = \omega/c$ in vacuum,

$$n(\omega) \equiv \frac{q}{q_0} = \frac{qc}{\omega} . \tag{3.21}$$

Equation (3.21) may be used to eliminate q from Eqs. (3.19) and (3.20) to obtain

$$n^2(\omega) = \frac{\kappa^2 c^2}{\omega^2} + \epsilon' \tag{3.22}$$

and

$$4\,\frac{\kappa^2 c^2}{\omega^2}\, n^2(\omega) = (\epsilon'')^2 , \tag{3.23}$$

where we took the square of Eq. (3.20). Solving Eq. (3.23) for $\kappa^2 c^2/\omega^2$, inserting into Eq. (3.22), and multiplying by $n^2(\omega)$ yields

$$n^4(\omega) - \epsilon' n^2(\omega) - \frac{\epsilon''^2}{4} = 0 , \qquad (3.24)$$

with the solution

$$n^2(\omega) = \frac{1}{2} [\epsilon' + \sqrt{(\epsilon')^2 + (\epsilon'')^2}] . \qquad (3.25)$$

Hence, the refractive index is

$$\boxed{n(\omega) = \sqrt{\frac{1}{2}\left[\epsilon' + \sqrt{(\epsilon')^2 + (\epsilon'')^2}\right]} \quad .}$$

refractive index (3.26)

From Eq. (3.23) we see that

$$2\kappa = \frac{\omega}{cn(\omega)} \epsilon''(\omega) . \qquad (3.27)$$

This is the *intensity absorption coefficient*, or simply, absorption coefficient,

$$\boxed{\alpha(\omega) = 2\kappa(\omega) = \frac{\omega}{cn(\omega)} \epsilon''(\omega) \quad .}$$

intensity absorption coefficient (3.28)

To understand this terminology, let us take the absolute square of Eq. (3.17) to obtain

$$|\mathbf{E}|^2 = |\mathbf{E}_0|^2 \, e^{-2\kappa z} = |\mathbf{E}_0|^2 \, e^{-\alpha z} . \qquad (3.29)$$

The radiation intensity I(z) is proportional to the absolute square of the field amplitude, allowing us to rewrite Eq. (3.29) as

$$\boxed{I(z) = I_0 \, e^{-\alpha z} \quad .}$$

Beer-Lambert's law (3.30)

This equation, which is referred to as the Beer-Lambert's law, states that the intensity of the incident radiation I_0 is attenuated exponentially in the medium. Note here that I_0 is the light intensity at the front surface *inside* the medium. The absorption coefficient is a measure of the propagation distance of the optical beam into the medium before the beam is dissipated to 1/e of its initial value [at $z = 1/\alpha(\omega)$ we get $\alpha(\omega)z = 1$, and $I/I_0 = 1/e$].

Rather than dealing with a refractive index and an absorption coefficient, one can also formally introduce a complex refractive index. The imaginary part of this complex index is then related to the absorption (see Chap. IV for more details).

3-2. Oscillator Model

To obtain a simple example for absorption coefficient and refractive index we use a simple oscillator model to describe the medium. This model was originally introduced in the early papers by Lorentz, Planck, and Einstein. It has been used and expanded in many ways (for example, by Drude) to explain some optical properties of simple metals. Here we derive some of Drude's results.

Let us assume that the crystal consists of charges (the electrons) which can be set in motion by an oscillating electric field of light, E(t), polarized in the x direction. Newton's second law gives the equation of motion of each electron in the ac field of light as

$$m_e \frac{d^2x(t)}{dt^2} + \frac{m_e}{\tau} \frac{dx(t)}{dt} = -eE(t) , \tag{3.31}$$

where x(t) represents the displacement of the electron from its equilibrium position. The second term in Eq. (3.31) describes a phenomenological damping term for the electrons, e.g., due to collisions with the ions. τ is the decay or relaxation time.

If we assume a monochromatic field with frequency ω,

$$E(t) \sim E_0 \, e^{-i\omega t} , \tag{3.32}$$

the displacement x(t), which is driven by that field, also oscillates accordingly,

$$x(t) \sim x_0 \, e^{-i\omega t} . \tag{3.33}$$

Substituting for E(t) and x(t) from Eqs. (3.32) and (3.33) into Eq. (3.31) and dividing both sides by $\exp(-i\omega t)$, we obtain

$$-m_e\omega^2 x_0 - m_e \frac{i\omega}{\tau} x_0 = -eE_0 \tag{3.34}$$

or

$$x_0 = \frac{e\tau}{m_e\omega} E_0 \frac{\omega\tau - i}{\omega^2\tau^2 + 1} . \tag{3.35}$$

This represents a particular solution of the second-order differential equation, Eq. (3.31), resulting from the driving term. The dipole moment -ex(t) times the electron density n gives the polarization density P_c of the conduction electrons,

$$P_c = -nex(t) = -\frac{ne^2\tau}{m_e\omega} E(t) \frac{\omega\tau - i}{\omega^2\tau^2 + 1} , \tag{3.36}$$

where we used Eqs. (3.33) and (3.35).

In addition to the electronic polarization, P_c, the applied electric field also induces a polarization, P_b, denoted as the background polarization. This background polarization is due to the displacement of bound particles, such as inner-shell electrons, about their equilibrium positions. The total polarization is, therefore,

$$P = P_b + P_c . \tag{3.37}$$

We define the *optical susceptibility*, $\chi(\omega)$, of the material as

$$\boxed{\chi(\omega) = \frac{P(\omega)}{E(\omega)} .}$$

optical susceptibility (3.38)

Since the displacement field **D** is given in terms of the polarization as

$$\mathbf{D} = \mathbf{E} + 4\pi\, \mathbf{P} = \mathbf{E}\, [\, 1 + 4\pi\chi(\omega)\,] , \tag{3.39}$$

and in terms of the dielectric function as

$$\mathbf{D} = \epsilon(\omega)\, \mathbf{E} , \tag{3.40}$$

a simple relation exists between susceptibility and dielectric function,

$$\epsilon(\omega) = 1 + 4\pi\chi(\omega) . \tag{3.41}$$

Using Eq. (3.37) we can write

$$\epsilon(\omega) = 1 + 4\pi \frac{P_b}{E} + 4\pi \frac{P_c}{E}$$

$$= 1 + 4\pi \frac{P_b}{E} - \frac{4\pi ne^2\tau}{m_e\omega} \frac{\omega\tau - i}{\omega^2\tau^2 + 1}$$

$$= \epsilon'(\omega) + i\epsilon''(\omega) . \tag{3.42}$$

For high frequencies, $\omega \rightarrow \infty$, the frequency-dependent term in Eq. (3.42) approaches zero and we obtain

$$\epsilon(\omega \rightarrow \infty) \equiv \epsilon_\infty = 1 + 4\pi \frac{P_b}{E} . \tag{3.43}$$

With this notation Eq. (3.42) becomes

$$\boxed{\epsilon(\omega) = \epsilon_\infty - \frac{4\pi ne^2\tau}{m_e\omega} \frac{\omega\tau - i}{\omega^2\tau^2 + 1} .}$$

oscillator (Drude) dielectric function (3.44)

Equation (3.44) can also be written as

$$\epsilon(\omega) = \epsilon_\infty - \frac{4\pi ne^2\tau^2}{m_e} \frac{1}{\omega^2\tau^2 + 1} + i \frac{4\pi ne^2\tau}{m_e\omega} \frac{1}{\omega^2\tau^2 + 1} . \tag{3.45}$$

Let us now analyze the behavior of this dielectric function in two limiting cases.

Low-Frequency Regime, $\omega << \frac{1}{\tau}$ or $\omega\tau << 1$

Conductivity measurements have revealed that the relaxation time τ is in the range of $\tau \simeq 10^{-13}$ s to 10^{-14} s for simple metals, so that the frequency $1/\tau$ corresponds to the infrared part of the spectrum. Since $\omega << 1/\tau$, we have $1/\omega >> \tau$ and

$$\frac{4\pi ne^2\tau}{m_e\omega} \frac{1}{\omega^2\tau^2 + 1} >> \frac{4\pi ne^2\tau^2}{m_e} \frac{1}{\omega^2\tau^2 + 1} >> \epsilon_\infty , \tag{3.46}$$

so that the dielectric function (3.45) becomes

$$\epsilon(\omega) \simeq i\,\frac{4\pi n e^2 \tau}{m_e \omega} = i\,\epsilon''(\omega), \tag{3.47}$$

where the term $\omega^2\tau^2$ in the denominator of Eq. (3.45) was ignored in comparison to 1. The dielectric function is, thus, purely imaginary in this limit, making the material highly absorptive [see Eq. (3.28)]. From Eq. (3.26) we then obtain the refractive index as

$$n(\omega) = \sqrt{\frac{1}{2}\,\epsilon''} = \sqrt{\frac{2\pi n e^2 \tau}{m_e \omega}}\,. \tag{3.48}$$

Inserting Eqs. (3.47) and (3.48) into Eq. (3.28), we get

$$\alpha(\omega) = \frac{\omega}{c}\,\sqrt{2\epsilon''} = \frac{\omega}{c}\sqrt{\frac{8\pi n e^2 \tau}{m_e \omega}}\,, \tag{3.49}$$

for the absorption coefficient.

High-Frequency Regime, $\omega >> 1/\tau$

In simple metals this regime corresponds to the ultraviolet part of the spectrum. Since $\omega\tau >> 1$, we obtain

$$\frac{4\pi n e^2 \tau^2}{m_e}\,\frac{1}{\omega^2\tau^2 + 1} >> \frac{4\pi n e^2 \tau}{m_e \omega}\,\frac{1}{\omega^2\tau^2 + 1}\,. \tag{3.50}$$

Therefore, we can ignore the imaginary part of the dielectric function in this regime and

$$\epsilon(\omega) \simeq \epsilon'(\omega) = \epsilon_\infty - \frac{4\pi n e^2 \tau^2}{m_e}\,\frac{1}{\omega^2\tau^2 + 1}\,. \tag{3.51}$$

Also, because $\omega\tau >> 1$, we ignore the 1 in the denominator of Eq. (3.51) with respect to $\omega^2\tau^2$ to get

$$\epsilon'(\omega) = \epsilon_\infty \left(1 - \frac{\omega_p^2}{\omega^2}\right), \tag{3.52}$$

where we have let

$$\omega_p^2 = \frac{4\pi n e^2}{\epsilon_\infty m_e} . \tag{3.53}$$

The frequency ω_p is called the *plasma frequency* for the electrons.

Since the dielectric function, Eq. (3.51), is purely real, we have vanishing absorption in this approximation, and the refractive index becomes

$$n(\omega) = \sqrt{\frac{1}{2}(\epsilon' + |\epsilon'|)} . \tag{3.54}$$

The dielectric function, Eq. (3.52), vanishes at the plasma frequency and becomes negative for frequencies below ω_p. For $\omega > \omega_p$, we have

$$n(\omega) = \sqrt{\epsilon_\infty \left(1 - \frac{\omega_p^2}{\omega^2}\right)} \qquad \text{for } \omega > \omega_p . \tag{3.55}$$

For frequencies less than the plasma frequency, $\epsilon' < 0$ and $|\epsilon'| = -\epsilon'$, and

$$n(\omega) = 0 \qquad \text{for } \omega < \omega_p , \tag{3.56}$$

from Eq. (3.54). It is noted that $n(\omega)$ should be real by our convention. Hence, a negative real part of a dielectric function corresponds to vanishing refractive index, which in turn corresponds to total reflection (see Prob. 3.3). However, for $\omega > \omega_p$, the dielectric function is real and positive, corresponding to unattenuated transverse electromagnetic waves, given by the solutions of the wave equation (3.12). In this regime the medium behaves like a nonabsorbing, transparent dielectric such as glass. Therefore, at high frequencies, when $\omega\tau >> 1$, the spectrum is divided into a highly transparent and a highly reflecting region. The dividing line is the plasma frequency ω_p. For the case of simple metals, the plasma frequency is in the ultraviolet; e.g., the plasma wavelength λ_p is $\simeq$ 810 Å for aluminum and $\lambda_p \simeq$ 2170 Å for sodium, corresponding to $\hbar\omega_p(\text{Al}) = 15.3$ eV and $\hbar\omega_p(\text{Na}) = 5.7$ eV, respectively.

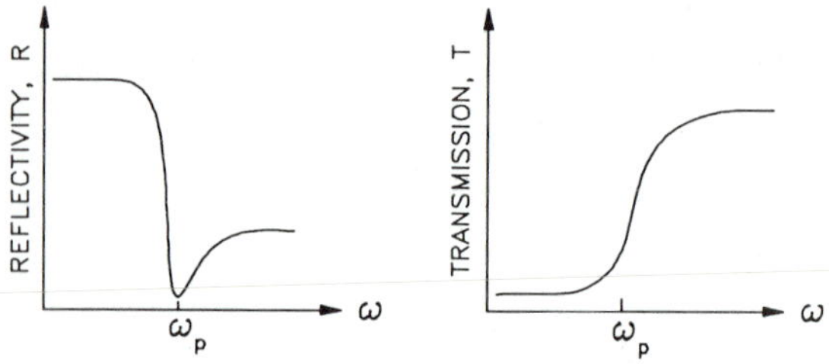

Figure 3.1. Schematical plot of the reflectivity R and transmission T of a simple metal in the vicinity of the plasma frequency ω_p as a function of frequency.

The discussion presented here shows why metals are highly reflective for visible frequencies. Simple metals such as aluminum and silver are widely used as good reflectors of light. Note the difference in the reflective qualities of a typical metal, as displayed in Fig. 3.1, for frequencies above and below ω_p.

3-3. Kramers-Kronig Transformations

We already mentioned that quantities such as the dielectric function and optical susceptibility have both real and imaginary components. Kramers-Kronig transformations allow us to determine the real part of the optical response function from the knowledge of the imaginary part at all frequencies, and vice versa. For example, the components of the optical susceptibility,

$$\chi(\omega) = \chi'(\omega) + i\chi''(\omega) , \tag{3.57}$$

are related by

$$\chi'(\omega) = \frac{2}{\pi} \, \mathrm{Pr} \int_0^\infty d\omega' \, \frac{\omega' \chi''(\omega')}{\omega'^2 - \omega^2} \tag{3.58}$$

and

$$\chi''(\omega) = -\frac{2\omega}{\pi} \, \mathrm{Pr} \int_0^\infty d\omega' \, \frac{\chi'(\omega')}{\omega'^2 - \omega^2} , \tag{3.59}$$

where Pr indicates that the principal value of the integral should be taken; i.e., the singular point, $\omega' = \omega$, should be omitted from the integration. For the example of Eq. (3.59), the principle value integral is defined as

$$\Pr \int_0^{\infty} d\omega' \frac{\chi'(\omega')}{\omega'^2 - \omega^2}$$

$$\equiv \lim_{\eta \to 0} \left[\int_0^{\omega - \eta} d\omega' \frac{\chi'(\omega')}{\omega'^2 - \omega^2} + \int_{\omega + \eta}^{\infty} d\omega' \frac{\chi'(\omega')}{\omega'^2 - \omega^2} \right], \tag{3.60}$$

assuming that this limit exists.

The relations in Eqs. (3.58) and (3.59) are called Kramers-Kronig relations. Figure 3.2 shows the typical spectrum for imaginary and real parts of a response function. For a simple resonance, the spectral shape of the imaginary part of the dielectric function often has a Lorentzian line shape, which varies with frequency ω like $1/(\omega - \omega_0)^2$, where ω_0 is the resonance frequency. For such a Lorentzian function, the real part of the response function behaves like $1/(\omega - \omega_0)$ far away from the resonance. As a consequence, the refractive index goes to zero much more slowly than the absorption far away from the resonance.

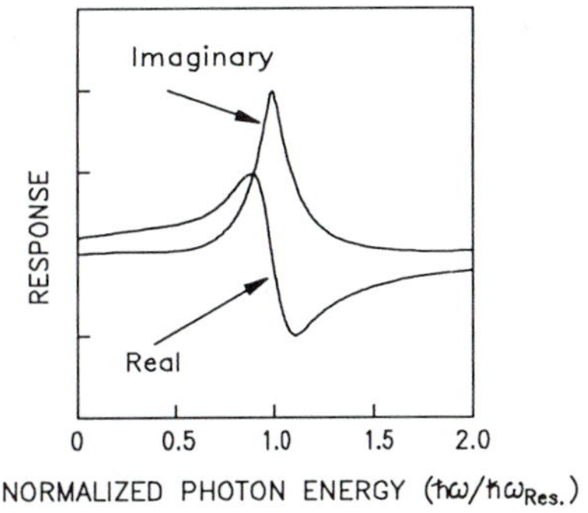

Figure 3.2. The real and imaginary parts of a typical response function.

3-4. Experimental Techniques to Obtain Optical Constants

In the previous sections of this chapter we repeatedly used quantities such as the complex dielectric function, refractive index, absorption coefficient, or optical susceptibility. In this section we now show how these quantities are related to experimentally measurable parameters.

Transparent Region, $\alpha \simeq 0$

The index of refraction $n(\omega)$ can be measured directly. One method is the prism method, which measures the deviation of an incident parallel beam impinging on a wedged plate of the material. Consider an incident beam perpendicular to the *exit* face of the prism, as shown in Fig. 3.3. The index of refraction is related to the incident and exit angles, γ and δ, through the relation (see Prob. 3.2),

$$\begin{aligned} n^2(\omega) &= 1 + \sin^2\delta + \frac{2\cos\gamma\,\sin\delta}{\sin\gamma} + \frac{\sin^2\delta\,\cos^2\gamma}{\sin^2\gamma} \\ &= 1 + \frac{2\cos\gamma\,\sin\delta}{\sin\gamma} + \frac{\sin^2\delta}{\sin^2\gamma}\,. \end{aligned} \tag{3.61}$$

Once γ is known, the index of refraction and its dispersion $n(\omega)$ can be obtained by measuring the deviation angle of the beam δ as a function of incident light frequency.

Another useful method relies on interferometry. Consider a parallel beam incident on a parallel plate functioning as a Fabry-Perot interferometer, as shown in Fig. 3.4. The directly transmitted beam I experiences a phase shift, $\Delta\phi$, after passing once through the plate.

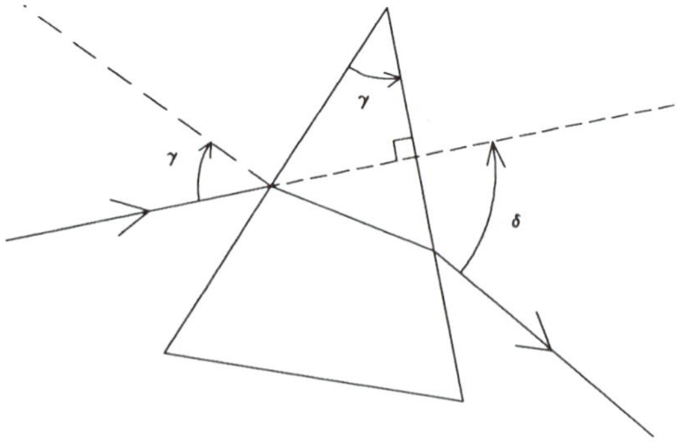

Figure 3.3. Measurement of the refractive index by the prism method.

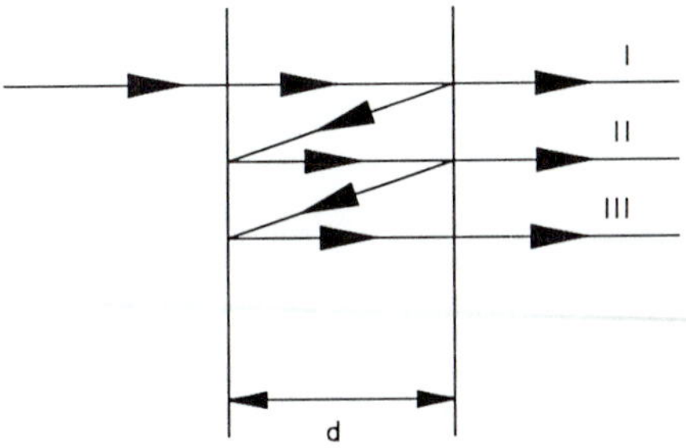

Figure. 3.4. Schematic drawing of multiple internal reflections of a beam transmitted through a parallel plate.

$$\Delta\phi = \frac{n(\omega)}{c}\,\omega d\ , \tag{3.62}$$

where d is the sample thickness and we assume near-normal incidence for simplicity. The output wave II, resulting from two internal reflections, has a phase change, $\Delta\phi' = 3\Delta\phi = 3n(\omega)\omega d/c$. Similarly, the wave III, with four internal reflections, has a phase shift of $5\Delta\phi$, etc. Constructive interference occurs if the phase difference between two consecutive rays is a multiple of 2π; i.e.,

$$\frac{2\ n(\omega)\ \omega d}{c} = 2\pi m \quad \text{with } m = 0,\ 1,\ 2,\ 3...$$

or

$$2n(\omega)d = m\lambda\ . \tag{3.63}$$

Similarly, destructive interference occurs if the two consecutive rays have a phase shift of π or odd multiples of π, i.e.9

$$\frac{2\ n(\omega)\ \omega d}{c} = \left(m + \frac{1}{2}\right) 2\pi \quad \text{with } m = 0,\ 1,\ 2,\ 3...$$

or

$$2n(\omega)d = \left(m + \frac{1}{2}\right)\lambda\ . \tag{3.64}$$

Therefore, by changing the wavelength λ one obtains interference fringes (see Fig. 3.5). The separation between successive maxima is given by

$$2n(\omega)d = m\,\lambda_1 \tag{3.65}$$

and

$$2n(\omega)d = (m + 1)\,\lambda_2\,, \tag{3.66}$$

and by eliminating m, we obtain

$$2n(\omega)d = \frac{1}{\left(\dfrac{1}{\lambda_2} - \dfrac{1}{\lambda_1}\right)}\,, \tag{3.67}$$

where it is assumed that the change of $n(\omega)$ between λ_1 and λ_2 is negligible. Hence, by measuring the separation between two maxima, the index of refraction may be obtained.

Absorbing Region, $\alpha > 0$

The linear absorption loss of a medium is expressed by the Beer-Lambert law, Eq. (3.30). The absorption coefficient $\alpha(\omega)$ cannot be obtained directly from a simple transmission experiment, since the reflection at the front and exit surfaces must be properly taken into account. For normal incidence, as in Fig. 3.4, the intensity of the beam reflected by the front surface is given by $I_R = I_0R$, where I_0 is the incident (external) intensity and R is the reflection coefficient given by (see Prob. 3.1)

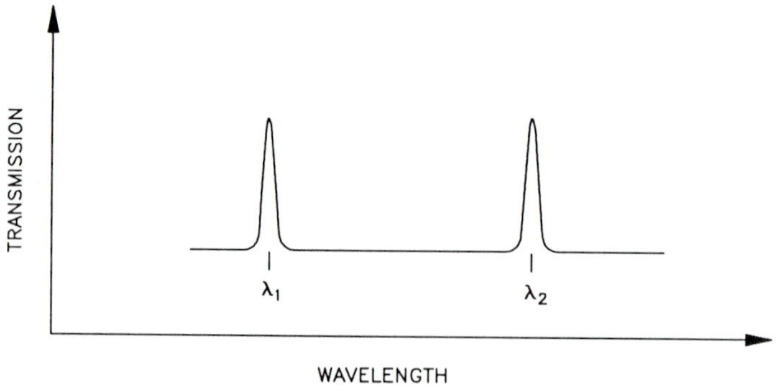

Figure 3.5. Schematic drawing of the transmission fringes that are typically observed in a parallel plate when the light frequency is varied.

$$R = \frac{(n(\omega) - 1)^2 + (\alpha c/2\omega)^2}{(n(\omega) + 1)^2 + (\alpha c/2\omega)^2} \, . \tag{3.68}$$

The transmission of the direct beam, using the Beer-Lambert law, is

$$I_0 = (1 - R)^2 \, I_0 e^{-\alpha d} \, , \tag{3.69}$$

because of the reflection at the front and exit faces. Similarly, for beams II and III in Fig. 3.4, we have

$$I_I = I_0 R^2 (1 - R)^2 \, e^{-3\alpha d} \, , \tag{3.70}$$

$$I_{II} = I_0 R^4 (1 - R)^2 \, e^{-5\alpha d} \, , \tag{3.71}$$

and so on. Summing all beams, I, II, III, ..., and ignoring the interference between the different beams (incoherent transmission), one obtains the total transmitted intensity,

$$\begin{aligned} I &= (1 - R)^2 \, I_0 e^{-\alpha d} \, [1 + R^2 e^{-2\alpha d} + R^4 e^{-4\alpha d} + \ldots] \\ &= I_0 \, \frac{(1 - R)^2 \, e^{-\alpha d}}{1 - R^2 \, e^{-2\alpha d}} \, , \end{aligned} \tag{3.72}$$

where use has been made of the relation

$$\sum_x s^x = \frac{1}{1 - s} \, , \qquad \text{with } |s| < 1 \, . \tag{3.73}$$

For $\alpha d > 1$, the second term in the denominator of Eq. (3.72) may be ignored compared with 1 and we get

$$I = I_0 (1 - R)^2 \, e^{-\alpha d} \, . \tag{3.74}$$

Therefore, a measurement of α requires knowledge of I/I_0, the reflection coefficient R, and the sample thickness d.

It is sometimes more convenient to avoid reflectivity measurements, since they are sensitive to the surface quality of the sample. The absorption coefficient can then be obtained by a relative measurement of transmission, I_1 and I_2, through two identical samples of thickness, d_1 and d_2, respectively. By forming the ratio of transmitted intensities I_1/I_2, we obtain

$$I_1/I_2 = e^{-\alpha(d_1 - d_2)} , \tag{3.75}$$

which requires only the difference of thicknesses and a knowledge of I_1/I_2. In the case where $\alpha d \sim 1$, a complication arises from interference effects due to reflections at the front and exit faces, as mentioned before. The analysis involves equations which are rather lengthy and will not be treated here.

In some cases (e.g., in many metals) it is impractical to perform transmission experiments because of the high reflection coefficient. It is then possible to obtain the refractive index and absorption coefficient by reflectance ellipsometry. The phase shift $\Delta\phi$ of the reflected wave depends on α, $n(\omega)$, the polarization direction, and the incidence angle. For instance, at normal incidence, $\theta = 0$, the phase change is

$$\tan \Delta\phi = \frac{2n(\omega)}{n(\omega)^2 + (c\alpha/2\omega)^2 - 1} . \tag{3.76}$$

Two measurements at oblique angles for both polarization directions provide sufficient information. For further details, the reader is referred to Ref. 3.1.

3-5. Plasma Oscillations and Plasmons

In Sec. 3-2 of this chapter we introduced the plasma frequency ω_p, which is the dividing frequency between propagation through the medium and high reflection at the surface. Physically, ω_p is the frequency of collective oscillations of the electron gas (plasma). To see this, let us consider a free electron plasma of mean density, n_0. The mean velocity of electrons at equilibrium is taken to be zero. Now let us assume that at time t the electron number density changes slightly from the mean value to

$$n(\mathbf{r}, t) = n_0 + n_1 , \tag{3.77}$$

where n_1 is a small quantity. As a result of a change in the electron density, an electric field is induced which is determined from Maxwell's equation,

$$\nabla \cdot \mathbf{E} = -4\pi e[n - n_0] , \tag{3.78}$$

where we dropped the arguments $\mathbf{r}$ and t in n for simplicity. Assuming that the mean electron velocity v_0 is zero, the actual velocity at time t is

$$v = v_0 + v_1 = v_1 \,. \tag{3.79}$$

Similarly, we assume that no field exists in the equilibrium state; i.e., $\mathbf{E} = 0$. Then

$$\mathbf{E} = \mathbf{E}_0 + \mathbf{E}_1 = \mathbf{E}_1 \,. \tag{3.80}$$

Now we use the continuity equation,

$$\nabla \cdot \mathbf{J} + \frac{\partial \rho_c}{\partial t} = 0 \,, \tag{3.81}$$

for the current density, $\mathbf{J} = -env$. Here ρ_c is the charge density,

$$\rho_c = -en \,. \tag{3.82}$$

Substituting for $\mathbf{J}$ and ρ_c in terms of electron density and velocity in Eq. (3.81), the continuity equation becomes

$$\nabla \cdot nv + \frac{\partial n}{\partial t} = 0 \,. \tag{3.83}$$

Substituting for n, v, and $\mathbf{E}$ from Eqs. (3.77), (3.79), and (3.80) into Eqs. (3.83) and (3.78), we find

$$\nabla \cdot (n_0 v_1) + \frac{\partial}{\partial t} n_1 = 0 \tag{3.84}$$

and

$$\nabla \cdot \mathbf{E}_1 = -4\pi e n_1 \,, \tag{3.85}$$

where terms proportional to $n_1 v_1$ have been ignored because n_1 and v_1 are small quantities and $\partial n_0/\partial t = 0$, since n_0 is a constant. Differentiating Eq. (3.84) with respect to time yields

$$\frac{\partial^2 n_1}{\partial t^2} + n_0 \nabla \cdot \frac{\partial v_1}{\partial t} = 0 \,. \tag{3.86}$$

We substitute for $\partial v_1/\partial t$ from Newton's equation of motion,

$$m_e \frac{\partial v_1}{\partial t} = -eE_1 \,, \tag{3.87}$$

to get

$$\frac{\partial^2 n_1}{\partial t^2} - \frac{en_0}{m_e} \nabla \cdot \mathbf{E}_1 = 0 \,. \tag{3.88}$$

Finally, substituting for $\nabla \cdot \mathbf{E}_1$ from Eq. (3.85) yields

$$\frac{\partial^2 n_1}{\partial t^2} + \frac{4\pi e^2 n_0}{m_e} n_1 = \frac{\partial^2 n_1}{\partial t^2} + \omega_p^2 \, n_1 = 0 \,. \tag{3.89}$$

(Note that we have let $\epsilon_\infty = 1$ in the substitution for ω_p^2.) Equation (3.89) describes the motion of a simple harmonic oscillator of frequency ω_p, showing that the electron density oscillates at the plasma frequency. This plasma oscillation is a collective excitation of the electron gas. Collective excitations are excitations that belong to the entire system. Note that the plasma oscillations are longitudinal excitations since their dispersion relation is given by Eq. (3.9). The quantum of the plasma oscillation is called *plasmon.* In the limit of the Drude oscillator model, the plasmons have no dispersion, indicating that the plasma frequency does not depend on the wave vector; it only depends on the electron density. According to quantum mechanics, the plasma oscillation only occurs with discrete quantized units of energy, $\hbar\omega_p$. Plasmons may be excited, for example, by inelastic electron scattering experiments where high-energy electrons are passed through a thin metal film. The incoming electrons lose some energy, equal to integer multiples of the plasmon energy, to excite the plasmons (see Fig. 3.6). Therefore, the plasmons are observed as multiple peaks in the electron energy loss spectra.

Plasmons may also be generated by excitation with x-rays or light. The plasmon excitation by light is observable using oblique incidence with the polarization of the vector **E** lying in the plane defined by the beam direction and the direction normal to the surface. In this way, the E-field component normal to the surface can excite longitudinal plasmons. It should be noted that light in the vacuum is a purely transverse electromagnetic wave (since $\nabla \cdot \mathbf{D} = \epsilon \nabla \cdot \mathbf{E} = 0$ and, thus, $\nabla \cdot \mathbf{E} = 0$). As such, it cannot excite a volume plasmon, which is a longitudinal wave. However, the presence of a surface allows the coupling of light into plasmons. This is possible because the solid-vacuum interface gives the light a longitudinal component. At the solid-vacuum interface, the dielectric function is discontinuous, changing from $\epsilon = \epsilon(\omega)$ in the solid to

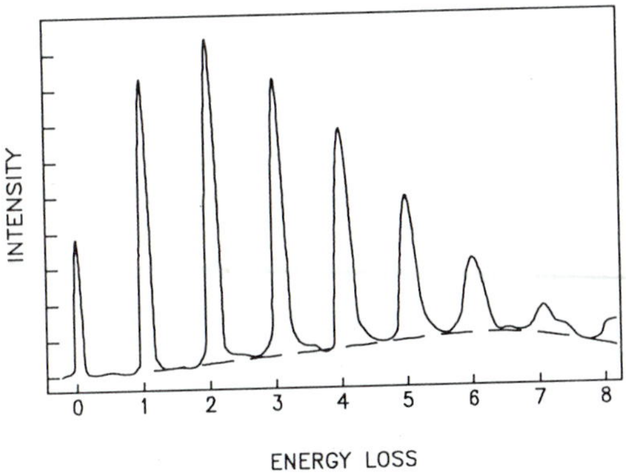

Figure 3.6. Spectrum of energy losses for 20-keV electrons passing through a 2580-Å Al foil. Increment of loss is 15 eV (after Ref. 3.2).

$\epsilon = 1$ in the vacuum. Thus, the dielectric function is space-dependent, $\epsilon = \epsilon(\mathbf{r})$, for which the Maxwell equation becomes $\nabla \cdot \mathbf{D} = \nabla \cdot \epsilon(\mathbf{r})\mathbf{E} = 0$, $\epsilon(\mathbf{r})\nabla \cdot \mathbf{E} + \mathbf{E} \cdot \nabla\epsilon(\mathbf{r}) = 0$, or $\nabla \cdot \mathbf{E} = -\mathbf{E} \cdot [\nabla\epsilon(\mathbf{r})]/\epsilon(\mathbf{r}) = -\mathbf{E} \cdot \nabla \ell n \epsilon(\mathbf{r}) \neq 0$. The nonzero $\nabla \cdot \mathbf{E}$ gives the light a longitudinal component, which allows the coupling of photons to plasmons. Plasmons are observed by a decrease of the transmittance or reflectance of the light from a thin metal film at the plasma frequency, because photon energy is fed into plasma oscillations. Figure 3.7 shows the experimentally observed transmittance of a 145-Å silver film. The resonance observed as a minimum in the transmittance at $\omega/\omega_p = 1$ corresponds to the volume plasmon excitation.

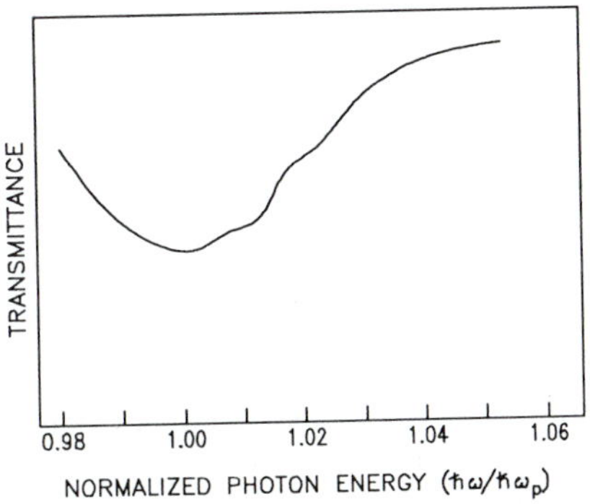

Figure 3.7. Transmittance of a 145-Å silver film for p-polarized light incident at an angle of 75^0, showing excitation of volume plasmons (after Ref. 3.3).

3-6. Surface Plasmons*

In addition to the possibility of optically exciting volume plasmons, the presence of a surface or interface between materials with different dielectric constants may also lead to specific surface-related excitations. For example, the interface between a medium with a positive dielectric constant and a medium with negative dielectric constant, such as metals, can give rise to special propagating electromagnetic waves called *surface plasma waves*, which stay confined near the interface. Such surface waves have a component of the electric field decreasing exponentially away from the interface into the directions ±z.

To obtain the conditions for the existence of surface plasma waves and their dispersion, consider the interface between two semi-infinite isotropic media with dielectric functions ϵ_1 and ϵ_2, as shown in Fig. 3.8. The z axis is perpendicular to the interface, which is the plane z = 0. The medium 1, occupying the half space z > 0, may be a dielectric or vacuum with $\epsilon_1 = 1$, and medium 2 may be a metal. The Maxwell equations for an electron plasma surface wave propagating along the x axis with frequency ξ and wave vector k_x, are of the form

$$\mathbf{E} = \mathbf{E}_1 \, e^{i(k_x x - \xi t)} \, e^{-\alpha_1 z} \qquad z > 0$$

or

$$\mathbf{E} = \mathbf{E}_2 \, e^{i(k_x x - \xi t)} \, e^{\alpha_2 z} \qquad z < 0 \, , \tag{3.90}$$

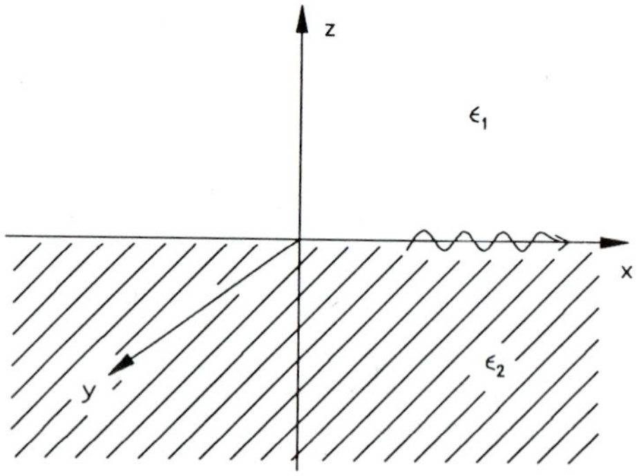

Figure 3.8. Excitation of a surface plasma wave at the interface between a metal with dielectric constant ϵ_2 and a dielectric or vacuum with dielectric constant ϵ_1.

with α_1 and $\alpha_2 > 0$. The z-dependent exponential terms ensure that the field decays away from the interface and is being confined at the surface. Only p-polarized TM (transverse magnetic) modes can satisfy the Maxwell equations together with the boundary conditions. An s-polarized TE (transverse electric) field that decays away from both sides of the interface cannot satisfy the boundary conditions at any wave vector (see Prob. 3.4). For TM waves the magnetic field is normal to the incident plane, containing the normal to the surface (z axis) and the propagation direction (x axis); i.e., **H** is along the y axis and **E** lies in the xz plane. Thus,

$$\mathbf{E} = (A, 0, B)\, e^{i(k_x x - \xi t)}\, e^{-\alpha_1 z} \qquad z > 0$$

or

$$\mathbf{E} = (C, 0, D)\, e^{i(k_x x - \xi t)}\, e^{\alpha_2 z} \qquad z < 0\,, \tag{3.91}$$

and

$$\mathbf{H} = (0, H_1, 0)\, e^{i(k_x x - \xi t)}\, e^{-\alpha_1 z} \qquad z > 0$$

or

$$\mathbf{H} = (0, H_2, 0)\, e^{i(k_x x - \xi t)}\, e^{\alpha_2 z} \qquad z < 0\,. \tag{3.92}$$

The boundary conditions that need to be satisfied at the interface are

$$\mathbf{E}_{\|}\Big|_{z=0} = E_x\Big|_{z=0} = \text{continuous}\,,$$

$$H_{\|}\Big|_{z=0} = H_y\Big|_{z=0} = \text{continuous}\,,$$

and

$$D_{\perp}\Big|_{z=0} = (\epsilon E)_{\perp}\Big|_{z=0} = \epsilon E_z\Big|_{z=0} = \text{continuous}\,. \tag{3.93}$$

The first two boundary conditions give

$$A = C$$

and

$$H_1 = H_2 = H\,. \tag{3.94}$$

From the Maxwell equations we know that

$$\nabla \times \mathbf{H} = \frac{1}{c}\frac{\partial \mathbf{D}}{\partial t} = \frac{\epsilon}{c}\frac{\partial \mathbf{E}}{\partial t}\,. \tag{3.95}$$

For the field components given in Eq. (3.92), we have

$$\left(-\frac{\partial H_y}{\partial z}, 0, \frac{\partial H_y}{\partial x}\right) = \frac{-i\xi\epsilon}{c}(E_x, E_y, E_z)\,. \tag{3.96}$$

The x component of Eq. (3.96) gives

$$\frac{\partial H_y}{\partial z} = \frac{i\xi\epsilon}{c}E_x\,. \tag{3.97}$$

When Eq. (3.97) is evaluated for media 1 and 2, we get

$$-\alpha_1 H = \frac{i\xi\epsilon_1}{c}A \qquad z > 0$$

and

$$\alpha_2 H = \frac{i\xi\epsilon_2}{c}A \qquad z < 0\,. \tag{3.98}$$

Dividing these two equations results in

$$\boxed{\frac{\alpha_1}{\alpha_2} = -\frac{\epsilon_1}{\epsilon_2}\,.}$$

condition for surface plasma mode (3.99)

Equation (3.99) indicates that $\epsilon_2 = -\epsilon_1\alpha_2/\alpha_1$, suggesting that for the medium 1, having $\epsilon_1 > 0$, the medium 2 needs to be a crystal with $\epsilon_2 < 0$ for the surface plasma wave to exist (since α_1 and α_2 are positive quantities). Metals have negative dielectric functions for frequencies $\xi < \omega_p$; thus, the interface between a metal and vacuum may support a surface plasma wave. Using the wave equations (3.12), we have

$$\nabla^2(E_x, E_y, E_z) = \frac{\epsilon}{c^2}\frac{\partial^2}{\partial t^2}(E_x, E_y, E_z)\,. \tag{3.100}$$

Substitution of Eqs. (3.91) and (3.94) into Eq. (3.100) gives

$$(-k_x^2 + \alpha_1^2)\,(A, 0, B) = \frac{-\epsilon_1}{c^2}\,\xi^2\,(A, 0, B) \qquad z > 0$$

and

$$(-k_x^2 + \alpha_2^2)\,(A, 0, D) = \frac{-\epsilon_2}{c^2}\,\xi^2\,(A, 0, D) \qquad z < 0\,, \tag{3.101}$$

resulting in

$$(-k_x^2 + \alpha_1^2) = \frac{-\epsilon_1}{c^2}\,\xi^2$$

and

$$(-k_x^2 + \alpha_2^2) = \frac{-\epsilon_2}{c^2}\,\xi^2\,, \tag{3.102}$$

or

$$\alpha_1^2 = k_x^2 - \frac{\epsilon_1}{c^2}\,\xi^2$$

and

$$\alpha_2^2 = k_x^2 - \frac{\epsilon_2}{c^2}\,\xi^2\,. \tag{3.103}$$

Dividing these two equations, we get

$$\frac{\alpha_1^2}{\alpha_2^2} = \frac{k_x^2 - \frac{\epsilon_1}{c^2}\xi^2}{k_x^2 - \frac{\epsilon_2}{c^2}\xi^2} . \tag{3.104}$$

Substituting for the left-hand side from Eq. (3.99), one obtains

$$\frac{\epsilon_1^2}{\epsilon_2^2} = \frac{k_x^2 - \frac{\epsilon_1}{c^2}\xi^2}{k_x^2 - \frac{\epsilon_2}{c^2}\xi^2} . \tag{3.105}$$

Rearranging the equation gives

$$\boxed{k_x^2 = \frac{\xi^2}{c^2}\frac{\epsilon_1\epsilon_2(\xi)}{\epsilon_1 + \epsilon_2(\xi)} ,}$$

dispersion relation for surface plasmons (3.106)

where the frequency dependence of the dielectric function of medium 2 is indicated as $\epsilon_2(\xi)$. The dispersion relation for the vacuum-metal interface with $\epsilon_1 = 1$ is

$$\xi = c\, k_x \left(\frac{\epsilon_2(\xi) + 1}{\epsilon_2(\xi)}\right)^{1/2} . \tag{3.107}$$

The dispersion relation of Eq. (3.106) is schematically plotted in Fig. 3.9. The frequency corresponding to the asymptotic value for large wave vectors, $k_x \rightarrow \infty$, is designated as the surface-plasma frequency. Examining Eq. (3.106) for large k_x values, we require

$$\epsilon_1 + \epsilon_2 = 0 \qquad \text{at } \xi = \omega_s . \tag{3.108}$$

The assignment of the surface plasma frequency ω_s for the condition $\epsilon_1 + \epsilon_2 = 0$ may be understood directly from Eq. (3.9). The total dielectric function of the system in this case is $\epsilon_1 + \epsilon_2$, and its vanishing magnitude at ω_s indicates that a longitudinal wave at frequency ω_s is sustained. Equation (3.108) at $\xi = \omega_s$ for a metal with dielectric function $\epsilon_2(\xi) = 1 - \omega_p^2/\xi^2$ gives

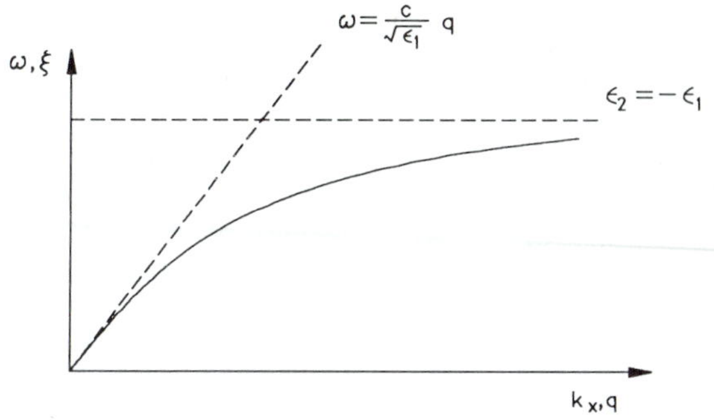

Figure 3.9. The dispersion curve for surface plasmons (the solid curve). The dispersion curve of a photon, $\omega = cq/\epsilon_1^{1/2}$, is also plotted on the same figure, as shown by the dashed line. The surface plasmon curve reaches an asymptotic value at $\epsilon_2 = -\epsilon_1$ when $k_x \rightarrow \infty$, which occurs at $\omega_s = \omega_p/(\epsilon_1 + 1)^{1/2}$.

$$1 - \frac{\omega_p^2}{\omega_s^2} = -\epsilon_1 \tag{3.109}$$

or

$$\omega_s = \frac{\omega_p}{\sqrt{\epsilon_1 + 1}} . \tag{3.110}$$

In Fig. 3.9 the dispersion relations for the photon with frequency ω and wave vector **q** in media 1 and 2 are also plotted; i.e., the frequency versus wave vector dependence of photon is plotted on the same figure as that of the surface plasmon. The surface plasmon and the photon dispersion curves do not cross each other anywhere. In other words, the wave number k_x of the surface plasmon wave is larger than the photon wave number $q = \omega\epsilon_1^{1/2}/c$ everywhere $k_x > \omega\epsilon_1^{1/2}/c$. This implies that the surface plasmon waves considered so far are nonradiative; i.e., they cannot be coupled with photons since the momentum (or wave vector) conservation that has to hold for any photon absorption or emission is not satisfied (this conservation law may be satisfied when higher-order processes such as Raman scattering are considered). Energy and momentum conservation is satisfied when there is a crossing of the dispersion curves of the photon and the surface excitation (or any other elementary excitation).

3-7. Optical Excitation of Surface Plasmons*

We pointed out in the previous section that the wave number k_x of the surface plasmon wave exceeds the photon wave number $\omega\epsilon_1^{1/2}/c$, resulting in a nonradiative surface mode. However, it is possible to couple light to surface plasmons by various techniques. To have coupling and excite a TM surface wave traveling along x direction, we require a longitudinal component of the incident electric field of the light to be along the x axis. Therefore, p-polarized light field must be used with its electric field lying in the plane defined by the light propagation vector and the direction normal to the interface (the xz plane).

Two of the most popular techniques are referred to as the "grating coupling" and the "prism coupling." In these schemes the wave vector of the light is increased because of the presence of the grating or the prism to match the surface plasmon wave vector. Figure 3.10 shows the grating-coupling method using a grating ruled on the surface. When the parallel component of the wave number of the incident light matches the surface plasmon wave number k_x, the light is coupled and the surface wave is excited. This occurs when the angle of incidence onto the grating satisfies the condition

$$\frac{\omega}{c}\sqrt{\epsilon_1}\sin\theta + \frac{2\pi m}{d} = k_x\,, \tag{3.111}$$

for a light beam with frequency ω and wave number $q = \omega\epsilon_1^{1/2}/c$. Here d is the period of the grating and m is an integer. In this technique, the light wave vector is increased by the Fourier component of the periodicity of the grating.

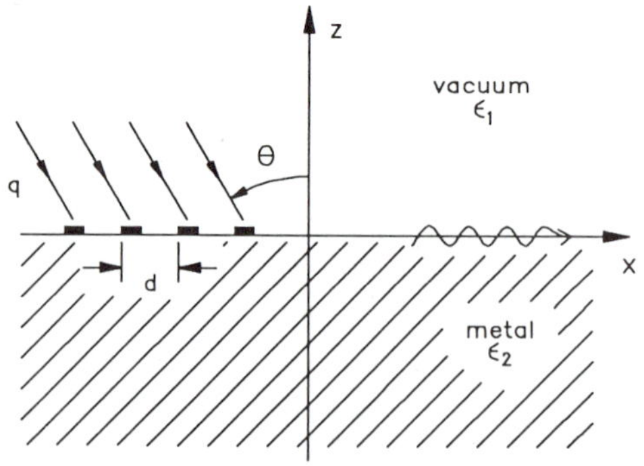

Figure 3.10. The grating-coupling scheme for excitation of surface plasmons.

The prism-coupling technique, shown schematically in Fig. 3.11, involves placing a prism with an index of refraction, $n_p(\omega) > 1$, very close to the surface of the metal. The light which is directed through the prism at an angle θ, larger than the critical angle of the prism for total internal reflection, has a wave vector parallel to the interface with magnitude $\omega\, n_p(\omega) \sin\theta/c$. Surface plasmons may be excited when this photon wave number matches the surface plasmon wave number k_x at an angle θ,

$$k_x = \frac{\omega}{c} \sqrt{\epsilon_p} \sin\theta \, , \tag{3.112}$$

where $\epsilon_p = n_p^2(\omega)$ is the dielectric constant of the prism. Therefore, by tuning the incident beam angle θ, we obtain coupling, as evidenced by the crossing of the dispersion curves of the photon and surface plasmon. Figure 3.12 shows the crossing for the prism-coupling case and the absence of crossing (and, hence, the absence of coupling of light to surface plasmons) when the prism is not used. The presence of the prism lowers the photon dispersion from $\omega = cq/\sin\theta$ to $\omega = cq/n_p(\omega) \sin\theta$ so that the two dispersion curves may cross, resulting in energy and momentum conservation. At the coupling angle for which the surface plasmon is excited, totally reflected light becomes attenuated as the incident light energy feeds into the surface mode. Experimentally, the intensity of the reflected wave is measured as a function of tuning angle θ. The dip in the reflected intensity indicates coupling to the surface plasmon. An example of the surface plasmon resonance is shown in Fig. 3.13.

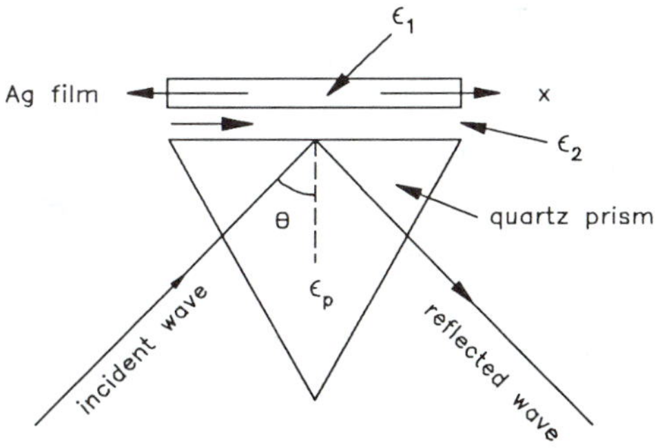

Figure 3.11. Surface plasmon excitation by prism-coupling technique.

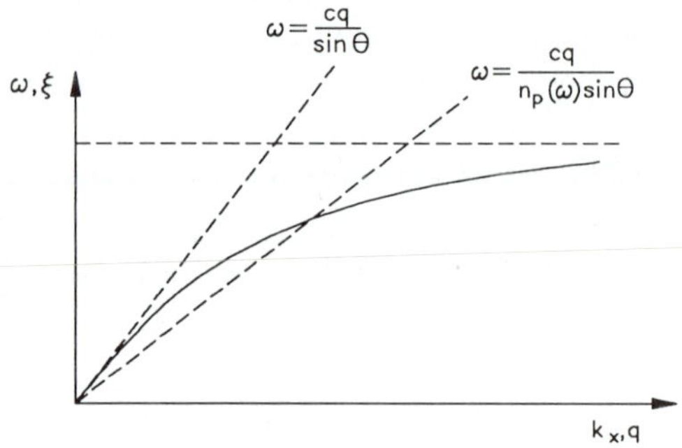

Figure 3.12. The crossing of the photon and surface plasmon dispersion curves.

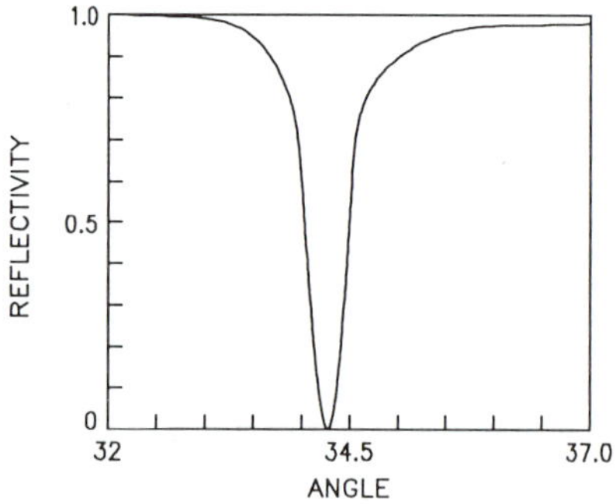

Figure 3.13. Reflectivity as a function of incident angle inside the prism for excitation of surface plasmons. The energy lost from the incident beam at resonance is fed to surface plasmons (after Ref. 3.4).

3-8. Problems

3.1. Compute the reflectance $R(\omega)$ of an electromagnetic wave in vacuum with complex dielectric constant $\epsilon(\omega) = \epsilon'(\omega) + i\epsilon''(\omega)$, and express $R(\omega)$ in terms of the refractive index and absorption coefficient. Show that when $n(\omega) \to 0$, the reflectance approaches 1.

3.2. Prove Eq. (3.61).

3.3. Show that the reflectivity of a material with negative dielectric function approaches unity.

3.4. Write the field components of a TE surface mode for the metal-vacuum interface, and prove that the interface cannot support a TE mode.

3.5. When a very intense laser irradiates a metal, it ablates material near the surface, resulting in a gas of ionized particles (a plasma) expanding into free space. At which densities of the plasma will light with $\lambda = 1\ \mu m$ be totally reflected?

3-9. References

1. T. S. Moss, *Optical Properties of Semiconductors* (Butterworths Scientific Publications, London, 1959).

2. L. Marton, J. Arol Simpson, H. A. Fowler, and N. Swanson, Phys. Rev. **126**, 182 (1962).

3. I. Lindau and P. O. Nilsson, Phys. Lett. A **31**, 352 (1970).

4. D. Sarid, R. T. Deck, A. E. Craig, R. K. Hickernell, R. S. Jameson, and J. J. Fasano, Appl. Opt. **21**, 3993 (1982).

Chapter IV
OPTICAL PROPERTIES OF PHONONS

In Chap. III we described a simple model for the analysis of optical properties associated with the electronic response of metals. This chapter concentrates on the optical properties associated with the ionic part of the material response. Since the mass of the ions is typically 10^3 times larger than that of the electrons, the ionic movement is usually restricted to small oscillations around the equilibrium position in the lattice. The oscillation modes of these lattice vibrations are called phonon modes. Phonons are almost entirely responsible for the thermal properties, such as heat capacity, thermal conductivity, and thermal expansion of insulators, and contribute significantly to the heat capacity of metals.

As we will see, some of these phonon modes interact directly with light and can be responsible for strong absorption and reflection by the medium. Other phonons, which do not absorb light directly, are nevertheless at the origin of light scattering. After an introduction describing the different phonons, we emphasize the optical properties associated with phonons, deriving the phonon dielectric function and introducing phonon-polaritons as mixed phonon-photon excitations in a crystal. We then discuss some aspects of light scattering and explain the concepts of Brillouin, Raman, and Rayleigh scattering. Finally, we discuss the experimental technique of coherent Raman spectroscopy of solids.

4-1. Optical and Acoustical Phonons

As a model for phonons, let us consider the one-dimensional diatomic lattice (a linear chain with two atoms per unit cell) shown in Fig. 4.1, where the atomic masses M_1 and M_2 are connected by ideal springs and the interatomic distance is a. This system obeys Hooke's law when the distortion (strain) in the medium is linearly related to the magnitude of the applied force (stress). We allow the lattice sites to vibrate, assuming that the movements of the ions from their equilibrium positions are small compared with their interatomic spacing. Under these conditions, a first approximation is to describe the motion of the atomic masses as a harmonic

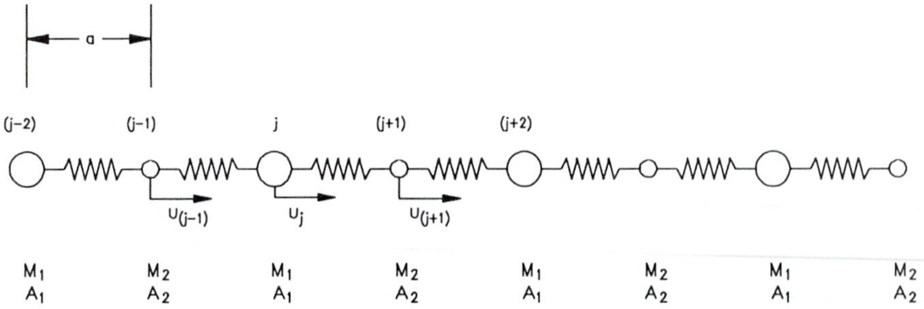

Figure 4.1. A one-dimensional diatomic lattice.

motion. We assume that the coordinates oscillate harmonically in time with constant amplitude and angular frequency. Furthermore, in the simplest approximation we only need to consider nearest-neighbor interactions, as they constitute the most significant forces acting on the ions. It should be noted that these approximations are not always appropriate; some phenomena, such as thermal expansion, require the inclusion of anharmonic terms. However, for the purposes of this chapter, we restrict ourselves to the harmonic approximation.

In order to determine the equation of motion of any two neighboring atoms, we let U_j be the displacement of the jth atom from its equilibrium position. The Hamiltonian of the system is the sum of the kinetic and potential energies, where the potential part is

$$\mathcal{E}_{pot} = \frac{1}{2} f \sum_{j=1}^{N} (U_j - U_{j-1})^2 , \tag{4.1}$$

and the kinetic part is

$$\mathcal{E}_{kin} = \sum_{j=1}^{N} p_j^2 / 2M_j . \tag{4.2}$$

In Eq. (4.1) f is the so-called spring constant which, as we will see later, is related to the vibrational frequency of the ion. The momentum of atom j is $p_j = M_j dU_j/dt$. The total Hamiltonian of the system is

$$H = \sum_{j=1}^{N} p_j^2 / 2M_j + \frac{1}{2} f \sum_{j=1}^{N} (U_j - U_{j-1})^2 . \tag{4.3}$$

If only nearest-neighbor interaction is considered, the Hamiltonian for the jth ion may be written as

$$H_j = \frac{1}{2} f (U_j - U_{j-1})^2 + \frac{1}{2} f (U_{j+1} - U_j)^2 + \frac{1}{2M_1} p_j^2 . \tag{4.4}$$

The equation of motion for the jth atom can be obtained from Hamilton's canonical equation,

$$\frac{dp_j}{dt} = - \frac{\partial H_j}{\partial U_j}$$

or

$$M_1 \frac{d^2 U_j}{dt^2} = -f (2U_j - U_{j-1} - U_{j+1}) . \tag{4.5}$$

Similarly, the equation of motion for atom $j + 1$ is

$$M_2 \frac{d^2 U_{j+1}}{dt^2} = -f (2U_{j+1} - U_j - U_{j+2}) . \tag{4.6}$$

The solutions for these equations may be obtained using

$$U_j = A_1 e^{i(\Omega t - Kx_j)} \tag{4.7}$$

and

$$U_{j+1} = A_2 e^{i(\Omega t - Kx_{j+1})} = A_2 e^{i(\Omega t - Kx_j)} e^{-iKa} , \tag{4.8}$$

where we used $x_{j+1} = x_j + a$. Also, different amplitudes were taken for the two neighboring atoms because the masses of the atoms are different. Similarly,

$$U_{j-1} = A_2 e^{i(\Omega t - Kx_j)} e^{iKa} \tag{4.9}$$

and

$$U_{j+2} = A_1 e^{i(\Omega t - Kx_j)} e^{-2iKa} . \tag{4.10}$$

Substituting from Eqs. (4.7) through (4.10) into Eqs. (4.5) and (4.6) we get

$$-M_1\Omega^2 A_1 = f(A_2 e^{-iKa} + A_2 e^{iKa} - 2A_1) \tag{4.11}$$

and

$$-M_2\Omega^2 A_2 e^{-iKa} = f(A_1 e^{-2iKa} + A_1 - 2A_2 e^{-iKa}) . \tag{4.12}$$

Rearranging Eqs. (4.11) and (4.12) and multiplying Eq. (4.12) by exp(iKa) gives

$$(M_1\Omega^2 - 2f)\, A_1 + 2f \cos Ka\, A_2 = 0 \tag{4.13}$$

and

$$2f \cos Ka\, A_1 + (M_2\Omega^2 - 2f)A_2 = 0 . \tag{4.14}$$

These equations can be satisfied so that A_1 and A_2 are not identically zero only if

$$\begin{vmatrix} M_1\Omega^2 - 2f & 2f \cos Ka \\ 2f \cos Ka & M_2\Omega^2 - 2f \end{vmatrix} = 0 , \tag{4.15}$$

which gives

$$\Omega^4 - 2f \left(\frac{M_1 + M_2}{M_1 M_2} \right) \Omega^2 + \frac{4f^2}{M_1 M_2} \sin^2 Ka = 0 . \tag{4.16}$$

The solution of Eq. (4.16) is

$$\Omega_\pm^2(K) = \frac{f(M_1 + M_2)}{M_1 M_2} \left[1 \pm \sqrt{1 - \frac{4M_1 M_2 \sin^2 Ka}{(M_1 + M_2)^2}} \right] . \tag{4.17}$$

Equation (4.17) describes the dispersion relation for the lattice vibrations (phonon dispersion relation). This dispersion can be plotted in the reduced zone scheme by noting that the reciprocal lattice vector for this diatomic array is π/a (the primitive unit vector in the direct lattice is 2a). The first Brillouin zone is the region between $-\pi/2a \leq K \leq \pi/2a$. For small K, sinKa approaches Ka, and Eq. (4.17) becomes

$$\lim_{K \to 0} \Omega_{\pm}^2(K) = f \frac{M_1 + M_2}{M_1 M_2} \left[1 \pm \sqrt{1 - \frac{4M_1 M_2 K^2 a^2}{(M_1 + M_2)^2}} \right] . \tag{4.18}$$

Expanding the square root, $\sqrt{1 - x} \simeq 1 - x/2$, one obtains the two distinct solutions

$$\Omega_+(0) = \lim_{K \to 0} \Omega_+(K) = \sqrt{\frac{2f(M_1 + M_2)}{M_1 M_2}} \tag{4.19}$$

and

$$\Omega_-(0) = \lim_{K \to 0} \Omega_-(K) = Ka \sqrt{\frac{2f}{M_1 + M_2}} . \tag{4.20}$$

At the zone boundary, where $K = \pi/2a$, we have $\sin^2 Ka = 1$,

$$\Omega_+(\pi/2a) = \sqrt{\frac{2f}{M_2}} , \tag{4.21}$$

and

$$\Omega_-(\pi/2a) = \sqrt{\frac{2f}{M_1}} . \tag{4.22}$$

For $M_1 > M_2$, where $\Omega_+(\pi/2a) > \Omega_-(\pi/2a)$, the dispersion relation is plotted in Fig. 4.2. The Ω_- solution is called *acoustic-phonon* dispersion, while the Ω_+ solution is the *optical-phonon* branch. Low-frequency phonons are acoustical phonons; their energy approaches zero as K approaches zero.

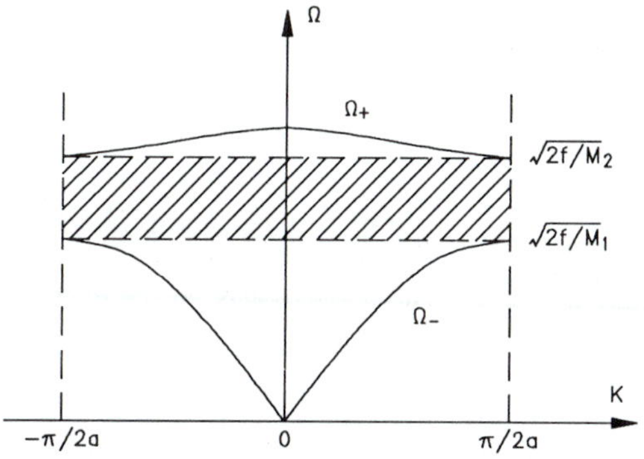

Figure 4.2. Dispersion curves for optical and acoustic phonons. The shaded region is the forbidden band. Waves created in this region are attenuated rapidly and cannot propagate.

Optical phonons have higher frequencies and are almost dispersionless; that is, their energy does not change much with K, as K is varied.

The distinction between acoustic phonon and optical phonon can be understood by examining the amplitude ratio A_1/A_2. From Eq. (4.13) we get

$$\frac{A_1}{A_2} = \frac{2f}{(2f - M_1\Omega_\pm^2)} \tag{4.23}$$

in the limit of small K. For the acoustical branch, we see that $A_1/A_2 \to 1$ for $K \to 0$, since $\Omega_- \to 0$. For the optical branch, however, Ω_+ is the constant given by Eq. (4.19) and the amplitude ratio becomes $A_1/A_2 = -M_2/M_1$. Therefore, for an acoustic mode at $K = 0$, the two types of atoms move in the same direction with the same amplitude, while for optical mode, at $K = 0$, vibrations of the two types of atoms are in opposite directions with the amplitudes being inversely proportional to the masses, as shown schematically in Fig. 4.3. This point can be further clarified by examining Eqs. (4.19) and (4.20), where the optical phonon frequency is given in terms of the reduced mass, M_r, of the two atoms; i.e., $1/M_r = 1/M_1 + 1/M_2$, in contrast to the acoustic phonon frequency, which varies with the total mass, $M_1 + M_2$. Thus, acoustic phonons are associated with the motion of the center of mass of the two atoms, whereas optical phonons are described by the oscillations of the relative coordinate of the two atoms. In ionic crystals, such as NaCl, the two different atoms in the unit cell have opposite charges. Their oscillations correspond to a set of dipoles. Therefore, the optical mode can be excited by an electric field associated with a light wave because the electric field tends to move the

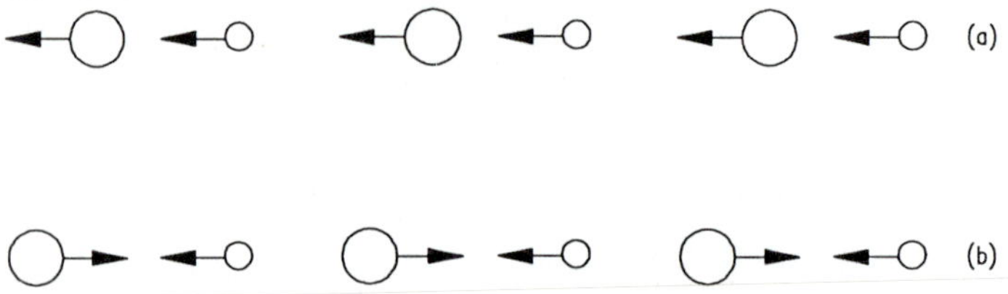

Figure 4.3. Schematics of (a) acoustic mode of vibration and (b) optical mode of vibration.

two oppositely charged ions in opposite directions. This is why they are called *optical* phonons. It must be emphasized, however, that there is not necessarily a dipole moment associated with each optical phonon. For example, in a covalent material like a crystal of germanium with two *identical* atoms per unit cell, optical phonons corresponding to vibrations of the two identical atoms in opposite directions do exist. However, these vibrations have no associated dipoles so that they cannot be excited by an incident light beam. One often says that such phonons have little optical activity.

For the one-dimensional diatomic chain treated so far there is one optical branch and one acoustic branch. For a three-dimensional solid, the same procedures can be followed in calculating phonon dispersion curves. A three-dimensional vibrational amplitude may be written as

$$\mathbf{U}_n = \hat{\mathbf{e}} U_0 \, e^{\, i(\Omega t \, - \, \mathbf{K}\cdot\mathbf{r})} \, , \tag{4.24}$$

where the wave vector, $\mathbf{K} = 2\pi/\lambda \; \hat{\mathbf{K}}$, specifies both the wavelength λ and direction of propagation $\hat{\mathbf{K}}$. The unit vector $\hat{\mathbf{e}}$ specifies the *polarization* of the waves. If $\hat{\mathbf{e}}$ is parallel to $\hat{\mathbf{K}}$, one has a longitudinal wave, while $\hat{\mathbf{e}}$ perpendicular to $\hat{\mathbf{K}}$ corresponds to a transverse wave. In general, the phonon wave in a solid is a mixture of transverse and longitudinal types.

For a mono-atomic solid, three equations of motion are obtained from the decomposition of $\hat{\mathbf{e}}U_0$ into U_{0x}, U_{0y}, and U_{0z}, which describe three acoustic modes. For a three-dimensional solid with two atoms per unit cell, there will be six branches. Three of them are acoustical and three are optical branches, as shown schematically in Fig. 4.4. From the three acoustic branches, two are transverse (TA) and one is longitudinal (LA). Similarly, two of the three optical branches are transverse (TO) and one is longitudinal (LO). In general, the following rule applies: A 3-D crystal with p atoms per unit cell has a total of 3p phonon branches, three of which are acoustical modes, and the remaining 3p-3 modes are optical.

The dispersion relations, energy as a function of wave vector, are not necessarily isotropic in wave vector space. The dispersion curves may have

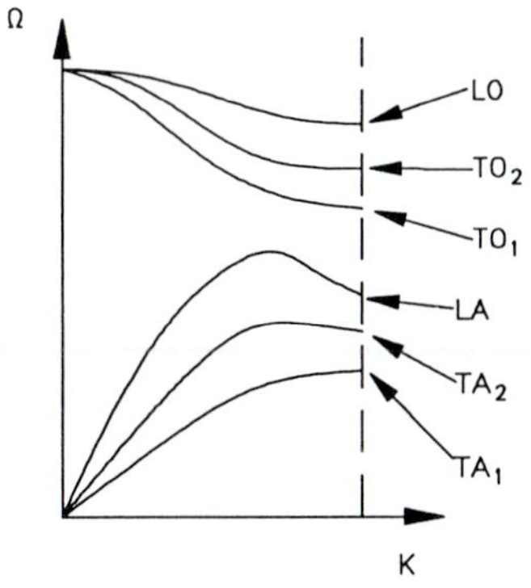

Figure 4.4. Typical phonon dispersion curves along a general direction in wave vector space for a three-dimensional diatomic solid such as Ge. Two TA and one LA branch are shown. Also, one LO and two TO branches are shown. Usually the longitudinal branch is higher than the transverse branch because the restoring forces associated with longitudinal oscillations are greater.

different profiles in different directions. Figure 4.5 shows the phonon dispersion curves along two different wave vector directions for aluminum, with one atom per unit cell, and germanium, with two identical atoms per unit cell. The right and left portions of the figure correspond to wave vector directions [100] and [111], respectively, showing that the phonon branches have different profiles in different directions. For example, in aluminum the TA branches are degenerate along [100] direction, while they are nondegenerate along [111] direction. Lattice dispersion curves are measured by inelastic neutron scattering methods.

The description that has been adopted so far in this chapter is a classical one. However, lattice vibrations, like electromagnetic radiation, occur in quantized energy units, as prescribed by quantum mechanics. The quantum mechanical treatment uses a Hamiltonian of the form (in one dimension),

$$H = -\frac{\hbar^2}{2M_1}\frac{d^2}{dx^2} + \frac{1}{2} M_1 \Omega^2 x^2 . \tag{4.25}$$

This is the Hamiltonian of a harmonic oscillator. The corresponding Schrödinger equation is

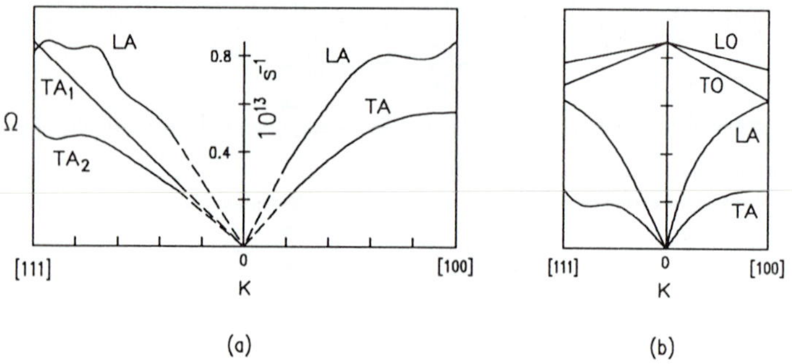

Figure 4.5. (a) Dispersion curves for acoustic phonon modes propagating along [100] (right portion) and [111] (left portion) direction in Al with an fcc structure (Ref. 4.1). (b) Dispersion curves of Ge in the [100] and [111] directions. In Ge, which has diamond structure, there are two identical atoms at positions (0, 0, 0) and (1/4, 1/4, 1/4), leading to six phonon branches. The TA (and TO) branch actually represents two degenerate branches.

$$\left[-\frac{\hbar^2}{2M_1}\frac{d^2}{dx^2}+\frac{1}{2}M_1\Omega^2x^2\right]\psi(x)=\mathcal{E}\,\psi(x)\ , \tag{4.26}$$

with the well-known solution of the form

$$\mathcal{E}_{n_K}=\left(n_K+\frac{1}{2}\right)\hbar\Omega \qquad \text{with}\quad n_K=0,\ 1,\ 2,\ \ldots \tag{4.27}$$

In analogy with the quantum theory of electromagnetic radiation, the vibrational energy $\mathcal{E}_{n_K}$ corresponds to the energy of the system with n-phonons. When $n_K = 0$, $\mathcal{E}_0 = 1/2\ \hbar\Omega$ (zero-point energy) is the energy of the system at a temperature of absolute zero. Equation (4.27) indicates that the energy of the lattice vibration is quantized with the quantum of energy being a phonon. For instance, thermal vibrations result in thermally excited phonon population.

In describing the interaction of phonons with one another or with electrons, it is often convenient mathematically to introduce phonon *creation* and *annihilation* operators, a^+ and a, such that

$$a^+|n_K\rangle=\sqrt{n_K+1}\ |n_K+1\rangle \tag{4.28}$$

and

$$a|n_K\rangle = \sqrt{n_K}\,|n_K - 1\rangle \ , \qquad (4.29)$$

where $|n_K\rangle$ denotes a state of the vibrating system containing n_K phonons, each having an energy $\hbar\Omega$. This means that the result of operating with a^+ on a state $|n_K\rangle$ containing n_K phonons is to change it to state $|n_K + 1\rangle$ containing $n_K + 1$ phonons.

4-2. Optical Excitation of Phonons

Light travels much faster than sound (acoustic-phonon mode) in a solid. The dispersion relation for light is $\omega = cq/n(\omega)$, where c is the speed of light, q is the photon wave number, and $n(\omega)$ is the refractive index of material. The dispersion relation of acoustic phonon is $\Omega = c_s K$ near $K = 0$, with $c_s \simeq 10^{-5}$ c being the typical speed of sound of a crystalline solid. The dispersion curves of phonons and photons are plotted separately for small wave vectors in Fig. 4.6. The region shown corresponds to $K,q \sim 10^{-3}\ \pi/a$, far from the first Brillouin zone boundary at $K,q \sim \pi/a$. The acoustic branch and the photon dispersion do not cross, indicating that there are no regions of the photon and acoustic phonon dispersion curves for which the energy and momentum conservation laws can be simultaneously satisfied. However, this is not true for the optical-phonon branch. At point P in Fig. 4.6, the optical phonon and light dispersion curves cross and, therefore, photons can be converted to optical phonons satisfying both momentum and energy conservation.

Ionic crystals exhibit interesting optical properties associated with optical phonons. The frequency of the optical phonon branch is $\Omega_o \sim (2f/M_1)^{1/2}$, where $f \sim 4 \times 10^3$ g s^{-2} (see Prob. 4.1), $M_1 \sim 10^{-23}$ g (e.g., for NaCl), giving $\Omega_o \sim 3 \times 10^{13}$ s^{-1}, corresponding to a wavelength of $\lambda_o \sim 10\ \mu$m. Thus, optical phonon frequencies lie in the infrared region of the spectrum. Characteristically, an ionic crystal such as NaCl exhibits strong optical reflection and absorption in the infrared region associated with optical phonons. Let us see how this happens.

Consider an ionic crystal with two ions per unit cell, such as NaCl with positively charged Na and negatively charged Cl ions. We apply a time-dependent light field of the form

$$E \sim E_o\, e^{-i\omega t} \qquad (4.30)$$

on the crystal. The electric field of the light moves the positive and negative charges to opposite directions. The E-field sets up a *forced*

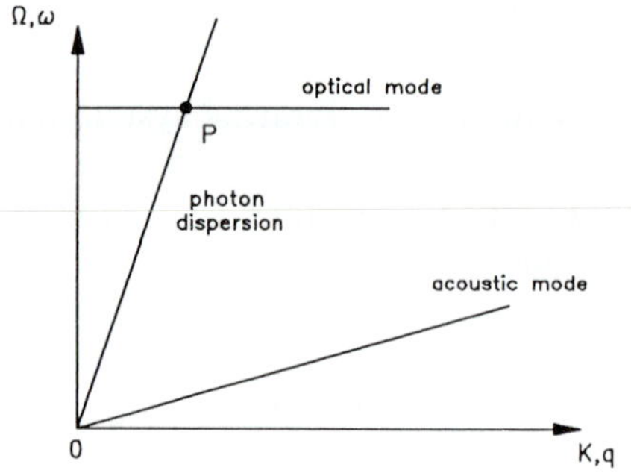

Figure 4.6. Dispersion for uncoupled photon and phonons.

vibration of atoms (optical mode). (It should be noted that in our treatment in Sec. 4-1 we considered free vibrations, with no external force applied.) The equation of motion for the two ions in our present case is similar to Eqs. (4.6) and (4.8), except for the additional force terms on the right-hand side,

$$M_1 \frac{d^2U_j}{dt^2} + M_1\gamma \frac{dU_j}{dt} - f(U_{j+1} + U_{j-1} - 2U_j) = -eE_0e^{-i\omega t} \tag{4.31}$$

and

$$M_2 \frac{d^2U_{j+1}}{dt^2} + M_2\gamma \frac{dU_{j+1}}{dt} - f(U_{j+2} + U_j - 2U_{j+1}) = eE_0e^{-i\omega t} . \tag{4.32}$$

The first (second) equation is written for the negatively charged (positively charged) ion. Note that we also added phenomenological damping terms, which are proportional to the first time derivative of the displacement.

To solve Eqs. (4.31) and (4.32), we again try solutions of the form

$$U_j = A_1 \, e^{-i\omega t} \tag{4.33}$$

and

$$U_{j+1} = A_2 \, e^{-i\omega t} , \tag{4.34}$$

where we assume that the vibrations oscillate at the frequency of the applied force. Substituting Eqs. (4.33) and (4.34) and their time derivatives into Eqs. (4.31) and (4.32), and rearranging terms, we obtain

$$[2f - M_1(\omega^2 + i\gamma\omega)]A_1 - 2f\,A_2 = -eE_0 \tag{4.35}$$

and

$$-2fA_1 + [2f - M_2(\omega^2 + i\gamma\omega)]A_2 = eE_0\ . \tag{4.36}$$

These are two algebraic equations for the two unknown quantities A_1 and A_2. The solution is

$$A_1 = \frac{-eE_0/M_1}{\Omega_+^2(0) - (\omega^2 + i\gamma\omega)} \quad \text{and} \quad A_2 = \frac{eE_0/M_2}{\Omega_+^2(0) - (\omega^2 + i\gamma\omega)}\ , \tag{4.37}$$

where we used Eq. (4.19) to introduce the optical-phonon frequency. The opposite signs of A_1 and A_2 indicate that the directions of displacement of the negative and positive ions are opposite to each other. Also, the above equations show that when the light frequency ω becomes equal to the optical-phonon frequency, a resonance effect should be observed, indicating that the vibration amplitudes become very large and that a strong absorption takes place.

As shown in Chap. III, the optical response of the crystal requires the knowledge of the dielectric constant. Therefore, we now calculate the dielectric function associated with the displacement of the ions. As in Chap. III, we divide the polarization into two parts,

$$P = P_i + P_b\ , \tag{4.38}$$

where P_i is the ionic contribution, which we want to compute in the following, and P_b is the electronic contribution to the total polarization, as treated in Chap. III. For N ionic pairs per unit volume (N unit cells per unit volume)

$$P_i = Ne(U_{j+1} - U_j)\ , \tag{4.39}$$

since $-e(U_j - U_{j+1})$ is the dipole moment for a pair of ions, and $(U_j - U_{j+1})$ is the relative displacement between ions. The dielectric function is then

$$\epsilon(\omega) = 1 + 4\pi\frac{P}{E} = 1 + \frac{4\pi P_b}{E} + \frac{4\pi P_i}{E}$$

$$= 1 + 4\pi\frac{P_b}{E} + \frac{4\pi Ne}{E_0}(A_2 - A_1)\ . \tag{4.40}$$

Substituting for A_2 and A_1 from Eq. (4.37) one gets

$$\epsilon(\omega) = 1 + 4\pi \frac{P_b}{E} + \frac{4\pi Ne^2}{M_r\Omega_+^2(0)} \frac{1}{1 - (\omega^2 + i\gamma\omega)/\Omega_+^2(0)} , \tag{4.41}$$

where $1/M_r = 1/M_1 + 1/M_2$ is the reduced mass of the two ions. For low frequencies, when $\omega << \Omega_+(0)$, we have

$$\epsilon(\omega \rightarrow 0) = 1 + 4\pi \frac{P_b}{E} + \frac{4\pi Ne^2}{M_r} \frac{1}{\Omega_+^2(0)} \equiv \epsilon_0 . \tag{4.42}$$

This is the static dielectric constant. For high frequencies, where $\omega >> \Omega_+(0)$, we get

$$\epsilon(\omega \rightarrow \infty) = 1 + 4\pi \frac{P_b}{E} \equiv \epsilon_\infty , \tag{4.43}$$

formally the same expression as Eq. (3.43). One should note that in principle, "$\omega \rightarrow 0$" or "$\omega \rightarrow \infty$" means that the frequencies are, respectively, much above or much below the resonance, $\Omega_+(0)$. In this sense, ϵ_0 and ϵ_∞ are the values of the dielectric function, which are assumed much below or above the analyzed resonance *if there is no other resonance in that spectral regime*. For the case that two or more resonances in a material are close in frequency, they cannot be modeled independently. The relevant physical processes responsible for the resonances must be treated simultaneously, including the possible interactions between the corresponding resonances.

Let us now come back to Eq. (4.42) where we insert the definition, Eq. (4.43), and rearrange the terms to obtain

$$\frac{4\pi Ne^2}{M_r} \frac{1}{\Omega_+^2(0)} = \epsilon_0 - \epsilon_\infty . \tag{4.44}$$

Using this result, Eq. (4.41) can be written as

$$\epsilon(\omega) = \epsilon_\infty + \frac{\epsilon_0 - \epsilon_\infty}{1 - (\omega^2 + i\gamma\omega)/\Omega_+^2(0)} . \tag{4.45}$$

This is the dielectric function in terms of directly measurable quantities, ϵ_0, ϵ_∞, and $\Omega_+(0)$. Note that at $\omega = \Omega_+(0)$, the dielectric function goes through a resonance which is broadened by the damping constant γ. At $\Omega_+(0)$, light with transverse polarization couples with TO phonons and we denote $\Omega_+(0) = \Omega_T$ as the frequency of the TO phonon near the center of the Brillouin zone. Thus, the dielectric function is

$$\epsilon = \epsilon_\infty + \frac{\epsilon_0 - \epsilon_\infty}{1 - (\omega^2 + i\gamma\omega)/\Omega_T^2} . \tag{4.46}$$

To get the longitudinal optical-phonon frequency $\omega = \Omega_L$, we recall that it is obtained by the condition $\epsilon(\omega) = 0$, as discussed in Chap. III (see Eq. 3.9). Equation (4.46) without the damping ($\gamma = 0$) gives

$$\epsilon = 0 = \epsilon_\infty + \frac{\epsilon_0 - \epsilon_\infty}{1 - \Omega_L^2/\Omega_T^2} \tag{4.47}$$

or

$$\boxed{\Omega_L = \Omega_T \sqrt{\frac{\epsilon_0}{\epsilon_\infty}} .}$$

Lyddane-Sachs-Teller relation (4.48)

This is called the Lyddane-Sachs-Teller (LST) relation, which expresses the ratio of LO and TO phonon frequencies in terms of the ratio of the low- and high-frequency dielectric constants. For NaCl, independently measured values are $\epsilon_0 = 5.62$, $\epsilon_\infty = 2.25$, $\Omega_L = 5.0 \times 10^{13}$ Hz, and $\Omega_T = 3.08 \times 10^{13}$ Hz, showing that Eq. (4.48) is well satisfied. Similarly, for LiF, $\epsilon_0 = 8.9$, $\epsilon_\infty = 1.9$, $\Omega_T = 5.8 \times 10^{13}$ Hz, and $\Omega_L = 12 \times 10^{13}$ Hz. Equations (4.42) and (4.43) indicate that $\epsilon_0 > \epsilon_\infty$ and, according to the LST relation, $\Omega_L > \Omega_T$.

The dielectric function of Eq. (4.46) is plotted in Fig. 4.7. For $\Omega_T < \omega < \Omega_L$, $\epsilon(\omega) < 0$, and the reflectivity becomes high, $R \to 1$, indicating that an incident wave whose frequency lies in the range $\Omega_T < \omega < \Omega_L$ gets totally reflected from the surface of the insulator. There is no solution for propagating waves in the crystal in this region. The behavior of the reflectivity of an incident wave is shown in Fig. 4.8.

In Chap. III we expressed the complex dielectric function, $\epsilon(\omega) = \epsilon'(\omega) + i\epsilon''(\omega)$, in terms of the absorption coefficient α and the refractive index $n(\omega)$, where both $\alpha(\omega)$ and $n(\omega)$ were chosen as real quantities. Equivalently, one can also introduce a complex refractive index, $\mathcal{n}(\omega)$, through the relation $\epsilon(\omega) = \mathcal{n}'^2(\omega) - \mathcal{n}''^2(\omega) + 2i\mathcal{n}'(\omega)\mathcal{n}''(\omega)$, and get

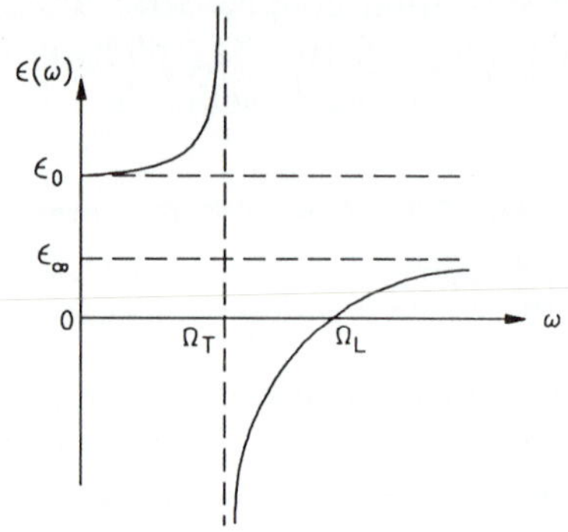

Figure 4.7. The plot of the dielectric function versus frequency.

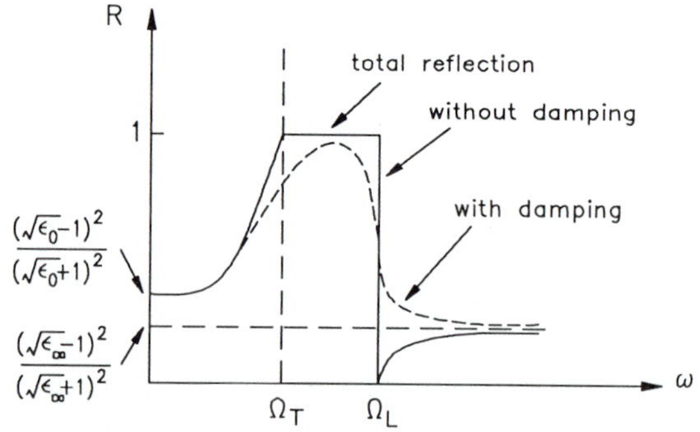

Figure 4.8. Schematic reflectivity versus frequency for an ionic insulator such as NaCl.

$$n'^2(\omega) - n''^2(\omega) = \epsilon_\infty + \frac{4\pi Ne^2}{M_r} \frac{\Omega_T^2 - \omega^2}{(\Omega_T^2 - \omega^2)^2 + \omega^2\gamma^2} \tag{4.49}$$

and

$$2n'(\omega)n''(\omega) = \frac{4\pi Ne^2}{M_r} \frac{\omega\gamma}{(\Omega_T^2 - \omega^2)^2 + \omega^2\gamma^2} . \tag{4.50}$$

The plot of $n'(\omega)$ and $n''(\omega)$ is shown in Fig. 4.9.

Figure 4.10 shows the observed infrared transmission in a thin film of NaCl. (The dip in transmission corresponds to a peak in absorption.) The peak of the transmission drop occurs at $\omega = \Omega_T$, as $\epsilon(\omega) \to \infty$ at this frequency. The phenomena of strong infrared absorption and reflection by the lattice is sometimes referred to as *reststrahlen* (German for residual rays).

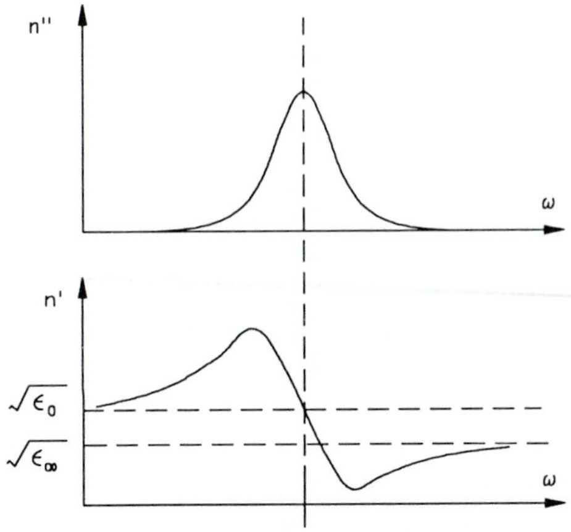

Figure 4.9. $n'(\omega)$ and $n''(\omega)$ versus frequency in the vicinity of TO phonon frequency.

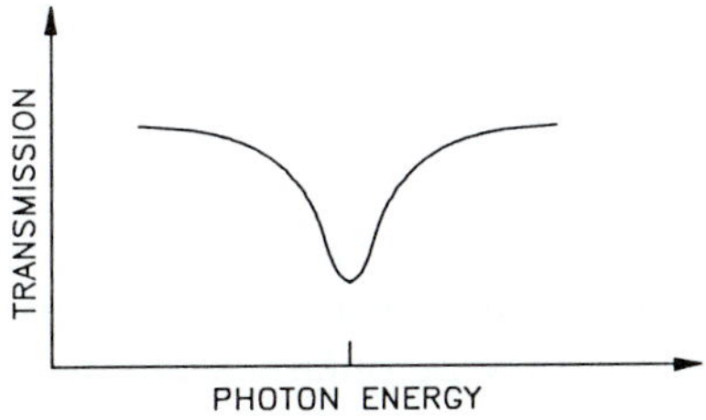

Figure 4.10. Transmission in the infrared in a NaCl thin film. The transmission minimum corresponds to a wavelength of $\simeq 60\ \mu m$ for NaCl.

4-3. Phonon Polaritons

As shown in Eq. (3.14), the dispersion relation for light inside a solid is modified from the case of free space, where $\epsilon(\omega) \equiv 1$. If, for simplicity, we ignore the damping, $\gamma \to 0$, we can write Eq. (4.46) as

$$\epsilon(\omega) = \epsilon_\infty \left[\frac{\Omega_L^2 - \omega^2}{\Omega_T^2 - \omega^2} \right], \tag{4.51}$$

using the LST relation, Eq. (4.48). The dispersion relation, Eq. (3.14), for this dielectric function becomes

$$\omega^2 \epsilon_\infty \left[\frac{\Omega_L^2 - \omega^2}{\Omega_T^2 - \omega^2} \right] = c^2 q^2 \ . \tag{4.52}$$

Equation (4.52) gives the dispersion relation for the light inside the crystal in the vicinity of the optical-phonon resonance. Figure 4.11 shows the plot of Eq. (4.52) by noting that when $q \to 0$, $\omega \to \Omega_L$ or $\omega \to 0$, and when $q \to \infty$, $\omega \to \Omega_T$ or $\omega \to \infty$. The dashed lines in Fig. 4.11 represent the dispersion curves of the noninteracting photon and phonon. The solid lines give the photon-phonon dispersion in the presence of interaction between the photon and phonon. We see that away from the point where the photon dispersion crosses the phonon dispersion (the point P of crossing of the two dashed lines), the solid and the dashed lines coincide, while they split apart in the neighborhood of the crossing point. The solid lines are called the *phonon-polariton dispersion* curves. They represent the mixed mode of photon and phonon. The name polariton was adopted as a short version of *polariz*ation wave and pho*ton*. They are mixtures of phonon and photon dispersions. In other words, when a photon impinges on a crystal with frequency in the vicinity of the TO phonon frequency, it excites a polariton. The curve with higher frequency is called the upper polariton branch. The low-frequency curve is called the lower polariton branch.

Since the phonon field is a mechanical field, phonon polaritons are mixtures of electromagnetic and mechanical fields. A simple physical picture for these polaritons may be obtained from their analogy to classical

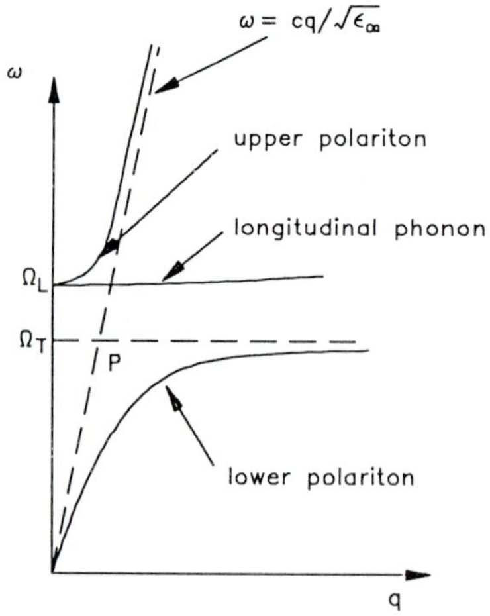

Figure 4.11. Schematics of phonon-polariton dispersion curves.

mixing of harmonic oscillators. Consider two oscillators of frequencies ω_1 and ω_2. Without coupling, the oscillators vibrate at these independent frequencies. However, if the oscillators are connected by a spring so that they couple together, they no longer vibrate at independent frequencies, and two different vibrational frequencies arise. In analogy, here the two independent oscillators are the pure photon and pure phonon which couple in the crystal to give two new modes; the polariton modes. Away from the photon-phonon crossing point, the two polariton modes behave like pure modes. For example, for large q values, the lower polariton branch behaves like a pure phonon mode, and for large ω, the upper polariton branch behaves like a pure photon mode.

Polaritons are always formed in the dielectric medium when the light dispersion crosses the dispersion of a transverse excitation to which it is coupled. We will discuss the so-called exciton polaritons in Chap. VI of this book.

4-4. Light Scattering

Photons may be scattered inelastically by both optic and acoustic phonons. In a scattering process, an incoming photon of frequency ω_i and wave vector $\mathbf{q}_i$ is converted into an outgoing photon of frequency ω_s and wave vector $\mathbf{q}_s$. The difference in energy, $\hbar(\omega_i - \omega_s)$, and momentum, $\hbar(\mathbf{q}_s - \mathbf{q}_i)$, is exchanged with the medium. When the scattered photon has a lower frequency than the incident photon, the exchanged energy creates a phonon, and the process is referred to as a *phonon emission* or a *Stokes process* [see Fig. 4.12(a)]. On the other hand, when the scattered photon has a higher frequency than the incident photon, a phonon is absorbed and the process is called *phonon absorption* or *anti-Stokes process* [see Fig. 4.12(b)]. When an optical phonon is involved in the scattering, the process is called *Raman scattering*. *Brillouin scattering* corresponds to photon-acoustic-phonon interaction. *Rayleigh scattering* is a process whereby the scattered photons fluctuate in frequency within a small width of ~ 10^{-9} to 10^{-4} cm^{-1}, centered at the incident photon frequency.

The scattered laser frequency is shifted with respect to the incident laser frequency, with the magnitude of the shift being the phonon energy. In the Raman spectra, this shift is determined by the optical phonon energy, which is in the range of approximately 5 meV to 250 meV. The energy shift is in the range of 6×10^{-3} meV to 0.1 meV or 0.05 to 1 cm^{-1} for acoustic phonons in Brillouin spectra. The Rayleigh spectrum, which is due to nonpropagating modes such as thermal diffusion or heat waves, is not shifted, but has a range in linewidths, as previously discussed. Figure 4.13 displays the three spectral ranges where both Stokes and anti-Stokes lines are schematically plotted. The line at the middle of each figure,

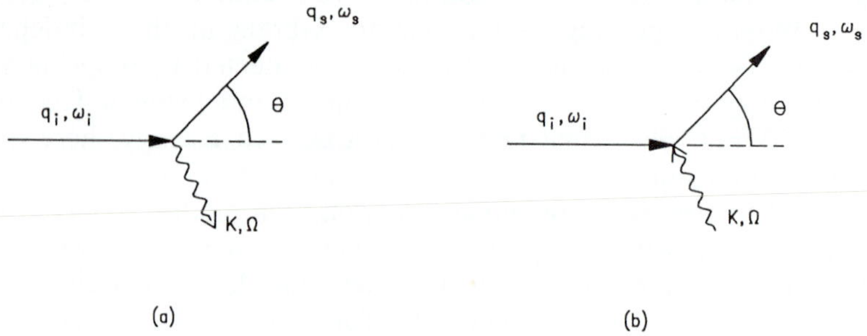

Figure 4.12. Schematics of a phonon scattering process; (a) phonon emission (Stokes) and (b) phonon absorption (anti-Stokes). Photons (phonons) are represented as straight (wiggled) lines.

denoted by zero, shows the spectral position of the incident monochromatic laser.

Selection rules have to be satisfied for the scattering to take place (see Chap. II). Conservation of momentum and energy provides one set of such selection rules. The conservation of momentum and energy for phonon absorption (anti-Stokes process) dictates

$$\hbar\omega_s = \hbar\omega_i + \hbar\Omega \tag{4.53}$$

and

$$\hbar\mathbf{q}_s = \hbar\mathbf{q}_i + \hbar\mathbf{K} \ . \tag{4.54}$$

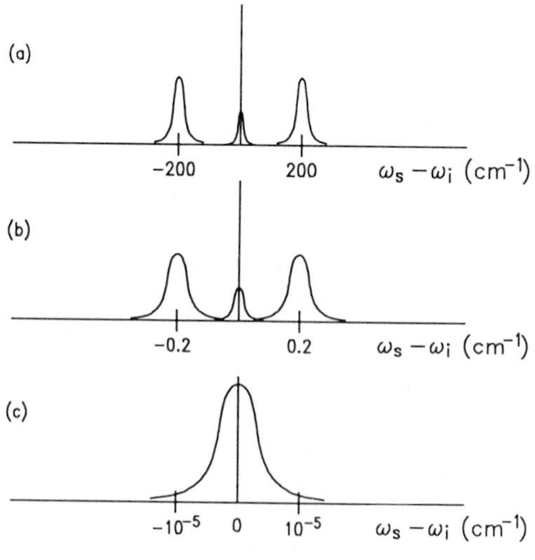

Figure 4.13. Schematic light scattering spectra for (a) Raman, (b) Brillouin, and (c) Rayleigh processes in the three spectral ranges.

For phonon emission (Stokes process) we have

$$\hbar\omega_s = \hbar\omega_i - \hbar\Omega \tag{4.55}$$

and

$$\hbar\mathbf{q}_s = \hbar\mathbf{q}_i - \hbar\mathbf{K}\ . \tag{4.56}$$

The conservation of momentum, as given by Eqs. (4.54) and (4.56), may be rearranged to

$$\mathbf{K} = \pm(\mathbf{q}_i - \mathbf{q}_s)\ . \tag{4.57}$$

Referring to Fig. 4.12, the magnitude of the vector equality of Eq. (4.57) may be written as

$$K^2 = q_i^2 + q_s^2 - 2q_i q_s \cos\theta\ , \tag{4.58}$$

where we denoted the angle between vectors $\mathbf{q}_s$ and $\mathbf{q}_i$ as θ. But, $q_i = n(\omega)\omega_i/c$ and $q_s = n(\omega)\omega_s/c$, and since $\Omega << \omega_i$ or ω_s (typically ω_i or $\omega_s \simeq 2$ eV and $\Omega \sim 10$ meV), we have $\omega_i \simeq \omega_s$ and, consequently $q_i \simeq q_s$, allowing us to simplify Eq. (4.58) as

$$K = 2q_i \sin(\theta/2)\ . \tag{4.59}$$

For Brillouin scattering, where acoustic phonons are involved, the phonon dispersion relation is given by $\Omega = c_s K$ or $K = \Omega/c_s$, where c_s is the sound velocity. Substituting for K in Eq. (4.59) yields

$$\Omega = 2\,\frac{c_s}{c}\,\omega_i\, n(\omega)\, \sin(\theta/2)\ . \tag{4.60}$$

Using the sound velocity of $c_s \sim 3 \times 10^5$ cm/sec and $\omega_i \sim 2$ eV, one gets $\Omega \sim 5 \times 10^{-2}$ meV or 0.4 cm^{-1}. This confirms that the acoustic phonon frequencies are small, and high resolution techniques are needed to detect them.

Figure 4.14 shows the representation of the Stokes and anti-Stokes processes in Brillouin scattering. In Fig. 4.14(a), an acoustic phonon with energy $\hbar\Omega_A$ and momentum $\hbar K_A$ is created as a result of the scattering process. This figure clarifies how acoustic phonons, which cannot be involved in the infrared absorption, may still participate in the scattering process. Phonons that are involved in the infrared absorption are called *IR or infrared active* modes, while those phonons that participate in the

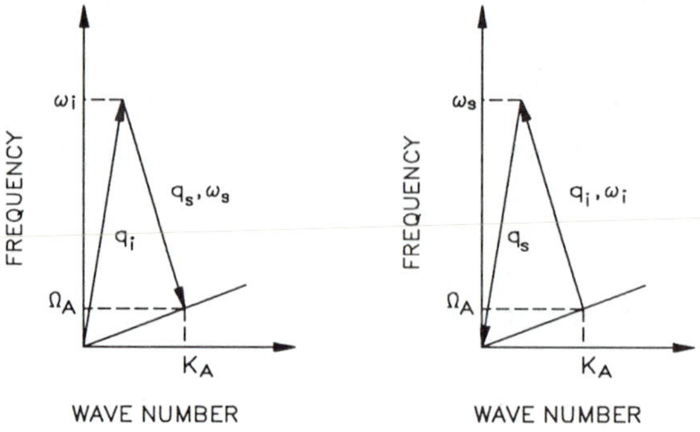

Figure 4.14. (a) Representation of conservation of momentum and energy for Stokes process in Brillouin scattering. (b) Vectorial representation for anti-Stokes process in Brillouin scattering where an acoustic phonon of energy $\hbar\Omega_A$ and momentum $\hbar K_A$ is absorbed to create the scattered photon with energy and momentum, $\hbar\omega_s$ and $\hbar \mathbf{q}_s$, respectively.

scattering process are referred to as *Raman active* modes. In a crystal with inversion symmetry, infrared active modes are Raman inactive, and vice versa. In a crystal without inversion symmetry, on the other hand, TO phonons may be both Raman and IR active, thus observable in infrared absorption as well as in scattering processes.

A second and more subtle set of selection rules that must be satisfied for the scattering to take place are the point-group selection rules, also mentioned in Chap. II. These originate from the fact that the crystal unit cell is invariant under only a certain set of symmetry operations, and in fact, it is this set of symmetry operations which defines the crystal point group. Similarly, the phonon eigenvector also has transformation properties. If these eigenvector transformation properties do not satisfy certain requirements, which depend on the crystal point group, then light will not be scattered. As an example, we consider the experimental configuration that is usually denoted by $\mathbf{q}_i(\epsilon_i\epsilon_s)\mathbf{q}_s$, determined by the incident photon wave vector $\mathbf{q}_i$, polarization unit vector ϵ_i, scattered photon wave vector $\mathbf{q}_s$, and polarization unit vector ϵ_s, all specified relative to the crystal principal axes. In a tetragonal crystal with no center of symmetry, the configuration y(zz)x is sensitive to phonon excitations involving motion by the ions parallel to the crystal z axis; the configuration y(xz)x is sensitive to excitations involving ionic motion in the plane perpendicular to the crystal z axis. Simply rotating the input polarization from x to z forbids the observation of some phonon modes, and allows the observation of other phonon modes. Crystal point-group selection rules are a simple yet powerful way to determine and distinguish between phonon modes of different symmetries. Group theoretical techniques, which are beyond the scope of this text, are essential for a more thorough understanding of these concepts.

Usually, the scattered photon is detected in a direction normal to the incident photon ($\theta \simeq \pi/2$). Figure 4.15 shows typical setups for Raman and Brillouin scattering experiments. A double monochromator is used in Raman scattering as a tunable spectral filter, while a Fabry-Perot interferometer (sometimes in connection with a double monochromator) is needed for Brillouin scattering, because the lower phonon frequencies involved require a better rejection of the incoming light and a higher detection resolution given by the Fabry Perot. Photon counting techniques are employed to detect the scattered photons. Figure 4.16 shows Raman scattering in an orthorhombic $KNbO_3$ crystal at 180° C. The line at zero frequency shift marks the position of the incident laser frequency, while the peak at an energy shift of $\simeq$ 280 cm^{-1} is due to optical phonon scattering. The acoustic phonon frequencies that are in the vicinity of the zero-line are not resolved in this figure. The results from a Brillouin scattering experiment of the same material are represented as an intensity versus frequency pattern (Fig. 4.17), where the frequency region in the neighborhood of the zero-line is displayed. The spectrum clearly exhibits identifiable peaks at $\simeq \pm 30$ GHz (1 cm^{-1}) from the center, corresponding to the Stokes and anti-Stokes lines.

Using angle-resolved scattering experiments, it is possible to directly measure the phonon-polariton dispersion curve near K ~ 0. Consider a crystal like GaP in which the transverse optical phonon is infrared as well as Raman active. By measuring the scattered photon frequency as a function of the scattering angle θ for small angles, one may construct the

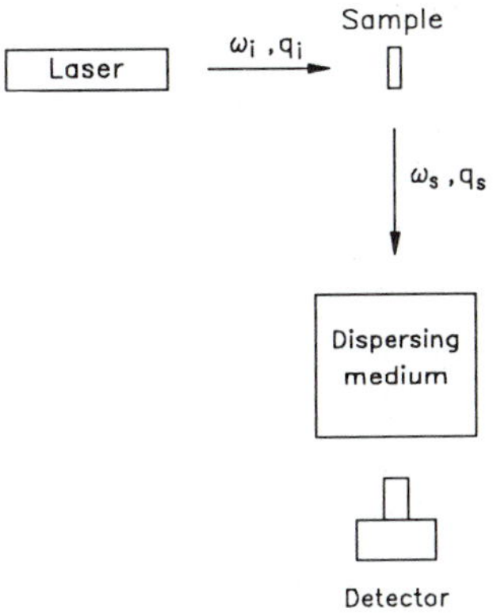

Figure 4.15. Schematic experimental setup for Raman and Brillouin scattering. The dispersing medium is typically a double monochromator or Fabry-Perot interferometer for the Raman and Brillouin scattering measurement.

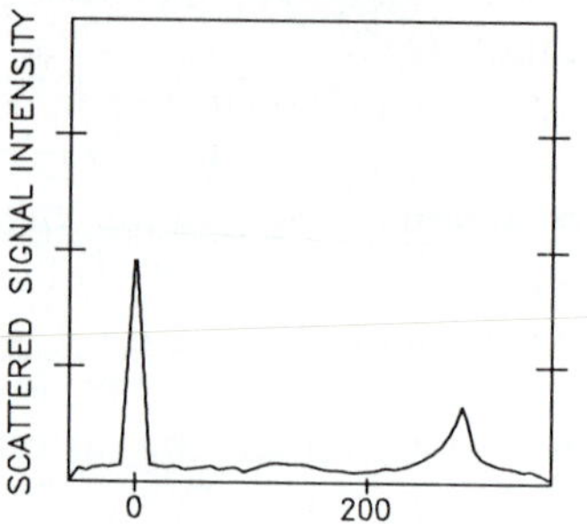

Figure 4.16. Raman spectrum of orthorhombic $KNbO_3$ (Ref. 4.3).

polariton dispersion curve, $\hbar\Omega = \omega_i - \omega_s$, versus $|\mathbf{K}| = |\mathbf{q}_s - \mathbf{q}_i|$.By varying the scattering angle, the wave vector of the phonon is changed, as can be simply seen from Eq. (4.59). Results of such a scattering experiment are shown in Fig. 4.18. Note that only the TO phonon shows the strong dispersion characteristic of a polariton, in contrast to the LO phonon, which is more or less dispersionless.

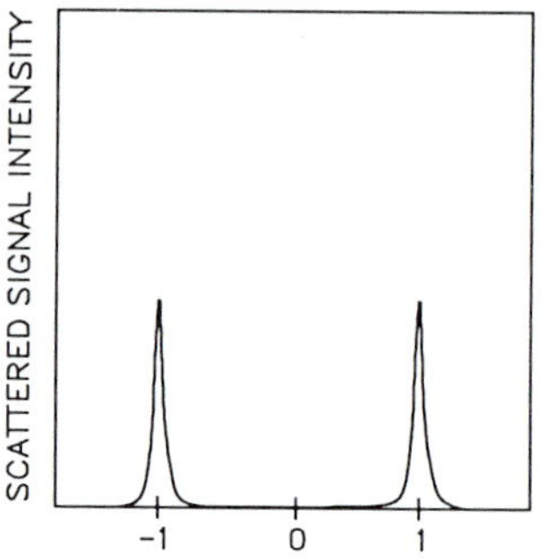

Figure 4.17. Brillouin spectrum of orthorhombic $KNbO_3$ showing Stokes and anti-Stokes lines (Ref. 4.3). The Rayleigh scattered light is removed by appropriate filters.

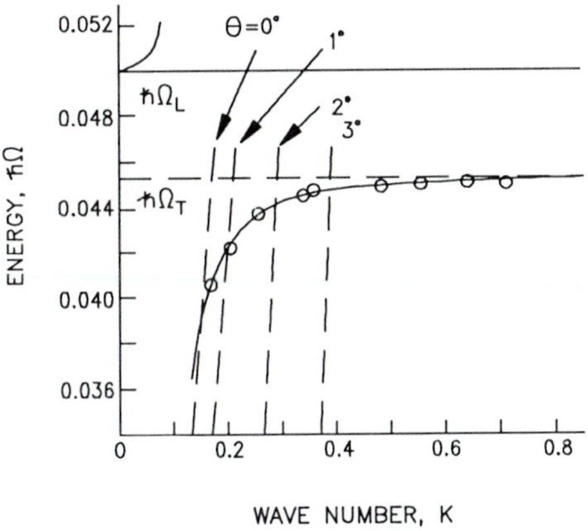

Figure 4.18. Phonon-polariton dispersion curve in a GaP crystal measured by Raman scattering. The horizontal axis denotes the wave number K represented by the corresponding photon energy $\hbar cK$ in eV (Ref. 4.4).

4-5. Coherent Raman Spectroscopy*

The scattering processes described in the previous section refer to spontaneous processes, whereby the light beam scatters from weak and random temperature fluctuations. Referring to Eq. (4.27), one sees that such spontaneous scattering processes can always occur, even if the population n_K of the phonons involved in the scattering process is small compared with unity, because the lattice fluctuates, even at $T = 0$. For example, the population of optical phonons, which is given by the Bose-Einstein distribution function as

$$n_K = \frac{1}{e^{\hbar\Omega/k_B T} - 1}, \tag{4.61}$$

has a value of $\simeq 0.5$ at room temperature for a phonon energy of $\hbar\Omega = 30$ meV. Because the probability of photon scattering is very small in most materials, the intensity of the scattered light, which is proportional to the phonon population, is extremely weak and requires high-sensitivity detection techniques such as the photon counting method. It is possible to circumvent this difficulty by using coherent Raman scattering techniques. The essence of these techniques is to *force* the oscillations of a particular mode of vibration with well-defined energy and wave vector. Two intense laser beams are adjusted such that their energy difference, $\hbar(\omega_1 - \omega_2)$, equals the phonon energy $\hbar\Omega$ and their momentum difference, $\hbar(\mathbf{q}_1 - \mathbf{q}_2)$, equals the phonon momentum $\hbar\mathbf{K}$. This forced oscillation builds up a large population of phonons, $n_K > 1$, which in turn produces a large scattering signal. The process for one such coherent scattering, known as coherent anti-Stokes Raman scattering (CARS), is schematically shown in Fig. 4.19. Using Eq. (4.53), it is clear that the scattered anti-Stokes signal occurs at

$$\hbar\omega_s = \hbar\omega_1 + \hbar\Omega = \hbar\omega_1 + (\hbar\omega_1 - \hbar\omega_2)$$
$$= \hbar(2\omega_1 - \omega_2) . \qquad (4.62)$$

The momentum of the scattered signal is at (using Eq. (4.54))

$$\hbar\mathbf{q}_s = \hbar\mathbf{q}_1 + \hbar\mathbf{K} = \hbar\mathbf{q}_1 + (\hbar\mathbf{q}_1 - \hbar\mathbf{q}_2)$$
$$= \hbar(2\mathbf{q}_1 - \mathbf{q}_2) . \qquad (4.63)$$

The intensity of the scattered anti-Stokes signal at $\hbar(2\omega_1 - \omega_2)$ is typically nine orders of magnitude stronger than a spontaneous anti-Stokes signal. The experimental geometry for a CARS process and the momentum conservation involved in this process are shown in Fig. 4.20. The fact that the CARS signal is generated in a particular direction, $\mathbf{q}_s = 2\mathbf{q}_1 - \mathbf{q}_2$, allows spatial separation of the signal from the intense incident laser beams. The CARS process may be viewed as a mixing of incident field ω_1 with the field oscillating at the difference frequency, $\omega_1 - \omega_2$ (i.e., $\omega_s = \omega_1 + (\omega_1 - \omega_2)$). Figure 4.21 shows the CARS signal observed in Si where a strong Raman line at 520 cm^{-1} is detected.

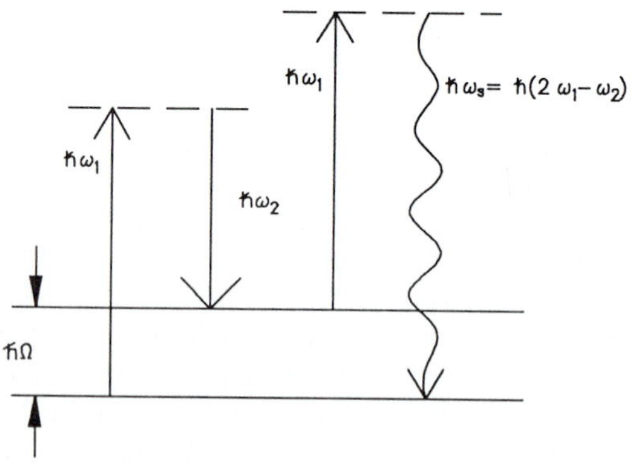

Figure 4.19. Energy diagram for CARS. $\hbar\Omega$ denotes the phonon energy.

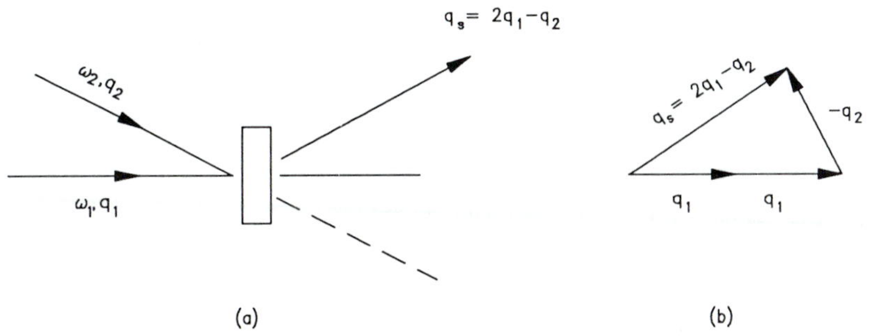

Figure 4.20. Experimental geometry for a CARS process (a) and momentum conservation for the CARS process (b).

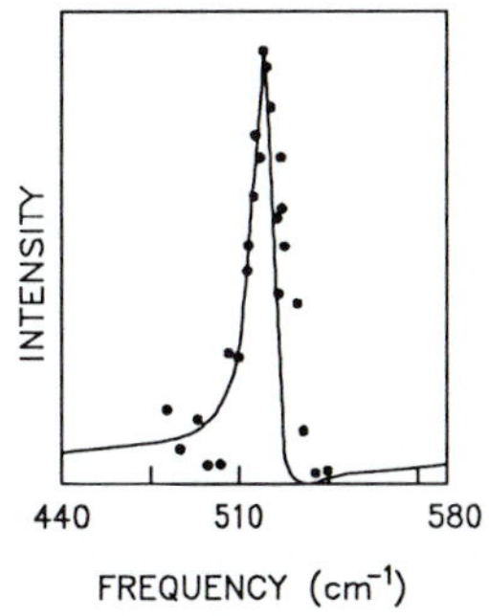

Figure 4.21. CARS signal for Si showing a strong Raman line with frequency of $\simeq$ 520 cm^{-1} (Ref. 4.6).

Other variations of coherent Raman scattering have been developed. For example, in a coherent Stokes Raman scattering (CSRS) the coherent scattered signal is generated with frequency $\omega_s = 2\omega_2 - \omega_1$ in the direction $\mathbf{q}_s = 2\mathbf{q}_2 - \mathbf{q}_1$. Higher-order Stokes and anti-Stokes signals such as $3\omega_2 - 2\omega_1$ may also be detected in the respective directions using the same experimental geometry. The reader is referred to specialized articles (see Refs. 4.5 and 4.7) for a description of these powerful spectroscopic tools for investigation of phonons and other excitations in solids using laser beams.

4-6. Problems

4.1. Write the force constant f in terms of Young's modulus. Using this relation, give an estimate for f. (Hint: You may use the relation $v_s = \sqrt{y/\rho}$, where ρ is the mass density, v_s is the velocity of sound, and y is Young's modulus.)

4.2. (a) Show that the dispersion relation for a mono-atomic one-dimensional solid is given by

$$\omega = 2\sqrt{f/M} \left| \sin \frac{1}{2} ka \right| , \tag{4.64}$$

where M is the atomic mass, f is the spring constant, and a is the interatomic spacing.

(b) Show that the density of states for the crystal of part (a) is given by

$$g(\omega) = \frac{1}{\pi a \omega_m} \left[1 - \frac{\omega^2}{\omega_m^2} \right]^{-1/2} , \tag{4.65}$$

where $\omega_m = 2\sqrt{f/M}$. The frequency $\omega = \omega_m$, at which the density of states goes through a resonance, is called a "Van-Hove Singularity."

(c) Plot $g(\omega)$.

4.3. Consider a crystal with two identical atoms per unit cell.

(a) What are the frequencies of acoustic phonons?

(b) What is the energy difference between acoustic and optical phonons at the zone boundary? Explain.

4.4. Consider a one-dimensional solid with three atoms per unit cell. How many modes of lattice vibrations exist for this solid? How many modes are optical and how many are acoustical? How many modes are associated with longitudinal and transverse waves?

4-7. References

1. R. F. Wallis, ed., *Lattice Dynamics* (Pergamon Press, New York, 1965).

2. H. Z. Cummins, in *Light Scattering in Solids*, edited by Blakanski, 1981.

3. J. P. Sokoloff, L. L. Chase, and D. Rytz, Phys. Rev. B **38**, 597 (1988).

4. C. H. Henry and J. J. Hopfield, Phys. Rev. Lett. **15**, 964 (1965).

5. M. D. Levenson and S. S. Kano, *Introduction to Laser Spectroscopy* (Academic Press, New York, 1988).

6. S. A. Akhmanov, S. V. Govorkov, N. I. Koroteev, and I. L. Shumay, in *Laser Optics of Condensed Matter*, edited by J. L. Birman, H. Z. Cummins, and A. A. Kaplyomskii (Plenum Press, New York, 1987).

7. Y. R. Shen, *The Principles of Nonlinear Optics* (Wiley, New York, 1984).

Chapter V
LINEAR OPTICAL PROPERTIES OF SEMICONDUCTORS: FREE ELECTRON-HOLE PAIRS

Crystalline materials are classified as insulators or semiconductors if their ground state is characterized by filled valence bands and empty conduction bands at low temperature, $T \simeq 0$. The ground state is defined as the state of the material when it is not excited, either by optical, electrical, thermal, or other excitation mechanisms. The distinction between semiconductors and insulators can be made by defining semiconductors as insulators with a relatively small bandgap, E_g. Electrons may be promoted from the valence band to the conduction band by photon absorption or by other means, such as thermal excitation, impact collisions, etc. For optical absorption to occur, the photon energy has to be larger than the bandgap energy of the semiconductor material. When the electron is lifted to the conduction band, it leaves behind a hole (missing electron) in the valence band. The charge assigned to the hole is that of the filled valence band minus an electron charge, i.e., a single positive charge. The electron in the conduction band with a negative charge (-e) and the hole in the valence band with a positive charge (+e) interact via the Coulomb potential, which is attractive for opposite charges. These Coulombic effects give rise to very important modifications of the band-edge absorption called excitonic effects, which will be described in Chap. VI. In this chapter, as an introduction, we treat optical properties and band-edge absorption in the absence of Coulomb interaction. The resulting spectra do not explain all experimental observations, since Coulomb effects are always present, but our discussion, nevertheless, yields some valuable insights into the physics of light-matter interaction.

5-1. Direct- and Indirect-Gap Semiconductors

It is common to distinguish two main groups of semiconductors having either direct gaps or indirect gaps. When the conduction band minimum occurs at the same point in **k** space as the valence band maximum, the semiconductor is called a direct-gap semiconductor. Examples are GaAs,

CdS, InSb, and many more compound III-V or II-VI materials made of elements from the third and fifth or second and sixth group, respectively, of the periodic table of elements [see Figs. 5.1(a) and 5.2(a)]. On the other hand, when the minimum of the conduction band occurs at a different point in **k** space than the maximum of the valence band, we would have an indirect-gap semiconductor. The most important examples of indirect semiconductors are the element semiconductors Si and Ge [see Figs. 5.1(b) and 5.2(b)].

The absorption or emission of photons is accompanied by conservation of momentum and energy. For direct transition, the conservation laws are given by [see Fig. 5.1(a)]

$$\boxed{\mathcal{E}_\mathrm{i} + \hbar\omega = \mathcal{E}_\mathrm{f}}$$

energy conservation (5.1)

and

$$\boxed{\hbar\mathbf{k}_\mathrm{i} + \hbar\mathbf{q} = \hbar\mathbf{k}_\mathrm{f} \ ,}$$

momentum conservation (5.2)

where $\mathcal{E}_\mathrm{i}$ and $\mathcal{E}_\mathrm{f}$ are the initial and final energies of the electron, $\mathbf{k}_\mathrm{i}$ and $\mathbf{k}_\mathrm{f}$ are the initial and final wave vectors of the electron, respectively, and $\hbar\omega$ and $\mathbf{q}$ are the energy and wave vector of the photon. Since $\mathbf{q}$ is small ($\sim 10^5\ \mathrm{cm}^{-1}$) in comparison to typical electron wave numbers in the

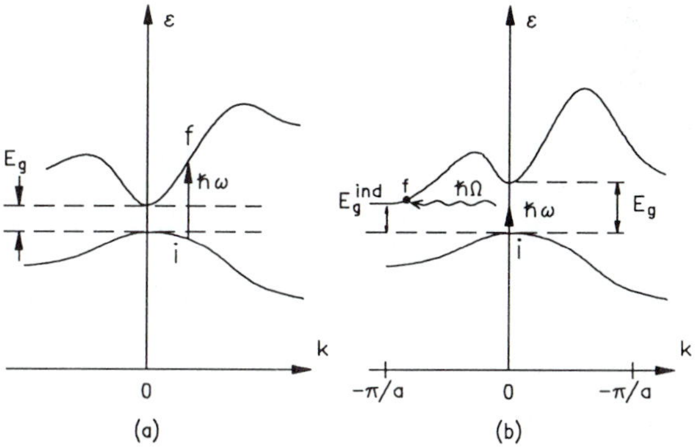

Figure 5.1. Schematics of (a) direct- and (b) indirect-gap semiconductors. The direct-gap energy is denoted as E_g, while the indirect-gap energy is E_g^{ind}.

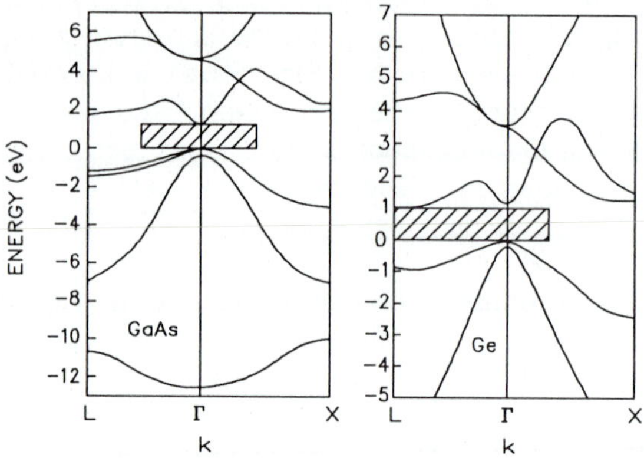

Figure 5.2. Electronic band structure of some semiconductors; (a) GaAs (direct-gap) and (b) Ge (indirect-gap). The hashed regions represent the regions of interest for optical measurements near the absorption edge (after Ref. 5.1). Γ, X, and L are points of high symmetry in the first Brillouin zone.

Brillouin zone, one often (not always!) ignores the photon momentum $\hbar q$ in comparison with the electron momentum $\hbar k_i$. In this case, the conservation laws become

$$\hbar\omega = \mathcal{E}_f - \mathcal{E}_i \gtrsim E_g \tag{5.3}$$

and

$$\mathbf{k}_f \simeq \mathbf{k}_i \ . \tag{5.4}$$

Equation (5.3) states that a semiconductor is transparent to light if the photon energy is less than the gap energy, but becomes absorbing above the threshold-gap energy E_g. Equation (5.4) states that only vertical transitions between valence and conduction bands are allowed if only photons are involved in the transition.

In the photon absorption process, the initial state of the electron is in the valence band and the final state is in the conduction band. If the zero of electron energy is chosen as the top of the valence band, the conduction and valence band energies can be written for parabolic bands (see Fig. 5.3) as

$$\mathcal{E}_c = E_g + \frac{\hbar^2 k^2}{2m_e} \tag{5.5}$$

and

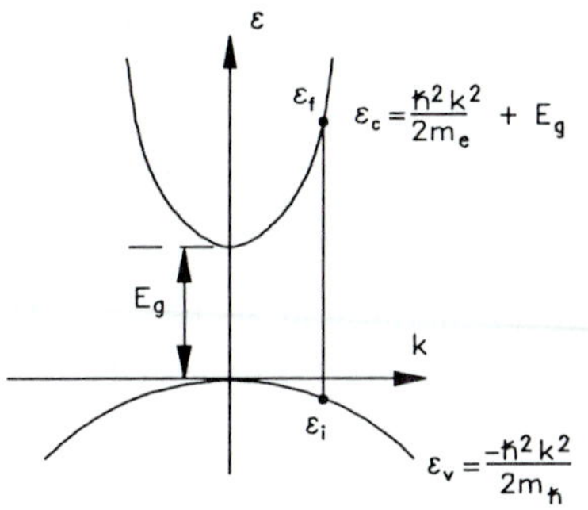

Figure 5.3. The single-particle energy bands for the valence band and conduction band near the center of the first Brillouin zone are schematically plotted as a function of wave vector for an idealized semiconductor with a direct bandgap. The quadratic dispersion law expresses the kinetic energy of an electron in the conduction band and a hole in the valence band, respectively. The vertical axis represents the electron energy.

$$\mathcal{E}_v = -\frac{\hbar^2 k^2}{2m_h}, \tag{5.6}$$

where $m_e = m_e^*$ is the electron effective mass, $m_h = m_h^*$ is the hole-effective mass, and $k_e = k_h = k$ because of the conservation of momentum, Eq. (5.4). The parabolic approximation for the bands is usually referred to as the effective-mass approximation. The conservation of energy for this direct transition is

$$\hbar\omega = \mathcal{E}_c - \mathcal{E}_v = E_g + \frac{\hbar^2 k^2}{2m_r}, \tag{5.7}$$

where we introduced the reduced electron-hole mass m_r as

$$\frac{1}{m_r} = \frac{1}{m_e} + \frac{1}{m_h}. \tag{5.8}$$

The conduction band minimum, which is labeled L in Fig. 5.2(b), has coordinates π/a_0 [1, 1, 1], while the valence band maximum, labeled Γ in Fig. 5.2(b), has coordinates [0, 0, 0] for the indirect-gap semiconductor Ge; a_0 is the length of the unit cell. In fact, as we have seen in Chap. II, points in the Brillouin zone differing by the reciprocal lattice vector G are equivalent. Therefore, there are four equivalent conduction band minima in Ge, with coordinates [1,1,1], [$\bar{1}$,1,1], [1,$\bar{1}$,1], and [1,1,$\bar{1}$]. These equivalent minima are often called valleys. (The point labeled X in Fig. 5.2 has coordinates π/a_0 [2,0,0], while point K (not shown) has coordinates of π/a_0 [0,3/2,3/2].) Therefore, the momentum difference between the conduction band minimum and valence band maximum prevents the momentum and energy conservation from being satisfied simultaneously if only photons are involved in the transition. Typically, the indirect transition needs an additional phonon to be completed. The conservation laws for indirect transitions state [see Fig. 5.1(b)]

$$\mathcal{E}_f = \mathcal{E}_i + \hbar\omega \pm \hbar\Omega \tag{5.9}$$

and

$$\mathbf{k}_f = \mathbf{k}_i \pm \mathbf{K} \ , \tag{5.10}$$

where $\hbar\Omega$ and $\mathbf{K}$ are the phonon energy and wave vector, respectively. The phonon supplies the required momentum transfer $\hbar K$ to allow the transition between the band extrema. For $\mathcal{E}_f - \mathcal{E}_i = E_g^{ind}$, where E_g^{ind} is the indirect-gap energy, the conservation of energy becomes

$$\hbar\omega \pm \hbar\Omega = E_g^{ind} \ , \tag{5.11}$$

where the + sign stands for phonon absorption (annihilation) and the - sign stands for phonon emission (creation).

It is worthwhile to note that it is possible also to have direct transitions in indirect-gap semiconductors at higher photon energies than the indirect transitions. They typically occur between the valence band maximum and in the conduction band around k = 0; i.e., at the Γ point of the Brillouin zone, as shown in Fig. 5.1(b).

The absorption coefficient associated with the fundamental absorption edge in a semiconductor with a direct gap is often quite large, having a magnitude of $\simeq 10^4$ cm^{-1} to 10^5 cm^{-1}. A semiconductor with a thickness of a few microns is, therefore, effectively opaque to photon energies near or above the bandgap, but is transparent to photons of energies less than the bandgap.

5-2. Free Electron-Hole-Pair Absorption

In Chaps. III and IV, when we discussed the optical properties associated with free electrons of mass m_0 in metals and lattice vibrations, we treated the interaction of light with matter classically, by writing the equation of motion as

$$m_0 \frac{d^2x}{dt^2} + m_0\gamma \frac{dx}{dt} + m_0\omega_0^2 x = -eE \,, \tag{5.12}$$

with E being the incident field of the form

$$E = E_0 e^{-i(\omega t - qx)} \,. \tag{5.13}$$

Then we solved for x and calculated the dielectric function from the induced polarization. The optical absorption and reflection were subsequently calculated using the dielectric function. To deal with the quantum mechanical properties of the medium excitations, we abandon the classical approach in this chapter and use quantum mechanics to compute the optical properties of semiconductors. However, we treat only the medium quantum mechanically and keep the notion of a classical light field. This *semiclassical* approach usually gives a correct treatment of absorption and induced emission properties, but it is not appropriate for the description of properties, such as spontaneous emission, which are associated with the true quantum character of the light field.

In this chapter and in Chap. VI, we deal with the linear regime. The incident light intensity is assumed to be weak enough that the response of the medium (the absorption and reflection coefficient) is independent of the light intensity. This is usually the case for light intensity up to a few W/cm^2.

In an electromagnetic field, an electron feels the Lorentz force, $\mathbf{F} = -e\mathbf{E} - e/c\, \boldsymbol{v} \times \mathbf{B}$. As discussed in Prob. 5.1, the corresponding Hamiltonian of the system is given by

$$H = \frac{1}{2m_0}\left(\mathbf{p} - \frac{e}{c}\mathbf{A}\right)^2 + W(\mathbf{r}) + e\phi \,, \tag{5.14}$$

where $\mathbf{p} = m_e \boldsymbol{v}$ is the electron momentum, m_e is the electron-effective mass, and $W(\mathbf{r})$ is the periodic potential of the lattice. The electric field strength E and the magnetic induction B are derived from a scalar potential, $\phi(\mathbf{r}, t)$, and a vector potential, $\mathbf{A}(\mathbf{r}, t)$, of the radiation field by the relations,

$$\mathbf{E} = -\nabla\phi - \frac{1}{c}\frac{\partial \mathbf{A}}{\partial t} \tag{5.15}$$

and

$$\mathbf{B} = \nabla \times \mathbf{A}\ . \tag{5.16}$$

However, for a given **E** and **B**, **A** and ϕ are not uniquely determined. An additional requirement (e.g., the Coulomb gauge, $\nabla\cdot\mathbf{A} = 0$) makes **A** unique. Equation (5.14) may be written as

$$H = \frac{\mathbf{p}^2}{2m_0} - \frac{e}{2m_0 c}(\mathbf{p}\cdot\mathbf{A} + \mathbf{A}\cdot\mathbf{p}) + \frac{e^2}{2m_0 c^2}\mathbf{A}^2 + W(\mathbf{r}) + e\phi\ . \tag{5.17}$$

The transition to quantum mechanics is made by replacing **p** with the operator $\hbar/i\ \nabla$ and making use of the commutation relation,

$$[x, p_x] \equiv xp_x - p_x x = i\hbar\ . \tag{5.18}$$

Using the chain rule for differentiation, we obtain for any function $f(\mathbf{r})$,

$$[f(\mathbf{r}), \mathbf{p}] = i\hbar\nabla f(\mathbf{r})\ , \tag{5.19}$$

which yields

$$\mathbf{p}\cdot\mathbf{A} = \mathbf{A}\cdot\mathbf{p} - i\hbar\nabla\cdot\mathbf{A} \tag{5.20}$$

for the special case of $f(r) = A$. Substituting for $\mathbf{p}\cdot\mathbf{A}$ from Eq. (5.20) into (5.17), we find

$$H = \frac{\mathbf{p}^2}{2m_0} - \frac{e}{m_0 c}\mathbf{A}\cdot\mathbf{p} + \frac{ie\hbar}{2m_0 c}\nabla\cdot\mathbf{A} + \frac{e^2}{2m_0 c^2}\mathbf{A}^2 + W(\mathbf{r}) + e\phi\ . \tag{5.21}$$

In this book, we always use the Coulomb gauge where $\nabla\cdot\mathbf{A} = 0$, and in the absence of any external free charges or currents we set $\phi = 0$, resulting in the simplified Hamiltonian

$$H = \frac{\mathbf{p}^2}{2m_0} + W(\mathbf{r}) - \frac{e}{m_0 c}\mathbf{A}\cdot\mathbf{p} + \frac{e^2}{2m_0 c^2}\mathbf{A}^2\ . \tag{5.22}$$

In the following, we ignore the last term in Eq. (5.22), which is usually much smaller than the previous term in the linear regime, since A^2 scales as the light intensity and we assume a weak intensity (see Prob. 5.4). Note also that this term does not involve the electron momentum operator p; thus, it does not alter the state of the electron in lowest order. Therefore, it does not contribute to linear light absorption. Hence,

$$H \simeq \frac{\mathbf{p}^2}{2m_0} + W(\mathbf{r}) - \frac{e}{m_0 c}\,\mathbf{A}\cdot\mathbf{p}\ , \tag{5.23}$$

which we write as

$$H = H_0 + H_{int}\ , \tag{5.24}$$

with

$$H_0 = \frac{\mathbf{p}^2}{2m_0} + W(\mathbf{r}) \tag{5.25}$$

as the system Hamiltonian in the absence of the radiation field, and

$$H_{int} = -\frac{e}{m_0 c}\,\mathbf{A}\cdot\mathbf{p} \tag{5.26}$$

as the interaction Hamiltonian in the momentum representation. Sometimes it is desirable to express the interaction Hamiltonian in the space representation instead of momentum representation. This may be accomplished by noting that the electric-field vector **E** and the vector potential **A** are related by $\mathbf{E} = -1/c\ \partial\mathbf{A}/\partial t$. Furthermore, the momentum **p** and displacement vector **r** are related by $\mathbf{p} = m_e\ \partial\mathbf{r}/\partial t$. Since the fields **E** and **A** and, consequently, the induced-electron displacement vector **r** vary with time like $e^{-i\omega t}$, the interaction Hamiltonian of Eq. (5.26) may be equivalently written in the space representation as

$$H_{int} = e\mathbf{r}\cdot\mathbf{E} = -\mathbf{d}\cdot\mathbf{E}\ , \tag{5.27}$$

where $\mathbf{d} = -e\mathbf{r}$ is the light-induced electric dipole moment.

We now use the interaction Hamiltonian of Eqs. (5.26) or (5.27) to calculate the probability for a field-induced transition between the energy eigenstates of the system, i.e., the absorption or emission rate. The Schrödinger equation for the wave function ψ' is

$$i\hbar\,\frac{\partial\psi'}{\partial t} = (H_0 + H_{int})\psi'\ . \tag{5.28}$$

The Hamiltonian H_0 has the eigenfunctions ψ_n and energy eigenvalues $\mathcal{E}_n = \hbar\omega_n$, so the Schrödinger equation in the absence of the radiation field,

$$H_0\psi_n = \mathcal{E}_n\psi_n = i\hbar \frac{\partial\psi_n}{\partial t} , \tag{5.29}$$

has the solution

$$\psi_n(\mathbf{r},t) = \phi_n(\mathbf{r}) \exp(-i\mathcal{E}_n t/\hbar) = \phi_n(\mathbf{r})\, e^{-i\omega_n t} . \tag{5.30}$$

We use first-order, time-dependent perturbation theory to calculate the transition rates due to the applied field. For this purpose, we expand the full wave function, $\psi'(\mathbf{r},t)$ in terms of $\psi_n(\mathbf{r},t)$, as

$$\psi'(\mathbf{r},t) = \sum_n a_n(t)\psi_n(\mathbf{r},t) = \sum_n a_n(t)\, \phi_n(\mathbf{r})\, e^{-i\omega_n t} . \tag{5.31}$$

The expansion coefficients $a_n(t)$ are related to the transition probability and, consequently, to the absorption or emission rates. The wave functions $\psi'(\mathbf{r},t)$ are normalized; i.e.,

$$1 = \int d\mathbf{r}\, \psi'^*(\mathbf{r},t)\, \psi'(\mathbf{r},t)$$

$$= \sum_{n,m} a_n^*(t)\, a_m(t)\, e^{i(\omega_n - \omega_m)t} \int d\mathbf{r}\, \phi_n^*(\mathbf{r})\, \phi_m(\mathbf{r}) . \tag{5.32}$$

Using the orthonormality of the eigenfunctions ϕ_n,

$$\int d\mathbf{r}\, \phi_n^*(\mathbf{r})\, \phi_m(\mathbf{r}) = \delta_{n,m} , \tag{5.33}$$

Eq. (5.32) becomes

$$1 = \sum_n a_n^*(t)\, a_n(t) = \sum_n |a_n(t)|^2 \ , \tag{5.34}$$

stating that the sum over all probabilities has to be unity. We assume that the electron at time $t = 0$ is in state ℓ, such that

$$a_n(0) = \delta_{n,\ell} = \begin{cases} 0 & (n \neq \ell) \\ 1 & (n = \ell) \end{cases} , \tag{5.35}$$

and at time t the incident field causes the electron to make a transition to state m, as shown in Fig. 5.4(a). We want to calculate the transition rate $[a_m^*(t)\, a_m(t)]/t$. Inserting Eq. (5.31) into Eq. (5.28) yields

$$\sum_n i\hbar \frac{da_n(t)}{dt} \phi_n(\mathbf{r})\, e^{-i\omega_n t} + \sum_n a_n(t)\, \hbar\omega_n\, \phi_n(\mathbf{r})\, e^{-i\omega_n t}$$

$$= \sum_n a_n(t)\, (H_0 + H_{int})\, \phi_n(\mathbf{r})\, e^{-i\omega_n t} \ . \tag{5.36}$$

Multiplying from the left by $\phi_m^*(\mathbf{r})$, integrating over space, using the orthonormality of the eigenfunctions, $\phi_n(\mathbf{r})$, Eq. (5.33), and noting that $H_0\phi_n = \hbar\omega_n\phi_n$ gives

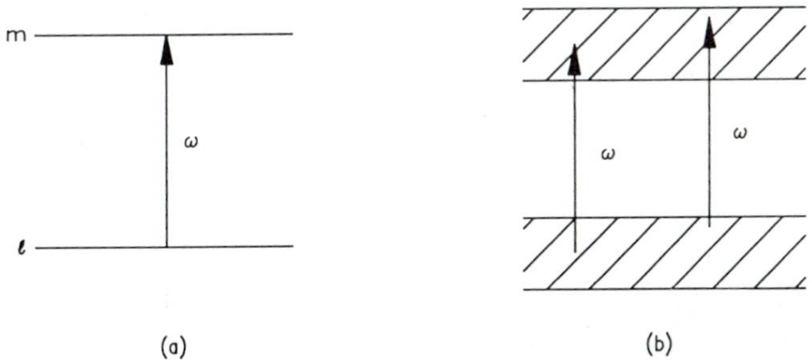

Figure 5.4. (a) Transition between two discrete energy levels. (b) Transition between two bands.

$$i\hbar \frac{da_m(t)}{dt} e^{-i\omega_m t} = \sum_n a_n(t)\, e^{-i\omega_n t} \langle\phi_m|H_{int}|\phi_n\rangle , \tag{5.37}$$

where we abbreviated

$$\int dr\, \phi_m^*(\mathbf{r})\, H_{int}\, \phi_n(\mathbf{r}) = \langle\phi_m|\, H_{int}\, |\, \phi_n\rangle . \tag{5.38}$$

Solving for $da_m(t)/dt$, we get

$$\frac{da_m(t)}{dt} = \frac{1}{i\hbar} \sum_n e^{i\omega_{mn} t}\, a_n(t)\, \langle\phi_m|H_{int}|\phi_n\rangle , \tag{5.39}$$

where $\omega_{mn} = \omega_m - \omega_n$. To solve Eq. (5.39) iteratively in first order, we simply insert the initial condition, Eq. (5.35), into the expression on the RHS of Eq. (5.39) to obtain

$$\frac{da_m^{(1)}(t)}{dt} = \frac{1}{i\hbar} e^{i\omega_{m\ell} t} \langle\phi_m|H_{int}|\phi_\ell\rangle . \tag{5.40}$$

This equation can be integrated to yield $a_m^{(1)}(t)$ and, thus, $a_m(t)$ as

$$a_m(t) = a_m(0) + a_m^{(1)}(t) + \ldots$$

$$= \delta_{m,\ell} + \frac{1}{i\hbar} \int_0^t dt' \langle\phi_m|H_{int}|\phi_\ell\rangle\, e^{i\omega_{m\ell} t'} + \ldots , \tag{5.41}$$

where the omitted terms contribute in higher order of the perturbation. The probability, $w_{m\ell}$, that the electron made a transition from the initial state ℓ to the final state $m \neq \ell$ is given by

$$w_{m\ell} = |a_m|^2 = \left| \frac{1}{i\hbar} \int_0^t dt' \langle \phi_m | H_{int} | \phi_\ell \rangle \, e^{i\omega_{m\ell} t'} \right|^2 . \tag{5.42}$$

To evaluate this expression, we use the interaction Hamiltonian, Eq. (5.27), assuming an incident monochromatic plane wave,

$$\begin{aligned} H_{int} &= -E_0 \cos(\mathbf{q}\cdot\mathbf{r} - \omega t)\, \mathbf{e}_q \cdot \mathbf{d} \\ &= -\frac{E_0}{2} \left(e^{i(\mathbf{q}\cdot\mathbf{r} - \omega t)} + e^{-i(\mathbf{q}\cdot\mathbf{r} - \omega t)} \right) \mathbf{e}_q \cdot \mathbf{d} , \end{aligned} \tag{5.43}$$

where $\mathbf{e}_q$ is the unit vector specifying the polarization of the incident field in the direction of the electric field. For a purely transverse field (Coulomb gauge), we have

$$\mathbf{e}_q \cdot \mathbf{q} = 0 . \tag{5.44}$$

The wavelength of light, $\lambda = 2\pi/q$, is large in comparison to the length a_0 of a lattice unit cell in the semiconductor (typically $a_0 \sim 5$ Å and $\lambda \sim 5000$ Å). Therefore, for typical variations of r on atomic scales, $q \cdot r \ll 1$, and we can expand

$$e^{i\mathbf{q}\cdot\mathbf{r}} = 1 + i\mathbf{q}\cdot\mathbf{r} + \ldots \tag{5.45}$$

It is usually sufficient to keep only the lowest-order term in this expansion, whose matrix element between the initial and final states, $|\phi_\ell\rangle$ and $|\phi_m\rangle$ in our case, does not vanish. Taking only the 1 in Eq. (5.45) is known as the *dipole approximation*. The next higher term, $\propto \mathbf{q}\cdot\mathbf{r}$, is the quadrupole term, which is usually not important in the visible part of the spectrum. However, quadrupole transitions become increasingly important in the far ultraviolet, where $\mathbf{q}^{-1} \sim a_0$.

Restricting the treatment to the dipole approximation, we obtain

$$\begin{aligned} \langle \phi_m | H_{int} | \phi_\ell \rangle &= -\frac{E_0}{2} \left(e^{-i\omega t} + e^{i\omega t} \right) \langle \phi_m | \mathbf{e}_q \cdot \mathbf{d} | \phi_\ell \rangle \\ &= -\frac{E_0}{2} \left(e^{-i\omega t} + e^{i\omega t} \right) d_{m\ell} , \end{aligned} \tag{5.46}$$

where $d_{m\ell} = \langle\phi_m|\mathbf{e}_q\cdot\mathbf{d}|\phi_\ell\rangle$ is the dipole matrix element for transition between states ℓ and m. Inserting Eq. (5.46) into Eq. (5.42) yields

$$w_{m\ell} = \left| \frac{d_{m\ell}E_0}{2\hbar} \int_0^t dt' \, (e^{-i\omega t'} + e^{i\omega t'})\, e^{i\omega_{m\ell}t'} \right|^2$$

$$= \left| \frac{d_{m\ell}E_0}{2\hbar} \left[\frac{e^{i(\omega_{m\ell} - \omega)t} - 1}{i(\omega_{m\ell} - \omega)} + \frac{e^{i(\omega_{m\ell} + \omega)t} - 1}{i(\omega_{m\ell} + \omega)} \right] \right|^2 . \tag{5.47}$$

The first term becomes important when its denominator approaches zero, $\omega_{m\ell} = \omega$ or $\omega_m - \omega_\ell = \omega$, i.e., for $\omega_m > \omega_\ell$, which is satisfied for the case of an absorption process. We should note here that the unphysical divergence of the transition probability at $\omega = \pm\,\omega_{m\ell}$ occurs in our treatment, since we did not include any damping (broadening) mechanism. The inclusion of broadening adds the linewidth term γ in the denominator; i.e., $1/i\omega \rightarrow 1/(i\omega + \gamma)$, eliminating all divergencies. Keeping this in mind, we notice that the second term in Eq. (5.47) becomes appreciable when $\omega_{m\ell} = -\omega$ or $\omega_m < \omega_\ell$, describing an emission process. For the case of light absorption, we can ignore the second term in comparison to the resonance of the first term, and the transition probability becomes

$$w_{m\ell} = \left| \frac{d_{m\ell}E_0}{2\hbar} \left[\frac{e^{i(\omega_{m\ell} - \omega)t} - 1}{i(\omega_{m\ell} - \omega)} \right] \right|^2 . \tag{5.48}$$

Now

$$e^{ix} - 1 = e^{ix/2}\,(e^{ix/2} - e^{-ix/2}) = 2ie^{ix/2}\sin(x/2) , \tag{5.49}$$

so

$$w_{m\ell} = a_m a_m^* = \left| \frac{d_{m\ell}E_0}{\hbar} \right|^2 \frac{\sin^2[(\omega_{m\ell} - \omega)t/2]}{(\omega_{m\ell} - \omega)^2} . \tag{5.50}$$

A similar expression can be obtained for the emission process by considering $\omega_m < \omega_\ell$ and using the second term of Eq. (5.47). We now take into account the fact that semiconductor transitions occur between energy bands and not between discrete levels, as shown schematically in Fig. 5.4(b). Therefore, we have to integrate over all valence- and conduction-band states using the concept of *joint* density of states.

The concept of density of states was discussed in Chap. II. Using the ingredients of Eq. (2.82), we write down the density of states in **k** space per unit volume of real space as

$$g(\mathbf{k})d\mathbf{k} = \frac{2}{8\pi^3}\, d\mathbf{k} \,, \tag{5.51}$$

where the factor of 2 in the numerator is due to the spin degeneracy for each **k** value and $d\mathbf{k} = 4\pi k^2 dk$. We also must consider the occupation probability of each band by electrons, which is given by the Fermi statistics. Here we consider the case of an unexcited semiconductor where the valence band is fully occupied and the conduction band is initially empty. The case of partially filled bands will be discussed in Chap. XII. Thus, the transition probability per unit volume becomes

$$w_{m\ell} = \frac{2}{8\pi^3} \left| \frac{d_{m\ell} E_0}{\hbar} \right|^2 \int_{-\infty}^{\infty} d\mathbf{k}\, \frac{\sin^2[(\omega_{m\ell} - \omega)t/2]}{(\omega_{m\ell} - \omega)^2} \,. \tag{5.52}$$

Noting that the integrand in Eq. (5.52) becomes very large for $\omega_{m\ell} = \omega$, one can show (see Prob. 5.6) that it may be approximated by a δ function at sufficiently long times (large ωt). That is,

$$\frac{\sin^2[(\omega_{m\ell} - \omega)t/2]}{(\omega_{m\ell} - \omega)^2} \simeq \frac{\pi t}{2}\, \delta(\omega_{m\ell} - \omega) \,. \tag{5.53}$$

Substituting Eq. (5.53) into Eq. (5.52) we get

$$w_{m\ell} = \frac{2}{8\pi^3} \left| \frac{d_{m\ell} E_0}{\hbar} \right|^2 \frac{\pi t}{2} \int_{-\infty}^{\infty} d\mathbf{k}\, \delta(\omega_{m\ell} - \omega) \,. \tag{5.54}$$

The transition probability per unit time, per unit volume is then

$$\frac{w_{m\ell}}{t} = \frac{2}{8\pi^3} \left| \frac{d_{m\ell} E_0}{\hbar} \right|^2 \frac{\pi\hbar}{2} \int_{-\infty}^{\infty} dk \; \delta(\hbar\omega_{m\ell} - \hbar\omega) \; , \tag{5.55}$$

where we used $\delta(ax) = 1/|a| \; \delta(x)$. The absorption coefficient of the semiconductor now can be calculated from Eq. (5.55) by noting that the absorption coefficient is given by multiplying the transition probability per unit time, per unit volume, by $\hbar\omega$ to get the energy absorbed per unit time, per unit volume and then dividing by the energy incident per unit time, per unit area. The incident energy is obtained from the time-averaged value of the Poynting vector, given by

$$S = \frac{n_b c}{8\pi} |E_0|^2 \; , \tag{5.56}$$

where n_b is the background index of refraction of the material that was discussed in Chap. III. The absorption coefficient is then

$$\alpha = \frac{w_{m\ell}}{t} \frac{\hbar\omega}{S} = \frac{\omega}{\pi n_b c} |d_{m\ell}|^2 \int dk \; \delta(\hbar\omega_{m\ell} - \hbar\omega)$$

$$= \frac{\omega}{\pi n_b c} |d_{m\ell}|^2 \int_0^{\infty} dk \; k^2 \; 4\pi \; \delta(\hbar\omega_{m\ell} - \hbar\omega) \; . \tag{5.57}$$

Note that the limit of integration in Eq. (5.57) is from 0 to ∞ because the magnitude of k in spherical coordinates varies from 0 to ∞. Since $dk = [dk/d(\hbar\omega_{m\ell})] \; d(\hbar\omega_{m\ell})$, we may rewrite Eq. (5.57) as

$$\alpha = \frac{\omega}{\pi n_b c} |d_{m\ell}|^2 \int_0^{\infty} d(\hbar\omega_{m\ell}) \frac{4\pi k^2 dk}{d(\hbar\omega_{m\ell})} \delta(\hbar\omega_{m\ell} - \hbar\omega) \; . \tag{5.58}$$

Equation (5.58) gives the absorption coefficient in terms of the density of states, $4\pi k^2 dk/d\omega_{m\ell}$, and interband transition matrix element, $d_{m\ell}$. The absorption will be evaluated in detail for direct and indirect transitions in the following two sections.

5-3. Direct Transitions

For a direct transition, the initial state ℓ in the valence band (ℓ = v) and the final state m in the conduction band (m = c) have nearly the same wave number, $k_v \simeq k_c = k$. We consider transitions occurring near extrema of the Brillouin zone, where the effective-mass approximation is valid. The absorption coefficient, using Eq. (5.58), becomes

$$\alpha = \frac{\omega}{\pi n_b c} |d_{cv}|^2 \int_0^\infty d(\hbar\omega_{cv}) \frac{4\pi k^2 dk}{d(\hbar\omega_{cv})} \delta(\hbar\omega_{cv} - \hbar\omega) . \tag{5.59}$$

The wave functions of the valence and conduction bands are then

$$\phi_{vk}(\mathbf{r}) = u_{vk}(\mathbf{r}) \frac{e^{i\mathbf{k}\cdot\mathbf{r}}}{\sqrt{V}}$$

and

$$\phi_{ck}(\mathbf{r}) = u_{ck}(\mathbf{r}) \frac{e^{i\mathbf{k}\cdot\mathbf{r}}}{\sqrt{V}} . \tag{5.60}$$

Using Eq. (5.46), the dipole matrix elements in Eq. (5.59) may be written as

$$d_{cv}(\mathbf{k}) = \langle\phi_{ck}|e_q \cdot \mathbf{d}|\phi_{vk}\rangle$$

$$= \int d\mathbf{r}\, \phi^*_{ck}(\mathbf{r})\, e_q \cdot \mathbf{d}\, \phi_{vk}(\mathbf{r}) . \tag{5.61}$$

Substituting for $\mathbf{d} = (e/im_0\omega)\,\mathbf{p}$ (see Prob. 5.7), where $\mathbf{p} = \hbar/i\, \nabla$, and substituting for $\phi_{ck}(\mathbf{r})$ and $\phi_{vk}(\mathbf{r})$ from Eqs. (5.60) gives

$$d_{cv}(k) = -\frac{e\hbar}{Vm_0\omega} \int d\mathbf{r}\, u^*_{ck}(\mathbf{r})\, e^{-i\mathbf{k}\cdot\mathbf{r}}\, e_q \cdot \nabla(u_{vk}(\mathbf{r})\, e^{i\mathbf{k}\cdot\mathbf{r}})$$

$$= -\frac{e\hbar}{m_0\omega} [e_q \cdot ik \langle u_{ck}|u_{vk}\rangle + \langle u_{ck}|e_q \cdot \nabla u_{vk}\rangle] , \tag{5.62}$$

where the first term, $\langle u_{ck}|u_{vk}\rangle = 1/V \int dr\, u^*_{ck}(\mathbf{r})\, u_{vk}(\mathbf{r}) = 0$, since the Bloch functions with the same wave vector belonging to two different bands are orthogonal. The second term may be rewritten, noting that $\mathbf{\nabla} = i/\hbar\, \mathbf{p}$. We get

$$d_{cv}(k) = \frac{e}{im_0\omega} \langle u_{ck}|\mathbf{e}_q \cdot \mathbf{p}|u_{vk}\rangle \ . \tag{5.63}$$

We expand the momentum matrix elements near k = 0 as

$$\langle u_{ck}|\mathbf{e}_q \cdot \mathbf{p}|u_{vk}\rangle \simeq \langle u_{c0}|\mathbf{e}_q \cdot \mathbf{p}|u_{v0}\rangle + k \frac{\partial}{\partial k} \langle u_{c0}|\mathbf{e}_q \cdot \mathbf{p}|u_{v0}\rangle \ . \tag{5.64}$$

If the interband transition is allowed at k = 0, we retain the first term on the RHS of Eq. (5.64). Equation (5.63) becomes

$$|d_{cv}(k)|^2 \simeq |d_{cv}(0)|^2 = \left(\frac{e}{m_0\omega}\right)^2 |\langle u_{c0}|\mathbf{e}_q \cdot \mathbf{p}|u_{v0}\rangle|^2$$

$$= \frac{e^2}{m_0^2\omega^2} |p_{cv}(0)|^2 \ , \tag{5.65}$$

where we let

$$p_{cv}(0) = \langle u_{c0}|\mathbf{e}_q \cdot \mathbf{p}|u_{v0}\rangle \ . \tag{5.66}$$

Using Eq. (5.7),

$$\hbar\omega_{cv} = \mathcal{E}_c - \mathcal{E}_v = E_g + \frac{\hbar^2 k^2}{2m_r} \ , \tag{5.67}$$

we obtain

$$\frac{1}{4\pi k^2} \frac{d\hbar\omega_{cv}}{dk} = \frac{\hbar^2}{4\pi k m_r} = \frac{1}{2\pi} \left(\frac{\hbar^2}{2m_r}\right)^{3/2} (\hbar\omega_{cv} - E_g)^{-1/2} \ . \tag{5.68}$$

Substituting Eqs. (5.65) and (5.68) into Eq. (5.59) yields

$$\alpha = \alpha_{free} = \frac{2e^2}{n_b c m_0^2 \omega} \left|p_{cv}(0)\right|^2 \left(\frac{2m_r}{\hbar^2}\right)^{3/2} \theta(\hbar\omega - E_g)\,(\hbar\omega - E_g)^{1/2} .$$

absorption coefficient for direct transitions allowed at k = 0 involving free electron-hole pairs (5.69)

The magnitude of $d_{cv}(0)/e$ is on the order of a few Angstroms, the distance between the atoms in the solid. For example, in GaAs $d_{cv}(0)/e \simeq 3$ Å. In Eq. (5.69) we used the Heaviside θ function, with

$$\theta(x) = \begin{cases} 0 & \text{for } x < 0 \\ 1 & \text{for } x > 0 \end{cases} , \tag{5.70}$$

since the density of states is zero below the bandgap. Equation (5.69) gives the absorption coefficient per unit length for direct transitions in a semiconductor in the absence of Coulomb interaction between electrons and holes. The subscript "free" in Eq. (5.69) stresses the absence of Coulomb interaction. Transitions for which the interband matrix elements $p_{cv}(0)$ are nonzero are called *first-class dipole-allowed* transitions. In a crystal with an inversion center, where parity of the Bloch states at k = 0 can be defined, such transitions are allowed between bands of opposite parity; i.e., between a valence band with p-like character and a conduction band with s-like character for the atomic part, $u_{k=0}(\mathbf{r})$, of the respective Bloch functions. For absorption near the band edge, we may approximate the prefactor ω in Eq. (5.69) as $\omega \simeq E_g/\hbar$ and obtain a square-root law for α as $\alpha \sim (\hbar\omega - E_g)^{1/2}$. This frequency dependence of α directly follows the density of states near the bandgap, and it holds when the matrix elements do not change with frequency. When the interband matrix element $p_{cv}(0)$ is zero, as is the case between bands of same parity (i.e., between s-like and d-like Bloch functions), we must consider the second term in RHS of Eq. (5.64). These transitions are called *second-class dipole transitions*. The extra factor k in Eq. (5.64) introduces a new frequency dependence of α that varies as $(\hbar\omega - E_g)^{3/2}$ (see Prob. 5.8).

5-4. Indirect Transitions

The momentum difference between conduction band minimum and valence band maximum in indirect semiconductors is too large to allow momentum conservation in an optical absorption process involving only a photon. For the interband transition to take place, the additional absorption or emission of a phonon is required, as shown schematically in

Fig. 5.5. Because of the simultaneous involvement of two quanta, a photon and a phonon, the transition is clearly of second order, and at least second-order perturbation theory must be used to calculate the transition probability. The corresponding transition rate is, therefore, much weaker than the dipole-allowed direct transitions, typically by several orders of magnitude.

In the following, we present a simple phenomenological derivation of the absorption coefficient for indirect transitions. We have to integrate over those states in both the conduction and valence band that may participate in the transition, since, in contrast to the case of direct transitions, the momentum in the initial and final state is not the same. As shown in Fig. 5.5, the initial- and final-state energies are given by

$$\mathcal{E}_c = E_g^{ind} + \mathcal{E} \tag{5.71}$$

and

$$\mathcal{E}_v = -\mathcal{E}' , \tag{5.72}$$

where E_g^{ind} is the indirect gap, and $\mathcal{E}$ and $\mathcal{E}'$ are assumed to be positive quantities. Conservation of energy requires

$$\mathcal{E}_v + \hbar\omega \pm \hbar\Omega = \mathcal{E}_c , \tag{5.73}$$

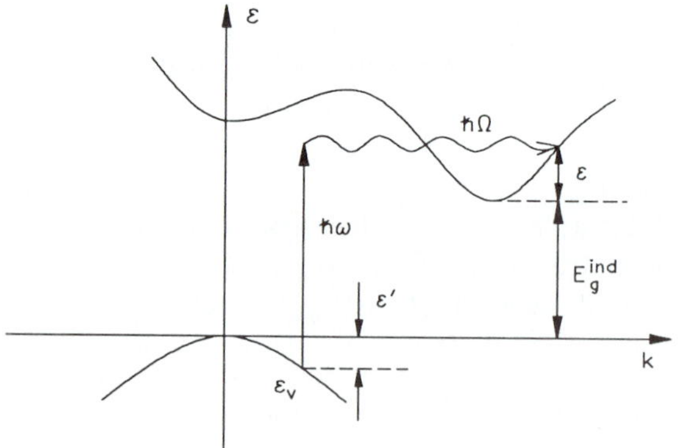

Figure 5.5. Schematic representation of an indirect transition. Here the photon (straight arrow) provides most of the energy, whereas the phonon (wiggled arrow) provides the required momentum.

where $\hbar\Omega$ is the energy of the phonon involved in the transition; the + sign is for phonon absorption and the - sign is for phonon emission. Substituting for $\mathcal{E}_c$ and $\mathcal{E}_v$ from Eqs. (5.71) and (5.72), we get

$$\mathcal{E} = \hbar\omega - E_g^{ind} - \mathcal{E}' \pm \hbar\Omega \,. \tag{5.74}$$

For a fixed $\mathcal{E}'$, the number of states in the interval $d\mathcal{E}$ is obtained from the density of conduction band states as

$$g_c(\mathcal{E}) = a_c\, \mathcal{E}^{1/2} = a_c\, (\hbar\omega - E_g^{ind} - \mathcal{E}' \pm \hbar\Omega)^{1/2} \,, \tag{5.75}$$

where we used Eq. (5.74) and $a_c = (1/2\pi)^2(2m_e/\hbar^2)^{3/2}$ from Eq. (2.89). To obtain the joint density of states, we have to multiply $g_c(\mathcal{E})$ with the valence band density of states, $a_v(\mathcal{E}')^{1/2}$, and integrate over $\mathcal{E}'$ from 0 to some maximum value, which we refer to as $\mathcal{E}'_m$,

$$g(\mathcal{E}) = a_c a_v \int_0 d\mathcal{E}'\, (\hbar\omega - E_g^{ind} - \mathcal{E}' \pm \hbar\Omega)^{1/2}\, \mathcal{E}'^{1/2} \,, \tag{5.76}$$

$\mathcal{E}'$ varies from 0 to $\mathcal{E}'_m$ with the $\mathcal{E}'_m$ being determined by setting $\mathcal{E} = 0$ in Eq. (5.74) (see Fig. 5.6):

$$\mathcal{E}'_m = \hbar\omega - E_g^{ind} \pm \hbar\Omega \,. \tag{5.77}$$

For any given $\hbar\omega \pm \hbar\Omega$, transitions can occur from the top of the valence band to states inside the conduction band if the conservation of energy is satisfied, i.e., if the energy difference between transitions is equal to $\hbar\omega \pm \hbar\Omega$ (see Fig. 5.6). Substituting $\mathcal{E}'_m = \hbar\omega - E_g^{ind} \pm \hbar\Omega$ in Eq. (5.76), we find

$$g(\mathcal{E}) = a_c a_v \int_0 d\mathcal{E}'\, (\mathcal{E}'_m - \mathcal{E}')^{1/2}\, \mathcal{E}'^{1/2} = \frac{a_c a_v \pi}{8}\, \mathcal{E}'^2_m\, \theta(\mathcal{E}'_m) \,, \tag{5.78}$$

where the θ function expresses the fact that the integral has a real solution only for $\mathcal{E}'_m > 0$. Furthermore, $\int_0^y dx\, (y - x)^{1/2}\, x^{1/2} = \pi y^2/8$ has been used. Hence, we obtain

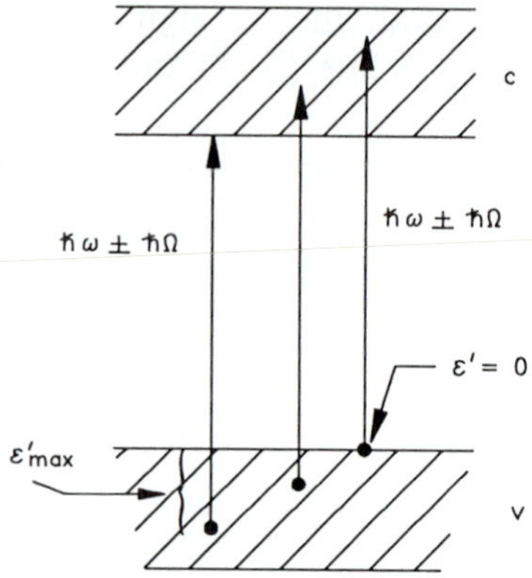

Figure 5.6. Schematics for determination of $\mathcal{E}'_m$.

$$g(\mathcal{E}) = \frac{a_c a_v \pi}{8} (\hbar\omega - E_g^{ind} \pm \hbar\Omega)^2 \, \theta(\hbar\omega - E_g^{ind} \pm \hbar\Omega) \ . \tag{5.79}$$

Therefore, the frequency dependence of the absorption coefficient near the bandgap, which follows the density of states, is given by

$$\alpha = A \, (\hbar\omega - E_g^{ind} \pm \hbar\Omega)^2 \, \theta(\hbar\omega - E_g^{ind} \pm \hbar\Omega) \ , \tag{5.80}$$

with the + sign for phonon absorption and the - sign for phonon emission. The constant A contains the transition matrix element, which is assumed to be independent of k, and is temperature dependent, primarily because of the involved phonon population factors. The square law of the absorption for indirect transitions is experimentally observed, for example, in Ge.

The temperature dependence of the absorption coefficient can be calculated using simple quantum mechanical arguments. We consider phonon absorption and emission processes separately. Phonon absorption eliminates one phonon from the system. If the initial state of the system, denoted by $|n_K\rangle$, contains n_K phonons, the final state has one less phonon, i.e., $|n_K - 1\rangle$. The total transition operator must, therefore, involve a phonon annihilation operator a, given by Eq. (4.29), and the total strength of the transition process is related to the matrix element,

$$|\langle n_K - 1|a|n_K\rangle|^2 = |\langle n_K - 1|\sqrt{n_K}|n_K - 1\rangle|^2 = n_K \ , \tag{5.81}$$

where we used $a|n_K\rangle = \sqrt{n_K}\,|n_K - 1\rangle$, Eq. (4.29), and the orthonormality of states; i.e., $\langle n_K - 1|n_K - 1\rangle = 1$. In Eq. (5.81), n_K is the occupation number for phonons that follows Bose-Einstein statistics,

$$n_K = \frac{1}{e^{\hbar\Omega/k_B T} - 1} . \tag{5.82}$$

The absorption coefficient can then be written as

$$\alpha_a = A' \frac{(\hbar\omega - E_g^{ind} + \hbar\Omega)^2}{e^{\hbar\Omega/k_B T} - 1}\,\theta(\hbar\omega - E_g^{ind} + \hbar\Omega) , \tag{5.83}$$

where A′ is a constant.

The final state, $|n_K + 1\rangle$, for the case of phonon emission contains one extra phonon, and the relevant transition operator contains a phonon creation operator $a^\dagger$. The transition matrix element is

$$\begin{aligned} |\langle n_K + 1|a^\dagger|\, n_K\rangle|^2 &= n_K + 1 = \frac{1}{e^{\hbar\Omega/k_B T} - 1} + 1 \\ &= \frac{1}{1 - e^{-\hbar\Omega/k_B T}} , \end{aligned} \tag{5.84}$$

where we used $a^\dagger|n_K\rangle = \sqrt{n_K + 1}\;|n_K + 1\rangle$ from Eq. (4.28). Consequently, the absorption coefficient for the case of phonon emission becomes

$$\alpha_e = \frac{A'(\hbar\omega - E_g^{ind} - \hbar\Omega)^2}{1 - e^{-\hbar\Omega/k_B T}}\,\theta(\hbar\omega - E_g^{ind} - \hbar\Omega) . \tag{5.85}$$

The factor $n_K + 1$ may be interpreted as originating from stimulated and spontaneous phonon emission, with n_K denoting the stimulated emission contribution (which depends on temperature), and 1 denoting the spontaneous emission contribution.

From Eqs. (5.83) and (5.85), we see that the absorption coefficient for indirect processes varies quadratically with photon energy $\hbar\omega$. Thus, a plot of $\alpha^{1/2}$ should be linear in $\hbar\omega$, as shown schematically in Fig. 5.7. The values of the indirect bandgap E_g^{ind} and of the phonon energy $\hbar\Omega$ may be obtained from the plot, as indicated in Fig. 5.7.

The slope of the straight lines in Fig. 5.7 depends on temperature according to Eqs. (5.83) and (5.85) for the phonon absorption and emission

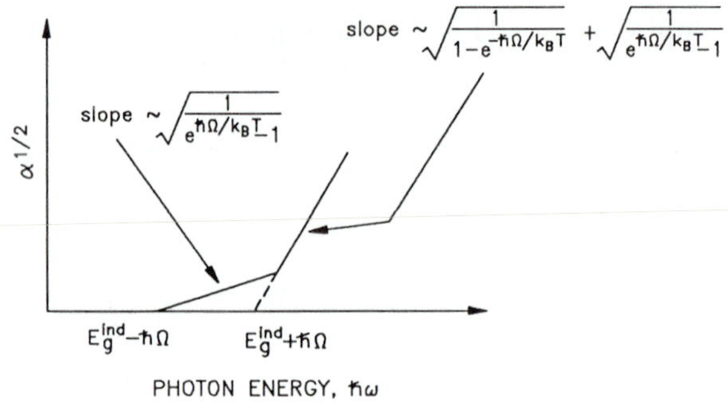

Figure 5.7. Schematics of energy dependence of $\alpha^{1/2}$ for indirect-gap materials.

processes. At a given temperature, the slope is smaller for the case of phonon absorption than for phonon emission because of the phonon occupation numbers. For example, in many semiconductors optical phonons typically have an energy of the order $\hbar\Omega \simeq 30$ meV, and at room temperature, where $k_BT \simeq 25$ meV, we have $\exp(\hbar\Omega/k_BT) \sim 3$, making the slope of $\alpha_a \sim 0.7$ and the slope of $\alpha_e \sim 1.7$. In general, when both phonon absorption and emission are present, the absorption coefficient is given by

$$\alpha = A' \left\{ \frac{(\hbar\omega - E_g^{ind} + \hbar\Omega)^2}{e^{\hbar\Omega/k_BT} - 1} + \frac{(\hbar\omega - E_g^{ind} - \hbar\Omega)^2}{1 - e^{-\hbar\Omega/k_BT}} \right\} \theta(\hbar\omega - E_g^{ind} \pm \hbar\Omega) \, .$$

absorption coefficient for indirect transitions (5.86)

5-5. Pressure and Temperature Dependence of the Bandgap

Energy gaps in semiconductors change with both pressure and temperature. The energy levels of a crystal are clearly functions of interatomic distances, so when temperature or pressure changes, the interatomic distance, the energy level, and E_g change as well. The bandgap energy may be written as

$$E_g = E_g^0 + \left(\frac{\partial E_g}{\partial P}\right)_T \Delta P + \left(\frac{\partial E_g}{\partial T}\right)_P \Delta T \, , \tag{5.87}$$

Table 5.1. Commonly used parameters for some semiconductors. Note that different values have been quoted for some of these parameters in the literature. The table only gives one of the measured values (Ref. 5.2).

	E_g (0^0K) eV	E_g (300^0K) eV	$\left(\frac{dE_g}{dT}\right)$ $\times 10^4$ (300^0K) eV/^{0}K	$\left(\frac{dE_g}{dP}\right)_T$ $\times 10^6$ eV/bar	Effective mass m_e^*	m_h^*	Refractive index	Static dielectric constant	Lattice constant Å
Si	1.166	1.11	-2.3	-1.5	m_ℓ 0.98 m_t 0.19	0.52	3.44	11.7	5.43
Ge	0.74	0.67	-3.7	5.0	m_ℓ 1.58 m_t 0.08	0.3	4.00	16.3	5.66
GaN	3.5	3.4	-4.8	4.2	0.2	0.8	2.4	12	a 3.18 c 5.16
GaP	2.4	2.25	-5.4	-1.7	0.13	0.67	3.37	10	5.450
GaAs	1.520	1.43	-5.0	11	0.07	0.5	3.4	12	5.653
GaSb	0.81	0.69	-4.1	12	0.045	0.39	3.9	15	6.095
InP	1.42	1.28	-4.6	4.6	0.07	0.40	3.37	12.1	5.8687
InAs	0.43	0.36	-3.3	5	0.028	0.33	3.42	12.5	6.058
InSb	0.235	0.17	-2.9	15	0.0133	0.18	3.75	18	6.4787
ZnO	--	3.2	-9.5	0.6	0.32	0.27	2.02	7.9	a 3.2496 c 5.2065
ZnSe	2.80	2.58	-7.2	6	0.17	--	2.89	8.1	5.667
ZnTe	2.39	2.2	-5	6	0.15	--	3.56	9.7	6.101
CdS	2.58	2.53	-5	3.3	0.20	0.7 ⊥c 5 //c	2.5	8.9	a 4.136 c 6.713
CdSe	1.85	1.74	-4.6	--	0.13	2.5 //† 0.4 ⊥	--	10.6	a 4.299 c 7.010
CdTe	1.60	1.50	-4.1	1.5	0.11	0.35	2.75	10.9	6.477
HgTe	-0.28	-0.15 0.14	+5.6	--	0.029	-0.3	3.7	20	6.42
CuCl	3.4	--	--	1.8	0.43	4.2	2.1	7.9	5.4

†Calculated value

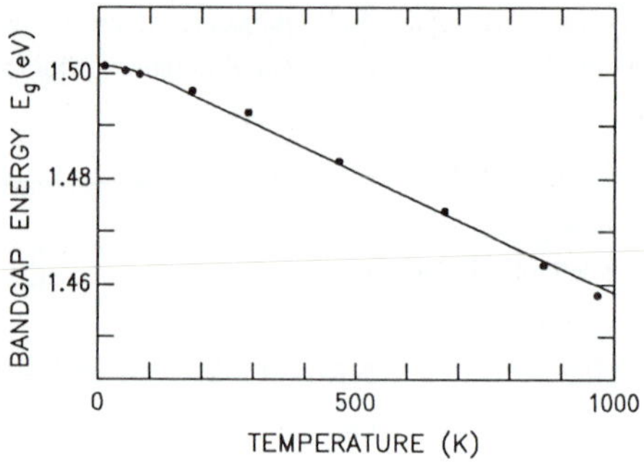

Figure 5.8. Temperature dependence of GaAs bandgap.

with E_g^0 being the bandgap energy at T = 0 in the absence of any external pressure. The changes of E_g with temperature are on the order of 0.1 meV/K, and the changes with pressure are on the order of 10^{-9} meV/dyn cm^{-2}. These numbers may be either positive or negative for various semiconductors. Usually, as temperature increases, the bandgap shrinks. For example, $(\partial E_g/\partial T)_P$ is negative for GaAs with a magnitude of -5×10^{-4} eV/°K. However, there are semiconductors such as CuCl whose bandgap energy increases as temperature rises. The value of $(\partial E_g/\partial P)_T$ for GaAs is 11×10^{-6} eV/bar with a positive sign, suggesting that the bandgap widens as pressure is applied to GaAs. Figure 5.8 gives the temperature dependence of the bandgap of GaAs. Table 5.1 summarizes values of $(\partial E_g/\partial P)_T$, $(\partial E_g/\partial T)_P$, and some other parameters for several semiconductors.

5-6. Problems

5.1. Prove that the Hamiltonian of an electron in a field expressed by potentials **A** and ϕ is given by

$$H = \frac{1}{2m_e}\left(\mathbf{p} - \frac{e}{c}\mathbf{A}\right)^2 + W(\mathbf{r}) + e\phi . \tag{5.88}$$

Hint: Start from the generalized momentum and note that the generalized momentum is $\mathbf{p} = \partial L/\partial v$, where L is the Lagrangian of the system.

5.2. Show that $[f(\mathbf{r}),\mathbf{p}] = i\hbar\nabla f(\mathbf{r})$. Hint: Multiply the commutator [. , .] by an arbitrary function g(r) and perform the differentiation, $[\nabla(f\cdot g) = f\nabla g + g\nabla f]$.

5.3. Show that $[H_0, r] = \dfrac{\hbar \mathbf{p}}{im_0}$ with $H_0 = -\dfrac{\hbar^2\nabla^2}{2m_0} + W(r)$. Hint: Use the same procedure as Prob. 5.2 and also note that

$\partial^2 f\cdot g/\partial x^2 = g\partial^2 f/\partial x^2 + 2\partial f\partial g/\partial x\partial x + f\partial^2 g/\partial x^2$.)

5.4. Show that the term involving A^2 in Hamiltonian, Eq. (5.22), is negligible compared to the term involving $A\cdot\mathbf{p}$ for light intensity, $I_0 \leq 10^6 W/cm^2$. For the electron momentum $p/\hbar$, assume a value of $10^5 cm^{-1}$. Hint: Use $\mathbf{E} = -1/c\ \partial\mathbf{A}/\partial t = i\omega/c\mathbf{A}$ and note that $E \sim 30\sqrt{I_0}$ with the units of E and I_0 being V/cm and W/cm^2, respectively.

5.5. Show that the solution to the Schrödinger equation,

$$\left[-\frac{\hbar^2}{2m_0}\nabla^2 + W(\mathbf{r})\right]\psi_n(\mathbf{r},t) = i\hbar\frac{\partial\psi_n(\mathbf{r},t)}{\partial t} = \mathcal{E}_n\,\psi_n(\mathbf{r},t)\ , \tag{5.89}$$

is given by $\psi_n(\mathbf{r},t) = \phi_n(\mathbf{r})\exp(-i\mathcal{E}_n t/\hbar)$.

5.6. Show that at sufficiently long times, Eq. (5.53) holds; that is,

$$\frac{\sin^2\left[\frac{1}{2}(\omega_{m\ell} - \omega)t\right]}{(\omega_{m\ell} - \omega)^2} \simeq \frac{\pi t}{2}\,\delta(\omega_{m\ell} - \omega)\ . \tag{5.90}$$

5.7. Show that the momentum and dipole operators, **p** and **d**, respectively, are related by

$$\mathbf{d} = \frac{e}{im_0\omega}\mathbf{p}\ . \tag{5.91}$$

5.8. For second-class dipole transitions where the first term in Eq. (5.64) is zero, the second term is kept. Calculate the absorption coefficient for such transitions and show that α varies with photon energy like $(\hbar\omega - E_g)^{3/2}$.

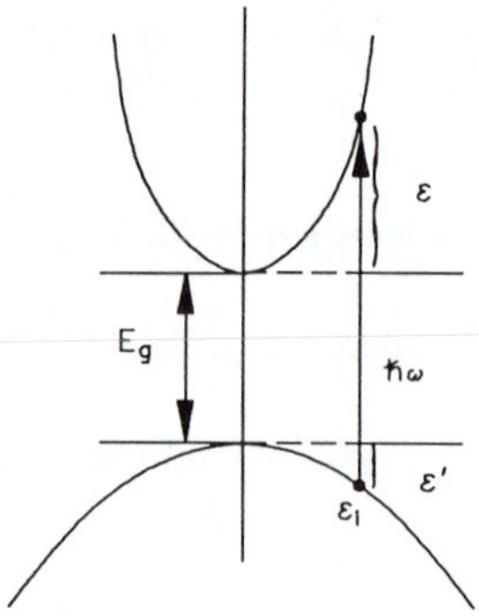

Figure 5.9. Schematic band structure for a direct semiconductor.

5.9. Calculate the joint density of states for direct transitions using Fig. 5.9 and following the procedures that were employed in Sec. 5-4. Hint: You should obtain the same result we got in Sec. 5-3; i.e., $g(\mathcal{E}) \sim (\hbar\omega - E_g)^{1/2}$.

5-7. References

1. M. L. Cohen and T. K. Bergstresser, Phys. Rev. A **141**, 789 (1966).

2. J. I. Pankove, *Optical Properties of Semiconductors* (Dover Publications Inc., New York, 1971).

Chapter VI
LINEAR OPTICAL PROPERTIES OF SEMICONDUCTORS: EXCITONS

In Chap. V we discussed the light-absorption process in semiconductors without including the Coulomb interaction between electrons and holes. This interaction, however, leads to considerable changes in the optical properties of the material, particularly around the absorption edge. When computing the minimum energy required to excite an electron-hole pair, it is necessary to include the Coulomb attraction between the conduction-band electron and the valence-band hole. We will see in this chapter that the inclusion of the Coulomb interaction results in the formation of bound electron-hole pairs, which are called *excitons*. Excitons are analogous to positronium atoms in which an electron is bound to a positron through the Coulomb interaction. In excitons, the electron is bound to the hole and the quasi-particle is electrically neutral. However, an exciton is free to move as an entity throughout the crystal, similar to a real atom in free space. Excitons are responsible for intense absorption lines below the bandgap energy in a region expected to be transparent according to the treatment of Chap. V. Less photon energy is required for the exciton transition compared with the bandgap transition because of the energy saved as a result of the binding of the electron-hole pair in the final state. The process of absorption of one photon corresponds to the direct *creation* of an exciton inside the crystal.

Here we will consider the so-called *Wannier excitons* more specifically. In Wannier excitons, the Bohr radius (i.e., the mean distance between electron and hole) is large in comparison to the length of the lattice unit cell. This condition is met in most II-VI, III-V, and column IV semiconductors. On the other hand, the case of *Frenkel excitons* is realized when the exciton Bohr radius is on the order of, or smaller than, an atomic unit cell. Frenkel excitons exist in wide-gap semiconductors/insulators and in some organic materials.

6-1. The Wannier Equation

To study the Coulomb modifications of the linear optical semiconductor properties in the Wannier approximation, it is sufficient to treat the case of a single excited electron-hole pair. The effect of the periodic atomic lattice, which gives rise to a periodic potential, is incorporated into the effective masses m_e and m_h of the electron and hole, respectively. We first examine the exciton wave function, since the Coulomb interaction leads to an attraction between the electron and the hole and, therefore, the electron tends to stay in the vicinity of the hole.

For the case of a direct-gap material, we write the exciton with wave function Ψ as a wave packet constructed from the linear combination of electron and hole Bloch functions,

$$\Psi(\mathbf{r}_e,\mathbf{r}_h) = \sum_{\mathbf{k}_e,\mathbf{k}_h} \Phi(\mathbf{k}_e,\mathbf{k}_h)\, \phi_{c\mathbf{k}_e}(\mathbf{r}_e)\, \phi_{v\mathbf{k}_h}(\mathbf{r}_h)$$

$$= \sum_{\mathbf{k}_e,\mathbf{k}_h} \Phi(\mathbf{k}_e,\mathbf{k}_h)\, u_{c\mathbf{k}_e}(\mathbf{r}_e)\, e^{i\mathbf{k}_e\cdot\mathbf{r}_e}\, u_{v\mathbf{k}_h}(\mathbf{r}_h)\, e^{i\mathbf{k}_h\cdot\mathbf{r}_h}\,, \qquad (6.1)$$

where $\Phi(\mathbf{k}_e,\mathbf{k}_h)$ is an expansion coefficient, depending on the electron and hole wave vectors, $\mathbf{k}_e$ and $\mathbf{k}_h$, respectively. $\phi_{c\mathbf{k}_e}(\mathbf{r}_e)$ describes the Bloch function for an electron in the conduction band with wave vector $\mathbf{k}_e$ at position $\mathbf{r}_e$. Similarly, $\phi_{v\mathbf{k}_h}(\mathbf{r}_h)$ represents the Bloch wave function for a hole in the valence band with wave vector $\mathbf{k}_h$ at position $\mathbf{r}_h$. In general, the atomic part of the Bloch wave functions (that is, $u_{c\mathbf{k}_e}$ and $u_{v\mathbf{k}_h}$) varies slowly with k_e and k_h. Since we deal with a direct-gap material with band extrema at the center of the Brillouin zone, we take the value of the atomic part of the Bloch wave function at $k_e = k_h = 0$,

$$\Psi(\mathbf{r}_e,\mathbf{r}_h) \simeq u_{c0}\, u_{v0} \sum_{\mathbf{k}_e,\mathbf{k}_h} \Phi(\mathbf{k}_e,\mathbf{k}_h)\, e^{i\mathbf{k}_e\cdot\mathbf{r}_e}\, e^{i\mathbf{k}_h\cdot\mathbf{r}_h}$$

$$\simeq u_{c0}\, u_{v0}\, \Phi(\mathbf{r}_e,\mathbf{r}_h)\,, \qquad (6.2)$$

where

$$\Phi(\mathbf{r}_e,\mathbf{r}_h) = \sum_{\mathbf{k}_e,\mathbf{k}_h} \Phi(\mathbf{k}_e,\mathbf{k}_h)\, e^{i\mathbf{k}_e\cdot\mathbf{r}_e}\, e^{i\mathbf{k}_h\cdot\mathbf{r}_h} \tag{6.3}$$

is called the exciton envelope function. Thus, the exciton wave function Ψ consists of an envelope function $\Phi(\mathbf{r}_e,\mathbf{r}_h)$, modulated on an atomic scale by the atomic part of the Bloch function of electron and hole, u_{c0} and u_{v0}, respectively. The envelope function $\Phi(\mathbf{r}_e,\mathbf{r}_h)$, which describes the electron-hole relative motion on a large scale compared to interatomic distances, obeys the two-particle Schrödinger equation,

$$\left(-\frac{\hbar^2}{2m_e}\nabla_e^2 - \frac{\hbar^2}{2m_h}\nabla_h^2 - \frac{e^2}{\epsilon_0|\mathbf{r}_e - \mathbf{r}_h|}\right)\Phi = \mathcal{E}\Phi\ , \tag{6.4}$$

where $-e^2/(\epsilon_0|\mathbf{r}_e - \mathbf{r}_h|)$ is the attractive Coulomb potential, and ϵ_0 is the background dielectric constant. The energy $\mathcal{E}$ is defined with respect to the bandgap energy E_g. The Schrödinger equation [Eq. (6.4)] is mathematically identical to that of the hydrogen atom, and we follow the standard procedures developed for this problem to obtain the energy eigenvalues.

First we transform Eq. (6.4) to relative and center-of-mass coordinates $\mathbf{r}$ and $\mathbf{R}$, respectively, using

$$\mathbf{r} = \mathbf{r}_e - \mathbf{r}_h\ , \tag{6.5}$$

$$\mathbf{R} = \frac{m_e\mathbf{r}_e + m_h\mathbf{r}_h}{m_e + m_h}\ . \tag{6.6}$$

The wave functions are then separated into functions which explicitly depend on either $\mathbf{R}$ or $\mathbf{r}$,

$$\Phi(\mathbf{R},\mathbf{r}) = g(\mathbf{R})\,\phi(\mathbf{r})\ . \tag{6.7}$$

Inserting Eqs. (6.5) - (6.7) into Eq. (6.4) yields

$$\left[-\frac{\hbar^2}{2M}\nabla_R^2 - \frac{\hbar^2}{2m_r}\nabla_r^2 - \frac{e^2}{\epsilon_0 r}\right] g(\mathbf{R})\,\phi(\mathbf{r}) = \mathcal{E}\, g(\mathbf{R})\,\phi(\mathbf{r})\ , \tag{6.8}$$

where $M = m_e + m_h$ and $1/m_r = 1/m_e + 1/m_h$, with m_r and M being the reduced mass and total mass of the electron-hole pair, respectively.

Furthermore, we introduced

$$\nabla_R^2 = \frac{\partial^2}{\partial X^2} + \frac{\partial^2}{\partial Y^2} + \frac{\partial^2}{\partial Z^2} \tag{6.9}$$

and

$$\nabla_r^2 = \frac{\partial^2}{\partial x^2} + \frac{\partial^2}{\partial y^2} + \frac{\partial^2}{\partial z^2} , \tag{6.10}$$

where $\mathbf{R} = X\mathbf{e}_x + Y\mathbf{e}_y + Z\mathbf{e}_z$ and $\mathbf{r} = x\mathbf{e}_x + y\mathbf{e}_y + z\mathbf{e}_z$, with $\mathbf{e}_i$ denoting the unit vector in the i direction. Defining $\mathcal{E} = \mathcal{E}_R + \mathcal{E}_r$ and dividing by $g(\mathbf{r})\phi(\mathbf{R})$ transforms Eq. (6.8) into

$$\left[-\frac{\hbar^2}{2M} \frac{\nabla_R^2 g(\mathbf{R})}{g(\mathbf{R})} - \frac{\hbar^2}{2m_r} \frac{\nabla_r^2 \phi(\mathbf{r})}{\phi(\mathbf{r})} - \frac{e^2}{\epsilon_0 r} \right] = \mathcal{E}_R + \mathcal{E}_r . \tag{6.11}$$

Since $\mathbf{r}$ and $\mathbf{R}$ are two independent variables, this equation can be separated into

$$-\frac{\hbar^2}{2M} \frac{\nabla_R^2 g(\mathbf{R})}{g(\mathbf{R})} = \mathcal{E}_R \tag{6.12}$$

and

$$-\frac{\hbar^2}{2m_r} \frac{\nabla_r^2 \phi(\mathbf{r})}{\phi(\mathbf{r})} - \frac{e^2}{\epsilon_0 r} = \mathcal{E}_r . \tag{6.13}$$

Equation (6.12) describes the center-of-mass motion, and Eq. (6.13) describes the relative motion, respectively. The solution of Eq. (6.12) gives the center-of-mass eigenfunction as

$$g(\mathbf{R}) = e^{i\mathbf{K}_c \cdot \mathbf{R}} , \tag{6.14}$$

where the center-of-mass wave number is given by

$$K_c^2 = \frac{2M\mathcal{E}_R}{\hbar^2} \quad \text{or} \quad \mathcal{E}_R = \frac{\hbar^2 K_c^2}{2M} . \tag{6.15}$$

Equation (6.14) shows that the center of mass of the electron-hole pair moves like a free particle of mass M and kinetic energy given by Eq. (6.15). Equation (6.13) can be written as

$$\boxed{\left[-\frac{\hbar^2}{2m_r}\nabla_r^2 - \frac{e^2}{\epsilon_0 r}\right]\phi(\mathbf{r}) = \mathcal{E}_r\phi(\mathbf{r})\ .}$$

Wannier equation (6.16)

Equation (6.16) is referred to as the Wannier equation; it is the Schrödinger equation for the excitonic envelope function describing the relative motion of the electron-hole pair. Since the Wannier equation is mathematically equivalent to the hydrogen Schrödinger equation for the relative motion of electron and proton, we refer to quantum mechanics textbooks (see Ref. 6.1) for the details of the solution. Here we merely quote the relevant results.

The energy eigenvalues are obtained as

$$\mathcal{E}_r \equiv \mathcal{E}_{n,\ell,m} = -\frac{m_r e^4}{2\hbar^2\epsilon_0^2}\left(\frac{1}{n^2}\right) \equiv -E_B\left(\frac{1}{n^2}\right), \tag{6.17}$$

where the integers $n = 1, 2, \ldots$; $\ell = 0, 1, \ldots n - 1$; and $m = -\ell, \ldots \ell$ are the main, orbital, and magnetic quantum numbers, respectively. In analogy with atomic spectroscopic notations, it is customary to refer to s-like excitons when $\ell = 0$, to p-like excitons when $\ell = 1$, etc. E_B is the exciton Rydberg energy, which is often written as

$$E_B = \frac{m_r e^4}{2\hbar^2\epsilon_0^2} = \frac{e^2}{2a_B\epsilon_0} = \frac{\hbar^2}{2m_r a_B^2}, \tag{6.18}$$

where

$$a_B = \frac{\epsilon_0\hbar^2}{m_r e^2} \tag{6.19}$$

is the exciton Bohr radius. Equation (6.17) specifies the binding energy of the electron-hole pair (exciton). The total exciton energy is

$$\mathcal{E}_n = E_g + \mathcal{E}_r + \mathcal{E}_R = E_g - E_B\left(\frac{1}{n^2}\right) + \frac{\hbar^2 K_c^2}{2M} \; . \tag{6.20}$$

The energetic positions of the lowest three excitonic states at $K_c = 0$ are displayed relative to the band edge in Fig. 6.1(a), showing clearly that the exciton resonances occur within the forbidden gap. The exciton dispersion (that is, exciton energy versus wave vector) is shown in Fig. 6.1(b). The zero of energy is the crystal ground state, which is the state of the crystal without any excitons. The wave vector K_c on the horizontal axis is the center-of-mass wave vector of the exciton. The parabolic shape of the exciton dispersion follows Eq. (6.15). Hence, we see that each exciton term, n = 1, 2, 3, ... etc., corresponds to a continuum of states, or to a kinetic energy band, not unlike free electrons.

Using Eqs. (6.2), (6.7), and (6.14), the exciton wave function becomes

$$\Psi(\mathbf{r},\mathbf{R}) = u_{c0}\, u_{v0}\, \phi(\mathbf{r})\, e^{i\mathbf{K}_c \cdot \mathbf{R}} \; . \tag{6.21}$$

The Wannier exciton wave function, Eq. (6.21), is a product of four terms. The first two terms contain the character of the atomic orbitals forming the valence and conduction bands. The symmetry of these functions determines the strength of the interband optical transition. The third term, $\phi(r)$, is the exciton envelope function describing the relative motion of the

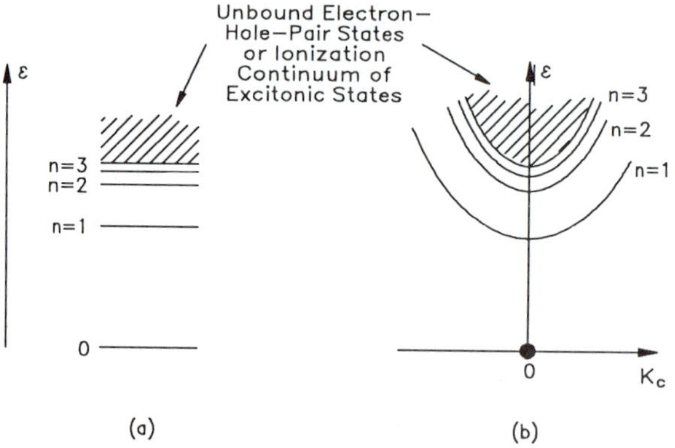

Figure 6.1. (a) Energy levels of excitons with zero center-of-mass kinetic energy, K = 0. (b) The exciton dispersion in a two-particle (electron-hole) excitation diagram of the entire crystal. The crystal ground state (zero energy, zero momentum) is the point at the origin. Different parabolas represent the kinetic energy bands associated with different terms of the excitonic series.

electron and hole on a large scale, over many atomic sites. The final factor, $\exp(i\mathbf{K}_c \cdot \mathbf{R})$, expresses the translational symmetry of the crystal, giving the exciton its running wave character. Figure 6.2 displays the schematic representation of excitons with s-like and p-like characters. The exciton of Fig. 6.2(a) has an s-like envelope function and is formed with p-like hole Bloch wave function, as represented by the ∞ symbol at the center, and an s-like electron wave function depicted by the circles surrounding the hole. The exciton of Fig. 6.2(b), which has a p-like envelope function, is also formed by a p-like hole and an s-like electron wave function.

For an n = 1 exciton, the eigenfunction of the Wannier equation [Eq. (6.16)] is given by

$$\phi_{1s}(r) = \frac{1}{\sqrt{\pi a_B^3}} \, e^{-r/a_B} , \tag{6.22}$$

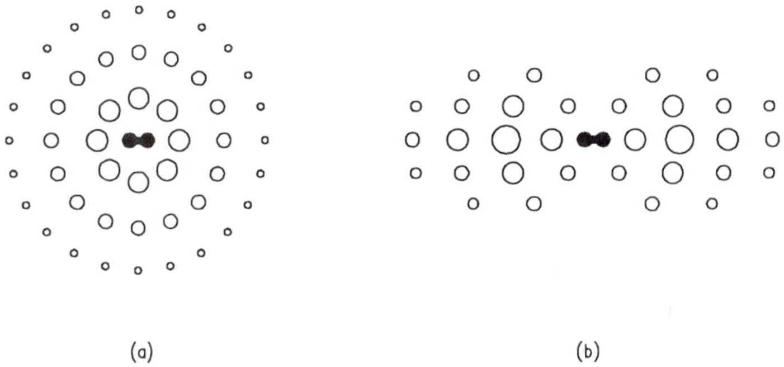

Figure 6.2. (a) Representation of an exciton with a p-like hole Bloch wave function, an s-like electron Bloch wave function, and an s-like envelope function. Such excitons are responsible for strong (first-class) transitions. (b) Representation of an exciton with p-like envelope function composed of electron wave function with s-like atomic character and a p-like hole Bloch wave function. The representation is in the hole reference frame, which sits at the center of each figure. The p-like character of the hole wave function is depicted by the ∞. The size of the circles represents the probability amplitude of the electron with s-like character (after Ref. 6.2).

where $1/\sqrt{\pi a_B^3}$ is the normalization factor, and a_B is the exciton Bohr radius [Eq. (6.19)].

The exciton binding energy is often difficult to determine with high precision experimentally, since a hydrogenic series of lines with more than two terms is rarely observed. A notable exception is Cu_2O, where several exciton transitions are seen, as shown in Fig. 6.3. Equation (6.18) shows that the exciton binding energy is inversely proportional to the exciton Bohr radius. The larger the binding energy, the smaller the Bohr radius. For example, for CdS, E_B = 27 meV and a_B = 28 Å, while for GaAs, E_B = 4.2 meV and a_B = 140 Å. Equation (6.18) also shows that the exciton binding energy is inversely proportional to the square of the background dielectric constant. For larger gap semiconductors such as CuCl, with $E_g \sim 3$ eV, the background dielectric constant is smaller and, thus, the binding energy is larger. For example, ϵ_0 = 8.3 and E_B = 18 meV for ZnSe, while ϵ_0 = 12 and E_B = 4.2 meV for GaAs. Figure 6.4 displays measured exciton binding energies for various direct-gap semiconductors. E_B decreases from I-VII semiconductors such as CuCl (where one element is from the first group of the periodic table and the other one from the seventh) to II-VI (e.g., CdSe) and III-V compounds (e.g., InSb).

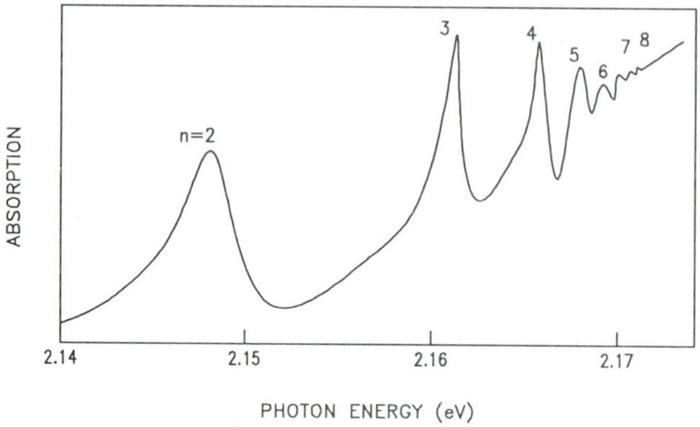

Figure 6.3. Optical absorption for the yellow exciton series of Cu_2O at 4.2 K (after Ref. 6.3).

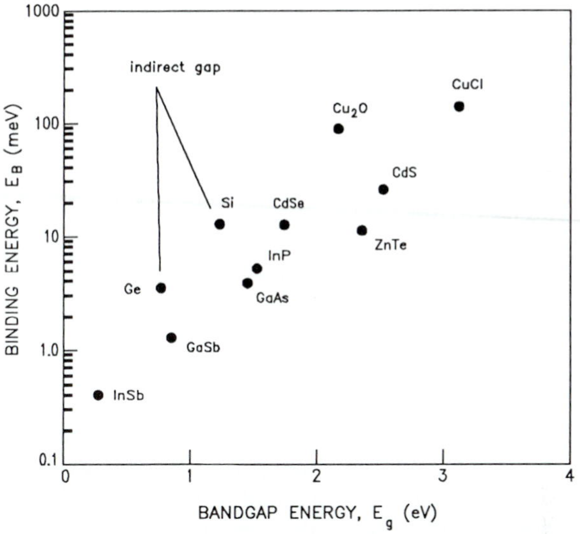

Figure 6.4. Experimental values for the exciton binding energy E_B versus bandgap energy.

6-2. Exciton Absorption

Exciton effects were seen in the absorption spectra of semiconductors as early as 1950. They could be explained on the basis of the theory for optical absorption by Wannier excitons, which was developed by Elliott in 1957. Before evaluating the exciton absorption spectrum, it is useful to discuss some of its general features by considering the conservation laws that must hold for the transition to take place. The conservation laws apply to energy, momentum, angular momentum, and parity (see Fig. 6.5).

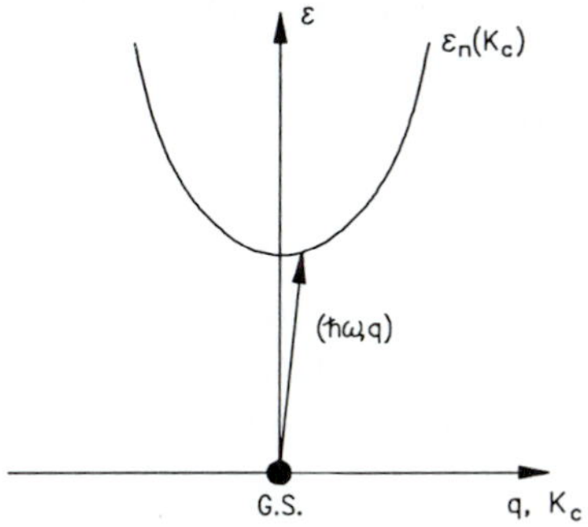

Figure 6.5. Schematic of exciton absorption. The initial state consists of the crystal ground state and a photon. The final state is an exciton state.

Energy Conservation

The exciton created by absorption of a photon must have the photon energy

$$\mathcal{E}_n(K_c) = \hbar\omega \; . \qquad (6.23)$$

Momentum Conservation

Similarly, the center-of-mass momentum of the exciton must be equal to the momentum of the absorbed photon,

$$\hbar\mathbf{K}_c = \hbar\mathbf{q} \; . \qquad (6.24)$$

For light in the visible part of the spectrum the photon wave number, $q = 2\pi/\lambda$, is negligibly small in comparison to the extension of the first Brillouin zone, and we usually assume $q \simeq 0$. Then, from Eq. (6.24), $K_c = 0$, leading to a vanishing of the center-of-mass kinetic energy of the exciton, the last term in Eq. (6.20). Therefore, one can see from Eq. (6.20) that although each term of the excitonic series consists of a band, only a series of sharp lines corresponding to an exciton with momentum, $\hbar K_c = 0$, is expected to appear in the absorption spectrum.

Angular Momentum Conservation

The photon carries an angular momentum of $\ell = 1$ in the dipole approximation. In the absorption process, the photon-angular momentum is transferred to the exciton. This conservation law may be understood by noting that the total angular momentum of the system must remain unchanged before and after the optical transition. Before the transition, the system consists of the crystal ground state plus a photon (see Fig. 6.5), while after the transition, an exciton state exists in the crystal. The crystal ground state is a state with no excited electron-hole pairs; thus, it is totally symmetric with no angular momentum. The angular momentum conservation may be expressed as

$$\Delta\ell = \pm 1 \; . \qquad (6.25)$$

This conservation law is similar to the usual dipole transition selection rule in atomic spectroscopy. This selection rule can be further split into a selection rule for the internal and envelope motions,

$$\Delta\ell = \Delta\ell_{int} + \Delta\ell_{env} = \pm 1 \; , \qquad (6.26)$$

where ℓ_{int} and ℓ_{env} correspond to the angular momenta for the internal and the exciton envelope motion, respectively. The part of the exciton wave function that describes the internal motion is the $u_{c0}u_{v0}$ component of Eq. (6.21); $\phi(\mathbf{r})$ describes the exciton envelope function in that equation. $\Delta\ell_{int}$ and $\Delta\ell_{env}$ can assume two possible sets of values for Eq. (6.26) to hold,

$$\Delta\ell_{int} = \pm 1, \qquad \Delta\ell_{env} = 0 \tag{6.27}$$

or

$$\Delta\ell_{int} = 0, \mp 2, \qquad \Delta\ell_{env} = \pm 1 \ . \tag{6.28}$$

In the case of Eq. (6.27), the transition is dipole-allowed between bands deriving from atomic orbitals with angular momenta differing by $\Delta\ell_{int} = \pm 1$; e.g., between a p-like valence band and an s-like conduction band. The envelope function of the exciton must be s-like to satisfy the conservation of the total angular momentum [Eq. (6.26)]. Such transitions give rise to strong absorption lines. They are the equivalent of the first-class transitions that we discussed in Chap. V. The exciton series starts with the $n = 1$ term with an s-like envelope function.

If the interband transition is dipole-forbidden between the valence band and conduction band at $k = 0$ (for example, if the valence band originates from s or d orbitals and the conduction band from s orbitals such that $\Delta\ell_{int} = 0$ or ± 2), the total angular momentum selection rule can still be satisfied by having $\Delta\ell_{env} = \pm 1$ [see Eq. (6.28)]. In that case, the dipole transition corresponds to the creation of excitons with p-like envelope function. Such transitions, which are referred to as the second-class transitions, are usually weaker than the first-class transitions. The exciton series starts with $n = 2$ since there is no solution to the two-particle Schrödinger equation for a 1p state.

Parity Conservation

In a crystal with inversion symmetry, a parity can be assigned to exciton states with $K_c = 0$. We recall that a state of positive parity remains invariant under reversal of coordinates,

$$\psi(\mathbf{r}) = \psi(-\mathbf{r}) \ , \tag{6.29}$$

whereas, a state of negative parity changes sign,

$$\psi(\mathbf{r}) = -\psi(-\mathbf{r}) \ . \tag{6.30}$$

The parity of the total system must also remain unchanged before and after the optical transition. The totally symmetric crystal ground state has a positive parity. The photon carries a negative parity, since the parity is equal to $(-1)^{\ell}$ and photons with $\ell = 1$ have a parity of -1. This can also be seen by inspecting the dipole-transition operator in the semiclassical picture, $e\mathbf{r}$, which has a negative parity. Therefore, the parity of the total system before the transition is negative. The state of the system after the absorption must also have a negative parity, leading to a negative parity requirement for the exciton. An exciton with overall negative parity can be constructed in different ways. The exciton wave function of Eq. (6.21) can have a negative parity, for instance, if the valence band Bloch function u_{v0} has a negative parity (from a p-like atomic state), and the conduction band Bloch function u_{c0} and the envelope function $\phi(\mathbf{r})$ are both of positive parity. This is the case for Fig. 6.1(a). The negative parity exciton wave function requirement can also be satisfied if both u_{c0} and u_{v0} are of positive parity (s-like) and the envelope function is of negative parity, or if all three wave functions are of negative parity, respectively. The angular momentum and parity conservation laws are treated more rigorously by using group theory. Such a treatment is outside the scope of this book.

1. Exciton Absorption in Direct-Gap Semiconductors

We now describe the mathematical form of the exciton absorption for the first-class transitions, which are dipole-allowed at the center of the Brillouin zone, k = 0. To compute the transition probability we must evaluate the interband matrix elements between all combinations of electron-hole states that make the exciton wave packet. Including the Coulomb effects, the transition probability w can be written as

$$w \propto \Big| \sum_{\mathbf{k}_e,\mathbf{k}_h} \Phi(\mathbf{k}_e,\mathbf{k}_h) \langle c,\mathbf{k}_e | \mathbf{e}_q \cdot \mathbf{d} | v,\mathbf{k}_h \rangle \Big|^2$$

$$= \Big| \sum_{\mathbf{k}_e,\mathbf{k}_h} \Phi(\mathbf{k}_e,\mathbf{k}_h) \int d\mathbf{r}\, \psi^*_{c\mathbf{k}_e}(\mathbf{r}_e)\, \mathbf{e}_q \cdot \mathbf{d}\, \psi_{v\mathbf{k}_h}(\mathbf{r}_h) \Big|^2 . \qquad (6.31)$$

In addition, we require

$$\mathbf{k}_e + \mathbf{k}_h = K_c \simeq 0 . \qquad (6.32)$$

Using Eqs. (5.69), (6.2), and (6.3), the Elliott's formula is obtained for the exciton absorption (see Ref. 2.5 for proof).

$$\alpha(\omega) = \frac{8\pi^2\omega|d_{cv}|^2}{n_b c} \sum_n |\phi_n(\mathbf{r}=0)|^2 \, \delta\left(\hbar\omega - E_g + \frac{E_B}{n^2}\right) .$$

Elliott formula for band-edge absorption (6.33)

The factor $|\phi_n(\mathbf{r} = 0)|^2$ is called the Sommerfeld or the Coulomb enhancement factor. It specifies the probability to find the electron at the same lattice site as the hole, since the envelope function $\phi_n(\mathbf{r})$ has to be evaluated at $\mathbf{r} = \mathbf{r}_e - \mathbf{r}_h = 0$. Since $\phi_n(\mathbf{r} = 0) \neq 0$ only for s wave functions, we see in Eq. (6.33) that only the s states contribute to the optical absorption.

Using the textbook results (see Ref. 2.5) for the hydrogen eigenfunctions, $\phi_n(\mathbf{r} = 0)$, and for the bound states and the ionization continuum, the Elliott formula can also be written as

$$\alpha(\omega) = \frac{e^2\omega|d_{cv}|^2}{n_b c\epsilon_0}\left(\frac{2m_r}{\hbar^2}\right)^2\left[E_B \sum_{n=1}^{\infty} \frac{4\pi}{n^3}\,\delta(\hbar\omega - E_g + E_B/n^2) + \theta(\hbar\omega - E_g)\,\frac{\pi e^Z}{\sinh Z}\right] , \tag{6.34}$$

where the unit-step function $\theta(\hbar\omega - E_g)$ is zero for $\hbar\omega < E_g$ and becomes 1 for $\hbar\omega > E_g$, and the variable Z is defined as

$$Z = \pi\sqrt{E_B/(\hbar\omega - E_g)} . \tag{6.35}$$

The remaining constants in Eq. (6.34) were defined in Chap. V. There are two terms in Eq. (6.34); the first term in the bracket describes the discrete excitonic absorption, and the second term gives the absorption for continuum states (i.e., absorption for $\hbar\omega > E_g$). Comparison of Eq. (6.34) with the free-carrier absorption of Eq. (5.69), where the Coulomb interaction was neglected, shows two major differences, as indicated in Fig. 6.6.

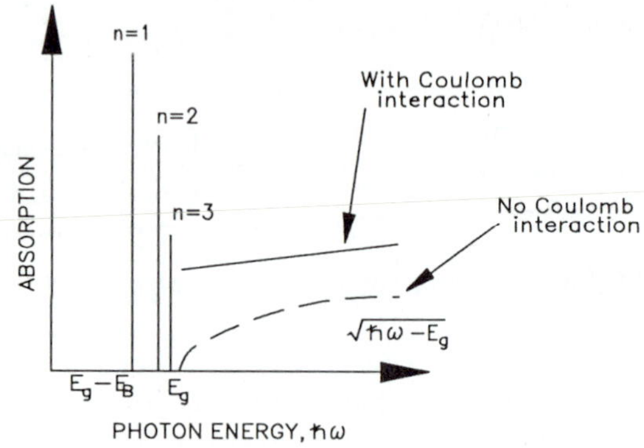

Figure 6.6. Schematic diagram of the absorption spectrum with inclusion of Coulomb interaction. The dashed line indicates the spectrum neglecting Coulomb interaction (after Ref. 6.4).

1. The absorption of a semiconductor near the band edge approaches a finite nonzero value as $\hbar\omega \rightarrow E_g$ from upper values, in contrast to the predictions of the free-electron theory stating $\alpha \rightarrow 0$ as $\hbar\omega \rightarrow E_g$. This can be seen from the continuum absorption term in Eq. (6.34), which can be written as (see Prob. 6.1)

$$\alpha_{\text{continuum}}(\omega) = \alpha_{\text{free}}(\omega)\ C(\omega)\ , \tag{6.36}$$

where

$$\alpha_{\text{free}}(\omega) \sim \frac{1}{Z}\ , \tag{6.37}$$

and

$$C(\omega) = \frac{Ze^{Z}}{\sinh(Z)} = \frac{2Z}{1 - e^{-2Z}} \tag{6.38}$$

describes the Coulomb enhancement of the continuum absorption. As $\hbar\omega \rightarrow E_g$, $\exp(-2Z) \rightarrow 0$ and $\alpha_{\text{continuum}} \rightarrow 2\alpha_{\text{free}} Z \sim 2(1/Z)Z \rightarrow$ const., showing that the continuum absorption assumes a constant value at the bandgap. The continuum absorption is strikingly different from the square-root law of the free-carrier absorption.

2. The most striking difference between the free-carrier absorption (with no Coulomb interaction between electrons and holes) and the true semiconductor absorption is that the free-carrier absorption is zero for photon energies below the bandgap, whereas the first term in Eq. (6.34) shows that the true absorption consists of discrete lines below the bandgap. In this equation, the different n values in the δ function give rise to discrete absorption resonances at $\hbar\omega = E_g - E_B$, $\hbar\omega = E_g - E_B/4$, $\hbar\omega = E_g - E_B/9$, etc., corresponding to exciton ground state 1s and higher excited states, 2s, ..., as displayed in Fig. 6.6. The magnitude of excitonic absorption lines decreases like $1/n^3$ so the 2s exciton with n = 2 has an absorption peak that is eight times smaller than the 1s exciton. Experimentally, however, only a few exciton states can be seen because of broadening mechanisms, such as scattering of electron-hole pairs by phonons. Figure 6.7 shows the computed band-edge absorption from Eq. (6.34), where the δ functions are replaced by broadened δ functions.

Figure 6.8 displays the measured linear absorption spectrum in GaAs at low temperatures. The upper curve shows the absorption for a 0.5-μm-thick GaAs platelet where the 1s exciton and the band edge are clearly observed. The lower curve was obtained for a 4.2-μm-thick sample. The increased absorbance, αd, results in the observation of the three lowest exciton states, 1s, 2s, and 3s.

It may happen that the interband dipole transition vanishes at k = 0; i.e.,

$$p_{cv}(0) = \langle u_{c0} | \mathbf{p} | u_{v0} \rangle = 0 \ . \tag{6.39}$$

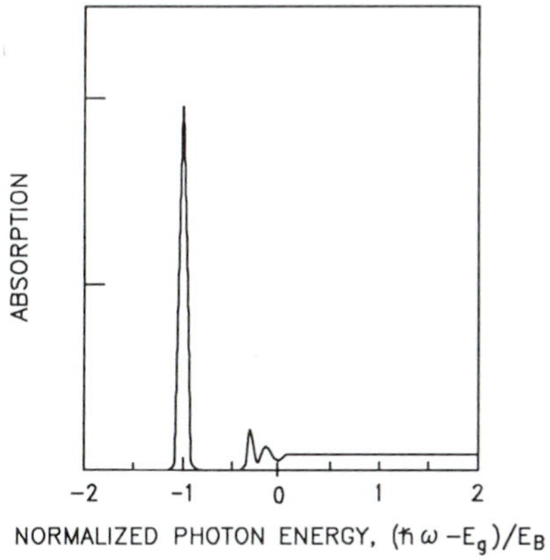

Figure 6.7. Computed band-edge absorption spectrum. The δ functions for the bound states in Eq. (6.33) are replaced by broadened δ functions.

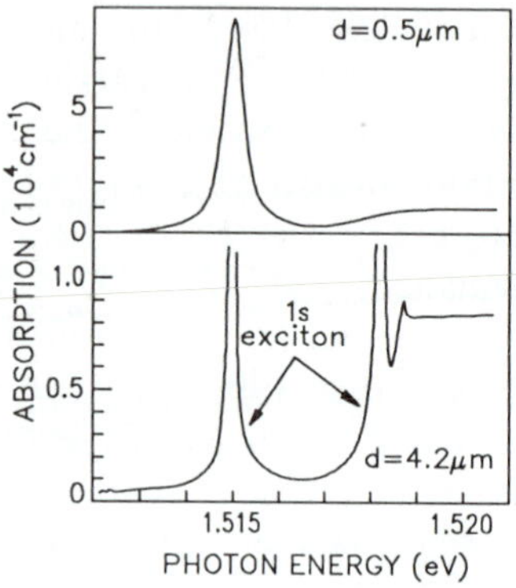

Figure 6.8. Measured band-edge absorption in GaAs (after Ref. 6.5).

This occurs, for instance, if the valence band is derived from s or d atomic orbitals and the conduction band originates from s atomic orbitals. A dipole transition is, nevertheless, possible with a final exciton state having a p-like envelope function. In that case, the second term of the RHS of Eq. (5.64) is introduced, resulting in second-class transition. The extra prefactor k in Eq. (5.64) leads to the derivative of the envelope function with respect to r at the origin, and the absorption coefficient α_n for the bound state n becomes

$$\alpha_n = \frac{4\pi}{m_e \hbar\omega} \left| \frac{\partial p_{cv}}{\partial k} \right|^2 \left| \frac{\partial \phi_n(r = 0)}{\partial r} \right|^2 . \tag{6.40}$$

For hydrogen functions, only p orbitals contribute to this type of transition with

$$\left| \frac{\partial \phi_n(0)}{\partial r} \right|^2 = \frac{n^2 - 1}{\pi a_B^5 n^5} \quad , \quad n = 2, 3, ... \tag{6.41}$$

Note that the n = 1 term is missing because for n = 1, ℓ = 0, and no p functions can exist. Thus, second-class transitions, including Coulomb interaction, lead to the creation of excitons with p-like envelope function. Second-class transitions are usually weaker than first-class transitions near the band edge, typically by two to three orders of magnitude. A well-known example of a second-class exciton transition is the yellow series of Cu_2O shown in Fig. 6.3, with n = 1 missing.

2. Exciton Absorption in Indirect-Gap Semiconductors

The transition to an exciton can also occur with the cooperation of a phonon, which provides the momentum, $\hbar K$. For instance, this type of transition becomes important in an indirect-gap material where the cooperation of phonons is necessary to provide the momentum mismatch between valence and conduction band extrema. The transition to a given exciton band n can take place either by absorption or by emission of a phonon (see Fig. 6.9). Phonon absorption and emission are proportional to n_K and $n_K + 1$, respectively, where n_K is the phonon occupation number (see Sec. 5-4) given by

$$n_K = \frac{1}{e^{\hbar\Omega/k_B T} - 1}, \tag{6.42}$$

with $\hbar\Omega$ as the phonon energy. The ratio of absorption due to phonon emission and phonon absorption is then [see Eq. (5.84) for phonon emission and Eq. (5.82) for phonon absorption]

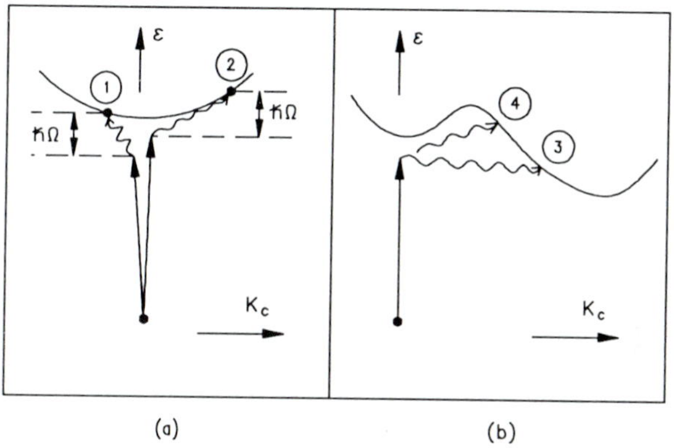

Figure 6.9. Phonon-assisted transition to an exciton term in a direct-gap (a) and indirect-gap (b) semiconductor. The photon is represented by a straight arrow, and the phonon is represented by a wiggled arrow. Final states (1) and (2) inside the same kinetic energy band of an exciton term are reached by simultaneous absorption of an optical phonon of same energy but different wave vector. Final states (3) and (4) in the indirect-gap material require a large phonon momentum, $\hbar K \sim 2\pi/a$, where a is the lattice constant. Here the process occurring via emission (3) or absorption (4) of a phonon is shown.

$$\frac{n_K + 1}{n_K} = \exp(\hbar\Omega/k_B T) , \tag{6.43}$$

so the emission process dominates at low temperatures. Phonon-assisted excitonic absorption consists of broad steps for each term of the excitonic series instead of sharp lines. This can be understood by inspecting the energy and momentum selection rules. We have

$$\hbar\omega \pm \hbar\Omega = \mathcal{E}_n \tag{6.44}$$

and

$$\hbar q \pm \hbar K = \hbar K_c , \tag{6.45}$$

where the + (-) sign refers to a transition in which a phonon of energy $\hbar\Omega$ and momentum $\hbar K$ is absorbed (emitted). As we have seen in Chap. IV, phonons have a continuous spectrum comprising wave vectors spanning over the first Brillouin zone. Therefore, any exciton state with wave vector K_c can be reached by the cooperation of a proper phonon with matching wave vector, $K \simeq K_c$ (Fig. 6.10). For instance, an indirect exciton absorption process involving optical-phonon emission has a threshold absorption at

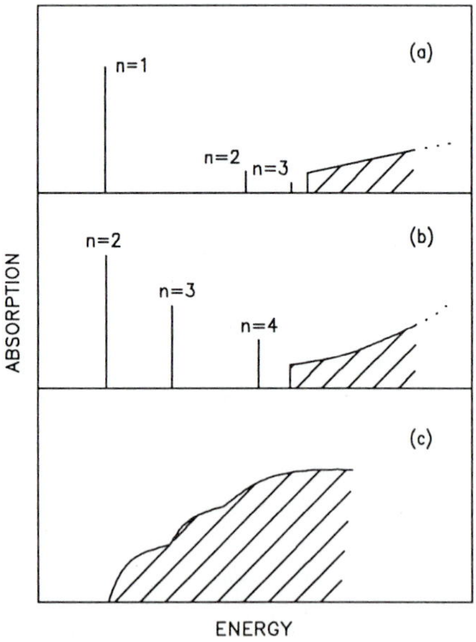

Figure 6.10. Schematic plot of the absorption spectra of (a) first-class, (b) second-class, and (c) phonon-assisted transitions.

$$\hbar\omega = E_g - E_B + \hbar\Omega . \tag{6.46}$$

The absorption coefficient increases with photon energy, reflecting the increased density of final excitonic states until the condition is reached,

$$\hbar\omega = E_g - E_B/4 + \hbar\Omega , \tag{6.47}$$

at which point the onset of phonon-assisted transition to the second exciton term, n = 2, leads to a new edge absorption. Figure 6.10(c) shows the schematic display of the absorption spectrum for the phonon-assisted transitions. For comparison, we also show in Fig. 6.10(a) and (b) the absorption spectra of first-class and second-class transitions, respectively.

Table 6.1. Photon energy dependence of linear absorption coefficient. The oscillator strength is the absorption coefficient integrated over the entire line profile for a given term n.

		without exciton effects	exciton effects included
First-class dipole transitions	$\hbar\omega < E_g$	$\alpha_{free} = 0$	exciton lines at $E_g - E_B/n^2$, n = 1, 2,.. oscillator strength $f_n \propto n^{-3}$, s excitons
	$\hbar\omega > E_g$	$\alpha_{free} \propto (\hbar\omega - E_g)^{1/2}$	$\alpha \propto \exp(Z)/\sinh(Z)$ with $Z = \pi[E_B/(\hbar\omega - E_g)]^{1/2}$
Second-class dipole transitions	$\hbar\omega < E_g$	$\alpha_{free} = 0$	exciton lines at $E_g - E_B/n^2$, n = 2, 3,... oscillator strength $f_n \propto (n^2 - 1)/n^5$, p excitons
	$\hbar\omega > E_g$	$\alpha_{free} \propto (\hbar\omega - E_g)^{3/2}$	$\alpha \propto (1 + Z^{-2}) \exp(Z)/\sinh(Z)$
Phonon-assisted transitions	$\alpha \propto (\hbar\omega - E_g \pm \hbar\Omega)^2$		$\hbar\omega < E_g$ absorption edges at $(\hbar\omega - E_g - E_B/n^2 \pm \hbar\Omega)^{1/2}/n^3$

Table 6.1 summarizes the frequency dependence of the absorption coefficient with and without the excitonic effects for the first-and second-class transitions. The behavior of the indirect-allowed transitions is also included. Of course, the expressions that describe the realistic absorptive behavior of bulk semiconductors are those that include the excitonic effects.

6-3. Exciton Luminescence*

An exciton in a crystal has a limited lifetime, since the exciton can dissociate into a free electron-hole pair, or it can recombine with the electron returning into the hole in the valence band. When this last process occurs with the emission of a photon, it is called luminescence.

Luminescence studies provide an interesting method of investigating several properties of the crystal, such as the excitonic effects and the band structure. They also help to evaluate the quality of the crystal. Impurities which act as substitutional atoms in the crystal, even in small quantities, give intense luminescence lines with characteristic wavelengths, allowing the determination of their nature and concentration. Controllable crystal purity is crucial in many optical applications, such as laser action, optical gain, and active optical devices. In this section, we first examine the "intrinsic" luminescence of a pure crystal, ignoring any impurity effect. This is also called the *free*-exciton luminescence, or recombination, to emphasize the fact that the decaying excitons are mobile particles, free to propagate throughout the crystal. We also assume the weak excitation regime (low exciton density) where excitons may be considered a gas of independent particles without mutual interactions. In Chap. XII we examine the exciton luminescence in the intermediate and high excitation regime, where exciton-exciton interaction leads to a new exciton chemistry with the appearance of more complex excitonic phases.

Since the exciton luminescence is the reverse process of absorption (destruction rather than creation of an exciton), the same conservation rules apply,

$$\hbar\omega_e = \mathcal{E}_n(K_c) \qquad \text{energy conservation} \tag{6.48}$$

and

$$\hbar\mathbf{q} = \hbar\mathbf{K}_c \qquad \text{momentum conservation ,} \tag{6.49}$$

where $\hbar\omega_e$ and $\hbar q$ are the energy and momentum of the emitted photon, respectively. K_c is the wave vector of the decaying exciton.

For emitted photons in the visible part of the spectrum, $\hbar q \simeq 0$, so only a small fraction of the total population, with momentum $\hbar K_c = \hbar q \sim 0$, is allowed to decay in a direct radiative process. Furthermore, we note that the corresponding emission wavelength is the same as that of the n = 1 exciton absorption. The emitted photon resulting from the recombination of an exciton in the bulk of the crystal can, therefore, be reabsorbed by the crystal, re-emitted in another decay process, reabsorbed, etc. A correct description of the whole process must be given in terms of the propagation of excitonic polaritons (see Sec. 6-5), the mixed exciton-photon modes, which can convert into external photons only when they reach a surface. We will see in the next section that excitons have a high probability of being captured by impurities on their way. The net result is that the intensity of luminescence of direct free-exciton decay is weak, because it involves only a small subset of the population and because of reabsorption.

Free-exciton luminescence can also occur in an indirect process, with the participation of optical phonons. In that case, energy and momentum conservation laws read

$$\hbar\omega_e = \mathcal{E}_n(K_c) \pm \hbar\Omega \tag{6.50}$$

and

$$\hbar q = \hbar K_c \pm \hbar K , \tag{6.51}$$

where + refers to the simultaneous absorption, and - refers to emission of an optical phonon. At low temperatures, $k_B T < \hbar\Omega$, the population of optical phonons is negligibly small, so that only exciton luminescence by emission of phonons needs to be considered.

We have seen in Chap. IV that optical phonons have little dispersion; i.e.,

$$\hbar\Omega(K) \simeq \text{const} . \tag{6.52}$$

In that case, both energy and momentum conservation laws can be satisfied simultaneously (see Fig. 6.11) for the entire population of decaying excitons. Therefore, the phonon-assisted exciton luminescence is usually stronger than the direct luminescence, even if it corresponds to a higher-order, less probable transition process.

Following the creation of an unbound electron-hole pair of total energy, $\mathcal{E}_1 = E_g + \Delta\mathcal{E}$, by photoexcitation in the band-to-band region, or by other means, interaction of the electron and hole with the lattice leads to a rapid relaxation of the excess energy in the form of successive emission of optical phonons. This energy relaxation is very fast compared to the exciton lifetime (typically less than a picosecond), as long as the

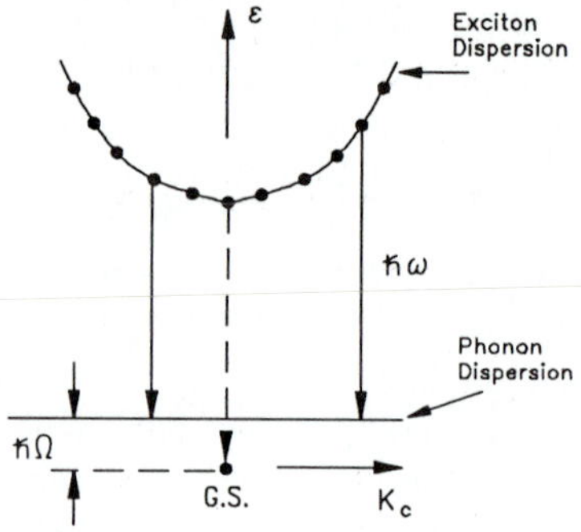

Figure 6.11. Schematic representation of the direct (dashed arrow) and phonon-assisted decay (full arrow). Note that only excitons near $K_c = 0$ can decay to the crystal ground state ($\mathcal{E} = 0$, $K_c = 0$).

excess energy, $\Delta\mathcal{E}$, is larger than the energy of longitudinal optical phonons. The created electron-hole pairs in a crystal kept at low temperatures ($k_B T < E_B$) tend, therefore, to accumulate in the form of a population of excitons in the n = 1 kinetic-energy band.

The further relaxation of the residual excess energy of the n = 1 exciton within its kinetic band occurs via emission of low-energy acoustic phonons. This process is slower, since the energy dissipated by each scattering is smaller, $\hbar\Omega_{ac} < \hbar\Omega_{opt}$. It typically requires a time of the order 10^{-9} sec or more to reach thermal equilibrium with the lattice. Therefore, if the exciton population lifetime is longer than 10^{-9} sec, one expects the energy distribution of the exciton population to be of the form

$$n(\mathcal{E}) = f(\mathcal{E})\, g(\mathcal{E}) \,, \tag{6.53}$$

where the density of states $g(\mathcal{E})$ and the statistical distribution function $f(\mathcal{E})$ are given by

$$g(\mathcal{E}) = \frac{1}{2\pi^2} \left(\frac{2M}{\hbar^2}\right)^{3/2} \mathcal{E}^{1/2} \tag{6.54}$$

and

$$f(\mathcal{E}) = \frac{1}{\exp[(\mathcal{E} - \mu)/k_B T] - 1} \,. \tag{6.55}$$

Since the exciton center-of-mass motion is that of a free particle of effective mass, $M = m_e + m_h$, the exciton density of states follows Eq. (2.89). The distribution $f(\mathcal{E})$ follows Bose-Einstein statistics because in the low-density regime, excitons belong to the class of Bose particles, as they consist of two Fermions, the electron and hole, each with half-integer spin so that the total spin is an integer. As long as the condition $na^3 << 1$ is satisfied, the internal Fermi structure can be ignored and the exciton can be treated like a point Bose particle. Mathematically, the Bose distribution function, Eq. (6.55), could diverge for $\mathcal{E} = \mu$ if μ would ever become positive. This divergence is related to the phenomenon of Bose-Einstein condensation, which is a condensation of Bosons into the momentum state $k = 0$. To understand the origin of the condensation, let us assume that the chemical potential for Bosons at a temperature T is negative; $\mu(T) < 0$. For decreasing T it can happen that $\mu \to 0$. [This may be found from the numerical solution of the integral equation, Eq. (6.61).] The temperature at which $\mu = 0$ is called the critical temperature, T_c,

$$\mu(T = T_c) = 0 \; . \tag{6.56}$$

If $\mu = 0$, then the $k = 0$ state becomes macroscopically populated, since

$$\mathcal{E}(k = 0) = 0 = \mu \; , \tag{6.57}$$

and, therefore,

$$f[\mathcal{E}(k = 0)] \to \infty \; . \tag{6.58}$$

Hence, the Bosons have a distribution given by Eq. (6.55) with

$$\mu < 0 \qquad \text{for } T > T_c \tag{6.59}$$

or

$$\mu = 0 \qquad \text{for } T \leq T_c \; . \tag{6.60}$$

The chemical potential μ for Bosons is always negative and is determined by the condition

$$n_{total} = \int_0^\infty d\mathcal{E} \; f(\mathcal{E}) \;\; g(\mathcal{E}) \; . \tag{6.61}$$

For low particle densities, μ is a large negative number [this may be seen from the numerical solution of the integral, Eq. (6.61)], and $-\mu/k_B T >> 1$. Thus, $f(\mathcal{E})$ can be approximated as

$$f(\mathcal{E}) = 4n\left(\frac{\hbar^2\pi}{2Mk_B T}\right)^{3/2} \exp(-\mathcal{E}/k_B T) . \tag{6.62}$$

Therefore, for low densities, Eq. (6.55) converts into the familiar Maxwell-Boltzmann function describing the velocity distribution of a classical gas. Substituting Eqs. (6.54) and (6.62) into Eq. (6.53) gives

$$n(\mathcal{E}) = 4n\left(\frac{\hbar^2\pi}{2Mk_B T}\right)^{3/2} g(\mathcal{E}) \exp(-\mathcal{E}/k_B T) . \tag{6.63}$$

The exciton distribution function $n(\mathcal{E})$ is plotted in Fig. 6.12 by the solid curve. The functions $g(\mathcal{E})$ and $f(\mathcal{E})$ are also plotted as dotted and dashed curves, respectively. The area under the distribution function that is marked by the hatched portion denotes the total number of excitons [Eq. (6.61)].

If the transition probability is independent of the phonon wave vector $\hbar K$, the luminescence lineshape is directly proportional to the density of excitons,

$$I(\hbar\omega) \propto \int n(\mathcal{E})\, \delta_\Gamma(\hbar\omega - \mathcal{E})\, d\mathcal{E} \sim \int g(\mathcal{E})f(\mathcal{E})\, \delta_\Gamma(\hbar\omega - \mathcal{E})\, d\mathcal{E} , \tag{6.64}$$

where $\delta_\Gamma(\hbar\omega - \mathcal{E})$ is the broadened lineshape function to relate $\hbar\omega$ and $\mathcal{E}$. For example, when the exciton distribution $n(\mathcal{E})$ follows Eq. (6.63), the luminescence lineshape $I(\hbar\omega)$ should have a Maxwellian shape.

An example of phonon-assisted free-exciton luminescence is shown in Fig. 6.13 for the case of Cu_2O at two lattice temperatures. The intensity of laser excitation was small to generate a small density of excitons of $< 10^{17}$ cm^{-3}. The large linewidth of the phonon-assisted emission peak is the result of the broad distribution of excitons, as excitons with different kinetic energies participate in the luminescence process (see Fig. 6.11). Note that the width of the emission line increases with temperature, in agreement with the Maxwell-Boltzmann distribution. The circles on the figure show a fit of the luminescence lineshape to a Maxwell-Boltzmann distribution. The good agreement between the experiment and the fit

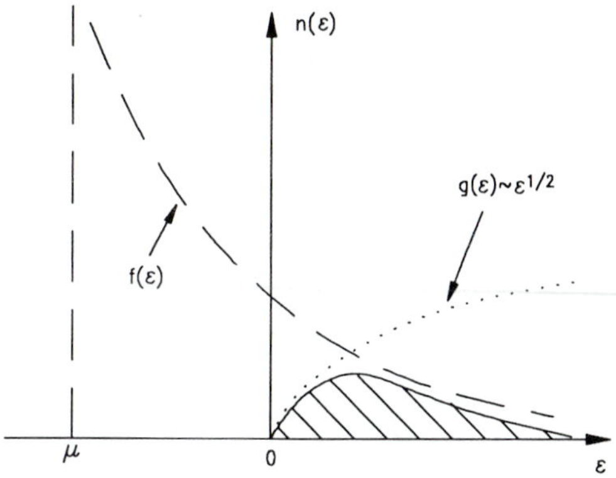

Figure 6.12. The distribution function of a gas of Bose particles such as excitons. The position of the chemical potential is determined by the requirement that the number of particles given by the hatched area in this figure equal the total number of particles. Temperature determines not only the shape of $f(\mathcal{E})$, but also its position through the position of μ. For values of $|\mu/k_B T| > 1$, the distribution $f(\mathcal{E})$ is indistinguishable from a classical Maxwell-Boltzmann distribution.

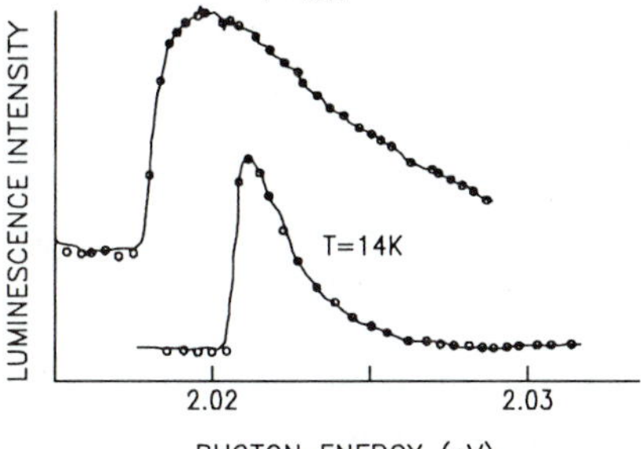

Figure 6.13. Optical phonon-assisted luminescence from n = 1 free exciton in Cu_2O at low densities, $n < 10^{17}$ cm^{-3}, at two temperatures. The circles denote the fit of the luminescence lineshape by a Maxwell-Boltzmann distribution function (after Ref. 6.6).

indicates that the kinetic energy of the decaying excitons actually follows a Maxwell-Boltzmann distribution. At higher densities, $n > 10^{17}$ cm^{-3}, the classical Maxwell-Boltzmann approximation no longer applies. The correct quantum-mechanical distribution function, Eq. (6.55), must be introduced. It leads to an increased occupation of low momentum states if compared to the classical regime, as shown in Fig. 6.14.

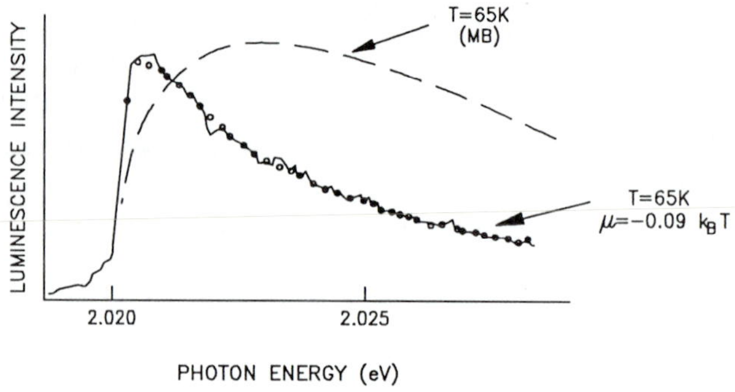

Figure 6.14. Phonon-assisted free-exciton luminescence at high densities, $n > 10^{18}$ cm^{-3}. The lineshape is fitted with a Bose-Einstein distribution function. The dashed curve shows a fit to a Maxwell-Boltzmann distribution, clearly demonstrating that such a fit cannot explain the observation (after Ref. 6.7).

6-4. Bound Excitons*

A real crystal is never perfect. Imperfections such as ion vacancies, interstitials, or substitutional atoms (either native or intentionally introduced) exist in densities ranging from $n_i < 10^{12}$ cm^{-3} in ultrapure crystals of Si or Ge, up to $n_i > 10^{17}$ cm^{-3} in doped semiconductors. Such imperfections can attract excitons that become localized at the defect site, becoming bound excitons. The binding energy of the exciton to the defect is often quite small, typically a few meV. Therefore, bound excitons are best observed at very low temperatures.

Excitons may be bound to a *donor*, which is a substitutional atom with a higher number of valence electrons compared with the host atom, or to an *acceptor*, a substitutional atom with a lower number of valence electrons. Donors contribute excess electrons to the crystal, while acceptors tend to capture electrons or equivalently donate holes. Donor or acceptor atoms may be electrically charged or neutral. When the donor atom has given away its initial extra valence electron, it becomes positively charged and is referred to as an ionized donor. Similarly, when an acceptor atom has captured an electron (or equivalently released a hole), it has a negative charge and is called an ionized acceptor. In contrast, a neutral donor or acceptor has no net charge, since it has kept its original number of valence electrons. Excitons may get bound to either an ionized donor or acceptor or a neutral donor or acceptor by forming complexes represented schematically in Fig. 6.15(a). In many crystals, the binding energy of the exciton to a neutral donor or acceptor is close to a tenth of the donor or

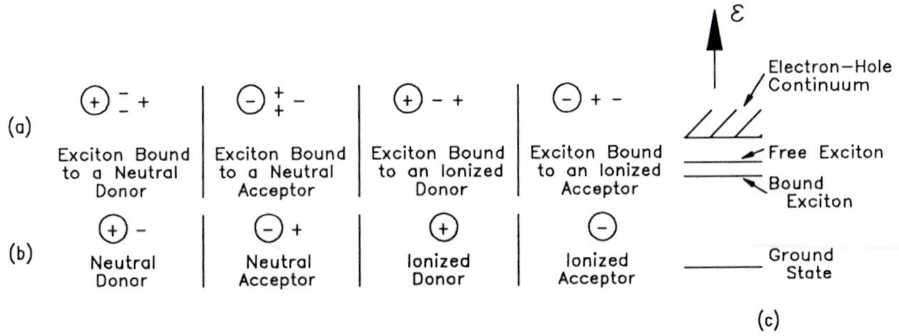

Figure 6.15. Commonly used schematic diagram to represent an exciton bound to neutral or ionized donors and acceptors. In the left part of (a) an exciton with a symbol of + - is shown with a neutral donor, represented by ⊕-, with a resulting complex that consists of the impurity atom ⊕, two electrons -, and a hole +. In the left part of (b) the impurity atom is shown without exciton (after Ref. 6.8). (c) The energetic position of the bound exciton relative to the free-exciton line.

acceptor ionization energy, which is the energy required to free the extra valence electron of a neutral donor, or the energy to free a hole (to accept an electron) in a neutral acceptor.

Another important class of defects is isoelectronic centers, which are substitutional atoms with the same valence electrons. Excitons can be bound to centers such as nitrogen (N) atoms substituting phosphor (P) in GaP, or oxygen (O) instead of tellurium (Te) in ZnTe.

The bound-exciton energy is lower than the n = 1 free-exciton energy by an amount equal to the binding energy to the impurity, as shown in Fig. 6.15(c). The bound-exciton energy, therefore, is

$$\mathcal{E}_{bx} = E_g - E_B - \mathcal{E}_D \,, \tag{6.65}$$

where $\mathcal{E}_D$ is the binding energy of the exciton to the impurity. Note the absence of the center-of-mass kinetic-energy term in the bound-exciton energy of Eq. (6.65) compared with the free-exciton energy of Eq. (6.20).

Bound excitons play a role disproportionate to their concentration, particularly in luminescence. To understand this, we first recall that the vast majority of free excitons cannot decay radiatively in a direct process (without the participation of phonons) because of the momentum conservation law. During their lifetime, these excitons propagate throughout the crystal with the possibility of becoming captured at an imperfection site. The time necessary to capture a free exciton can be evaluated as follows.

The inverse capture time (the capture rate) is proportional to the density n_i of defects, to a capture cross section σ, which is of the order of the square exciton radius a_B^2, and to the mean thermal velocity of the exciton $\bar{v}$.

$$\frac{1}{\tau_c} = n_i \, \sigma \, \bar{v} \, . \qquad (6.66)$$

Even in a very pure crystal with $n_i \sim 10^{15}$ cm^{-3}, Eq. (6.66) indicates that the capture time at T = 2 K is short, less than 10^{-9} s for an exciton with Bohr radius $a_B \geq 50$ Å. This capture lifetime is much smaller than a typical exciton lifetime in an indirect-gap semiconductor, which is $\gtrsim 10^{-6}$ sec. Therefore, impurities act as very efficient capture centers. Once localized at an imperfection site, bound excitons may recombine radiatively in a direct-emission process. The prohibition due to the momentum conservation law no longer applies, because the localization of the electron-hole pair in real space implies, by Fourier transformation, an extension in momentum or k space. Therefore, the direct luminescence of bound excitons, without the participation of phonons, can even be observed in an indirect-gap material such as silicon (see Fig. 6.16).

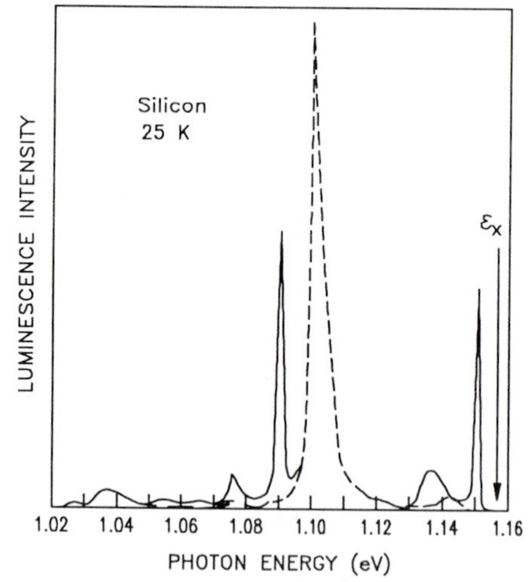

Figure 6.16. The photoluminescence of silicon at 25 K. The dashed curve, which is the luminescence of a pure crystal, has a main peak not at the intrinsic exciton energy, $\mathcal{E}_X = 1.155$ eV, but rather at $\mathcal{E}_X - \hbar\Omega = 1.10$ eV, one phonon energy below exciton. The solid curve is the spectrum from a crystal containing 8×10^{16} arsenic atoms/cc. There is now a zero phonon bound-exciton line at 1.149 eV, as well as its phonon replica (after Ref. 6.9) at 1.09 eV.

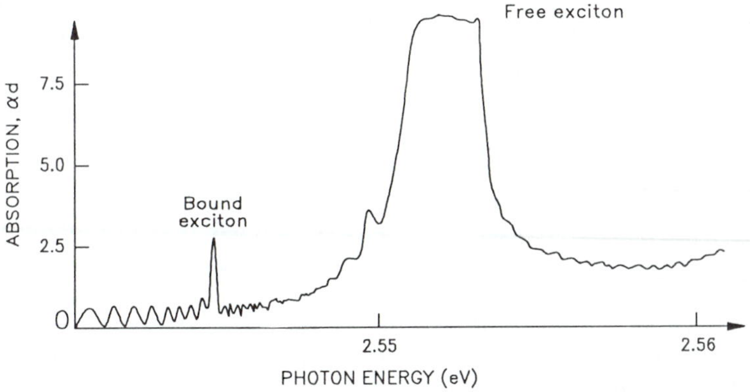

Figure 6.17. Free- and bound-exciton absorption lines in CdS at 2 K. The large peak at ≃ 2.552 eV is assigned to the free exciton (the so-called A-exciton in CdS), and the smaller peak at ≃ 2.545 eV is the bound-exciton resonance (after Ref. 6.10).

Bound-exciton luminescence can be recognized by the sharpness of the lines located below the free exciton. Since bound excitons are localized around the impurities, there is no thermal broadening corresponding to an increase in the width of the energy distribution, as is the case for free excitons. As long as the binding energy of the exciton to the impurity is smaller than the exciton binding energy, bound-exciton lines disappear before free exciton lines as the crystal temperature is raised, because thermal energy dissociates the bound exciton first. The nature of the impurity center (neutral or ionized donor or acceptor) can be identified by the application of a magnetic field with a characteristic Zeeman splitting for each impurity center. Bound-exciton lines can also be observed in absorption. The absorption coefficient is smaller than the intrinsic free-exciton absorption line, because it is proportional to the density of defects in the crystal, which is usually a small percentage of the density of crystal atoms. An example of a bound exciton seen in absorption is shown in Fig. 6.17.

6-5. Exciton Polaritons*

The correct description of an optically active exciton must include the polariton concept. We noted in Chap. IV that phonon polaritons arise when the phonon dispersion crosses the photon dispersion. Similarly, in those frequency regions where the exciton and photon dispersions cross, the proper states of the system are exciton polaritons, a mixture of electronic polarization wave and optical wave. The basic concept for exciton polaritons formation follows the same steps as for phonon polaritons. Thus, at very small k values, $k < 10^5$ cm^{-1}, the polariton dispersion curve is a straight line, representing the dispersion curve of the light. In the crossing region, $k \sim 10^5$-10^6 cm^{-1}, the polariton is a mixture of the exciton and photon. Here the exciton and photon strongly couple. In the interaction process, the exciton field plays the role of the classical polarization field that couples with the electromagnetic field of the photon. Eigenstates of the coupled exciton-photon system are mixed states of excitons and photons. The eigenvalues of these mixed states are the exciton polaritons. Therefore, the propagating modes are the polaritons in the crystal, and the speed of propagation is the group velocity of the polariton. The equation describing the exciton-polariton dispersion in the effective-mass approximation is given by (see Ref. 6.11)

$$\left[\mathcal{E}(K_c)\right]^4 - A\left[\mathcal{E}(K_c)\right]^2 + B = 0 , \tag{6.67}$$

where

$$A = (\hbar^2 c^2 K_c^2/\epsilon_\infty) + E_{XT}^2(K_c)(1 + 4\pi\beta^0) , \tag{6.68}$$

$$B = (\hbar^2 c^2 K_c^2/\epsilon_\infty) + E_{XT}^2(K_c) , \tag{6.69}$$

$$2\pi\beta^0 \simeq [E_{XL}(0) - E_{XT}(0)]/E_{XT}(0) , \tag{6.70}$$

and

$$E_{XT}(K_c) = E_{XT}(0) + (\hbar^2/2M)K_c^2 . \tag{6.71}$$

Here M is the effective-exciton mass, $E_{XT}(0)$ and $E_{XL}(0)$ are the transverse and longitudinal exciton energies at $K_c = 0$, and $E_{XT}(K_c)$ is the wave vector dependent transverse exciton energy. Equation (6.67) is plotted schematically in Fig. 6.18, showing exciton-polariton dispersion curves. It should be noted that, as for the photon-phonon coupling case, only the transverse exciton with a dipole moment normal to its wave vector interacts with a photon, which is a purely transverse mode.

The exciton-polariton dispersions may be measured directly by resonant two-photon Raman scattering.

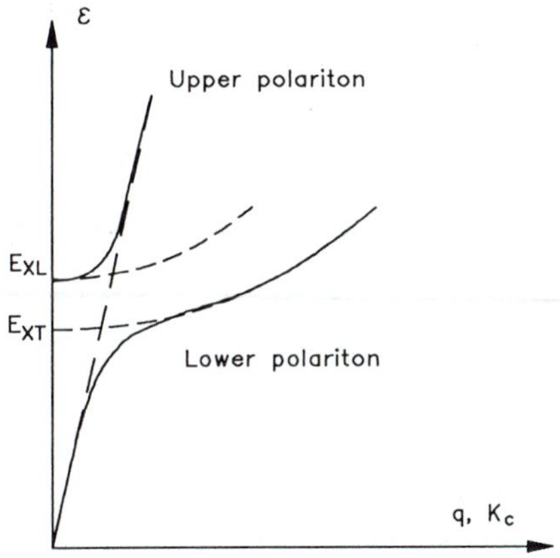

Figure 6.18. The dispersion of excitonic polaritons. The dashed lines are the photon and transverse exciton dispersions in the absence of coupling.

6-6. Problems

6.1. Using Eq. (5.69), show that the continuum absorption term in Eq. (6.34) can be written as

$$\alpha_{\text{continuum}} = \alpha_{\text{free}}\, C(\omega) , \tag{6.72}$$

where

$$C(\omega) = \frac{Ze^Z}{\sinh Z} . \tag{6.73}$$

6.2. Show that Eq. (6.34) may be written as

$$\alpha(\omega) = \frac{2|d_{cv}|^2}{\hbar n_b c} \frac{1}{a_B^3} \frac{\hbar\omega}{E_B} \left[\theta(\hbar\omega - E_g) \frac{\pi \exp Z}{\sinh Z} \right.$$

$$\left. + E_B \sum_{n=1}^{\infty} \frac{4\pi}{n^3} \delta\left(\hbar\omega - E_g + \frac{E_B}{n^2}\right)\right] . \tag{6.74}$$

This equation indicates that the excitonic absorption strength varies with the exciton Bohr radius, like $1/a_B^3$, i.e., the larger the exciton Bohr radius, the smaller the exciton absorption coefficient. This behavior has been observed experimentally for several semiconductors. For example, the exciton absorption is larger for CuCl with $a_B \simeq 7$ Å than that in GaAs with $a_B \simeq 140$ Å.

6.3. Compute the time required to capture an exciton of radius, $a_B = 100$ Å, for a donor concentration, $n_i = 10^{16}\ cm^{-3}$, and a lattice temperature, T = 2K. (Hint: Express the average speed as $M\bar{v}^2/2 = 3/2\ k_B T$ and use GaAs mass parameters.)

6-7. References

1. L. Schiff, *Quantum Mechanics* (McGraw Hill Book Company, New York, 1955).
2. R. S. Knox, *Theory of Excitons* (Academic Press, New York, 1963).
3. S. Nikitine, Philos. Mag. **4**, 1 (1959).
4. F. Bassani, *Electronic States and Optical Transitions in Solids* (Pergamon, Oxford, 1975).
5. G. W. Fehrenbach, W. Schäfer, J. Treusch, and R. G. Ulbrich, Phys. Rev. Lett. **49**, 1281 (1982).
6. A. Mysyrowicz, J. B. Grun, A. Bivas, R. Levy, and S. Nikitine, Phys. Lett. A **25**, 286 (1967).
7. A. Mysyrowicz, D. Hulin, and A. Antonetti, Phys. Rev. Lett. **43**, 1123 (1979).
8. D. G. Thomas and J. J. Hopfield, Phys. Rev. Lett. **7**, 316 (1961).
9. J. R. Haynes, Phys. Rev. Lett. **4**, 361 (1960).
10. M. Dagenais and W. F. Sharfin, J. Opt. Soc. Am. B 2, 1179 (1985).
11. D. G. Thomas and J. J. Hopfield, Phys. Rev. **128**, 2135 (1962).

Chapter VII
OPTICAL PROPERTIES OF SOME IMPORTANT BULK SEMICONDUCTORS

In this chapter the linear optical properties of some representative bulk semiconductors are described. We start with the direct-gap semiconductor GaAs, since this crystal provides a model case for Wannier excitons with a relatively simple band structure. We then discuss Cu_2O because of the appearance of a large variety of excitonic transitions in this semiconductor, in very good agreement with the theory. This allows us to clearly illustrate processes that are often masked or difficult to resolve in other semiconductors because of the more complex band structure and the much weaker exciton binding energy. We then examine CuCl, also a model case for excitons. This is a I-VII compound consisting of one element from the first row and one element from the seventh row of the periodic table. We then describe representatives of II-VII materials and, finally, indirect-gap Si and Ge elemental semiconductors from the fourth column of the periodic table.

7-1. Gallium Arsenide - GaAs

Gallium arsenide is a III-V compound semiconductor with the Ga and As belonging to the third and fifth column of the periodic table, respectively. This material is gaining increasing importance in modern technology. Transistors based on GaAs have higher speeds than those of Si. Laser diodes and optically active nonlinear elements using bulk GaAs or GaAs quantum wells operate at room temperature. It is anticipated that circuitry based on GaAs technology will allow interfacing between conventional electronics and optoelectronics in the near future. GaAs has a cubic structure without inversion symmetry, belonging to a point group T_d, which is also known as zinc blende structure. The arrangement of atoms in the zinc blende structure is shown in Fig. 7.1, together with that of diamond. The two structures differ in that the diamond structure has only one type of atom. Thus, diamond has two identical atoms per unit cell, while GaAs has two different atoms per unit cell with the atoms being

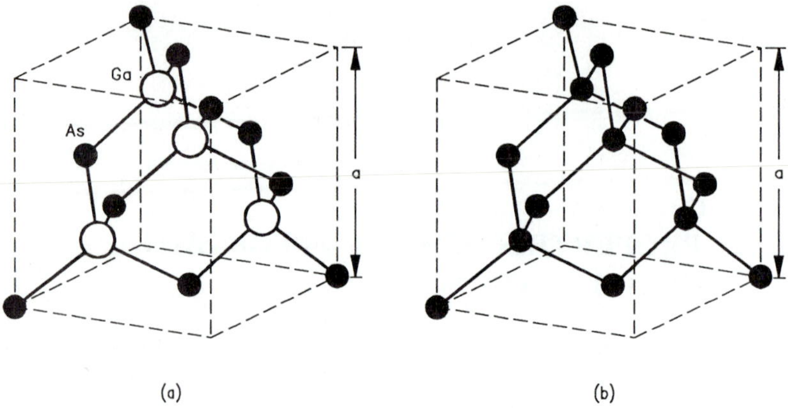

Figure 7.1. (a) Zinc blende lattice. (b) Diamond lattice.

at positions (0,0,0) and (1/4,1/4,1/4). The structure consists of two face-centered cubic lattices separated from each other along the [111] axis by a quarter of diagonal length. The lattice constant of GaAs is a = 0.5653 nm. The band structure is shown in Fig. 7.2. Notice that the conduction band minima at points L and X of the Brillouin zone have nearly the same energy as the minimum of the conduction band at point Γ. Thus, GaAs is nearly an indirect-gap semiconductor. (In the related compound crystal GaP, the minima at X and L are further decreased, resulting in an indirect-gap material. Physical effects, such as the Gunn effect, and applications, such as solid-state microwave emitters, rely on this peculiarity of GaAs.)

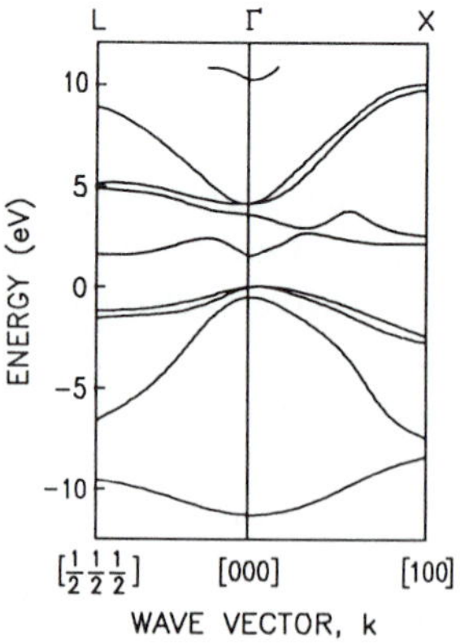

Figure 7.2. Band structure of GaAs (after Ref. 7.1).

An expanded view of the band structure near k = 0 is schematically shown in Fig. 7.3 with the bandgap energy at the Γ point being E_g = 1.519 eV at low temperatures. This band structure is similar for many III-V compound semiconductors and has the following origin (see Ref. 7.2). Ga, with 31 electrons, has the electronic structure $1s^2 2s^2 2p^6 3s^2 3p^6 3d^{10} 4s^2 4p^1$, consisting of three outer shell electrons, while As, with 33 electrons, has the atomic configuration $1s^2 2s^2 2p^6 3s^2 3p^6 3d^{10} 4s^2 4p^3$, with five outer shell electrons. When we form GaAs, the atomic levels of the individual constituents are mixed because the electrons of Ga feel the potential of As, and vice versa. Therefore, a total of eight outer shell electrons for the compound hybridize to form bonds between the two atoms. The hybridization leads to splitting of the atomic energy levels into two levels, referred to as the bonding and antibonding levels. Figure 7.4 schematically shows this energy splitting where the outer shell electron levels 4s and 4p have split into the bonding (lower) and antibonding (upper) levels. The lower energy bonding levels are completely filled with electrons, while the antibonding levels are empty. The s levels are spin degenerate (i.e., twofold degenerate), while the p levels are sixfold degenerate (three p orbitals each with two electrons having the same energy). In crystalline GaAs, with many unit cells, the energy levels broaden to form bands, resulting in valence bands from the bonding levels and conduction bands from the antibonding levels. Thus, if we ignore spin-orbit interaction, the top of the valence band at k = 0 in GaAs would be sixfold degenerate due to the degeneracy of the atomic orbitals. In this case we have three valence bands, each doubly degenerate because of spin. As spin-orbit interaction, we refer to the situation where the orbital angular momentum ℓ and spin s of the electrons are not independent, but have to be coupled to a total angular momentum,

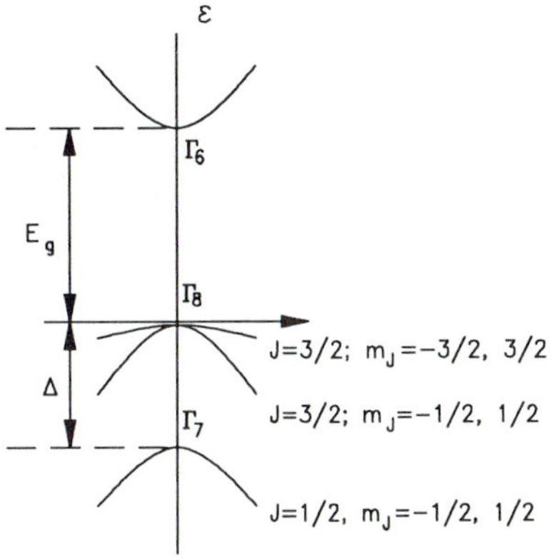

Figure 7.3. An expanded band structure of GaAs near k = 0, including spin-orbit coupling.

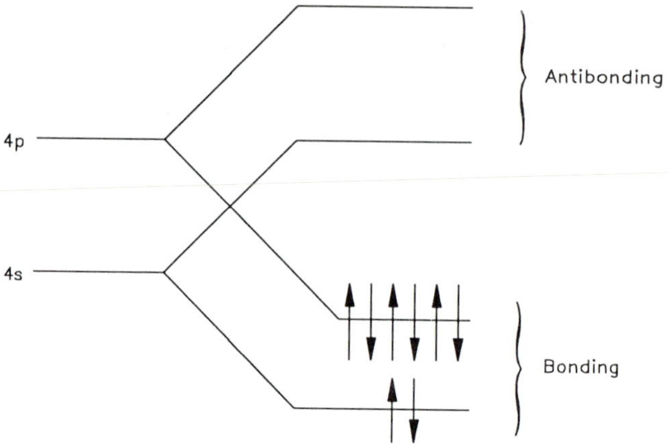

Figure 7.4. Atomic energy level splitting for a binary complex like GaAs.

J = $\boldsymbol{\ell}$ + **s**. Hence, the electronic states cannot be classified according to the spin or orbital angular momentum quantum numbers s,m_s and ℓ,m_ℓ, respectively. They have to be classified according to the total angular momentum J,m_J. Therefore, s,m_s and ℓ,m_ℓ are no longer "good quantum numbers." Coupling of spin and orbital angular momentum leads to further reduction of the valence-band degeneracy. Thus, one of the three valence bands is shifted down in energy, as schematically shown in Fig. 7.3. The total angular momenta of the valence bands are 3/2 and 1/2, since $\ell = 1$ (for p orbitals, $\ell = 1$) and s = 1/2. The top two bands are referred to as the heavy-hole (J = 3/2, m_J = -3/2, 3/2) and light-hole (J = 3/2, m_J = -1/2, 1/2) bands. The lower band is called the split-off band (J = 1/2, m_J = -1/2, 1/2), resulting from the spin-orbit coupling with the energy separation of $\Delta = 0.341$ eV. These bands have point-group symmetries noted in Fig. 7.3 with the top two valence bands having the Γ_8 symmetry and the split-off band having the Γ_7 symmetry. Note that here we use the point-group symmetry notations, Γ symbols, only for band labeling purposes, with the use of the letter Γ indicating that the band extrema lie at the center of the Brillouin zone. The group theoretical implications are ignored.

The ordering of the valence bands, with the Γ_8 band above the Γ_7 band, is the normal ordering. The degeneracy of the Γ_8 band is removed away from k = 0. The energy dependence of the heavy- and light-hole bands versus k shows anisotropy; i.e., the effective hole mass depends on the orientation of k with respect to the crystal axis. This effect is often called warping. Furthermore, the hole bands deviate markedly from a quadratic dispersion law for energies, $\mathcal{E}_h(k) \sim \Delta$, the spin-orbit coupling energy. The dispersion $\mathcal{E}(k)$ of the valence bands is expressed in terms of the so-called Luttinger parameters, j_1, j_2, and j_3. To a good approximation, near k = 0, they read (see Ref. 7.2)

$$\mathcal{E}_\ell = \frac{\hbar^2 k^2}{2m_{\ell h}} \tag{7.1}$$

and

$$\mathcal{E}_h = \frac{\hbar^2 k^2}{2m_{hh}} , \tag{7.2}$$

with

$$\frac{m_0}{m_{\ell h}} = j_1 + 2j_2 \tag{7.3}$$

and

$$\frac{m_0}{m_{hh}} = j_1 - 2j_2 , \tag{7.4}$$

for energies of the light-hole, $\mathcal{E}_\ell$, and heavy-hole, $\mathcal{E}_h$, bands for k along [100] direction; and

$$\frac{m_0}{m_{\ell h}} = j_1 + 2j_3 \tag{7.5}$$

and

$$\frac{m_0}{m_{hh}} = j_1 - 2j_3 , \tag{7.6}$$

for k along [111] direction, with the following values for the Luttinger parameters for GaAs,

$j_1 = 6.9 \ (7.49)$,

$j_2 = 2.4 \ (2.41)$,

and

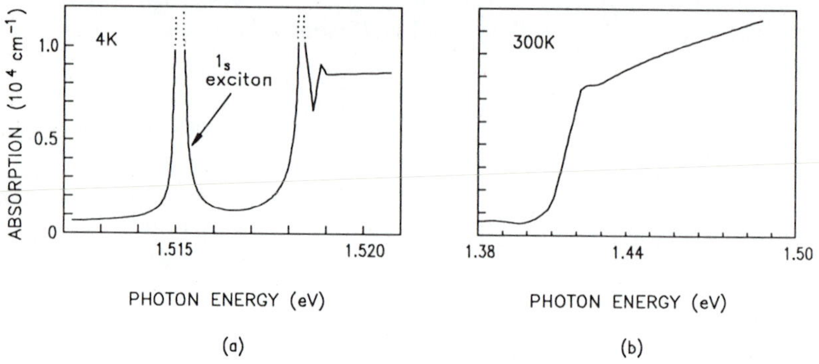

Figure 7.5. Absorption spectrum of GaAs at (a) low temperatures (after Ref. 7.4) and (b) room temperature (after Ref. 7.5).

$$j_3 = 2.6 \quad (3.28) \tag{7.7}$$

(after Ref. 7.3).

Figure 7.5 shows an absorption spectrum of GaAs at room and low temperatures. The large absorption coefficients $\alpha \sim 10^5\text{-}10^6\ \text{cm}^{-1}$ are typical of dipole-allowed first-class transitions. The peak at 1.515 eV at low temperatures corresponds to the n = 1 exciton. In thin crystals of very high quality, the n = 2 term is also observed. The exciton binding energy is E_B = 4.2 meV. At room temperature, with $k_B T > E_B$, the exciton structure broadens and merges with the band-to-band continuum due to the very short exciton lifetime of 10^{-13}s, limited by thermal dissociation into unbound electron-hole pairs.

7-2. Cuprous Oxide - Cu_2O

Cu_2O is a red semiconductor that crystallizes in the cubic structure, having the inversion symmetry and belonging to the symmetry point group O_h. Although Cu_2O was considered a material for electrical rectification in the early days of semiconductor physics, it has little technical importance now. However, it serves as a model case for the study of excitonic effects because of the large variety of excitonic transitions observed that are in good agreement with theory.

Figure 7.6 shows the arrangement of the atoms in the direct lattice. Oxygen atoms occupy the sites of a body-centered cubic lattice, whereas the copper atoms are arranged on a face-centered cubic lattice, with a lattice constant of a = 4.26 Å.

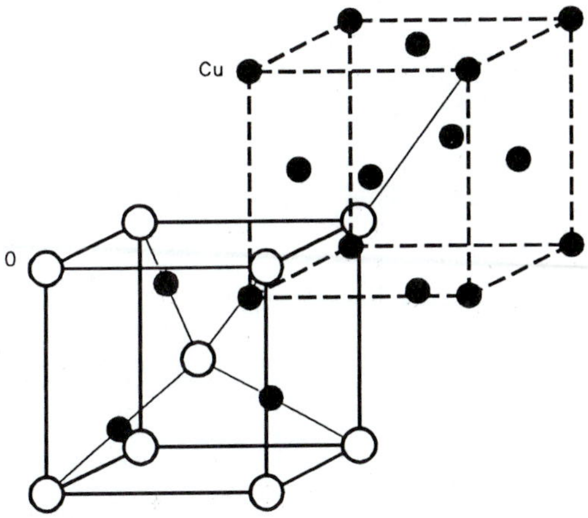

Figure 7.6. Arrangement of atoms in the crystal, Cu_2O.

Figure 7.7 schematically shows the electronic bands responsible for the optical transitions in the visible and ultraviolet. The conduction band is formed out of Cu 4s orbitals. The valence band originates from Cu 3d orbitals that are split, due to the spin-orbit interaction.

For Cu_2O, the extrema of the valence and conduction bands are located at the center of the Brillouin zone, the Γ point. The parity of the bands can be defined in Cu_2O since the material has an inversion center. The valence bands have a positive parity. The lowest conduction band, Γ_6^+, also has a positive parity. The second conduction band, Γ_8^-, has a negative parity. The resulting band structure near k = 0 is shown in Fig. 7.8. Note the reverse ordering of the spin-orbit split upper valence bands Γ_7^+ and Γ_8^+.

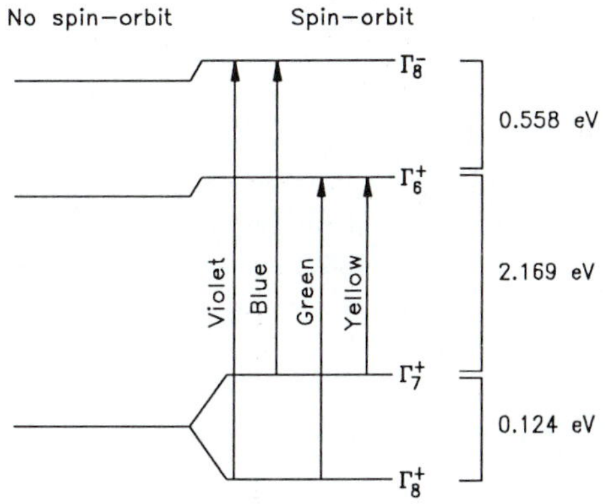

Figure 7.7. Schematic energy spectrum of valence and conduction bands in Cu_2O showing the four-exciton series (after Ref. 7.6).

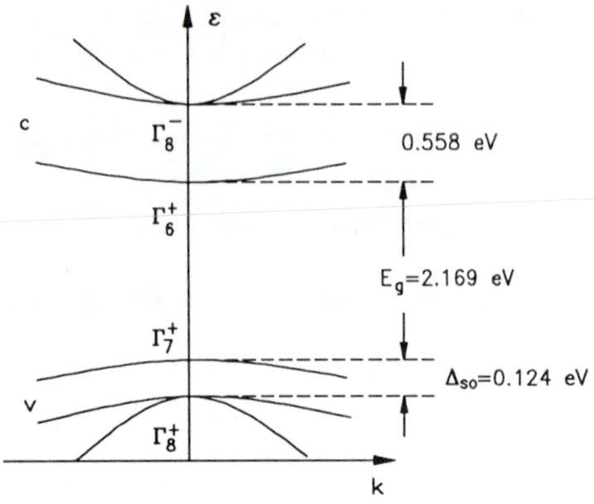

Figure 7.8. Band structure of Cu_2O. All band extrema are at the center of the Brillouin zone at $k = 0$.

In linear absorption, four groups of excitonic lines are observed which correspond to the interband transitions between the two valence bands and the two conduction bands. They are often referred to as the yellow, green, blue, and violet excitons, from the spectral region where they appear, as shown schematically in Fig. 7.7.

The yellow and green excitonic absorption results from transitions between the two upper valence bands and the lowest conduction band. As many as seven lines are observed for the yellow series, as shown in Fig. 7.9, which obey the relation

$$E = E_g - \frac{E_B}{n^2} \qquad \text{for } n = 2, 3, \dots , \tag{7.8}$$

with $E_g = 2.172$ eV at 2 K and $E_B = 96.7$ meV. Since both the valence and conduction bands have the same positive parity, due to their original d and s atomic character, the interband transition is forbidden at $k = 0$ in the dipole approximation ($\Delta\ell = 2$). Second-class dipole transitions are allowed, with the final excitons having a p-like envelope function. The total change of angular momentum is

$$\Delta\ell = \pm 1 \quad \text{with } \Delta\ell_{int} = \pm 2 \text{ and } \Delta\ell_{env} = \pm 1 . \tag{7.9}$$

We note that the $n = 1$ exciton, which must have an s-envelope function, cannot satisfy the transition selection rule and is therefore missing.

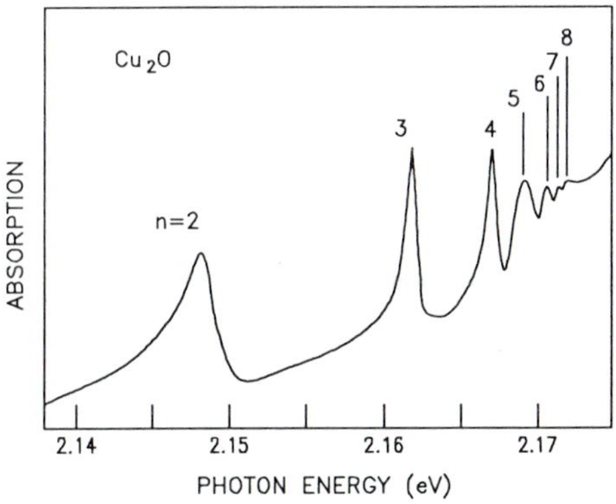

Figure 7.9. Absorption spectrum of the yellow excitonic series of Cu_2O at 2 K (after Ref. 7.7).

Second-class dipole transitions are relatively weakly allowed. In Cu_2O, the absorption coefficient of the yellow excitons, which appear in the 2.1-2.2 eV spectral range, is of the order of 10^3-10^4 cm^{-1}. The n = 1 exciton becomes observable as a very weak absorption line in samples with a few mm thickness, as it is allowed in the quadrupole approximation [see Eq. (5.45)]. Optical transitions to the n = 1 exciton level can also occur with the cooperation of optical phonons. The participating optical phonon, of negative parity, supplies the required parity and momentum to complete the transition to the different k states of the n = 1 exciton band. Thus, in addition to the sharp n = 1 line which is weakly seen, a smooth absorption edge, typical of indirect-gap semiconductors, is observed in the direct-gap semiconductor Cu_2O.

At higher photon energies, above 2.5 eV, strong absorption lines appear, corresponding to first-class transitions for s excitons from the two upper valence bands and the second conduction band. The negative parity of the second conduction band allows the first-class transitions from positive-parity valence bands. Very thin samples, less than one micron, are required to observe the absorption line profile of these dipole-allowed excitons.

The shape of the linear absorption of Cu_2O spanning the spectral region, comprising transitions between the two upper valence bands and the two lowest conduction bands, is depicted schematically in Fig. 7.10. In most other crystals with direct bandgap structure, first-class transitions with strong exciton absorption occur between the highest valence band and the lowest conduction band, consequently determining the onset of absorption.

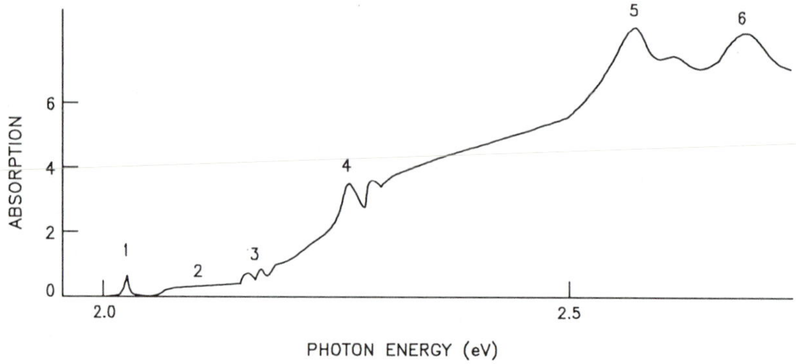

Figure 7.10. Profile of absorption spectrum of Cu_2O at low temperatures. Starting from lower energies, the following excitonic transitions are possible. (1) n = 1 yellow exciton, which is weakly observed as a quadrupole transition; (2) n = 1 yellow exciton by phonon-assisted transition; (3) n = 2, 3, ... yellow p excitons by second-class transition; (4) n = 2, 3, ... green p excitons by second-class transition; (5) n = 1, 2, ... blue s excitons by first-class transition; and (6) n = 1, 2, ... violet s excitons by first-class transition.

7-3. Cuprous Chloride - CuCl

Cuprous chloride is a good example of a I-VII compound semiconductor. Such crystals generally have large energy gaps, $E_g > 3$ eV, and are transparent to visible radiation. Below 400°C the crystal structure of CuCl is zinc blende, like GaAs with T_d point-group symmetry. The lattice constant of CuCl at room temperature is a = 0.5406 nm. A schematic band structure is shown in Fig. 7.11. Extrema of the lowest conduction band and the highest valence band are located at the center of the Brillouin zone, at the Γ point.

Because of the lack of inversion symmetry, no parity can be assigned to the bands at k = 0. The conduction band is derived from s orbitals of the Cu 4s levels. The highest valence bands, Γ_7 and Γ_8, are derived principally from 3d orbitals of the Cu ions, with about 23% admixture of the p-orbitals of the chlorine ion. The energy separation Δ between the two upper valence bands due to spin-orbit coupling is 68 meV. The ordering of the valence bands in CuCl is different from that of typical zinc blende materials, such as the related compound CuBr or GaAs (see Fig. 7.11), with the twofold degenerate Γ_7 band above the fourfold Γ_8 band. This is due to the strong d-like character of the valence bands, which inverts the sign of the spin-orbit splitting. Thus, the band structure of

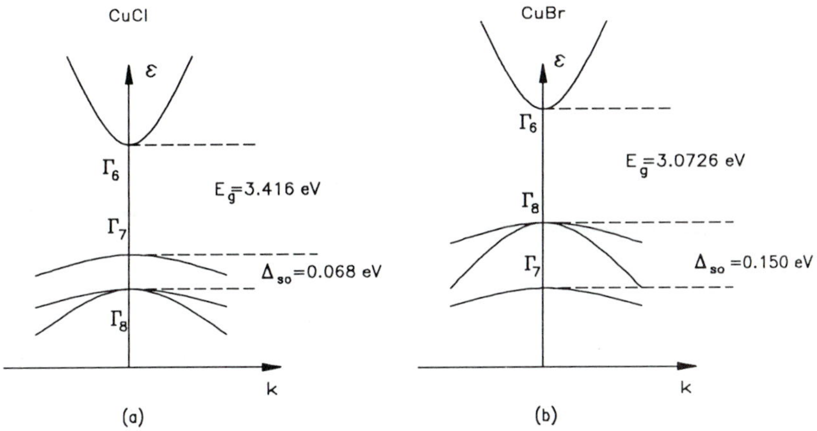

Figure 7.11. Schematic band structure of CuCl (a) and CuBr (b) near the Γ point, at the center of the Brillouin zone. CuBr shows the normal ordering of the upper valence bands, with fourfold degenerate Γ_8 band above the twofold degenerate Γ_7 band. Away from k = 0, the Γ_8 band splits into a heavy-hole and light-hole band.

CuCl is very similar to the case of Cu_2O discussed above, except for the fact that the one-photon interband matrix element $\langle c|p|v\rangle$ is dipole-allowed at k = 0 between the highest valence band and the lowest conduction band, because of the partial p-like character of the valence band. This leads to strong exciton absorption lines, as in Fig. 7.12. Due to the large absorption coefficients $\alpha \sim 10^5$ cm^{-1}, near the band edge, thin films of thickness, $d < 0.1$ μm, are required for linear transmission spectroscopy. Excitons from the two upper valence bands and the lowest conduction band are often referred to as Z_3 and Z_{12} excitons, respectively. The two strong absorption lines at 3.205 eV and 3.2754 eV correspond to the n = 1 excitons from the Z_3 and Z_{12} series, respectively. The n = 2 and n = 3 term of the Z_3 series are also observed at 3.366 eV and 3.385 eV. The three terms do not exactly follow a hydrogenic series, $E_g - E_B/n^2$, because the n = 1 exciton radius 0.7 nm is comparable to the lattice constant. In that case, the static dielectric constant ϵ_0, appropriate for $a_B >> a$, has to be replaced by a dielectric constant of intermediate value close to the high-frequency value, and the exciton is intermediate between a Wannier-type and Frenkel-type exciton. The binding energy of the n = 1, Z_3 exciton is 0.210 eV.

Because of the strong coupling with light, the n = 1 exciton displays a pronounced polariton character. The polariton dispersion curve of the n = 1, Z_3 polariton of CuCl is shown in Fig. 7.13.

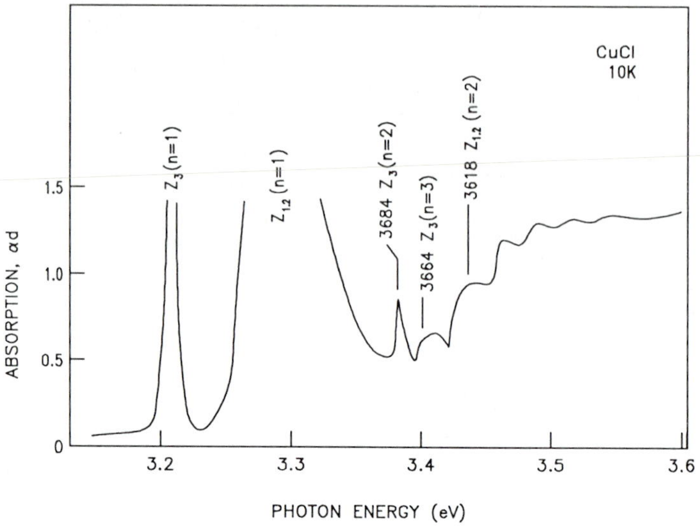

Figure 7.12. Linear absorption spectrum of CuCl at 2 K (after Ref. 7.8). Lines n = 1, 2, and 3 of the Z_3 exciton series, as well as exciton lines n = 1 and n =2 of the Z_{12} series, are observed.

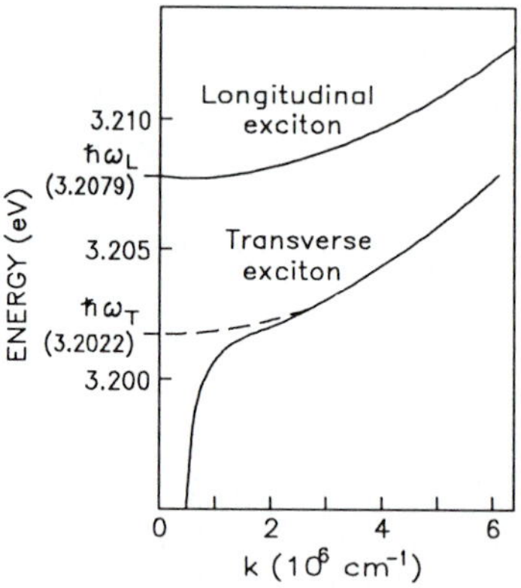

Figure 7.13. Polariton dispersion curves in CuCl (after Ref. 7.9).

7-4. Cadmium Sulfide - CdS

This yellow compound material, representative of the II-VI family, has been extensively studied for its excitonic properties (see Ref. 7.10). The crystal structure is wurzite, with the point-group symmetry notation C_{6v}. It is hexagonal with an optical c axis and inversion symmetry in the xy

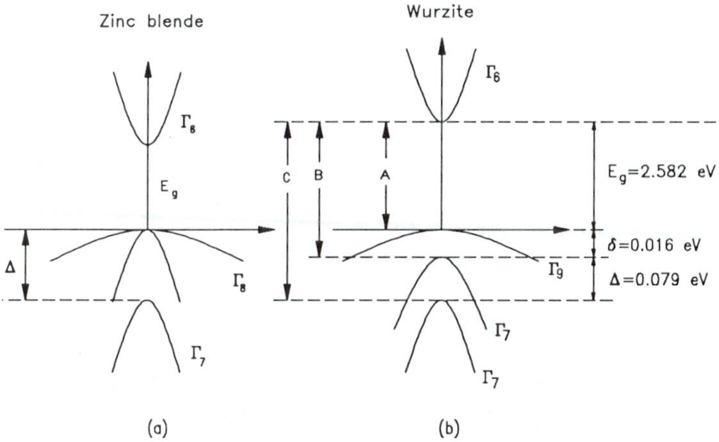

Figure 7.14. Schematic band structure near $\mathbf{k} = 0$ for (a) a zinc blende structure with usual spin-orbit coupling and (b) a Wurzite structure with additional splitting, δ, of the upper valence band due to the crystal field.

plane. This can be viewed as a trigonal distortion of the zinc blende structure along [111] direction. Assuming normal ordering of the spin-orbit split valence bands, Γ_8 and Γ_7, this trigonal distortion acts like a potential field, which is referred to as the crystal field. This lowering of symmetry removes the degeneracy of the Γ_8 band, leading to the band structure shown in Fig. 7.14. Transitions between the three valence bands to the conduction band Γ_6 in CdS are traditionally labeled A, B, and C.

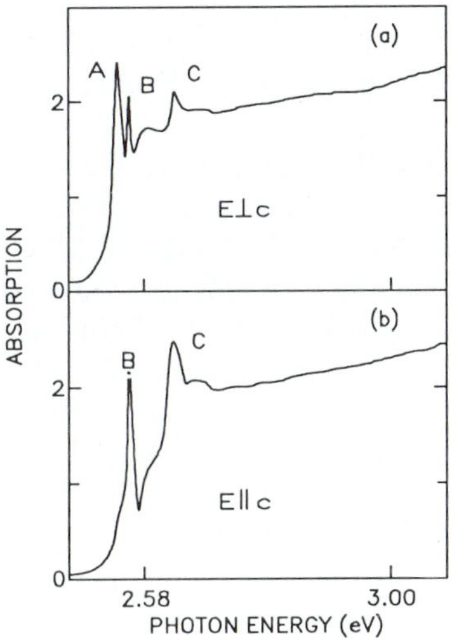

Figure 7.15. Excitonic absorption spectrum of CdS for (a) $\mathbf{E} \perp \mathbf{c}$ and (b) $\mathbf{E} \parallel \mathbf{c}$.

Excitons from the A, B, and C series are dipole-allowed, independent of the polarization direction in the xy plane. On the other hand, for light propagation perpendicular to the crystal optic axis, **k** $\perp$**c**, the crystal shows anisotropic behavior. If the incident electric field is polarized perpendicular to the crystal c axis, **E** $\perp$**c**, excitons of the A, B, and C series are allowed as before. However, for **E** || c, only s-like excitons of the B and C series are allowed. This anisotropic absorption is shown in Fig. 7.15. The exciton binding energy for the A and B series is 27.4 meV.

7-5. Silicon (Si) and Germanium (Ge)

These crystals, particularly Si, are at the heart of modern electronics, forming the constituent of most transistors and other elements of integrated circuits. They are totally opaque to visible light, giving them a metallic aspect. The optical properties near the absorption edge are determined by the indirect bandgap structure, making the onset of absorption smooth and rather weak in absolute terms. The optical properties of Si and Ge are also exploited in efficient infrared light detectors.

Si and Ge belong to the diamond structure (point group O_h with inversion center), which is similar to the zinc blende structure except that there is only one type of atom surrounded by four nearest neighbors forming a tetrahedron (see Fig. 7.1). Lattice parameters are a = 0.543 μm for Si and a = 0.3658 Å for Ge.

Silicon

The band structure of silicon is shown in Fig. 7.16. The valence band has its maximum at the Γ point of the Brillouin zone. It shows the normal ordering of bands with Γ_8^+, the fourfold degenerate valence band, lying above the split-off band Γ_7^+. However, the spin-orbit coupling is very weak, resulting in an energy splitting, Δ = 0.0441 eV. For this reason, the symmetry points of silicon are often labeled according to the single group representation, which ignores spin effects. As in the previous cases, the Γ_8 valence bands separate from k = 0 into a light- and heavy-hole band. The bands show warping; i.e., the surfaces of constant energy $\mathcal{E}_v(k)$ show deviation from spherical symmetry. The Luttinger parameters describing this anisotropy in k space are

$$j_1 = 4.285, \quad j_2 = 0.339, \quad j_3 = 1.446 , \tag{7.10}$$

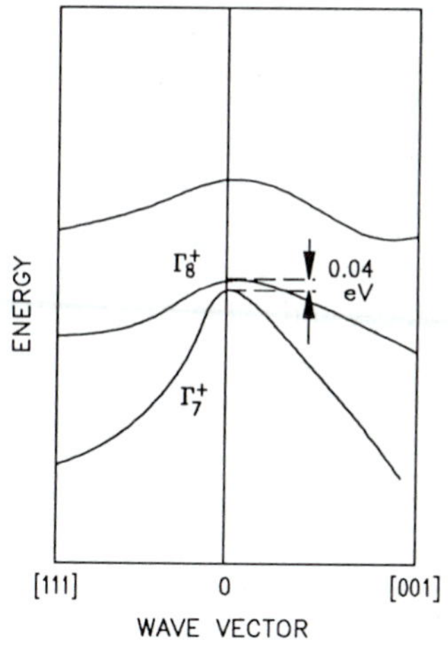

Figure 7.16. Schematic band structure of Si.

resulting in average effective hole masses

$$m_{hh} = 0.537\ m_0\ ,$$

$$m_{\ell h} = 0.153\ m_0\ ,$$

and

$$m_{so} = 0.234\ m_0. \tag{7.11}$$

The conduction band minima lie along the [100] axis with six equivalent minima, about 0.85 ($2\pi/a$) toward the symmetry point Δ. The surfaces of constant energy have the form of elongated ellipsoids pointing in the [100] directions. The effective electron mass values are

$$\text{longitudinal mass } m_{\|} = 0.1905\ m_0$$

and

$$\text{transverse mass } m_{\perp} = 0.9163\ m_0\ . \tag{7.12}$$

Germanium

Ge is also an indirect-gap material (see Fig. 7.17). The conduction band has four equivalent minima at the end points of the axis along [111] in the Brillouin zone. Here also the effective electron mass is strongly anisotropic. Surfaces of constant energy have the shapes of ellipsoids of revolution with the major axis pointing along the [111] direction. The effective electron masses along the [111] direction are

$$m_{e\perp} = 0.0815\ m_0$$

and

$$m_{e\|} = 1.588\ m_0\ . \tag{7.13}$$

The valence band is analogous to the case of Si, except for the larger spin-orbit splitting between bands Γ_8 and Γ_7, with $\Delta = 0.296$ eV. The Luttinger parameters j_1, j_2, and j_3 are

$$j_1 = 13.35\ ,$$

$$j_2 = 4.25\ ,$$

and

$$j_3 = 5.69\ , \tag{7.14}$$

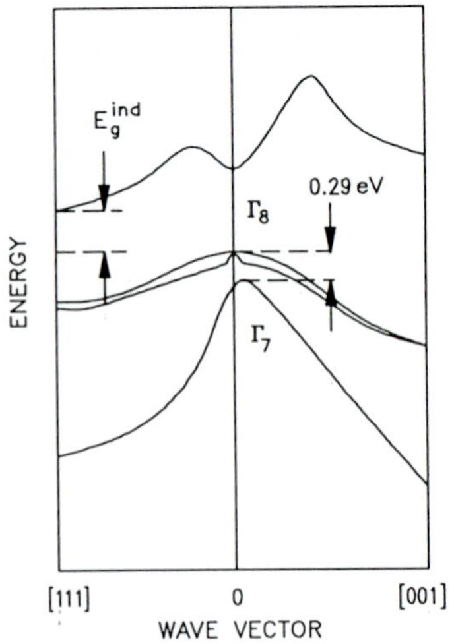

Figure 7.17. Schematic band structure of Ge.

leading to effective hole masses

$$m_{hh} = 0.284\ m_0\ ,$$

$$m_{\ell h} = 0.0438\ m_0\ ,$$

and

$$m_{so} = 0.095\ m_0\ . \tag{7.15}$$

The indirect bandgap energy is $E_g = 0.744$ eV at T = 2 K, and the exciton binding energy is $E_B = 4.15$ meV. The linear absorption spectrum of a 10-μm-thick sample of Ge at 30 K near the exciton resonance is shown in Fig. 7.18. The exciton resonance resulting from the direct transition between the upper valence band and the Γ minimum of the conduction band is clearly seen. The square root of the absorption spectra of Ge, in the vicinity of the indirect gap, recorded at different temperatures is shown in Fig. 7.19. The smooth, temperature-dependent absorption edge characteristic of indirect phonon-assisted transitions is seen as expected (see Sec. 5-4).

Crystals of germanium and silicon can be grown in ultrapure form with concentrations of defects less than 10^{13} cm^{-3}. Since the exciton radiative recombination process is an indirect transition (the reverse of absorption), and thus forbidden in first approximations, n = 1 exciton lifetimes are long, of the order of 10^{-5}s in ultrapure crystals. Exciton transport over macroscopic distances in such samples can be observed, as in the case of Cu_2O. Therefore, excitons are sensitive to nonradiative decay centers, even if these centers have low density. These are the reasons that Ge and Si are poor light emitters.

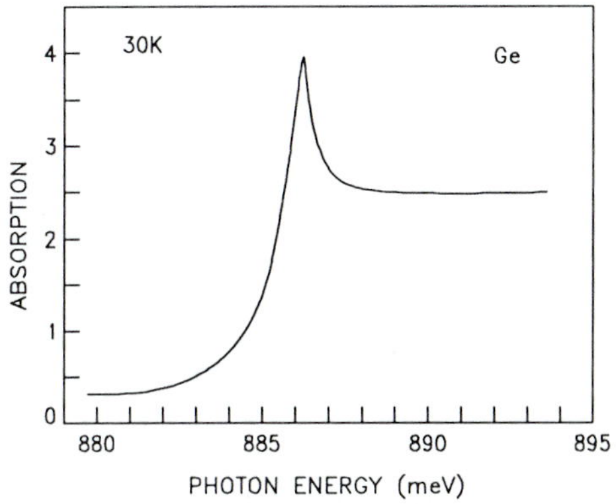

Figure 7.18. Absorption spectrum of Ge at 30 K showing the structure connected with the onset of direct transitions (after Ref. 7.11).

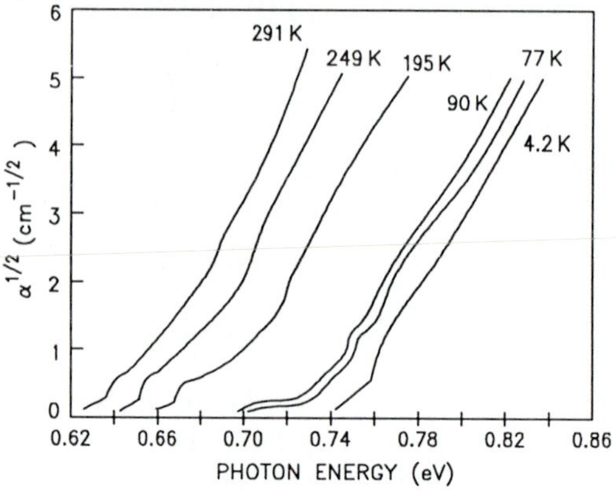

Figure 7.19. Absorption spectrum of Ge showing onset of indirect transitions at various temperatures (after Ref. 7.12).

7-6. References

1. F. Bassani and P. Parravicini, *Electronic States and Optical Transitions in Solids* (Pergamon Press, Oxford, 1975).

2. G. Bastard, *Wave Mechanics Applied to Semiconductor Heterostructures* (Les Editions de Physique, Paris, 1988).

3. C. Hermann, in *Optical Orientation*, edited by Meir and Zakharchevya (North-Holland, 1984).

4. G. W. Fehrenbach, W. Schaefer, J. Treusch, and R. G. Ulbrich, Phys. Rev. Lett. **49**, 1281 (1982).

5. Y. H. Lee, A. Chavez-Pirson, S. W. Koch, H. M. Gibbs, S. H. Park, J. Morhange, A. Jeffery, N. Peyghambarian, L. Banyai, A. C. Gossard, and W. Wiegmann, Phys. Rev. Lett. **57**, 2446 (1986).

6. M. Ueta, H. Kanzaki, K. Kobayashi, Y. Toyozawa, and E. Hanamura, *Excitonic Processes in Solids* (Springer-Verlag, Berlin, 1986).

7. S. Nikitine, in *Optical Properties of Solids*, edited by Nudelman and Mi (Plenum, New York, 1969), pp. 197-237.

8. Y. Kato, T. Goto, T. Fujii, and M. Ueta, J. Phys. Soc. Jpn. **36**, 175 (1974).

9. T. Mita, K. Sotome, and M. Ueta, Solid State Commun. **33**, 1135 (1980).

10. J. J. Hopfield and G. D. Thomas, Phys. Rev. **122**, 35 (1961).

11. H. Schweizer, A. Forchel, A. Hangleifer, S. Schmitt-Rink, J. P. Löwenau, and H. Haug, Phys. Rev. Lett. **51**, 698 (1983).

12. T. P. McLean and R. London, Phys. Chem. Sol. **13**, 1 (1960).

Chapter VIII

QUASI-TWO-DIMENSIONAL SEMICONDUCTORS: QUANTUM WELLS AND SUPERLATTICES

Modern growth techniques such as Molecular Beam Epitaxy (MBE) and Metal Organic Chemical Vapor Deposition (MOCVD) make it possible to manufacture ultrathin semiconductor structures of high quality. Ultrathin refers to a film thickness that is comparable to the exciton Bohr radius. Semiconductor films of thicknesses less than 100 Å are referred to as quantum-confined quasi-two-dimensional systems. One class of such artificially grown quasi-two-dimensional systems is the quantum-well structures. A *single quantum well* consists of an ultrathin layer ($d \leq 100$ Å) of a semiconductor sandwiched between thin layers of a larger gap semiconductor with matching lattice constant [see Figs. 8.1(a) and 8.2]. A much studied example described in this chapter is a thin layer of GaAs, with bandgap energy $E_g \sim 1.5$ eV, between layers of the ternary alloy

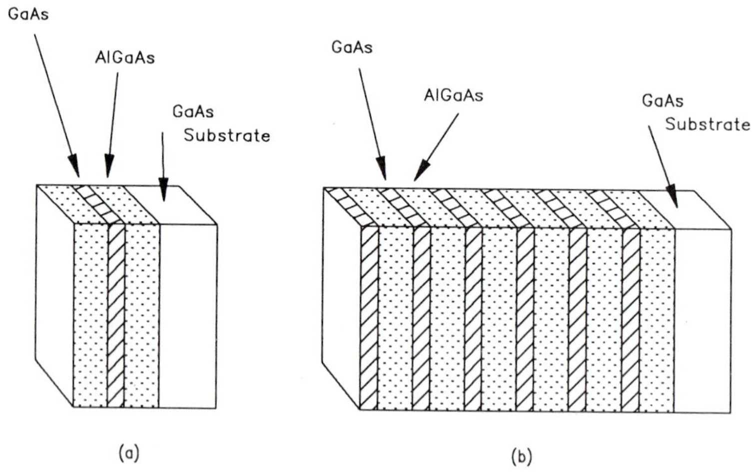

Figure 8.1. Schematics of (a) a single quantum well and (b) a multiple quantum well.

$Al_xGa_{1-x}As$, where x is the molar fraction of aluminum with $0 < x < 1$. Pure AlAs is an indirect-gap semiconductor, but for $x < 0.4$, $Al_xGa_{1-x}As$ has a direct gap of somewhat less than 2 eV, depending on the Al concentration. When the growth of GaAs-$Al_xGa_{1-x}As$ layers continues to include many layers of GaAs separated from each other by thin $Al_xGa_{1-x}As$ layers, a *multiple quantum well* (MQW) results. Typically, the structure may consist of 100 periods of 100-Å-thick-GaAs layers separated by AlGaAs layers of slightly larger width, say 150 Å [see Fig. 8.1(b)]. Ideally, multi quantum wells display the optical properties of a single quantum well, except that the optical density of the structure along the growth axis z is multiplied by N, the total number of periods.

Another important class of thin film structures are superlattices, where both the GaAs and AlGaAs sizes are ultrathin, for instance, 100 periods of 50-Å-thick-GaAs wells between 25-Å-thick-AlGaAs barriers. In superlattices the electronic wave functions from neighboring wells overlap to make the properties of the structure differ from those of multiple quantum wells. The name superlattice expresses the fact that a new spatial periodicity is impressed upon the natural lattice periodicity of the constituent semiconductors along the growth axis z.

The availability of quantum wells and superlattices with high structural quality is of considerable importance. There is a fundamental interest in the understanding of semiconductor systems with reduced dimensionality. The recently discovered quantum Hall effect is one example of the new effects that can be observed. These structures also offer the possibility, to some extent, of tailoring the optical properties according to the needs. For

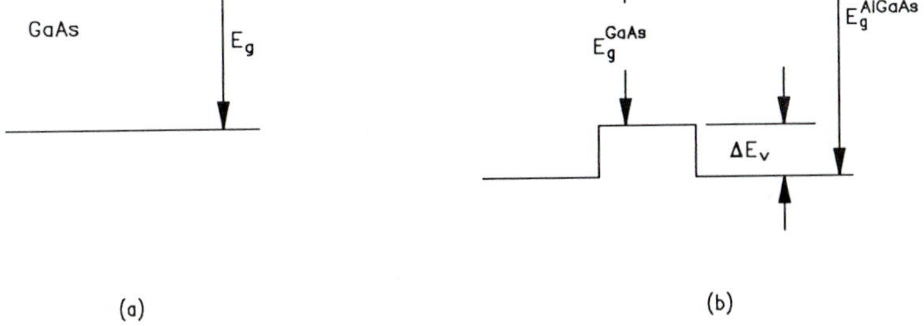

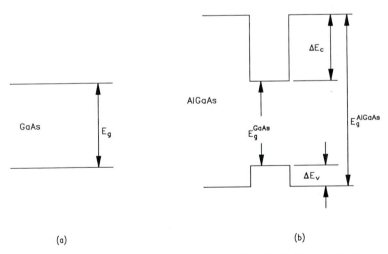

Figure 8.2. Band structure in real space for (a) a bulk GaAs and (b) a single quantum well of GaAs sandwiched between two AlGaAs barriers.

instance, it is possible to shift the onset of absorption of GaAs MQWs from infrared (λ = 8200 nm) to red ($\lambda \sim$ 6500 Å) simply by changing the width of the wells. Some of the many applications of quantum wells or superlattices are described in later chapters.

In this chapter we examine the optical properties of quantum wells and superlattices, with particular emphasis on GaAs-AlGaAs systems. We first consider an isolated single quantum well. To further simplify the problem, we model it by a free electron in a rectangular well, with barriers of infinite height. We then examine the more realistic case of finite barrier height. The effect of the Coulomb interaction on optical absorption in two-dimensional quantum wells will be described. Different types of quantum wells, including type-I, type-II, and n-i-p-i structures, will be discussed.

8-1. Electronic States for Infinite Potential Barriers

We consider a single quantum well consisting of a thin semiconductor film of thickness L_z. We model it as an electron confined in an infinitely deep potential well. The z axis is chosen to be along the direction perpendicular to the film. Thus, the electron motion is restricted in the z direction. We calculate the corresponding energy levels with the confinement potential V(z),

$$V(z) = \begin{cases} 0 & \text{for } -L_z/2 < z < L_z/2 \\ \infty & \text{for } |z| > L_z/2 \end{cases} , \tag{8.1}$$

shown schematically in Fig. 8.3. The carriers can move freely in the xy plane, and the corresponding energy bands are assumed to be parabolic and nondegenerate. We treat the problem in the effective mass approximation with the electron or hole having the effective masses m_e or m_h, respectively. The Schrödinger equation, in the absence of the electron-hole Coulomb interaction, may be written as

$$\left[-\frac{\hbar^2}{2m_i}\nabla_i^2 + V(z_i)\right]\psi_i(x_i, y_i, z_i) = \mathcal{E}_i\psi_i(x_i, y_i, z_i) , \tag{8.2}$$

where i = e (electron) or h (hole), m_i is the effective mass for the corresponding particle, and $\mathcal{E}_i$ and $\psi_i(x_i, y_i, z_i)$ are the energy eigenvalues and wave functions. For simplicity of notation, we drop the subscript i from now on, and Eq. (8.2) becomes

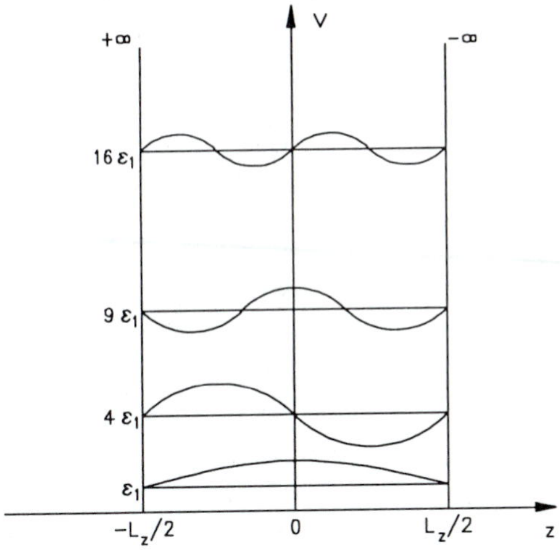

Figure 8.3. Schematic representation of an infinitely deep potential well. The lowest four single-particle energy levels and wave functions for a particle in this potential well are shown. The ground state is an even function, and the next higher energy state is an odd function, etc. The ground-state energy is given by $\mathcal{E}_1 = \pi^2\hbar^2/2m_i L_z^2$.

$$\left[-\frac{\hbar^2}{2m}\nabla^2 + V(z)\right]\psi(x,y,z) = \mathcal{E}\,\psi(x,y,z)\,. \tag{8.3}$$

The wave function may be separated into $\phi(x,y)$, describing the motion of the particle in the plane of the film, and $\varsigma(z)$, for the motion normal to the film,

$$\psi(x,y,z) = \phi(x,y)\,\varsigma(z)\,. \tag{8.4}$$

Knowing that

$$\nabla^2 = \frac{\partial^2}{\partial x^2} + \frac{\partial^2}{\partial y^2} + \frac{\partial^2}{\partial z^2} = \nabla_\perp^2 + \frac{\partial^2}{\partial z^2}\,, \tag{8.5}$$

where

$$\nabla_\perp^2 = \frac{\partial^2}{\partial x^2} + \frac{\partial^2}{\partial y^2}\,, \tag{8.6}$$

Eq. (8.3) becomes

$$\left[-\frac{\hbar^2}{2m}\left(\nabla_\perp^2 + \frac{\partial^2}{\partial z^2}\right) + V(z)\right]\phi(x,y)\zeta(z) = (\mathcal{E}_\perp + \mathcal{E}_z)\,\phi(x,y)\zeta(z)\,, \qquad (8.7)$$

where we let $\mathcal{E} = \mathcal{E}_\perp + \mathcal{E}_z$. Operating $\nabla_\perp^2$ on $\phi(x,y)$, and $\partial^2/\partial z^2$ on $\zeta(z)$, and dividing by $\phi(x,y)\,\zeta(z)$, we can separate Eq. (8.7) into two equations for the xy motion,

$$-\frac{\hbar^2}{2m}\nabla_\perp^2\,\phi(x,y) = \mathcal{E}_\perp\,\phi(x,y)\,, \qquad (8.8)$$

and for the z motion,

$$\left[-\frac{\hbar^2}{2m}\frac{\partial^2}{\partial z^2} + V(z)\right]\zeta(z) = \mathcal{E}_z\,\zeta(z)\,. \qquad (8.9)$$

Equation (8.8) has the well-known plane-wave solutions with the corresponding energy eigenvalues,

$$\mathcal{E}_\perp = \frac{\hbar^2}{2m}(k_x^2 + k_y^2)\,. \qquad (8.10)$$

Taking the zero of energy on the top of the valence band, the energy eigenvalues for an electron and a hole in the plane of the film are obtained from Eq. (8.10) as

$$\mathcal{E}_e(k) = E_g + \frac{\hbar^2}{2m_e}(k_x^2 + k_y^2) \qquad (8.11)$$

and

$$\mathcal{E}_h(k) = -\frac{\hbar^2}{2m_h}(k_x^2 + k_y^2)\,, \qquad (8.12)$$

showing that the energy in the plane of the film exhibits the usual parabolic dispersion due to the kinetic energy of the particle moving freely in the xy plane. However, the solution of Eq. (8.9) for the motion perpendicular to the film is distinctly different, resulting from the interface potential V(z), given by Eq. (8.1). Equation (8.9) describes the familiar particle-in-a-box problem in quantum mechanics with well-known solutions. For $-L_z/2 < z < L_z/2$, where $V(z) = 0$, Eq. (8.9) becomes

$$-\frac{\hbar^2}{2m}\frac{\partial^2}{\partial z^2}\zeta(z) = \mathcal{E}_z\,\zeta(z) \tag{8.13}$$

or

$$\frac{\partial^2\zeta(z)}{\partial z^2} + k_z^2\,\zeta(z) = 0\ , \tag{8.14}$$

where we defined

$$k_z^2 = \frac{2m\mathcal{E}_z}{\hbar^2}\ . \tag{8.15}$$

The solution to Eq. (8.14) is given by

$$\zeta(z) = A\sin k_z z + B\cos k_z z\ , \tag{8.16}$$

where A and B are constants. Because of the infinite potential well, the particle cannot move out of the well and the wave function outside the well vanishes. Since wave functions have to be continuous everywhere, $\zeta(z)$ must approach zero at the boundaries,

$$\zeta\left(z = \frac{L_z}{2}\right) = \zeta\left(z = -\frac{L_z}{2}\right) = 0\ . \tag{8.17}$$

Note that typically the derivative of the wave function $(1/m)\,\partial\zeta(z)/\partial z$ is also required to be continuous at the boundary for the continuity of the probability current to hold. However, this boundary condition is relaxed in this case because the derivative of the wave function is not defined outside the well, as a result of the infinite well depth.

Since the potential is symmetric, the wave functions also have a definite symmetry; i.e., they must be either even or odd. We consider both of these possibilities.

Even States

The wave function for even states satisfies the relation

$$\varsigma(z) = \varsigma(-z) \ . \tag{8.18}$$

Since sin $k_z z$ is an odd function, A = 0 in Eq. (8.18) and

$$\varsigma^+(z) = B \cos k_z z \ , \tag{8.19}$$

where we use the wave function superscript $^+$ to denote even parity. The normalization of the wave function requires

$$\langle \varsigma^+(z) | \varsigma^+(z) \rangle = 1, \tag{8.20}$$

giving

$$B = \sqrt{\frac{2}{L_z}} \ , \tag{8.21}$$

and Eq. (8.19) becomes

$$\varsigma^+(z) = \sqrt{\frac{2}{L_z}} \cos k^+{}_z z \ . \tag{8.22}$$

The boundary condition, Eq. (8.17), requires

$$\cos \frac{k_z^+ L_z}{2} = 0 \ , \tag{8.23}$$

which is satisfied if

$$\frac{k_z^+ L_z}{2} = (j_e - 1/2)\pi \qquad \text{for } j_e = 1, 2, 3, \ldots \tag{8.24}$$

or

$$k_z^+ = \frac{2\pi(j_e - 1/2)}{L_z} \ . \tag{8.25}$$

The even eigenfunctions and energies may now be obtained by substituting Eq. (8.25) into Eqs. (8.22) and (8.15),

$$\zeta^{+}(z) = \sqrt{\frac{2}{L_z}} \cos \frac{2\pi(j_e - 1/2)}{L_z} z \tag{8.26}$$

and

$$\mathcal{E}_z^{+} = \frac{4(j_e - 1/2)^2 \pi^2 \hbar^2}{2m\, L_z^2} \qquad j_e = 1, 2, 3, \ldots \tag{8.27}$$

Odd States

For odd wave functions, we require

$$\zeta(z) = -\zeta(-z) \; . \tag{8.28}$$

In this case we have to put B = 0 in Eq. (8.16). The resulting normalized wave function takes the form,

$$\zeta^{-}(z) = \sqrt{\frac{2}{L_z}} \sin k_z^{-} z \; . \tag{8.29}$$

The application of boundary condition leads to

$$\sin \frac{k_z^{-} L_z}{2} = 0 \tag{8.30}$$

or

$$\frac{k_z^{-} L_z}{2} = j_0 \pi \qquad \text{for } j_0 = 1, 2, 3, \ldots \tag{8.31}$$

or

$$k_z^{-} = \frac{2 j_0 \pi}{L_z} \; . \tag{8.32}$$

The odd wave functions and energies using Eq. (8.32) become

$$\zeta^-(z) = \sqrt{\frac{2}{L_z}} \sin \frac{2j_0\pi}{L_z} z \tag{8.33}$$

and

$$\mathcal{E}_z^- = \frac{4j_0^2\pi^2\hbar^2}{2mL_z^2} . \tag{8.34}$$

From Eqs. (8.27) and (8.34) it is clear that the lowest energy, or the ground-state energy, is given by $j_e = 1$ in Eq. (8.27)

$$\mathcal{E}_{1z}^+ = \frac{\pi^2\hbar^2}{2mL_z^2} , \tag{8.35}$$

with the accompanying wave functions

$$\zeta_1^+(z) = \sqrt{\frac{2}{L_z}} \cos \frac{\pi z}{L_z} . \tag{8.36}$$

The next higher energy state is the odd state, with $j_0 = 1$ and

$$\mathcal{E}_{2z}^- = 4 \frac{\pi^2\hbar^2}{2m\, L_z^2} , \tag{8.37}$$

with

$$\zeta_2^-(z) = \sqrt{\frac{2}{L_z}} \sin \frac{2\pi}{L_z} z . \tag{8.38}$$

The two next higher energy states are those with $j_e = 2$ in Eq. (8.27) and $J_0 = 2$ in Eq. (8.34), respectively,

$$\mathcal{E}_{3z}^+ = 9 \frac{\pi^2\hbar^2}{2m\, L_z^2} \tag{8.39}$$

and

$$\mathcal{E}^-_{4z} = 16\,\frac{\pi^2\hbar^2}{2m\,L_z^2}\,. \tag{8.40}$$

Thus, the ground state in the potential well is an even function, the next higher state is an odd function, etc., with alternating even and odd. The energy levels and the wave functions for the lowest four levels are shown in Fig. 8.3. Summing up the contributions in the xy plane and the z direction, the total energy of an electron or a hole becomes

$$\boxed{\mathcal{E} = \frac{\hbar^2}{2m}\left(\frac{j^2\pi^2}{L_z^2} + k_x^2 + k_y^2\right) \qquad j = 1, 2, 3, \ldots ,}$$

energy eigenvalues for an electron or hole confined in a potential well of infinite height (8.41)

where j is the subband quantum number. The term subband expresses the fact that even if the electron wave function has a discrete solution along z, there is still a quasi continuum of solutions in the xy plane, forming a subband (this will be shown more clearly in Fig. 8.10). We also note that the lowest electron conduction subband $j = 1$ is shifted to higher energies with respect to the bottom of the three-dimensional conduction band by $\Delta\mathcal{E} = \pi^2\hbar^2/2mL_z^2$ in Fig. 8.3. This reflects the confinement of the particle in the two-dimensional well.

For the case that is treated here, where the bands are parabolic and nondegenerate, the expressions for energy eigenvalues and wave functions are identical for electrons and holes, except for the different effective masses. Energy splitting between successive hole-energy levels (subbands) are generally smaller because of the larger hole effective mass. Figure 8.4 schematically displays the single-particle energy levels and wave functions for the electrons and the holes. There are an infinite number of bound states, as j can take all positive integer values.

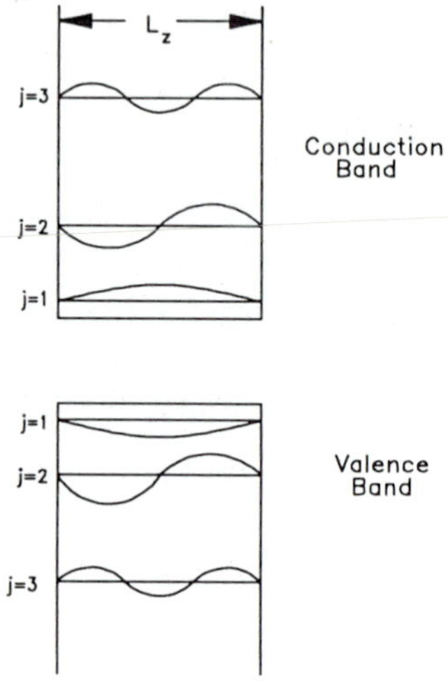

Figure 8.4. The single-particle energies for electrons and holes in an infinite potential well.

8-2. Electronic States for Finite Potential Barriers

We now consider a semiconductor heterostructure consisting of a single layer of material A sandwiched between two layers of material B, which has a larger bandgap energy than that of material A. An example of such a system may be the one given before where material A is GaAs with a bandgap energy of ~ 1.5 eV, and material B is $Al_xGa_{1-x}As$ with $E_g \simeq 2$ eV. The difference between the bandgap energies of GaAs and $Al_xGa_{1-x}As$ of ~ 0.5 eV provides a finite potential well, confining the excited electron-hole pairs in the GaAs layer.

To treat this problem we let the potential well have a depth of V_0, and for simplicity, we again assume that the bands are parabolic and nondegenerate. This is not a good assumption for the GaAs-$Al_xGa_{1-x}As$ case because the top valence band in GaAs is fourfold degenerate (see Chap. VII), making the treatment more complex. However, the approach employed here retains most of the physics and keeps the algebra to a minimum.

The square well potential of finite depth is displayed in Fig. 8.5. This potential is defined by

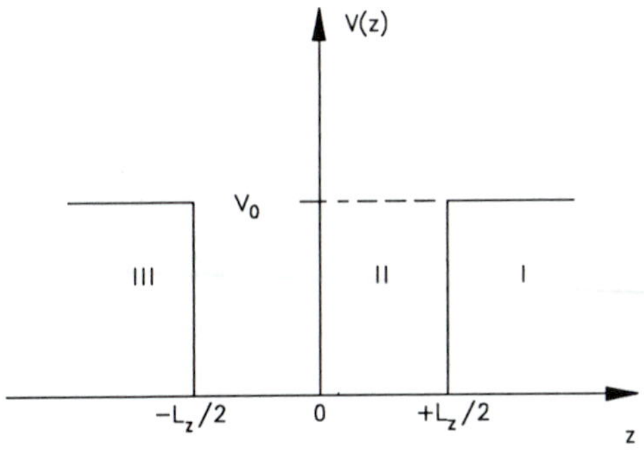

Figure 8.5. A square well potential of width L_z and depth V_0.

$$V(z) = \begin{cases} 0 & -L_z/2 < z < L_z/2 \\ V_0 & |z| > L_z/2 \end{cases} . \tag{8.42}$$

The Schrödinger equation describing the motion of the particles in the plane of the layer does not change compared with the infinite potential well case treated in Sec. 8-1 with the energy eigenvalues given by Eq. (8.10). However, the motion normal to the confining layer is changed. This motion is represented by a Schrödinger equation similar to Eq. (8.9),

$$-\frac{\hbar^2}{2m}\frac{\partial^2\zeta(z)}{\partial z^2} + V(z)\zeta(z) = \mathcal{E}_z\zeta(z) , \tag{8.43}$$

where we again drop the subscript i = e,h and keep in mind that the solutions are applicable to both electrons and holes as long as the proper effective mass is chosen. For the solution, we now have to consider the regions I, II, and III shown in Fig. 8.5. In region II, V(z) vanishes and Eq. (8.43) reduces to

$$\frac{\partial^2\zeta(z)}{\partial z^2} + \frac{2m\mathcal{E}_z}{\hbar^2}\zeta(z) = 0 . \tag{8.44}$$

The eigenfunction of this equation is

$$\zeta(z) = A \sin k_z z + B \cos k_z z \qquad \text{in region II} , \tag{8.45}$$

where

$$k_z^2 = \frac{2m\mathcal{E}_z}{\hbar^2} . \tag{8.46}$$

In regions I and III, $V(z) = V_0$ and Eq. (8.43) becomes

$$\frac{\partial^2 \zeta(z)}{\partial z^2} - \frac{2m}{\hbar^2}(V_0 - \mathcal{E}_z)\,\zeta(z) = 0 \tag{8.47}$$

or

$$\frac{\partial^2 \zeta(z)}{\partial z^2} - K_z^2\,\zeta(z) = 0 , \tag{8.48}$$

with

$$K_z^2 = \frac{2m}{\hbar^2}(V_0 - \mathcal{E}_z) . \tag{8.49}$$

We are mainly interested in the states that are bound inside the quantum well. The corresponding energy, $\mathcal{E}_z < V_0$, and K_z^2 is positive in Eq. (8.48). The eigenfunction of Eq. (8.47) is an exponential of the form $\exp(\pm K_z z)$. Since the wave functions have to be normalized, we require them to be finite for $|z| \to \infty$. Therefore, the solutions in regions I and III are

$$\zeta(z) = C\,e^{-K_z z} \qquad \text{in region I} \tag{8.50}$$

and

$$\zeta(z) = D\,e^{K_z z} \qquad \text{in region III} . \tag{8.51}$$

Again we make use of the symmetry of the wave functions because the potential is symmetric, and consider the even and odd states.

Even States

The condition $\zeta(z) = \zeta(-z)$ requires $D = C$ and $A = 0$ i.e.,

$$\zeta^+(z) = \begin{cases} B\cos k_z^+ z & -L_z/2 < z < L_z/2 \\ C\,e^{-K_z^+ z} & z > L_z/2 \\ C\,e^{K_z^+ z} & z < -L_z/2 . \end{cases} \tag{8.52}$$

Furthermore, the continuity of the wave function and its derivative at the boundary lead to

$$B \cos \frac{k_z^+ L_z}{2} = C\, e^{-K_z^+ L_z/2} \tag{8.53}$$

and

$$-k_z^+ B \sin \frac{k_z^+ L_z}{2} = -K_z^+ C\, e^{-K_z^+ L_z/2} \ . \tag{8.54}$$

Dividing Eq. (8.54) by Eq. (8.53) gives

$$k_z^+ \tan \frac{k_z^+ L_z}{2} = K_z^+ \ , \tag{8.55}$$

and substituting for k_z and K_z from Eqs. (8.46) and (8.49), we get

$$\boxed{\sqrt{\mathcal{E}_z^+}\ \tan\left(\sqrt{\frac{m\mathcal{E}_z^+}{2\hbar^2}}\ L_z\right) = \sqrt{V_0 - \mathcal{E}_z^+}\ .}$$

transcendental equation for energy eigenvalues for even states (8.56)

The solution of this equation gives the energy eigenvalues $\mathcal{E}_z^+$ for the even states. However, the roots of this transcendental equation cannot be determined algebraically; therefore, we rely on a graphical solution. For this purpose we plot both sides of Eq. (8.56) as a function of $\mathcal{E}_z^+$ with the intersections of the two curves giving the solutions. Figure 8.6 shows such a procedure, where the dashed-dotted curve shows the plot of the right-hand side of Eq. (8.56), while the full lines are the plot of the left-hand side, exhibiting the behavior of the multivalued tangent function. For the particular V_0 chosen, there are two intersections between the dashed and full lines, showing that two even bound states exist. It is evident from this figure that the number of intersections depends on the number of zeros of the left-hand side of Eq. (8.56). These zeros occur when

$$\mathcal{E}_z^+ = 0,\ \frac{2\hbar^2\pi^2}{mL_z^2}\ ,\ \frac{8\hbar^2\pi^2}{mL_z^2}\ , \ldots \tag{8.57}$$

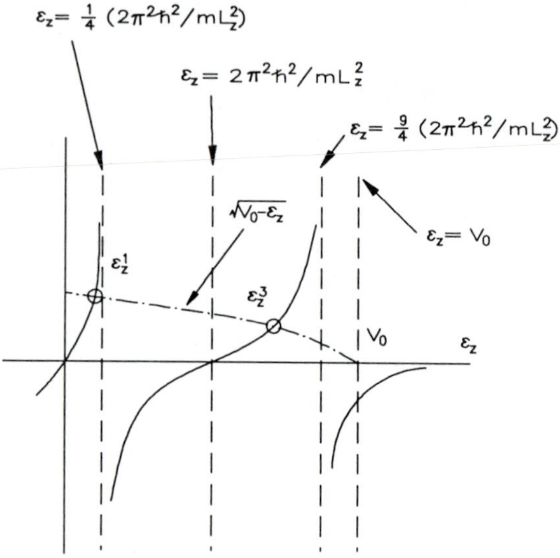

Figure 8.6. Graphical solution of Eq. (8.56) giving the allowed even bound-state energies. The dotted-dashed curve is the right-hand side, and the solid curves represent the left-hand side of Eq. (8.56).

The lowest bound-state energy is labeled $\mathcal{E}_z^1$, and the next higher energy is labeled $\mathcal{E}_z^3$, etc. Figure 8.6. shows that the number of bound states also depends on V_0. If V_0 is decreased below $2\pi^2\hbar^2/mL_z^2$, only one bound state remains, whereas an increase of V_0 leads to the occurrence of more and more bound states. As long as V_0 is positive, there is always at least one bound even state.

Odd States

For this case, $\varsigma(z) = -\varsigma(-z)$. It is left as an exercise (see Prob. 8.1) to show that the wave functions in regions I, II, and III become

$$\varsigma^-(z) = \begin{cases} A \sin k_z^- z & \text{for } -L_z/2 < z < L_z/2 \\ Ce^{-K_z^- z} & \text{for } z > L_z/2 \\ -Ce^{K_z^- z} & \text{for } z < -L_z/2 \,. \end{cases} \tag{8.58}$$

The transcendental equation for the energy eigenvalues in this case becomes

$$-\sqrt{\mathcal{E}_z^-}\ \mathrm{ctn}\sqrt{\frac{m\mathcal{E}_z^-}{2\hbar^2}}\ L_z = \sqrt{V_0 - \mathcal{E}_z^-}\ .$$

transcendental equation for energy eigenvalues for odd states (8.59)

This equation is plotted in Fig. 8.7 for the same value of V_0 as in Fig. 8.6. Again, there are two allowed bound states which are labeled $\mathcal{E}_z^2$ and $\mathcal{E}_z^4$. The comparison of Figs. 8.6 and 8.7 reveals that $\mathcal{E}_z^1 < \mathcal{E}_z^2 < \mathcal{E}_z^3 < \mathcal{E}_z^4$. The number of odd bound states also depends on V_0, and there are no odd bound states for $V_0 < \pi^2\hbar^2/2mL_z^2$. Therefore, as V_0 is increased from zero, initially there is only one bound state that has even parity. The next bound state is odd, then a second even state becomes allowed, and so on.

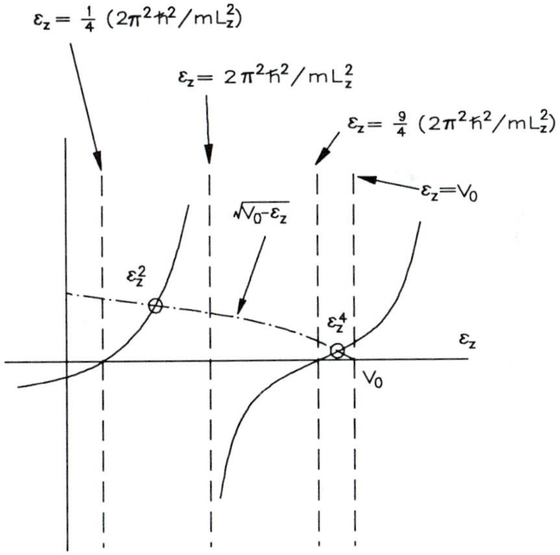

Figure 8.7. Graphical solution of Eq. (8.59) giving the allowed odd bound-state energies. The dashed-dotted curve is the right-hand side and the solid curves represent the left-hand side of Eq. (8.59).

Figure 8.8 shows the wave functions for a potential V_0 that allows four bound states. As can be seen from Eqs. (8.52) and (8.58), the wave functions are oscillatory within the well (region II) but penetrate through the barrier into regions I and III where they decay exponentially. From Fig. 8.8 one can see that as V_0 is lowered, $\mathcal{E}_z^4$ becomes unbound first and merges with the continuum. Next the $\mathcal{E}_z^3$ state is ionized, and so on. Equations (8.27) and (8.34) show that the quantized bound-state energies vary with L_z like $1/L_z^2$. Thus, as L_z is reduced, the bound-state energies shift to higher values. This phenomenon of blue shift due to quantum confinement is called *quantum size effect.*

The expressions derived in this section under the assumption of nondegenerate bands apply to both electrons and holes, as long as the proper effective masses are taken into account. Figure 8.9 shows the first two bound-state energies and wave functions for electrons and holes in the conduction and valence bands. The penetration of the wave functions into the barrier layers is referred to as the tunneling of the carriers. We will see later that such a tunneling is at the origin of the coupling between adjacent wells in superlattices.

In summary, the single-particle wave functions are plane waves with parabolic energy bands for the motion in the plane of the quantum-well layers. The motion normal to the layers is quantized with wave functions that are sinusoidal in the layer and exponentially decaying outside. They are characterized by a confinement quantum number j and a confinement energy scaling, like j^2 and $1/L_z^2$.

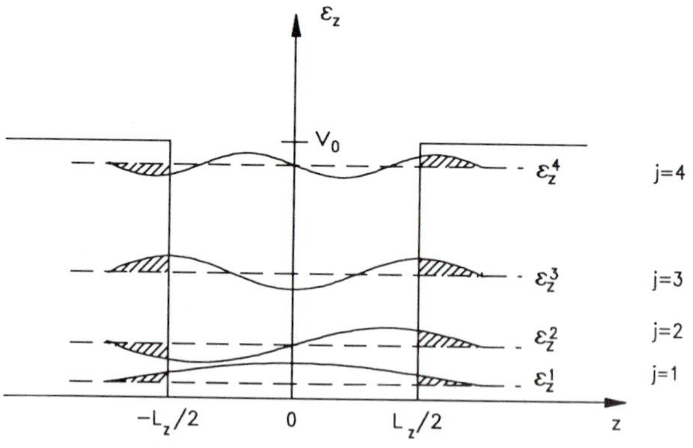

Figure 8.8. Even and odd bound-state energies and wave functions for a single particle in a square potential well of finite depth V_0.

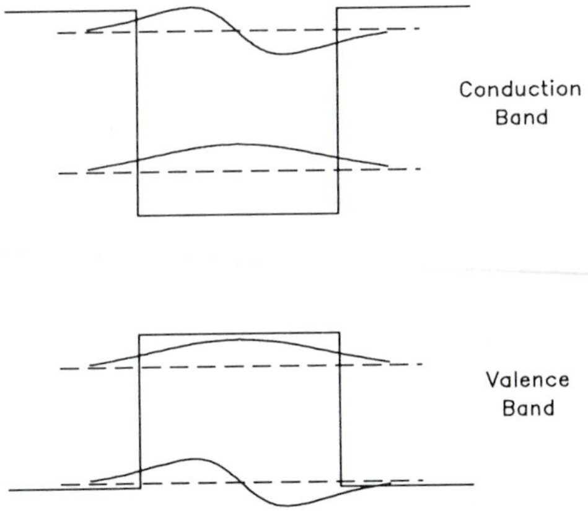

Figure 8.9. The bound-state energy levels for a single electron in the conduction band and a single hole in the valence band.

8-3. Density of States in Two Dimensions

It was shown in Secs. 8-1 and 8-2 how the motion of the particles confined in a two-dimensional semiconductor is quantized along the z direction normal to the layer. The quantized wave vectors k_z are integer multiples of π/L_z. We now consider the density of state behavior and optical absorption in such a system.

The single-particle density of states in two dimensions, which we denote here as $g^{2d}(\mathcal{E})$, gives the number of states per unit energy, per unit area in the plane of the layer. Similar to our procedure in Chap. II, we can write

$$g^{2d}(\mathcal{E})\, d\mathcal{E} = g^{2d}(k_\perp)\, d^2k_\perp, \tag{8.60}$$

where $k_\perp$ is the wave vector in the plane of the layer. The density of states in momentum space, $g^{2D}(k_\perp)$, is written as

$$g^{2d}(k_\perp) = \left(\frac{1}{2\pi}\right)^2 . \tag{8.61}$$

Also, for integrands which depend only on the magnitude of $k_\perp$

$$d^2k_\perp = 2\pi k_\perp dk_\perp. \tag{8.62}$$

Substituting Eqs. (8.61) and (8.62) into Eq. (8.60) and multiplying by a factor of 2 to take into account the spin degeneracy, we get

$$g^{2d}(\mathcal{E}) = \frac{2}{(2\pi)^2}\, 2\pi k_\perp \frac{dk_\perp}{d\mathcal{E}} . \tag{8.63}$$

The total energy of the free particle in an infinite potential well is

$$\mathcal{E} = \frac{\hbar^2}{2m}(k_x^2 + k_y^2 + k_z^2) = \frac{\hbar^2}{2m}(k_x^2 + k_y^2) + \frac{\hbar^2}{2m}\left(\frac{j\pi}{L_z}\right)^2 \tag{8.64}$$

or

$$\mathcal{E} = \frac{\hbar^2}{2m}k_\perp^2 + \frac{\hbar^2}{2m}\left(\frac{j\pi}{L_z}\right)^2 , \tag{8.65}$$

where $j = 1, 2, 3, \ldots$ and m is the effective mass of the particle ($m = m_e$ for the electron, and $m = m_h$ for the hole). Equation (8.65) leads to

$$d\mathcal{E} = \frac{\hbar^2 k_\perp}{m} dk_\perp . \tag{8.66}$$

Substitution into Eq. (8.63) gives

$$g^{2d}(\mathcal{E}) = \frac{m}{\pi\hbar^2} . \tag{8.67}$$

More generally, this equation may be written as

$$g^{2d}(\mathcal{E}) = \frac{m}{\pi\hbar^2}\,\theta(\mathcal{E} - \mathcal{E}_j) , \tag{8.68}$$

where the usual step function $\theta(\mathcal{E} - \mathcal{E}_j)$ has been included to emphasize the point that the states start from $\mathcal{E}_j$. The step function is equal to 1 for $\mathcal{E} > \mathcal{E}_j$ and zero for $\mathcal{E} < \mathcal{E}_j$. Equation (8.68) specifies that the two-dimensional density of states is a constant, independent of energy, in contrast to the square-root energy dependence of the three-dimensional density of states given by Eq. (2.89). Thus, the two-dimensional parabolic band of the form $\mathcal{E}_\perp = \hbar^2 k_\perp^2/2m$ leads to a constant density of states given by Eq. (8.68). Since the total energy of the particle in an infinite potential well, as given by Eq. (8.65), varies with the subband quantum number j, for each j value we have a different parabolic energy band in $k_\perp$ direction. Consequently, the energy levels in the $k_\perp$ direction consist of a series of parabolic bands separated by $\hbar^2 j^2 \pi^2/2mL_z^2$, as shown in Fig. 8.10(a). Each of these two-dimensional parabolic bands brings about a constant density of states [see Fig. 8.10(b)]. Hence, the total single-particle density of states at an energy $\mathcal{E}$ is equal to Eq. (8.68); i.e.,

$$\boxed{g^{2d}(\mathcal{E}) = \frac{m}{\pi\hbar^2} \sum_j \theta(\mathcal{E} - \mathcal{E}_j)} \qquad j = 1, 2, 3, \dots, \tag{8.69}$$

density of states in a two-dimensional system

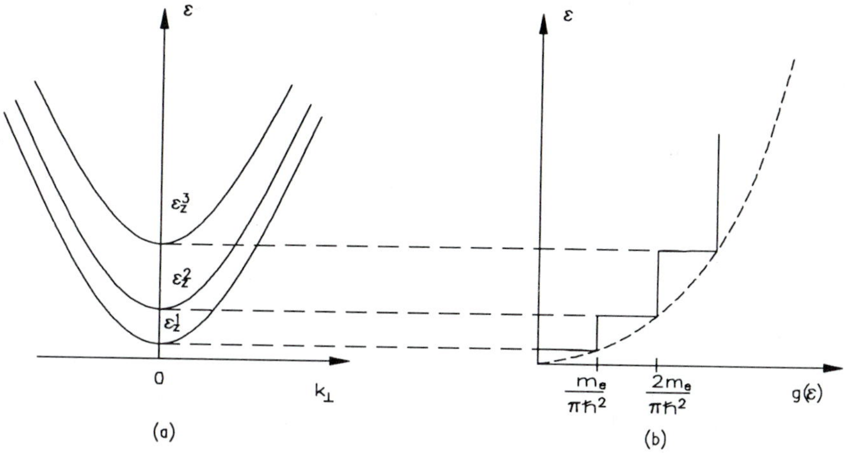

Figure 8.10. (a) Two-dimensional parabolic bands in $k_\perp$ direction. The bands correspond to the discrete $\mathcal{E}_z$ values. (b) The single-particle density of states (full line) for the bands of part (a). The dashed curve is the three-dimensional single-particle density of states expressed in two-dimensional units, $g^{(3)}(\mathcal{E})L_z$.

where $\mathcal{E}_j = (\hbar^2/2m)(j\pi/L_z)^2$. The total density of states, therefore, has a staircase shape, as shown in Fig. 8.10(b). For comparison, Fig. 8.10(b) also displays (dashed curve) the usual square-root three-dimensional density of states in two-dimensional units, $g^{(3)}(\mathcal{E})L_z$. Note that the steps occur at the quantized $\mathcal{E}_z$ energy levels, and the two-dimensional and the three-dimensional density of states (in two-dimensional units, i.e., $g^{(3)}(\mathcal{E})L_z$) coincide at each step (see Prob. 8.3).

Equation (8.69) describes the electron or hole density of states with m being the effective mass of the corresponding particle. Similar to the procedure we adopted in Chap. V, the joint density of states in two dimensions, which is relevant for optical transitions, may be obtained by replacing m with the reduced mass of the electron-hole pair, m_r. This can be verified by noting that

$$\mathcal{E}_e = E_g + \frac{\hbar^2}{2m_e} k_\perp^2 + \frac{\hbar^2}{2m_e}\left(\frac{i\pi}{L_z}\right)^2 \qquad i = 1, 2, 3, \ldots \tag{8.70}$$

and

$$\mathcal{E}_h = -\frac{\hbar^2}{2m_h} k_\perp^2 - \frac{\hbar^2}{2m_h}\left(\frac{j\pi}{L_z}\right)^2 \qquad j = 1, 2, 3, \ldots . \tag{8.71}$$

Thus,

$$\mathcal{E}_e - \mathcal{E}_h = E_g + \frac{\hbar^2 k_\perp^2}{2m_r} + \frac{\pi^2\hbar^2 i^2}{2m_e L_z^2} + \frac{\pi^2\hbar^2 j^2}{2m_h L_z^2}, \tag{8.72}$$

$$\frac{\partial(\mathcal{E}_e - \mathcal{E}_h)}{\partial k_\perp} = \frac{\hbar^2 k_\perp}{m_r}, \tag{8.73}$$

and

$$g^{2d}(\mathcal{E}) = \frac{2}{(2\pi)^2}\, 2\pi\, k_\perp \frac{m_r}{\hbar^2 k_\perp}, \tag{8.74}$$

or

$$g^{2d}(\mathcal{E}) = \frac{m_r}{\pi\hbar^2} . \tag{8.75}$$

The total joint density of states is then

$$\boxed{g^{2d}(\mathcal{E}) = \frac{m_r}{\pi\hbar^2} \sum_{i,j} \theta(\mathcal{E} - E_g - \mathcal{E}_i^e - \mathcal{E}_j^h) ,}$$

joint density of states in a two-dimensional system (8.76)

where $\mathcal{E}_j^{e,h} = (\hbar^2/2m_{e,h})(j\pi/L_z)^2$, and E_g is the three-dimensional bandgap energy. The major difference between the joint density of states (and also the single-particle density of states) in two dimensions and three dimensions may be realized by examining Fig. 8.11. The three-dimensional density of states is zero at the lowest energy level, E_g. However, the two-dimensional density of states has a finite value at the lowest two-dimensional energy, $\mathcal{E}_z^1$. This is an important consequence of the confinement in two dimensions, and results in finite optical absorption at the lowest energies, in contrast to the behavior of the three-dimensional free-carrier absorption that is zero at the lowest energy, E_g.

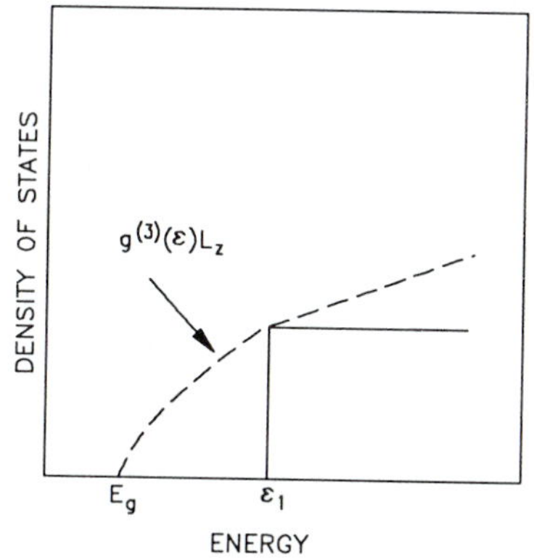

Figure 8.11. Total joint density of states in two dimensions (solid curve) and in three dimensions (dashed curve).

8-4. Excitons in Two Dimensions

The procedure that we have adopted in the previous chapters may be used to calculate the absorption spectra in two-dimensional semiconductors near the band edge. The total Hamiltonian for electron-hole pairs in a quantum well is

$$H = -\frac{\hbar^2}{2m_e}\nabla_e^2 - \frac{\hbar^2}{2m_h}\nabla_h^2 + V_{confinement} + V_{Coulomb}, \tag{8.77}$$

where the first two terms are the kinetic energies of the electron and the hole, and $V_{confinement}$ and $V_{Coulomb}$ are the confinement potential and the electron-hole Coulomb potential, respectively. The confinement potential is the sum of the confinement energies of the electrons, ΔE_c, and the hole, ΔE_v, as shown in Fig. 8.2. Substituting for the Coulomb and confinement terms and for ∇_e^2 and ∇_h^2 in terms of their x, y, and z coordinates, we get

$$H = -\frac{\hbar^2}{2m_e}\left(\frac{\partial^2}{\partial x_e^2} + \frac{\partial^2}{\partial y_e^2} + \frac{\partial^2}{\partial z_e^2}\right) - \frac{\hbar^2}{2m_h}\left(\frac{\partial^2}{\partial x_h^2} + \frac{\partial^2}{\partial y_h^2} + \frac{\partial^2}{\partial z_h^2}\right)$$

$$+ V_0(z_{e,h}) - \frac{e^2}{\epsilon_0|\mathbf{r}_e - \mathbf{r}_h|}, \tag{8.78}$$

where $V_0(z_{e,h}) = \Delta E_c + \Delta E_v = V_{confinement}$ and ϵ_0 is the static dielectric constant of the medium. We keep the $\partial^2/\partial z_e^2$ and $\partial^2/\partial z_h^2$ terms in Eq. (8.78) and transform the xy motion into the center of mass and the relative coordinates using the procedures adopted in Chap. VI. (Note that this procedure is only valid in the "strong confinement" limit where the electron-hole-pair Coulomb attraction is not too strong to perturb the motion in the z direction.) The electron-hole-pair Hamiltonian becomes

$$H = -\frac{\hbar^2}{2m_e}\frac{\partial^2}{\partial z_e^2} - \frac{\hbar^2}{2m_h}\frac{\partial^2}{\partial z_h^2} - \frac{\hbar^2}{2M_{xy}}\left(\frac{\partial^2}{\partial X^2} + \frac{\partial^2}{\partial Y^2}\right) - \frac{\hbar^2}{2m_{xy}}\left(\frac{\partial^2}{\partial x^2} + \frac{\partial^2}{\partial y^2}\right)$$

$$+ V_0(z_{e,h}) - \frac{e^2}{\epsilon_0 r}, \tag{8.79}$$

where M_{xy} and m_{xy} are the in-plane total and reduced masses of the electron-hole pair, as given by $M_{xy} = (m_e + m_h)_{xy}$,

$1/m_{xy} = (1/m_e + 1/m_h)_{xy}$, and $\mathbf{r} = \mathbf{r}_e - \mathbf{r}_h$. Furthermore, X and Y are the center-of-mass coordinates and x and y are the relative coordinates (see Chap. VI for definitions). The Schrödinger equation associated with this Hamiltonian needs to be solved, obtaining the energy eigenvalues and eigenfunctions. The Schrödinger equation may be written as

$$\left[-\frac{\hbar^2}{2m_e}\frac{\partial^2}{\partial z_e^2} - \frac{\hbar^2}{2m_h}\frac{\partial^2}{\partial z_h^2} - \frac{\hbar^2}{2M_{xy}}\left(\frac{\partial^2}{\partial X^2} + \frac{\partial^2}{\partial Y^2}\right)\right.$$

$$\left. - \frac{\hbar^2}{2m_{xy}}\left(\frac{\partial^2}{\partial x^2} + \frac{\partial^2}{\partial y^2}\right) + V_0(z_{e,h}) - \frac{e^2}{\epsilon_0 r}\right]\psi(\mathbf{r}) = \mathcal{E}\,\psi(\mathbf{r}) \,, \qquad (8.80)$$

where the wave function $\psi(\mathbf{r})$ can be separated into the z and the xy components as

$$\boxed{\psi(\mathbf{r}) = \Phi_n^{xy}(r_{xy})\,\zeta_{ei}(z_e)\,\zeta_{hj}(z_h)\,,}$$

excitonic wave function in a quantum well
with infinite barriers (8.81)

where $\Phi_n^{xy}(r_{xy}) = u_{c0}u_{v0}\phi_n^{xy}(r_{xy})e^{iK_c^{xy}\cdot R_{xy}}$. u_{c0} and u_{v0} are the Bloch functions, ϕ_n^{xy} is the exciton envelope function in the xy plane, and the exponential term expresses the center-of-mass motion of the exciton in the xy plane with $\mathbf{R} = (m_e\mathbf{x}_e + m_h\mathbf{x}_h)/(m_e + m_h)$. The subscripts i and j describe the subband quantum numbers for the quantized motion in the z direction given by Eqs. (8.70) and (8.71) for infinite potential wells and $r_{xy} = r_{xy}^e - r_{xy}^h$. The solution of the two-dimensional Schrödinger equation for the relative motion in the xy plane can be found in Ref. 2.5. It yields a series of bound excitonic lines at positions

$$\mathcal{E}_n^{2d} = E_g + \frac{\hbar^2\pi^2 j^2}{2m_r L_z^2} - \frac{E_B}{(n_j - 1/2)^2} \quad \text{where } n_j = 1, 2, 3, \ldots \,. \qquad (8.82)$$

Here E_B is the three-dimensional exciton Rydberg given by Eq. (6.18), $E_B = e^4 m_r/2\epsilon_0^2\hbar^2$. The second term is the confinement energy for the electron-hole pair with j = 1, 2, 3, ... The third term, resulting from the inclusion of the Coulomb interaction, is the two-dimensional exciton binding energy. Equation (8.82) states that each j value has its own excitonic series. We will get back to this point shortly. A comparison between the two-dimensional binding energy and the three-dimensional exciton binding energy, Eq. (6.18), where $E_B^{3d} = E_B/n^2$, reveals that the substitution $n \rightarrow n - 1/2$ transforms the three-dimensional results to two dimensions. We see from Eq. (8.82) that the lowest two-dimensional exciton energy associated with $n = 1$ has a magnitude four times larger than the three-dimensional exciton ground state; i.e.,

$$E_B^{2d} = \frac{E_B}{(n - 1/2)^2} , \tag{8.83}$$

and, therefore, for $n = 1$,

$$E_{B,n=1}^{2d} = 4E_B = 4E_{B,n=1}^{3d} . \tag{8.84}$$

Thus, the exciton ground state is farther away from the bandgap in two dimensions compared with three dimensions. In analogy with Eq. (6.19), we may introduce a two-dimensional exciton Bohr radius through the relation

$$E_B^{2d} = \frac{\hbar^2}{2m_r(a_B^{2d})^2} . \tag{8.85}$$

Using Eqs. (8.84) and (6.19) we see that

$$a_B^{2d} = \frac{a_B^{3d}}{2} ; \tag{8.86}$$

i.e., the two-dimensional Bohr radius is half as big as the three-dimensional value.

So far we have assumed implicitly that the dielectric constants in the well and the barriers are the same, $\epsilon_w = \epsilon_b = \epsilon_0$. This condition is rather well fulfilled in the GaAs-AlGaAs structures, for instance, where $\epsilon_w = 13.1$, $\epsilon_b = 11.6$ for $x \sim 0.5$. However, an interesting situation occurs for $\epsilon_b < \epsilon_w$, or more generally, $\epsilon_w > \epsilon_I, \epsilon_{III}$, where ϵ_I and ϵ_{III} are the dielectric constants of the regions I and III, as shown in Fig. 8.5. This may be the case, for example, for an ultrathin semiconductor film of thickness, $d < a_B$, on an insulating substrate, where a_B is the three-dimensional exciton Bohr radius. In that case, the exciton binding energy is strongly enhanced, because the Coulomb interaction between the electron and hole occurs primarily through media I (the dielectric) and media III (free space) with less effective screening. In the truly two-dimensional limit we get

$$E_B^{2d} = \left(\frac{2\epsilon_w}{\epsilon_I + \epsilon_{III}}\right)^2 \frac{E_B}{(n - 1/2)^2} .$$

two-dimensional exciton binding energy (8.87)

The increase of exciton binding energy due to this effect, which is referred to as the local field effect, can be considerable. For instance, if $\epsilon_I \simeq \epsilon_{III} = \epsilon_w/10$, it enhances the binding energy by two orders of magnitude.

8-5. Optical Absorption in Two Dimensions

The solution of the two-dimensional Schrödinger equation for the relative motion in the xy plane also gives the wave functions, which may be substituted in Elliott's formula, to calculate the optical susceptibility and, consequently, the optical absorption. For the case of first-class transitions (dipole-allowed interband transition at k = 0), the result of these calculations leads to

$$\alpha^{2d}(\omega) \sim |d_{cv}|^2 \sum_{i,j,n} |\langle \zeta_{ei}(z_e)|\zeta_{hj}(z_h)\rangle|^2 \, |\phi_n^{xy}(r_{xy} = 0)|^2 \, \delta(\hbar\omega - \mathcal{E}_n^{2d}) , \tag{8.88}$$

where $\phi_n^{xy}(r_{xy} = 0)$, $\zeta_{ei}(z_e)$, and $\zeta_{hj}(z_h)$ are defined in Eq. (8.81), and $\mathcal{E}_n^{2d}$ is given by Eq. (8.82). $\phi_n^{xy}(r_{xy} = 0)$ is similar to the two-dimensional hydrogen atom wave functions. For example, for n = 1, the two-dimensional exciton with s-like envelope function in the xy plane is of the form

$$\phi_{1s}^{xy}(r_{xy}) = \sqrt{\frac{8}{\pi a_B^2}} \, e^{-2r_{xy}/a_B} , \tag{8.89}$$

where $a_B = a_B^{3d} = \epsilon_0\hbar^2/m_r e^2$ is the three-dimensional exciton Bohr radius. For any n, we get

$$|\phi_{ns}^{xy}(r_{xy} = 0)|^2 = \frac{1}{8(n - 1/2)^2} |\phi_{1s}^{xy}(r_{xy} = 0)|^2 . \tag{8.90}$$

[Note that in three dimensions $|\phi_{ns}^{3d}(r = 0)|^2 = (1/n^3)|\phi_{1s}^{3d}(r = 0)|^2$.] Again, we should emphasize that Eq. (8.88) is valid when the thickness of the film is very small (e.g., for narrow quantum wells). In Eq. (8.88) the functions $\zeta_{ei}(z_e)$ and $\zeta_{hi}(z_h)$ are orthornormal functions and, consequently, the overlap integral becomes

$$\langle\zeta_{ei}(z_e)|\zeta_{hj}(z_h)\rangle = \int_{-\infty}^{\infty} dz\, \zeta_{ei}(z_e)\, \zeta_{hj}(z_h) = 0 \quad \text{for } i \neq j , \tag{8.91}$$

forcing $\alpha^{2d}(\omega) = 0$ for $i \neq j$ or $\Delta j \neq 0$. This is one of the absorption selection rules in two-dimensional systems stating that optical absorption occurs between quantized states in the valence and conduction bands only when $\Delta j = 0$. In other words, the allowed transitions are those that occur between the quantized electron and hole states with the same quantum number, j. Therefore, only the term $i = j$ in the joint density of states, Eq. (8.76), contributes to the optical absorption in two dimensions, which results in transitions such as $\mathcal{E}_{1z}^h \rightarrow \mathcal{E}_{1z}^e$, $\mathcal{E}_{2z}^h \rightarrow \mathcal{E}_{2z}^e$, $\mathcal{E}_{3z}^h \rightarrow \mathcal{E}_{3z}^e$ becoming allowed. The allowed transitions are shown in Fig. 8.12. The other selection rules pertain to the transition dipole moment between the states of the valence and conduction bands. They are similar to the three-dimensional case; i.e., the interband matrix element between the valence and conduction Bloch states should not vanish for the transition to take place; i.e., $p_{cv}(0) \neq 0$. Finally, the momentum conservation rule requires $\hbar K_{xy} = \hbar q \simeq 0$.

For $i = j$, Eq. (8.88) becomes

$$\alpha^{2d}(\omega) \sim \sum_{j,n} |\phi_n^{xy}(r_{xy} = 0)|^2\, \delta(\hbar\omega - \mathcal{E}_n^{2d}) . \tag{8.92}$$

Therefore, the free-carrier absorption, consisting of a series of steps corresponding to different j values, is modified by the Coulomb

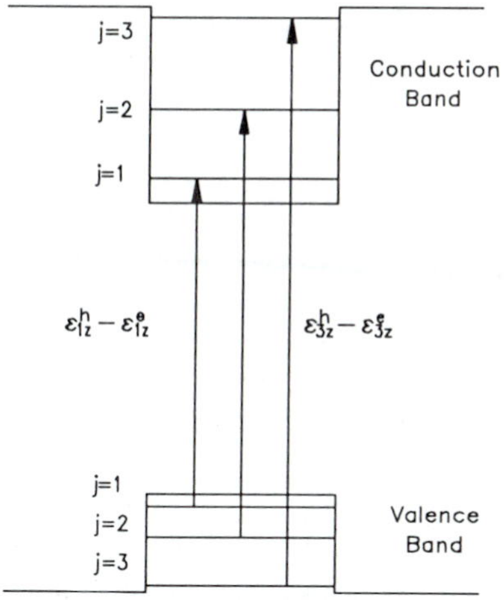

Figure 8.12. Possible electron-hole transitions in a quantum well. Allowed transitions are those with the same quantum numbers for the electron and the hole, such as $\mathcal{E}^h_{1z}$ to $\mathcal{E}^e_{1z}$, etc.

interaction. The inclusion of Coulombic effects causes each step to acquire its own excitonic series. Figure 8.13 schematically shows the accumulative two-dimensional absorption spectrum, and for comparison, the free-carrier step-like density of states. In Sec. 8-6 we give some examples of experimentally observed quasi-two-dimensional absorption spectra.

The relative separation between the electron and the hole in the exciton state in two dimensions is reduced as compared to three dimensions because of the confinement. Thus, the term $|\phi^{xy}_n\,(r_{xy} = 0)|^2$ in Eq. (8.92), which gives the probability that the electron and the hole are in the same unit cell, is larger in two dimensions. This results in stronger excitonic resonances (larger exciton oscillator strengths) that are accompanied by larger exciton binding energies, as we discussed earlier.

A detailed analysis of Coulombic effects on the two-dimensional optical absorption at each step may be done by examining Eq. (8.92) further. For a particular j, that is, for a particular step, Eq. (8.92) may be reduced to (see Ref. 2.5)

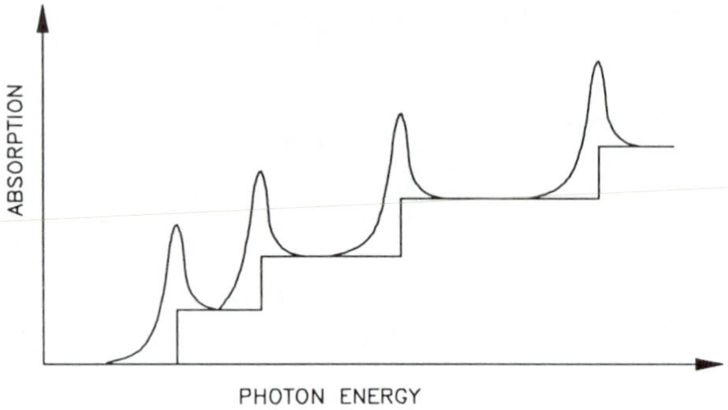

Figure 8.13. Schematic absorption spectra of a two-dimensional semiconductor. The staircase function is the free-carrier absorption. The spectrum with absorption peaks shows the influence of Coulomb effects where each step acquires its own excitonic series.

$$\alpha^{2d}(\omega) = \frac{2\pi\omega|d_{cv}|^2}{n_0 c L_z}\left(\frac{2m_r}{\hbar^2}\right)\left[\sum_{n=1}^{\infty}\frac{4E_B}{(n-1/2)^3}\,\delta\left(\hbar\omega - E_g^{2d} + \frac{E_B}{(n-1/2)^2}\right) + \theta(\hbar\omega - E_g^{2d})\frac{e^{\Delta}}{\cosh\Delta}\right],$$

two-dimensional absorption system (8.93)

where $\Delta = \pi/\sqrt{(\hbar\omega - E_g^{2d})/E_B}$ and E_g^{2d} is given by $E_g^{2d} = E_g + \hbar^2\pi^2 j^2/2m_r L_z^2$. Equation (8.93) is the two-dimensional Elliott formula. The first term in the bracket corresponds to the excitonic lines at positions $\hbar\omega = E_g + \hbar^2\pi^2 j^2/2m_r L_z^2 - E_B/(n - 1/2)^2$. The magnitude of the exciton absorption lines varies with n like $1/(n - 1/2)^3$. Thus, the strength of absorption for the exciton decreases more rapidly with n in two dimensions than with n in three dimensions. For instance, the ratio $\alpha_{2d}(n = 2/n = 1) = 1/27$, rather than $\alpha_{3d}(n = 2/n = 1) = 1/8$. The second term in the bracket gives the Coulomb enhanced continuum absorption in two dimensions. The corresponding two-dimensional Coulomb enhancement factor is now

$$C^{2d}(\omega) = \frac{e^{\Delta}}{\cosh \Delta} . \tag{8.94}$$

When $\hbar\omega \rightarrow E_g^{2d}$, $(\hbar\omega - E_g^{2d}) \rightarrow 0$, and $\Delta = \pi/\sqrt{(\hbar\omega - E_g^{2d})/E_B} \rightarrow \infty$, making $C^{2d}(\omega) = 2e^{\Delta}/(e^{\Delta} + e^{-\Delta}) = 2/(1 + e^{-2\Delta}) \rightarrow 2$. Thus, the absorption at the two-dimensional band edge is twice the free-carrier-continuum absorption, as a result of the Coulombic effects. For $\hbar\omega >> E_g^{2d}$, $\hbar\omega - E_g^{2d} \rightarrow \infty$, and $\Delta \rightarrow 0$ and $C^{2d}(\omega) \rightarrow 1$, indicating that for photon energies far above the two-dimensional band edge, the absorption approaches the free-carrier-continuum absorption. Figure 8.14 shows the calculated excitonic and the Coulomb enhanced continuum absorption by the solid lines. Only the n = 1 exciton ground state is displayed as a sharp δ function. The free-carrier absorption for two and three dimensions is also shown, for comparison, by the dashed curves. It can be seen that the Coulomb enhanced continuum absorption at E_g^{2d} is twice as large as the free-carrier absorption in three dimensions. The two-dimensional 1s exciton is better resolved spectrally as a result of its four-times-larger binding energy.

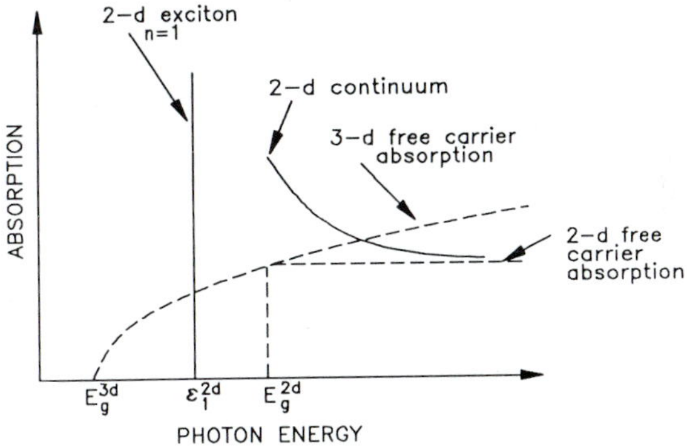

Figure 8.14. Schematic comparison of the free-carrier and excitonic two-dimensional absorption spectra. The free-carrier spectra are shown by the dashed curves, while the excitonic effects are included in the full curves.

8-6. GaAs-AlGaAs Multiple Quantum Wells

The heterostructures that have been the most extensively studied so far are the artificially prepared GaAs-$Al_xGa_{1-x}As$ MQWs. The multilayer structure consists of layers of GaAs sandwiched between layers of AlGaAs, grown on a [100] plane of a GaAs substrate. The concentration x of Al is typically between 0.2 to 0.3. For simplicity, we drop the subscript x in the following discussions. The thicknesses of alternating GaAs and AlGaAs are controlled to a precision of a few tens to a few hundreds of angstroms. The GaAs layers form the "wells" with thickness L_z where the excited carriers are confined, and the AlGaAs layers are the "barriers" with thickness L_b. For MQWs, the barrier thickness is wide enough to prevent penetration of the wave functions from neighboring wells.

GaAs and AlGaAs are almost perfectly lattice matched, meaning that the lattice constant in both materials is nearly the same. They are both direct-gap semiconductors with zinc blende structures and band extrema at $k = 0$. The bandgap of AlGaAs is larger than that of GaAs by about 0.5 eV. This band off-set is distributed between the valence band and conduction band with a larger discontinuity at the conduction band, $\Delta E_c > \Delta E_v$ (see Fig. 8.2). The ratio of $\Delta E_c/\Delta E_v$ is not precisely known. Various measurements have resulted in different values for this ratio, ranging from $\Delta E_c/(\Delta E_v + \Delta E_c) \sim 0.6$ to 0.85.

Quantum wells are often referred to as quasi-two-dimensional semiconductors because they retain some properties of a three-dimensional system. For instance, a limited number of quantized states exist for the motion in z direction. The conduction band electron energy levels and eigenfunctions are similar to those we derived in Sec. 8-2. The hole energies and wave functions in the valence band are more complicated, and the procedures of Sec. 8-2 do not directly apply for them. This is mainly due to the fact that the valence band at $k = 0$ is degenerate. As discussed in Chap. VII, the bulk GaAs valence bands around $k = 0$ are described by basis functions with total angular momentum $J = 3/2$ for the upper two bands, which are referred to as the heavy- and light-hole bands, and $J = 1/2$ for the lower valence band, known as the spin-orbit split-off band, as shown in Fig. 8.15. In bulk GaAs the $J = 3/2$ upper valence bands are fourfold degenerate at $k = 0$, since the projection of angular momentum along the z axis, which is taken as the quantization axis, can assume values $J_z = \pm 1/2$ and $\pm 3/2$. The hole mass is no longer single valued, but takes different values for the different bands. The hole dispersion near the center of the Brillouin zone becomes

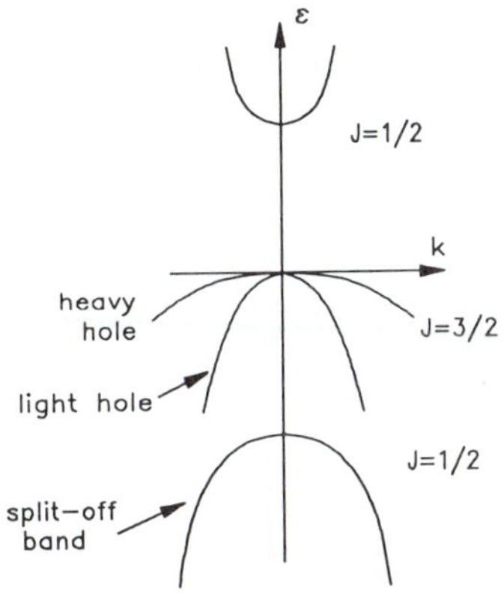

Figure 8.15. Schematic band structure in the vicinity of k = 0 for bulk GaAs. The lower-energy valence bands with J = 3/2 are called heavy-hole and light-hole bands, while the higher-energy valence band with J = 1/2 is referred to as the split-off band.

$$\mathcal{E}_{hh} = -\frac{\hbar^2 k_z^2}{2\left(\dfrac{m_0}{\gamma_1 - 2\gamma_2}\right)} - \frac{\hbar^2 k_\perp^2}{2\left(\dfrac{m_0}{\gamma_1 + \gamma_2}\right)} \qquad \text{for } J_z = \pm\frac{3}{2} \tag{8.95}$$

and

$$\mathcal{E}_{\ell h} = \frac{\hbar^2 k_z^2}{2\left(\dfrac{m_0}{\gamma_1 + 2\gamma_2}\right)} - \frac{\hbar^2 k_\perp^2}{2\left(\dfrac{m_0}{\gamma_1 - \gamma_2}\right)} \qquad \text{for } J_z = \pm\frac{1}{2}\,, \tag{8.96}$$

where m_0 is the free-electron mass and γ_1 and γ_2 are the Luttinger parameters that we introduced in Chap. VII, which in GaAs have the values $\gamma_1 \simeq 6.9$ and $\gamma_2 \simeq 2.4$. The hole masses, $m_{z,hh} = m_0/(\gamma_1 - 2\gamma_2)$, $m_{\perp,hh} = m_0/(\gamma_1 + \gamma_2)$, $m_{z,\ell h} = m_0/(\gamma_1 + 2\gamma_2)$, and $m_{\perp,\ell h} = m_0/(\gamma_1 - \gamma_2)$, are referred to as the *heavy-hole* and the *light-hole* mass, respectively, noting that $m_{hh} > m_{\ell h}$. These two masses are degenerate in bulk GaAs. However, this degeneracy at the zone center is lifted due to the quantum confinement. Any interaction potential may lift the degeneracy of the system. In the case studied here, the confinement potential acts as a perturbation that lifts the degeneracy of the valence band and results in the heavy-hole and the light-hole masses. This can easily be seen by noting that the different masses of the heavy hole and light holes lead to different confinement energies and, consequently, to a removal of the degeneracy. The values for the effective masses are $m_{z,hh} \simeq 0.5m_0$, $m_{\perp hh} \simeq 0.11m_0$, $m_{z,\ell h} \simeq 0.086\ m_0$, and $m_{\perp\ell h} \simeq 0.23$, showing the "mass reversal effect"; i.e., $m_{z,hh} > m_{z,\ell h}$ but $m_{\perp\ell h} > m_{\perp hh}$.

Excitonic absorption spectra in GaAs-$Al_xGa_{1-x}As$ MQWs have been extensively studied both at low and room temperatures. Excitonic lines appear at the onset of each of the two-dimensional interband transitions, i.e., around every step in the joint density of states. Figure 8.16 displays typical absorption spectra of two MQWs and their comparison with a three-dimensional bulk GaAs absorption at 2 K. The top trace is for a sample with GaAs well size of 4000 Å, showing the usual absorption spectrum for a high-purity bulk GaAs. As we discussed in Chap. VI, this spectrum is characterized by a sharp exciton line just below the band edge and a structureless Coulomb enhanced continuum absorption. The bottom two traces display absorption for two MQW structures with well sizes of 210 Å and 140 Å, respectively. In distinct contrast to the bulk trace, the MQW spectra feature multiple absorption peaks above the GaAs bandgap. The lowest energy excitonic peaks in MQWs are blue shifted, compared with the bulk exciton. The shift is larger for smaller well sizes, clearly exhibiting the quantum-confinement induced static blue shift (quantum size effect). Furthermore, the lowest exciton transition for the L_z = 140-Å

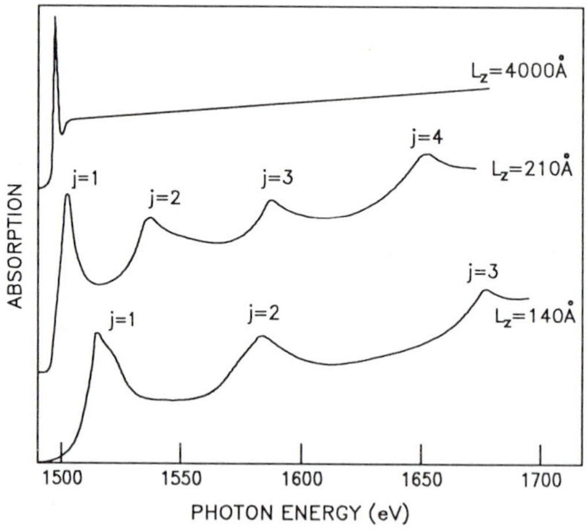

Figure 8.16. Low-temperature absorption spectra of three samples. The two lower spectra show the absorption for two MQW samples with well sizes of 140 Å and 210 Å. The top spectrum is for a sample with GaAs thickness of 4000 Å that effectively describes the three-dimensional bulk behavior (after Ref. 8.3).

sample with thickness $L_z = 140$ Å has two peaks. The two peaks in this doublet are due to transitions from the heavy-hole and the light-hole states in the valence band to the conduction band, as shown schematically in Fig. 8.17. In bulk material, the heavy- and light-hole bands are degenerate at $k = 0$, leading to a single exciton resonance. As already discussed, this degeneracy is lifted in the presence of a quantum-confinement potential. The resulting splitting between light- and heavy-hole bands increases as the well size is reduced. For the sample with thickness $L_z = 210$ Å in Fig. 8.17, the splitting is not large enough to be resolvable.

The lowest-energy exciton lines in MQWs are broader than those of the bulk material, as seen in Fig. 8.16. The larger broadening is due to fluctuations in the well sizes of the MQW samples. During the growth process, the well sizes in MQWs cannot be kept absolutely constant, leading to quantum wells with small thickness fluctuations, i.e., $L_z \pm \Delta L_z$. Since the transition energies between the quantum-confined states vary with L_z like $1/L_z^2$, the fluctuations in L_z lead to fluctuations in transition energies, $\mathcal{E}_j \rightarrow \mathcal{E}_j \pm \Delta\mathcal{E}_j$. The fluctuations in transition energies consequently give an additional width to each transition line, making them inhomogeneously broadened. A typical value of the broadening of the exciton resonance in high-quality undoped wells with thickness of 100-Å sizes is $\simeq 2$ meV at low temperature, to be compared with 0.1 meV in good bulk samples. The inhomogeneous broadening increases for smaller well sizes.

Because of the larger exciton binding energy in low-dimensional materials, the excitonic absorption lines remain resolved even at room temperature in MQWs, in contrast to bulk GaAs. Figure 8.18 shows the room-temperature absorption spectrum of a GaAs-$Al_xGa_{1-x}As$ MQW

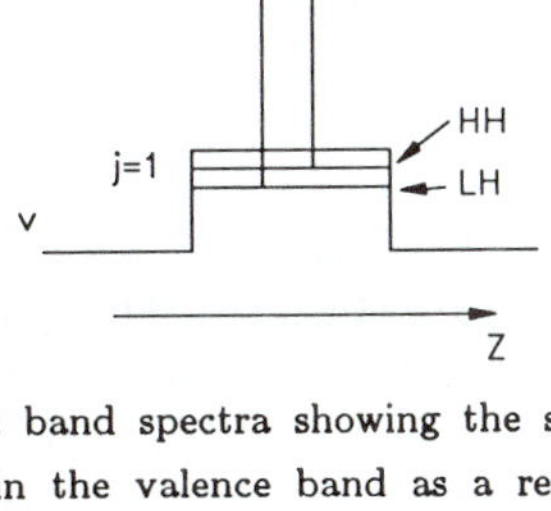

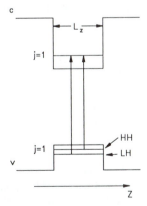

Figure 8.17. Schematic band spectra showing the splitting of the heavy- and light-hole states in the valence band as a result of confinement in small-well-size samples.

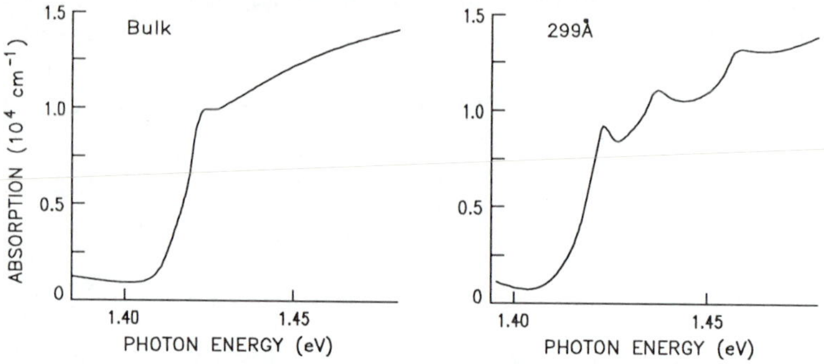

Figure 8.18. Room-temperature absorption spectrum in a GaAs-$Al_xGa_{1-x}As$ MQW with well size of 299 Å. For comparison, the optical absorption of a bulk GaAs is also shown. The appearance of excitonic modified steps is clear in the MQW sample (after Ref. 8.5).

consisting of 61 layers of 299-Å-thick GaAs between 98-Å-thick $Al_{0.36}Ga_{0.64}As$ layers with a total GaAs thickness of 1.8 μm. For comparison, we also show the absorption spectrum of a 2.05-μm-thick bulk GaAs. The MQW sample clearly displays resolved excitonic features, while the exciton is not resolved in the bulk sample. There are two important reasons for this effect. First, the enhanced exciton oscillator strength and the larger binding energy in MQWs lead to a larger absorption coefficient, making exciton resonances more resolvable at room temperature. However, this alone cannot explain the observed spectra, because other bulk semiconductors such as CdS, with even larger binding energy and stronger exciton binding energy, do not exhibit resolvable room-temperature excitonic absorption. The second factor is the unchanged phonon broadening in MQWs compared with the bulk. In bulk semiconductors the exciton binding energy increases as the bandgap increases through the dependence of the binding energy on the dielectric constant and effective masses. However, the exciton-LO phonon coupling strength responsible for the exciton line broadening in polar semiconductors also increases as the bandgap increases. Therefore, in semiconductors such as CdS with larger bandgap and larger exciton binding energy, broadening effects prevent the clear observation of exciton lines at room temperature. In MQWs the larger binding energy is not a result of the larger bandgap, but is due to the reduced dimensionality. Thus, the larger exciton binding energy is not coupled with enhanced phonon broadening. Consequently, excitons are clearly observed at room temperature. The splitting of the lowest-energy exciton into the heavy- and light-hole components is also evident at room temperature. Figure 8.19 shows the heavy- and light-hole excitons in a 76-Å sample.

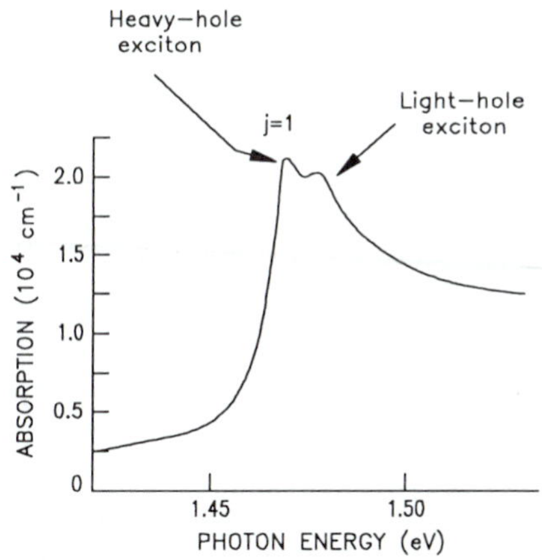

Figure 8.19. Room-temperature absorption of a MQW of GaAs-$Al_xGa_{1-x}As$ with well thickness of 76 Å. Only the lowest exciton transition is displayed. The appearance of the heavy- and light-hole excitons as a result of removal of the degeneracy is clearly shown (after Ref. 8.5).

In summary, the GaAs-$Al_xGa_{1-x}As$ MQWs have properties intermediate between two dimensions and three dimensions. The excitonic binding energy is larger, the Bohr radius is smaller, and the exciton absorption peak is larger for MQWs. For example, the exciton binding energy and Bohr radius of a 100-Å MQW are 9 meV and 65 Å, respectively, compared with a 4.2-meV binding energy and 140-Å Bohr radius in bulk GaAs.

8-7. Absorption Anisotropy in GaAs MQWs*

The transition selection rules in MQWs may be obtained by considering the transition matrix elements in the three-dimensional case. The most important selection rule is the one we described in Sec. 8-4 where the allowed transitions are those between the confined valence and conduction states with the same quantum number j. This selection rule holds for the infinite-well approximation and is a result of orthogonality of the envelope wave functions for the motion in z direction. However, for finite potential wells, this selection rule is not strictly obeyed and transitions with $\Delta j \neq 0$ are also weakly allowed. Such transitions, which are referred to as *forbidden transitions*, have been observed in MQWs.

The strength of the transitions between the heavy-hole-to-conduction state and the light-hole-to-conduction state strongly depends on the

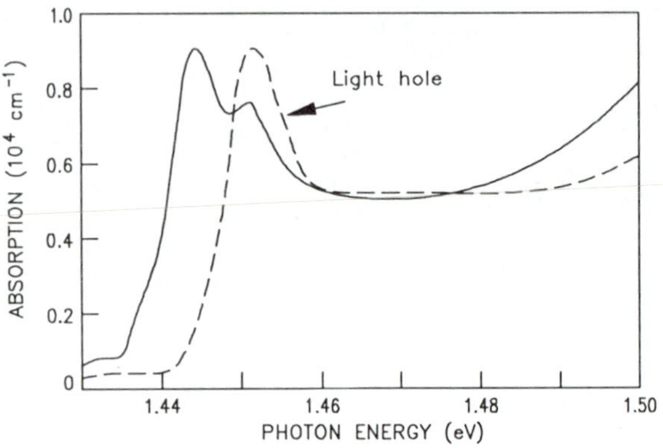

Figure 8.20. Absorption anisotropy in quantum wells. The solid curve (dashed curve) represents the room-temperature absorption for light polarization parallel (normal) to the layers (after Ref. 8.6).

polarization of the incoming photon. Calculations of the transition dipole moments yield that in the case of free-electron-hole absorption (no exciton effects) for light polarized in the plane of the layers (TE polarization), the ratio of the transition strength of the heavy-hole exciton to the light-hole exciton is 3:1. That is, $\alpha_{hh}/\alpha_{\ell h} = 3/1$, where $\alpha_{hh}(\alpha_{\ell h})$ refers to the absorption coefficient for the heavy-hole exciton (light-hole exciton). However, this ratio is $\alpha_{hh}/\alpha_{\ell h} = 0/1$ for light polarized along the z direction (TM polarization), suggesting that the heavy-hole exciton transition is not allowed for light polarized normal to the layers. Such an absorption property has interesting implications for waveguide devices made with GaAs-$Al_xGa_{1-x}As$ MQWs, as will be explained later in this book.

Figure 8.20 shows the room-temperature absorption spectra for an MQW for light polarized along and normal to the layers. The dashed curve gives the absorption for light polarized normal to the layers, and the solid curve is for light polarization along the layers. The lack of the heavy-hole exciton in the dashed curve is clearly demonstrated. Even though the ratio of 3:1 for $\alpha_{hh}/\alpha_{\ell h}$ has not always been seen for the TE polarization, the absorption behavior qualitatively agrees with the described selection rules.

8-8. Type-I and Type-II Quantum Wells*

One important aspect of heterostructures is the band alignment at the interface. As can be seen from Fig. 8.21, several possibilities exist for a given band offset, $\Delta E_g = E_g^B - E_g^A$, between materials A and B, forming a quantum well. In type-I quantum wells [see Fig. 8.21(a)], we have the following relation

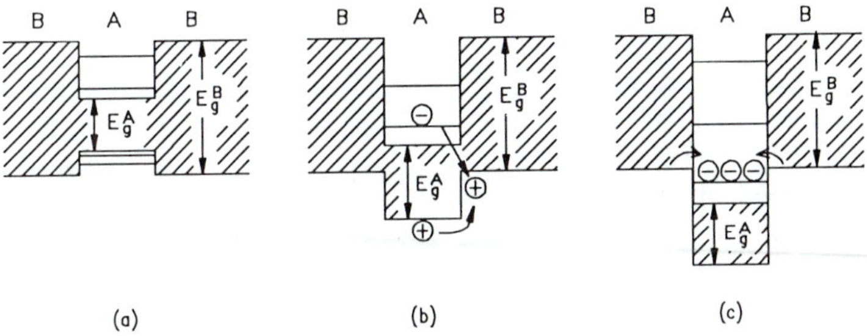

Figure 8.21. Schematic representation of a (a) type-I semiconductor quantum well and (b and c) two examples of type-II structures.

$$E_v^A - E_v^B = \Delta E_v > 0 , \tag{8.97}$$

$$E_c^A - E_c^B = \Delta E_c < 0 , \tag{8.98}$$

and

$$\Delta E_v , \Delta E_c < E_g^B - E_g^A . \tag{8.99}$$

This is the type of quantum well that we have described so far. We note that the parameter $Q = \Delta E_c / \Delta E_v$, specifying the respective valence and conduction band alignments, needs to be determined to completely characterize the structure. At present, there is no rigorous theoretical rule specifying the Q value of a heterostructure. It is generally assumed that most of the band offset occurs between the conduction bands.

There are several variations of type-II quantum wells. Here we consider only two of these variations. The first example of a type-II quantum well is shown in Fig. 8.21(b). The following conditions exist for such a structure,

$$E_c^A - E_c^B = \Delta E_c < 0 , \tag{8.100}$$

$$E_v^A - E_v^B = \Delta E_v < 0 , \tag{8.101}$$

$$\Delta E_v , \Delta E_c < E_g^B , \tag{8.102}$$

and

$$\Delta E_v, \Delta E_c > E_g^B - E_g^A . \tag{8.103}$$

The zero of energy is chosen at the top of the valence band of material A.

The optical properties of type-II quantum wells differ radically from those of type I because of the *spatial* misalignment between the lowest conduction band, located in region A, and the highest valence band in region B. This can be understood simply by considering that following optical excitation of an electron-hole pair, the hole at the top of the valence band in material A is unstable against relaxation to the lowest energy state. The subsequent recombination of the electron-hole pair by emission of a photon occurs through the spatial overlap of the wave function of the electron in material A and the hole in material B. The overlap is small but nonzero due to the exponential tail of the wave function in the barriers of finite height. This results in two characteristic features. First, the energy of the emitted photon is shifted to lower energies, compared with absorption photon energy, by a quantity,

$$\Delta E = \Delta E_v . \tag{8.104}$$

Second, the luminescence has a long lifetime because the dipole matrix element $d_{cv}(0)$ is small due to the spatial separation of the electron and hole wave functions. The material system for such a type-II quantum well may be the system of GaAs-AlAs.

Another example of a type-II quantum well is shown in Fig. 8.21(c). For this structure the following conditions prevail,

$$E_v^A - E_v^B = \Delta E_v < 0 , \tag{8.105}$$

$$E_c^A - E_c^B = \Delta E_c < 0 , \tag{8.106}$$

and

$$\Delta E_c > E_g^B . \tag{8.107}$$

The lowest conduction subband in material A lies energetically below the upper valence band of material B. As a consequence, electrons from the full valence band of material B transfer into the conduction subband of material A, leading to a conducting (semi-metallic) heterostructure. One example of such a quantum well is GaSb/InAs, which may provide a good candidate for infrared optical detectors.

8-9. Semiconductor Superlattices and Minibands*

Structures in which the $Al_xGa_{1-x}As$ barriers are thick enough that the wave functions in the neighboring wells do not overlap are referred to as MQWs. In contrast, periodic structures that have thin barriers so the electronic wave functions from neighboring wells overlap are called *superlattices*. The thin barriers allow tunneling of particles from one well across the barrier to another well. This brings about new effects, leading to new devices and interesting transport properties.

To describe the properties of superlattices, it is instructive to first consider a coupled double-well structure where two GaAs wells of depths V_0 are separated by a narrow $Al_xGa_{1-x}As$ barrier, as shown in Fig. 8.22. For a thin barrier, the exponentially decaying wave function through the barrier can have a finite value in wells 1 and 2. In the absence of wave function overlap between the wells, the quantized energy levels in each well are doubly degenerate for the double-well structure with values $\mathcal{E}_1$, $\mathcal{E}_2$, $\mathcal{E}_3$, etc. The wave function overlap between the wells removes the degeneracy and splits each degenerate energy level in each well into two states with energies $\mathcal{E}_1 \pm \Delta V$, and so on for other transitions. Here ΔV is the interaction energy due to the wave function overlap, $\Delta V = \langle \psi_1 | V_0 | \psi_2 \rangle$, where ψ_1 and ψ_2 are the unperturbed wave functions of the single wells 1 and 2. Thus, the energy levels $\mathcal{E}_1$, $\mathcal{E}_2$, $\mathcal{E}_3$, etc., become doublets.

For a structure with three wells the quantized energy levels become triplets. Similarly, for a structure with four wells one gets quadruplet energy levels, and so on. In general, for N wells each degenerate level splits into N levels and, consequently, forms energy bands having 2N states when the spin degeneracy is included. The tight-binding model is the simplest way to treat this N-well problem. The Bloch-like wave function that may be used for this tight-binding approximation is written as

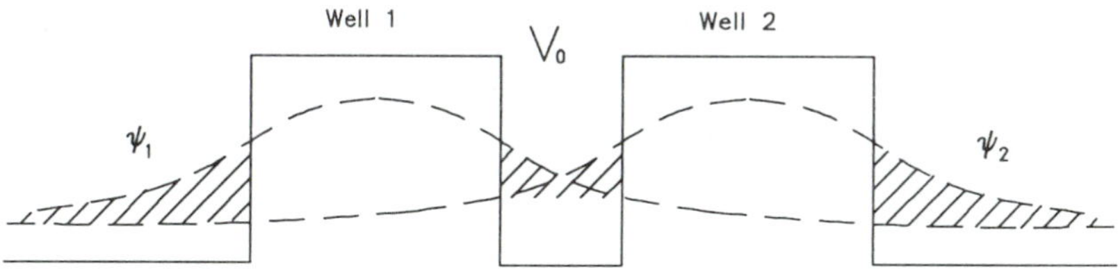

Figure 8.22. Schematic representation of a coupled double-well heterostructure.

$$\psi_k^j(z) = \frac{1}{\sqrt{N}} \sum_n e^{ikn(L_z + L_b)} \zeta_j[z - n(L_z + L_b)] , \tag{8.108}$$

where $L_z + L_b$ is the period of the structure (see Fig. 8.23), k is the Bloch wave vector, and $\zeta_j[z - n(L_z + L_b)]$ is the jth wave function of the quantum well centered at $z = n(L_z + L_b)$. Assuming the nearest neighbor interaction between the wells, the energy of the jth quantized level is found to be a band of the form (see Chap. II)

$$\mathcal{E}_j(k) = \mathcal{E}_j - A_j - 2B_j \cos k(L_z + L_b) , \tag{8.109}$$

where $\mathcal{E}_j$ is the unperturbed quantized energy level, and A_j and B_j are given by

$$A_j = -\int_{-\infty}^{\infty} dz\, \zeta_j(z - L_z - L_b)\, V(z)\, \zeta_j(z - L_z - L_b) \tag{8.110}$$

and

$$B_j = -\int_{-\infty}^{\infty} dz\, \zeta_j(z)\, V(z)\, \zeta_j(z - L_z - L_b) . \tag{8.111}$$

Examination of Eq. (8.109) reveals that $\mathcal{E}_j(k)$ are cosine energy bands with a bandwidth given by $4B_j$, as plotted in Fig. 8.24, where the energy band in the first Brillouin zone is drawn. The Bloch wave vector k can only take discrete values that are integer multiples of $2\pi/N(L_z + L_b)$, as obtained from the periodic boundary condition ($e^{ikNd} = 1$). Therefore, the total number of allowed k values is

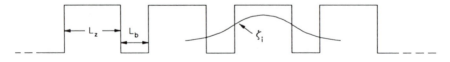

Figure 8.23. Schematic representation of the valence bands of a superlattice along the growth direction z. The wave function $\zeta_i[z - n(L_z - L_b)]$ is plotted for one of the wells.

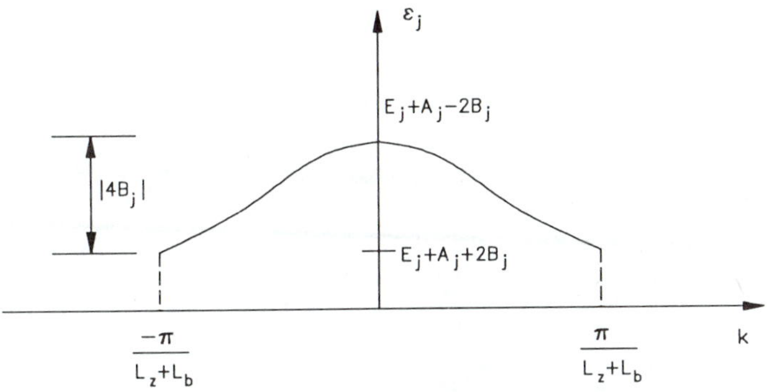

Figure 8.24. The tight-binding energy band for a superlattice. The parameter B_j is assumed to be positive in the band plotted here.

$$\frac{2\pi/(L_z + L_b)}{2\pi/N(L_z + L_b)} = N , \tag{8.112}$$

and the total number of electrons that can occupy this superlattice band is 2N, considering the spin degeneracy. The superlattice energy bands formed this way are also referred to as the *minibands*.

More precise calculations of the superlattice bands have been carried out, and the result of one such calculation is shown in Fig. 8.25. The four lowest quantized energy levels, $\mathcal{E}_1$, $\mathcal{E}_2$, $\mathcal{E}_3$, and $\mathcal{E}_4$, and their formation into bands as the barrier width (and the well width, since $L_z = L_b = a$ in this case) is reduced, allows tunneling and interaction between wells. Energy bands are formed that are separated by energy gaps. As the barrier sizes reduce, the interaction between the wells becomes stronger and the bandwidth increases. The effective unit cell in the superlattice case is a combination of a well and a barrier with total thickness of $L_z + L_b$ that is repeated in the structure.

In summary, artificially grown semiconductor heterostructures have provided a unique type of materials for studying new effects. In addition to the new physics, quantum wells have also provided interesting new devices that will be discussed in later chapters.

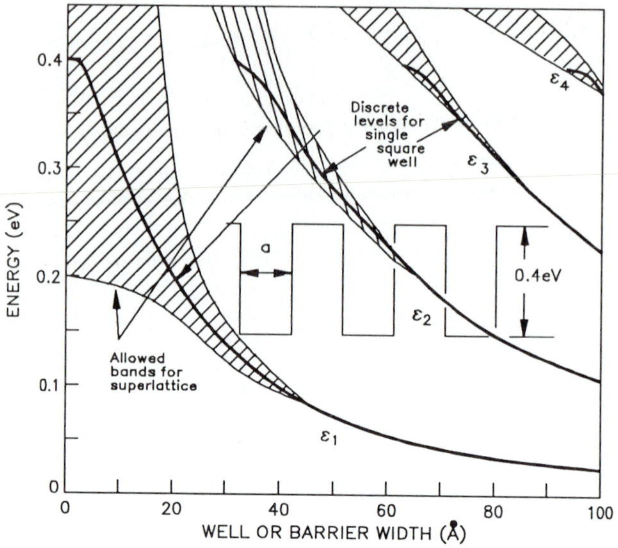

Figure 8.25. Calculated superlattice energy bands and energy gaps for the four lowest electron quantized levels, $\mathcal{E}_1$, $\mathcal{E}_2$, $\mathcal{E}_3$, and $\mathcal{E}_4$. The bands are formed when the barrier sizes are small, allowing interaction between wells (after Ref. 8.7).

8-10. Semiconductor Doping Superlattices (n-i-p-i Structures)*

Semiconductor doping superlattices or n-i-p-i structures are periodic structures of n-type, intrinsic (undoped), p-type, intrinsic materials (a description of n and p doping is given in Chap. XVII). Figure 8.26(a) shows the schematic for such a structure in real space (see Refs. 8.8 and 8.9). A donor may transfer an electron to an acceptor, making both types of impurities charged. The ionized impurities cause a build-up of a periodic space charge potential in the doping layers, which modulates the conduction and valence band edges. The modulation of the electronic bands has the strength of 2 V_0. The space charge potential is parabolic, as shown in Fig. 8.26(b), in contrast to the rectangular potentials in compositional superlattices. In compositional superlattices such as GaAs-AlGaAs, the superlattice potential originates from different bandgaps of the components. The energy gap of the n-i-p-i structure, E_g^{eff}, is the energy separation between the top of the valence band in the p layer and the bottom of the conduction band in the n layer,

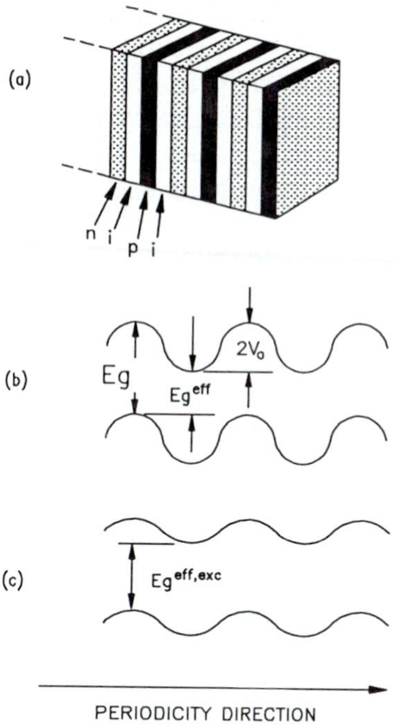

Figure 8.26. (a) Schematic structure of a n-i-p-i superlattice. (b) Band diagrams for no additional injected carriers. (c) Band diagram with additional injected carriers.

$$E_g^{eff} = E_g - 2\,V_0\,, \tag{8.113}$$

where E_g is the bandgap of the host material. The built-in potential (electric field) makes the electrons go to the bottom of the conduction band in the n layer and the holes accumulate at the top of the valence band in the p layer. Therefore, the electrons and holes are spatially separated, which results in longer recombination lifetimes.

If additional carriers are injected in the material, either by light absorption or through the application of electrical currents, the impurity space charge potential is partially compensated, since electrons and holes partially neutralize the positively charged donors and negatively charged acceptors, respectively. Thus, by increasing the excess carrier concentration, the space charge potential is reduced and the effective n-i-p-i bandgap increases, approaching the bandgap of the host material E_g, as schematically displayed in Fig. 8.26(c). This effect makes the doping superlattices tunable in bandgap and recombination lifetime.

The description given above represents a simple picture where the energies of the electronic states (valence-band and conduction-band states) are merely shifted by the built-in space charge field. The confinement of the carriers in the superlattice direction was ignored. However, for thin layers the quantum-confinement energy becomes significant, and the motion of particles in the direction normal to the layers becomes quantized. This leads to the formation of subbands in the conduction and valence bands. In contrast to compositional superlattices, the parabolic potentials in doping superlattices give rise to subbands with energies and spacings that follow harmonic oscillator levels.

So far we have only considered simple n-i-p-i structures in which the modulation is due to dopants. In hetero-n-i-p-i superlattices there is a compositional variation in addition to doping modulation (see Ref. 8.10). For example, one hetero-n-i-p-i superlattice uses a larger-gap material (such as AlGaAs) as the doping layers and a smaller-gap material (such as GaAs) as the intrinsic layer between the doped layers. In these structures there is a potential associated with the spatial variation of the band structure of the two different materials, in addition to the space charge potential. The hetero-n-i-p-i structures combine the unique features of the compositional and doping modulated superlattices.

8-11. Problems

8.1. Using a procedure similar to the one we adopted in Sec. 8-2, show that the wave functions in regions I, II, and III for odd states in a finite potential are given by

$$\varsigma(z) = \begin{cases} A \sin kz & -L_z/2 < z < L_z/2 \\ C\, e^{-k'z} & z > L_z/2 \\ -C\, e^{k'z} & z < -L_z/2\,. \end{cases} \tag{8.114}$$

Furthermore, show that the transcendental equation for this case becomes

$$-\sqrt{\mathcal{E}_z}\, \text{ctn} \sqrt{\frac{m\mathcal{E}_z}{2\hbar^2}}\, L_z = \sqrt{V_0 - \mathcal{E}_z}\,. \tag{8.115}$$

8.2. Using the expressions given in Sec. 8-2, show that the four lowest energy eigenvalues of the finite well potential reduce to those in the infinite well potential in the limit of $V_0 \to \infty$.

8.3. Show that the evaluation of the three-dimensional density of states ($g^{(3)}(\mathcal{E})L_z$) at discrete $\mathcal{E}_z$ energy levels, $\mathcal{E}_z^1$, $\mathcal{E}_z^2$, $\mathcal{E}_z^3$, ..., would result in the two-dimensional density of states values, $m/\pi\hbar^2$, $2m/\pi\hbar^2$, $3m/\pi\hbar^2$, etc., as shown in Fig. 8.10.

8.4. Assume that the effective mass of a particle in a heterostructure consisting of materials A and B is different in the two materials and is given by m_A and m_B. Also assume that the potential well has a finite value V_0 and that material A has a smaller bandgap energy than material B. Using the continuity of the probability current, $[1/m(z)][\partial\zeta(z)/\partial z]$, as a boundary condition (instead of the continuity of wave function derivatives), calculate the transcendental equation for the even states.

8-12. References

1. D. S. Saxon, *Elementary Quantum Mechanical* (Holden-Day, San Francisco, 1968).

2. L. Schiff, *Quantum Mechanics* (McGraw Hill Book Company, New York, 1955).

3. R. Dingle, in *Festkörperprobleme* (Advances in Solid State Physics), Volume XV, edited by H. J. Queisser (Pergamon/Viewweg, Braunschweig, 1975) p. 21.

4. G. Bastard, *Wave Mechanics Applied to Semiconductor Heterostructures* (Les Editions de Physique, Paris, 1988).

5. S. H. Park, J. F. Morhange, A. D. Jeffery, R. Morgan, A. Chavez-Pirson, H. M. Gibbs, S. W. Koch, N. Peyghambarian, M. Derstine, A. C. Gossard, J. H. English, and W. Wiegmann, Appl. Phys. Lett. **52**, 1201 (1988).

6. J. S. Weiner, D. S. Chemla, D. A. B. Miller, H. Haus, A. C. Gossard, W. Wiegmann, and C. A. Burrus, Appl. Phys. Lett. **47**, 6641 (1985).

7. L. Esaki, in *Recent Topics in Semiconductor Physics*, edited by H. Kamimura and Y. Toyozawa (World Scientific, Singapore, 1983) pp. 1-71.

8. G. H. Döhler, H. Kunzel, D. Olego, K. Ploog, P. Ruden, H. Stolz, and G. Abstreiter, Phys. Rev. Lett. **47**, 864 (1981).

9. P. Ruden and G. H. Döhler, Phys. Rev. B 27, 2538 (1983).

10. A. Kost, E. Garmire, A. Danner, and P. D. Dapkus, Appl. Phys. Lett. 52, 837 (1988).

Chapter IX*
QUASI-ONE- AND ZERO-DIMENSIONAL SEMICONDUCTORS: QUANTUM WIRES AND QUANTUM DOTS

The effects observed in quantum wells have led to a surge of interest in the study of even lower dimensional semiconductors with enhanced quantum-confinement properties. It has become customary to use the term "quantum wires" for structures in which the motion of the electron-hole pairs is restricted in two space dimensions. Hence, quantum wires represent quasi-one-dimensional materials. As "quantum dots" or "quantum boxes," we denote all those semiconductor microstructures that confine the laser-excited electron-hole pairs in all three space dimensions (quasi-zero-dimensional semiconductors). Attempts to fabricate quantum wires have been made by etching small stripes out of quantum well materials or by indirect methods to effectively confine the carriers to quasi-one dimension. A number of laboratories have attempted to fabricate quantum dots using various techniques, including colloidal suspension of semiconductor particles, electron-beam lithography, and semiconductor microcrystallites in glass matrices.

In this chapter we describe the density of states and free-carrier absorption in quantum wires and quantum dots, and analyze the effects of Coulomb interaction on the optical properties of such quantum structures.

9-1. Density of States in One-Dimensional Semiconductors

It is assumed throughout this chapter that the wire structure is along the y axis so that the motions along the x and z directions are quantized. The dimensions of the semiconductor along the x and z axes are assumed to be L_x and L_z, and the structure extends to infinity along the y axis. The single-particle Schrödinger equation (for free carriers) may be written as

$$-\frac{\hbar^2}{2m}\nabla^2\psi(x, y, z) = \mathcal{E}\,\psi(x, y, z) \tag{9.1}$$

or

$$-\frac{\hbar^2}{2m}\left(\frac{d^2}{dx^2}+\frac{d^2}{dy^2}+\frac{d^2}{dz^2}\right)\psi(x, y, z) = \mathcal{E}\,\psi(x, y, z)\ , \tag{9.2}$$

where m is the effective mass of the particle that can be an electron or a hole. Separating the wave function and the energy for motions along x, y, and z directions, we get

$$\psi(x, y, z) = \chi(x)\,\phi(y)\,\varsigma(z) \tag{9.3}$$

and

$$\mathcal{E} = \mathcal{E}_x + \mathcal{E}_y + \mathcal{E}_z\ . \tag{9.4}$$

Substitution in Eq. (9.2) and separation of variables leads to three equations for the motions along x, y, and z axes:

$$-\frac{\hbar^2}{2m}\frac{d^2\chi(x)}{dx^2} = \mathcal{E}_x\,\chi(x)\ , \tag{9.5}$$

$$-\frac{\hbar^2}{2m}\frac{d^2\phi(y)}{dy^2} = \mathcal{E}_y\,\phi(y)\ , \tag{9.6}$$

and

$$-\frac{\hbar^2}{2m}\frac{d^2\varsigma(z)}{dz^2} = \mathcal{E}_z\,\varsigma(z)\ . \tag{9.7}$$

The motions along the x and z directions are quantized and the wave functions follow the infinite-well boundary conditions; they vanish at the boundaries L_x and L_z. The solutions of Eqs. (9.5) and (9.7) are the familiar particle-in-a-box solutions, while Eq. (9.6) gives a parabolic energy band in the y direction. Similar to the results of Chap. VIII, we get

$$\mathcal{E}_x = \mathcal{E}_{j_x} = \frac{\hbar^2}{2m}\left(\frac{j_x\pi}{L_x}\right)^2 \qquad j_x = 1, 2, 3, \ldots\ , \tag{9.8}$$

$$\mathcal{E}_z = \mathcal{E}_{j_z} = \frac{\hbar^2}{2m}\left(\frac{j_z \pi}{L_z}\right)^2 \qquad j_z = 1, 2, 3, \ldots , \tag{9.9}$$

and

$$\mathcal{E}_y = \frac{\hbar^2}{2m} k_y^2 . \tag{9.10}$$

The total energy is then

$$\mathcal{E} = \frac{\hbar^2}{2m}\left(\frac{j_x \pi}{L_x}\right)^2 + \frac{\hbar^2}{2m}\left(\frac{j_z \pi}{L_z}\right)^2 + \frac{\hbar^2}{2m} k_y^2 . \tag{9.11}$$

Thus, the energy levels $\mathcal{E}_x$ and $\mathcal{E}_z$ are quantized, and $\mathcal{E}_y$ is continuous. The density of states, denoted as $g^{1d}(\mathcal{E})$, can be written as

$$g^{1d}(\mathcal{E})\, d\mathcal{E} = g^{1d}(k_y)\, dk_y , \tag{9.12}$$

where

$$g^{1d}(k_y) = \frac{1}{2\pi} . \tag{9.13}$$

Including the spin degeneracy, $g^{1d}(\mathcal{E})$ becomes

$$g^{1d}(\mathcal{E}) = \frac{2}{2\pi} \frac{1}{d\mathcal{E}/dk_y} . \tag{9.14}$$

Using Eq. (9.11), we have

$$\frac{d\mathcal{E}}{dk_y} = \frac{\hbar^2 k_y}{m} = \left(\frac{2\hbar^2}{m}\right)^{1/2} \sqrt{\mathcal{E} - \frac{\hbar^2}{2m}\left(\frac{j_x \pi}{L_x}\right)^2 - \frac{\hbar^2}{2m}\left(\frac{j_z \pi}{L_z}\right)^2} . \tag{9.15}$$

Substituting into Eq. (9.14) leads to

$$g^{1d}(\mathcal{E}) = \frac{1}{\pi}\left(\frac{m}{2\hbar^2}\right)^{1/2} \frac{1}{\sqrt{\mathcal{E} - \frac{\hbar^2}{2m}\left(\frac{j_x\pi}{L_x}\right)^2 - \frac{\hbar^2}{2m}\left(\frac{j_z\pi}{L_z}\right)^2}} \tag{9.16}$$

or

$$g^{1d}(\mathcal{E}) = \frac{1}{\pi}\left(\frac{m}{2\hbar^2}\right)^{1/2} \frac{1}{\sqrt{\mathcal{E} - \mathcal{E}_{j_x} - \mathcal{E}_{j_z}}}\,. \tag{9.17}$$

The total single-particle density of states is obtained by summing over all possible j_x and j_z values.

$$g_{sp}^{1d}(\mathcal{E}) = \frac{1}{\pi}\left(\frac{m}{2\hbar^2}\right)^{1/2} \sum_{j_x', j_z'} \frac{1}{\sqrt{E - \mathcal{E}_{j_x'} - \mathcal{E}_{j_z'}}}\,. \tag{9.18}$$

The joint density of states is (see Prob. 9.1)

$$g_{joint}^{1d}(\mathcal{E}) = \frac{1}{\pi}\left(\frac{m_r}{2\hbar^2}\right)^{1/2} \sum_{\substack{j_x', j_z' \\ j_x, j_z}} \frac{1}{\sqrt{\mathcal{E} - E_g - \mathcal{E}^e_{j_x} - \mathcal{E}^e_{j_z} - \mathcal{E}^h_{j_x'} - \mathcal{E}^h_{j_z'}}}\,.$$

joint density of states (9.19)

Therefore, the density of states in one dimension diverges like $\left(\mathcal{E} - E_g - \mathcal{E}^e_{j_x} - \mathcal{E}^e_{j_z} - \mathcal{E}^h_{j_x'} - \mathcal{E}^h_{j_z'}\right)^{-1/2}$. Figure 9.1 shows the one-dimensional joint density of states (dotted curve) where two of the diverging resonances are plotted. For comparison we have also plotted the two- and three-dimensional density of states by dashed and full curves, respectively.

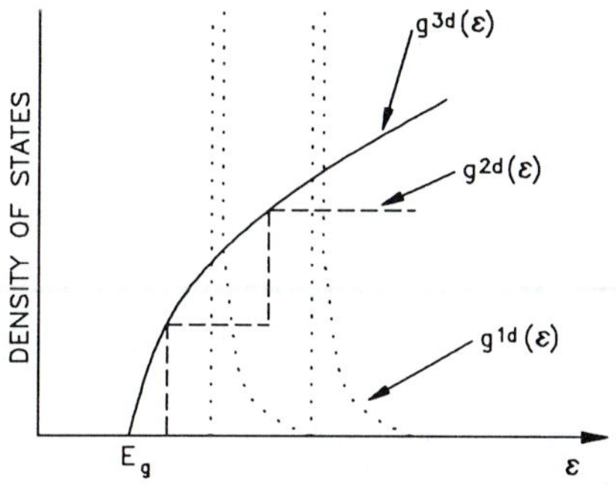

Figure 9.1. The joint density of states in one dimension (dotted curve) and its comparison with two- (dashed curve) and three-dimensional (solid curve) densities of states.

9-2. Optical Absorption in Quantum Wires

The quantum wire wave functions and energies may be calculated by modeling the structure as a cylindrical wire with a finite radius R in the xz plane and with the unconfined coordinate along the y axis. The electrons and holes are confined in a cylindrical potential and the wave functions vanish at the boundaries. The electron-hole-pair Schrödinger equation in this quasi-one-dimensional structure is then

$$\left[-\frac{\hbar^2}{2m_e}\nabla_e^2 - \frac{\hbar^2}{2m_h}\nabla_h^2 + V_e(\mathbf{r}_e) + V_h(\mathbf{r}_h) + V_{Coulomb}\right]\psi(\mathbf{r}_e, \mathbf{r}_h)$$

$$= [H_{kin}^e(\mathbf{r}_e) + H_{kin}^h(\mathbf{r}_h) + V_e(\mathbf{r}_e) + V_h(\mathbf{r}_h) + V_{Coulomb}]\,\psi(\mathbf{r}_e,\mathbf{r}_h)$$

$$= \mathcal{E}\,\psi(\mathbf{r}_e, \mathbf{r}_h)\,, \tag{9.20}$$

where $V_e(\mathbf{r}_e)$ and $V_h(\mathbf{r}_h)$ are the confining potentials of the wire. To solve Eq. (9.20) we assume that the wave function $\psi(\mathbf{r}_e, \mathbf{r}_h)$ is separable and has cylindrical symmetry about the y axis. Changing to cylindrical coordinates, $\mathbf{r} \rightarrow (\rho, y)$, we have

$$[H^{eh}_{eff}(\rho) + H^{e}_{kin}(y_e) + H^{h}_{kin}(y_h) + V_{Coulomb}]\, \psi(\rho, y)$$

$$= \mathcal{E}\psi(\rho, y)\ , \qquad (9.21)$$

where

$$H^{eh}_{eff}(\rho) = H^{e}_{kin}(\rho) + H^{h}_{kin}(\rho) + V_e(\rho) + V_h(\rho)\ , \qquad (9.22)$$

$$H^{e}_{kin}(y_e) = -\frac{\hbar^2}{2m_e}\frac{\partial^2}{\partial y_e^2}\ , \qquad (9.23)$$

and

$$H^{h}_{kin}(y_h) = -\frac{\hbar^2}{2m_h}\frac{\partial^2}{\partial y_h^2}\ , \qquad (9.24)$$

taking into account that the confinement potential has only radial dependence. Separating $\psi(\rho, y)$ into functions of ρ and y, we get

$$\psi(\rho, y) = \phi^{eh}(\rho)\ \zeta(y_e, y_h)\ , \qquad (9.25)$$

with

$$H^{eh}_{eff}(\rho)\ \phi^{eh}(\rho) = (\mathcal{E}^e + \mathcal{E}^h)\ \phi^{eh}(\rho)\ , \qquad (9.26)$$

where $\phi^{eh}(\rho) = \phi_e(\rho_e)\ \phi_h(\rho_h)$, with $\phi_e(\rho_e)$ and $\phi_h(\rho_h)$ being the single-particle ground-state wave functions, which are the solutions of

$$\left[-\frac{\hbar^2}{2m_i}\frac{1}{\rho_i}\frac{\partial}{\partial\rho_i}\rho_i\frac{\partial}{\partial\rho_i} + V_i(\rho_i)\right]\phi_i(\rho_i) = \mathcal{E}_i\phi_i(\rho_i)\ , \qquad (9.27)$$

with i = e or h. Substituting Eq. (9.25) into Eq. (9.21), multiplying from the left by $\phi^{*eh}(\zeta)$, and integrating over ρ for both electrons and holes, we obtain

$$\left[-\frac{\hbar^2}{2m_e} \frac{\partial^2}{\partial y_e^2} - \frac{\hbar^2}{2m_h} \frac{\partial^2}{\partial y_h^2} + V_{eff}(y_e, y_h) \right] \zeta(y_e, y_h)$$

$$= (\mathcal{E} - \mathcal{E}^e - \mathcal{E}^h)\, \zeta(y_e, y_h) \,, \tag{9.28}$$

where we made use of Eq. (9.26) and let

$$\int d\rho\; \phi^{*eh}(\rho)\, \phi^{eh}(\rho) = 1 \quad . \tag{9.29}$$

Also, we defined

$$V_{eff}(y_e, y_h) = \int d\rho\; \phi^{*eh}(\rho)\, V_{Coulomb}\, \phi^{eh}(\rho) \,, \tag{9.30}$$

which is the effective potential for the quasi-one-dimensional exciton problem. $\mathcal{E}^e$ and $\mathcal{E}^h$ are the single particle energies from Eq. (9.26). An approximate expression for $V_{eff}(y = |y_e - y_h|)$ is given by

$$V_{eff}(y) \sim \frac{e^2}{\epsilon_0 (y + \gamma R)} \,, \tag{9.31}$$

where γ is a fitting parameter, R is the radius of the wire, and ϵ_0 is the background dielectric constant of the quantum wire and the surrounding medium. γ can be varied to give a good approximation to the true $V_{eff}(y)$. Writing Eq. (9.28) in terms of the center-of-mass, $Y = (m_e y_e + m_h y_h)/(m_e + m_h)$, and the relative, $y = (y_e - y_h)$, coordinates, the relative-coordinate equation becomes

$$-\frac{\hbar^2}{2m_r} \frac{\partial^2 \zeta(y)}{\partial y^2} + V_{eff}(y)\, \zeta(y) = (\mathcal{E} - \mathcal{E}^e - \mathcal{E}^h)\, \zeta(y) \,, \tag{9.32}$$

where $\mathcal{E} - \mathcal{E}^e - \mathcal{E}^h = E_B$ is the one-dimensional exciton binding energy. The center-of-mass equation yields the usual parabolic energy solution as those we derived for the bulk and quantum well cases. The solution of Eq. (9.32) gives the one-dimensional exciton binding energy, as shown in Fig. 9.2. The figure shows that the exciton binding energy increases with decreasing radius R. For $R = a_B$, where a_B is the exciton Bohr radius, the exciton binding energy is 2.7 times the three-dimensional exciton Rydberg.

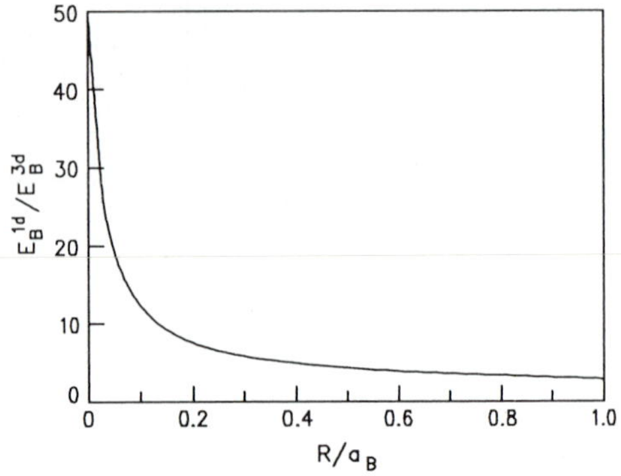

Figure 9.2. One-dimensional exciton binding energy normalized to the three-dimensional exciton Rydberg, E_B^{1d}/E_B^{3d}, as a function of normalized quantum wire radius R/a_B (after Ref. 9.1).

The absorption spectrum of quasi-one-dimensional quantum wires has been calculated recently. Since the result is expressed in terms of special functions (confluent hypergeometric functions, Whittaker functions), we do not give the final results here (see Ref. 9.3 for details). In Fig. 9.3 we show a computed absorption spectrum for a single two-band model and one confinement level, indicating that only one strong exciton resonance is visible. The $(\hbar\omega)^{-1/2}$ singularities of the free-particle density of states are completely suppressed. The strong Coulomb interaction moves the oscillator strength out of the continuum states into the exciton resonance. Practically the entire oscillator strength is accumulated in the ground-state exciton.

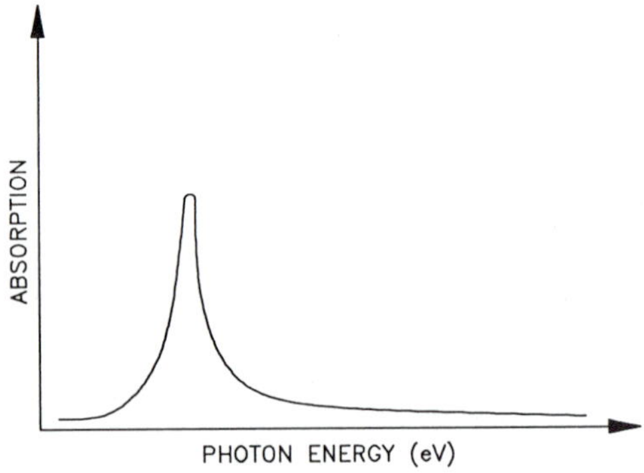

Figure 9.3. Computed exciton absorption spectrum for a quasi-one-dimensional quantum wire (after Ref. 2.5).

Experimental results of GaAs quantum wires are shown in Fig. 9.4. The photoluminescence excitation spectrum, which is the amount of luminescence obtained as a function of incident (excitation) photon energy, basically reflects absorption of the sample, since it is directly proportional to the density of excited carriers. The comparison of the quantum-wire spectra for different wire widths and the reference quantum well spectrum shows a size-dependent blue shift of the heavy-hole exciton in the wires. Furthermore, the resonance in the wires is split into two, corresponding to the two energetically lowest quantum-confined subbands. The splitting increases with decreasing wire width.

Generally, in quantum wires it is difficult to relate the shift of the absorption peak directly to the confinement energy, since one has the balance between two potentially large effects. One is the blue shift due to the quantum confinement (see, e.g., Fig. 9.1), and the other effect is the red shift due to the very large exciton binding energy in quasi-one-dimensional systems. Only a detailed analysis of these effects allows a comparison between the theory and experiments.

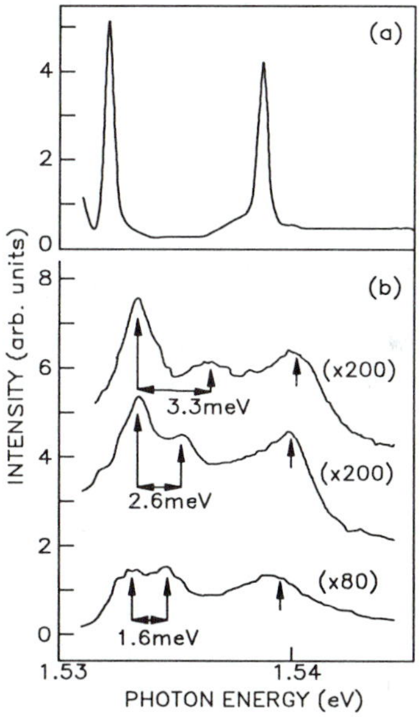

Figure 9.4. Photoluminescence excitation spectra at 5 K for a GaAs quantum well (a) and for quantum-well wires (b) of different thicknesses, L_x. L_x is 60 Å, 70 Å, and 100 Å, respectively, from top to bottom traces in part (b). The spectra are shifted vertically for clarity (after Ref. 9.4).

9-3. Density of States in Zero-Dimensional Semiconductors

Quasi-zero-dimensional semiconductor structures that confine the electron-hole pairs in all three dimensions are usually referred to as *quantum dots*. We discuss quantum dots which have a spherical shape with a characteristic length given by the radius R of the sphere. The radius of the quantum dot is usually on the order of a few tens of Angstroms, comparable to the exciton Bohr radius. Examples of such systems are semiconductor microcrystallites in glass matrices, as well as semiconductor particles in a solution (colloids).

The density of states for quantum dots may be obtained by noting that quantization of particle motion occurs in all three, x, y, and z, directions. If we let j_x, j_y, and j_z be the quantum numbers for the quantized energies in the respective directions, where j_x, j_y, and j_z = 1, 2, 3, ..., the joint density of states becomes

$$g^{0d}(\mathcal{E}) = 2 \sum_{j_x,\, j_y,\, j_z} \delta\left(\mathcal{E} - E_g - \mathcal{E}^e_{j_x, j_y, j_z} - \mathcal{E}^h_{j_x, j_y, j_z}\right).$$

zero-dimensional electron-hole density of states (9.33)

The factor of 2 is included for the spin degeneracy. Equation (9.33) states that the density of states in zero dimension is a series of δ functions corresponding to the quantized energy levels. The δ function resonances appear on the high-energy side of the bulk band edge due to confinement. Therefore, the free-carrier absorption in quantum dots, which is directly

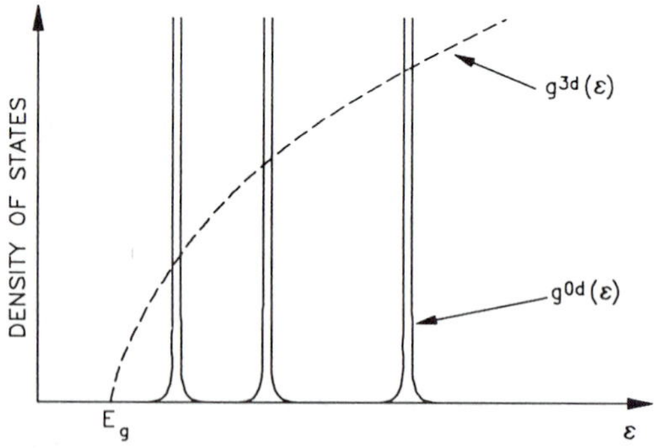

Figure 9.5. Schematic representation of zero-dimensional density of states (solid lines). For comparison the three-dimensional density of states is also shown by the dashed curve.

related to the joint density of states, also consists of sharp resonances on the high-energy side of the bulk bandgap. Figure 9.5 schematically shows the zero-dimensional density of states.

9-4. Optical Absorption in Quantum Dots

To calculate the optical absorption in quasi-zero-dimensional semiconductors, we consider spherical semiconductor microcrystallites with radii R and background dielectric constant ϵ_2 embedded in another material with background dielectric constant ϵ_1. This would closely model the system of semiconductor crystallites in glass which has been extensively studied.

For microcrystals with radius R in the range $a \ll R \simeq a_B$, where a is the lattice constant of the semiconductor and a_B is the exciton Bohr radius, the single-electron properties are determined by the periodic lattice. Hence, the quantum dot has a macroscopic size in comparison to the unit cell, but it is small compared with all other macroscopic scales. In such microcrystallites, which are usually categorized under *mesoscopic* structures, the effective-mass approximation is assumed. The electrons and holes are, thus, assumed to have the effective masses m_e and m_h, respectively, of the bulk material. Optically excited electron-hole pairs are influenced by the small size of the microcrystals, leading to quantum-confinement effects.

For the case of cubic quantum dots (quantum boxes), the single-particle Schrödinger equation (9.1) can be treated exactly as in our discussion of quantum wires in Sec. 9-1. The only, but significant, difference is that now Eq. (9.6) also has to be solved under quantum-confinement conditions, as Eqs. (9.5) and (9.7). Consequently, for the quantum box problem with infinite confinement potential, Eq. (9.10) has to be replaced by

$$\mathcal{E}_y = \frac{\hbar^2}{2m}\left(j_y \frac{\pi}{L_y}\right)^2 \qquad j_y = 1, 2, 3, \ldots , \tag{9.34}$$

leading to a total energy,

$$\mathcal{E} = \frac{\hbar^2\pi^2}{2m}\left[\left(\frac{j_x}{L_x}\right)^2 + \left(\frac{j_y}{L_y}\right)^2 + \left(\frac{j_z}{L_z}\right)^2\right] . \tag{9.35}$$

For the more experimentally relevant case of spherical quantum dots, the single particle Schrödinger equations for the electron and the hole in the absence of Coulomb interaction can be written as

$$-\frac{\hbar^2}{2m_i} \nabla^2 \zeta_i(\mathbf{r}) = \mathcal{E}_i \, \zeta_i(\mathbf{r}) \,, \tag{9.36}$$

where i = e or h. The boundary condition of ideal quantum confinement dictates that

$$\zeta_i(\mathbf{r}) = 0 \qquad \text{for } r = R \,. \tag{9.37}$$

The solution of the Schrödinger equation (9.36) with the spherical boundary condition, Eq. (9.37), is

$$\boxed{\zeta_i(\mathbf{r}) = \sqrt{\frac{1}{4\pi R^3}} \; \frac{j_\ell\left(\alpha_{n\ell}\,\frac{r}{R}\right)}{j_{\ell+1}(\alpha_{n\ell})} \; Y_\ell^m(\theta,\phi) \,,}$$

electron wave function in spherical quantum dot with infinite confinement potential (9.38)

where j_ℓ is the ℓth order spherical Bessel function, $Y_\ell^m(\theta,\phi)$ are the spherical harmonics, $\alpha_{n\ell}$ is the nth root of the ℓth order Bessel function (Ref. 9.2). The quantum numbers for the particle are n, ℓ, and m. The boundary condition (9.37) is satisfied if

$$j_\ell\left(\alpha_{n\ell}\,\frac{r}{R}\right)\Bigg|_{r=R} = 0 \tag{9.39}$$

or

$$j_\ell(\alpha_{n\ell}) = 0 \,. \tag{9.40}$$

Equation (9.40) is satisfied for

$$\alpha_{10} = \pi,\ \alpha_{11} = 4.4934,\ \alpha_{12} = 5.7635,\ \alpha_{20} = 6.2832,$$

$$\alpha_{21} = 7.7253,\ \alpha_{22} = 9.0950,\ \alpha_{30} = 9.4248, \text{ etc.} \tag{9.41}$$

The denominator in Eq. (9.38) comes simply from the normalization of the wave function. Inserting Eq. (9.38) into Eq. (9.36) gives the discrete energy eigenvalues as

$$\boxed{\mathcal{E}^{i} = \frac{\hbar^2}{2m_i}\left(\frac{\alpha_{n\ell}}{R}\right)^2 .}$$

electron energy levels in spherical quantum dot with infinite confinement potential (9.42)

It is customary to refer to the $n\ell$ eigenstates as 1s, 1p, 1d, etc., where s, p, d, etc., correspond to $\ell = 0, 1, 2, \ldots$, respectively ($\alpha_{10} = \alpha_{1s}$, $\alpha_{11} = \alpha_{1p}$, etc.). Note the somewhat unusual notation for atomic spectroscopists, for whom a 1p state would not be possible. The difference here comes from the fact that we are not dealing with a Coulomb potential, but a spherical confinement potential.

The lowest energy states are those with the lowest $\alpha_{n\ell}$ values. Examination of Eq. (9.41) points to these lowest energies as those with α_{1s}, α_{1p}, α_{1d}, etc. Taking the zero of energy at the top of the valence band, the electron and hole energy levels are then given by [using Eq. (9.42)]

$$\mathcal{E}^{e} = E_g + \frac{\hbar^2}{2m_e}\left(\frac{\alpha_{n_e\ell_e}}{R}\right)^2 \tag{9.43}$$

and

$$\mathcal{E}^{h} = -\frac{\hbar^2}{2m_h}\left(\frac{\alpha_{n_h\ell_h}}{R}\right)^2 . \tag{9.44}$$

The lowest two energy levels are plotted schematically in Fig. 9.6. We see from this figure that the usual three-dimensional band structure is drastically modified and has become a series of quantized single-particle states.

As we mentioned before, the single-particle spectrum does not correspond to the optical absorption spectrum, since the electron-hole Coulomb effects are excluded. The Schrödinger equation for one electron-hole pair is written as

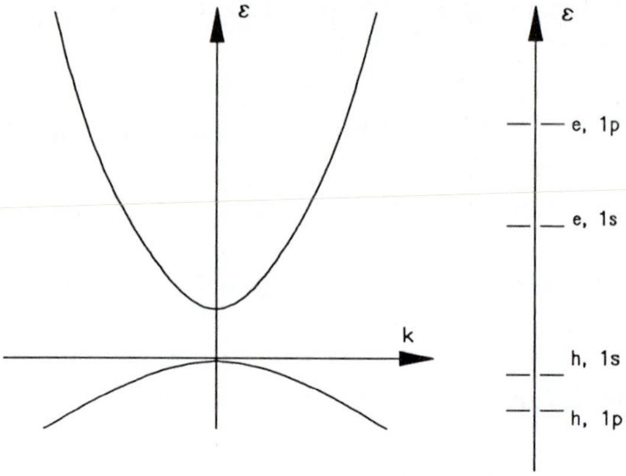

Figure 9.6. Schematic plot of the single-particle energy spectrum in bulk semiconductors (left) and in small quantum dots (right).

$$\left[-\frac{\hbar^2}{2m_e}\nabla_e^2 - \frac{\hbar^2}{2m_h}\nabla_h^2 + V_c\right]\phi(\mathbf{r}) = \mathcal{E}\,\phi(\mathbf{r})\,, \tag{9.45}$$

with the usual spherical boundary condition $\phi(r = R) = 0$. V_c is the Coulomb potential. The Schrödinger equation (9.45) with the boundary condition can be solved analytically in the absence of the Coulomb interaction, resulting in

$$\mathcal{E} = \mathcal{E}^e + \mathcal{E}^h = E_g + \frac{\hbar^2}{2m_e}\left(\frac{\alpha_{n_e \ell_e}}{R}\right)^2 + \frac{\hbar^2}{2m_h}\left(\frac{\alpha_{n_h \ell_h}}{R}\right)^2 \tag{9.46}$$

and

$$\phi(\mathbf{r}_e,\mathbf{r}_h) = \zeta(\mathbf{r}_e)\,\zeta(\mathbf{r}_h)\,, \tag{9.47}$$

where

$$\zeta(\mathbf{r}) = \sqrt{\frac{1}{4\pi R^3}}\;\frac{j_\ell\left(\alpha_{n\ell}\,\frac{r}{R}\right)}{j_{\ell+1}(\alpha_{n\ell})}\;Y_\ell^m(\theta, \phi)\,. \tag{9.48}$$

Equation (9.46) shows that the absorption is blue shifted with respect to the bulk bandgap E_g. The shift varies with crystal size R, like $1/R^2$, being larger for smaller sizes. Furthermore, this equation states that the energy spectrum consists of a series of lines corresponding to the electron-hole transitions. Figure 9.7 exhibits the schematic representation of the one-electron-hole-pair states. The selection rules for the dipole-allowed interband transitions are $\Delta\ell = 0$ in the absence of Coulomb interaction. For example, the $\mathcal{E}_{1s-1s}$-transition, where electron and hole are both of 1s-type, is allowed.

When the Coulomb interaction is included, the problem can no longer be solved analytically and a numerical approach is needed (see Ref. 2.5 for details). The absolute value of the one- and two-pair states is only weakly shifted by Coulomb effects, since the kinetic energy terms dominate for dots with $R \simeq a_B$. The electron-hole-pair wave functions are, however, modified enough to strongly influence the nonlinear effects. The selection rules stated earlier are no longer valid, and transitions with $\Delta\ell \neq 0$ become weakly allowed.

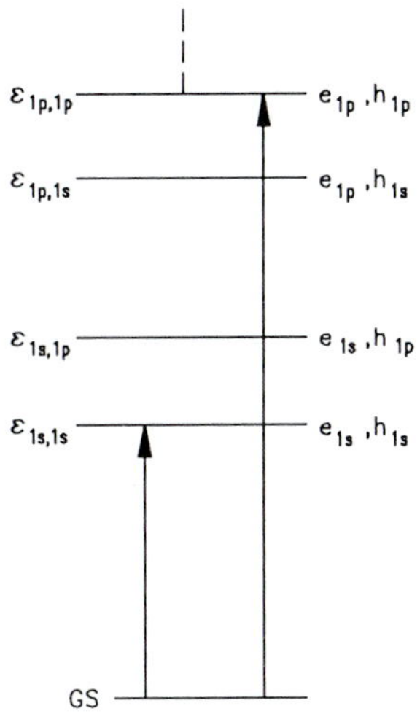

Figure 9.7. Schematic representation of the one-electron-hole-pair transitions in a semiconductor quantum dot. The notation e_{1s}, h_{1p}, etc. refers to the electron being in 1s state, the hole being in 1p state, etc.

9-5. Semiconductor Quantum Dots in Glass

There has been a growing interest in the search for systems that exhibit three-dimensional confinement effects and the understanding of their behavior. A number of laboratories have attempted to fabricate quasi-zero-dimensional structures using various techniques. It has been shown that special glasses doped with CdS, CdSe, CdTe, CuCl, or CuBr crystallites can be fabricated, clearly exhibiting quantum confinement. The microcrystallites in these glasses form out of the super-saturated solid solution of the basic constituents originally brought into the glass melt. The crystallites are more or less randomly distributed in the glass matrix. It has been reported that crystallite growth follows the Lifshitz-Slyozov growth law, $R \propto t^{1/3}$, where R is the crystallite size and t is the duration of the heat treatment during which the crystallites actually grow. Average crystallite sizes from approximately ten Å to several hundred Å have been obtained.

The linear absorption spectra of two samples of CdS quantum dots in glass at room temperature are shown in Fig. 9.8. The spectrum labeled "bulk" refers to a glass with semiconductor microcrystallites large enough to retain the three-dimensional bulk properties. This spectrum is typical of bulk CdS absorption spectrum at room temperature with a sharp band edge and structureless Coulomb-enhanced continuum absorption at higher photon energies. The average crystallite size becomes smaller for lower heat-treatment temperatures. The spectrum labeled "QD" refers to a sample with small crystal sizes. The quantum-confinement effects are clearly observable in this sample. The absorption has shifted to higher

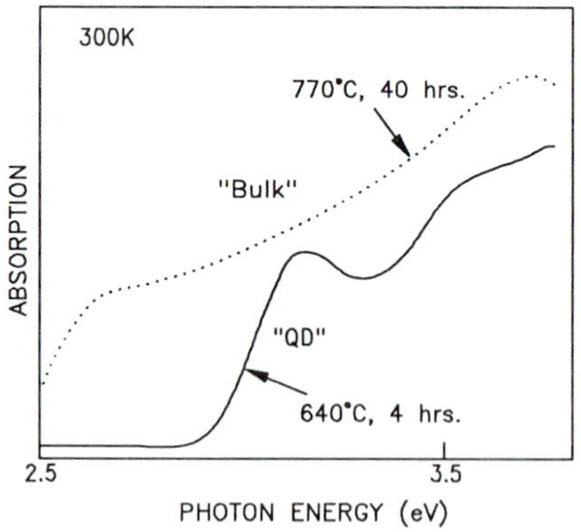

Figure 9.8. The linear absorption spectra at room temperature of two samples of CdS microstructures in glass. The crystal sizes in the sample labeled "bulk" are large enough to have bulk behavior. In the "QD" sample the crystals have very small sizes.

energies as expected for the confinement effect. Furthermore, discrete quantum-confined electron-hole-pair states appear in the "QD" sample.

Figure 9.9 displays the absorption spectra for similar samples at a temperature of T = 10 K. The lower temperature reduces the phonon broadening and, consequently, the transitions become narrower. However, the main features of the spectra have been retained. The quantum-confined transitions are more clearly observed. In this figure, two samples with different quantum dot sizes are displayed.

The transition lines in Fig. 9.9 are much broader than transitions in bulk materials. The width of the transitions is a result of homogeneous and inhomogeneous broadening mechanisms. The homogeneous component is due to phonon and other scattering mechanisms. The inhomogeneous broadening comes from the fact that the crystallites do not have the same radii. Each radius has its own transition frequency ($\mathcal{E} \sim 1/R^2$), making the effective linewidth broad. This broadening can be taken into account theoretically. Using a density-matrix approach, the optical susceptibility and, consequently, optical absorption can be computed. The absorption coefficient of a single quantum dot as a function of photon frequency in such a calculation is given by

$$\alpha(\omega) = \frac{4\pi\omega}{\hbar c\sqrt{\epsilon_2}} \sum_i |d_{oi}|^2 \frac{\gamma_i}{\gamma_i^2 + (\omega_i - \omega)^2} , \tag{9.49}$$

quantum-dot absorption spectrum (9.49)

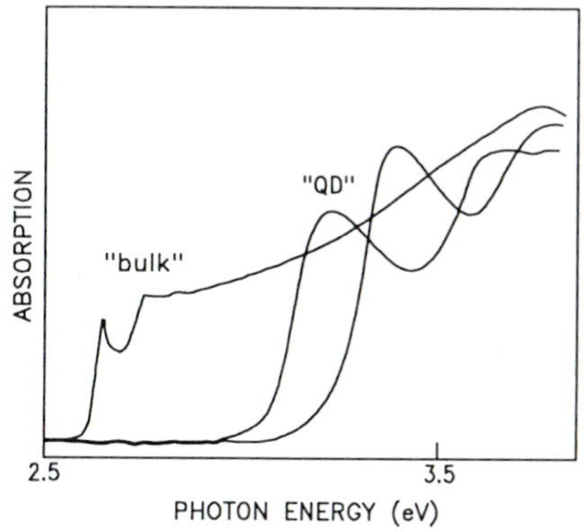

Figure 9.9. The linear absorption spectra at T = 10 K of CdS crystallites in glass. The two spectra labeled "QD" refer to two samples with small crystal sizes in the quantum-confinement regime. The "bulk" sample has large crystal sizes, exhibiting bulk CdS properties.

where ϵ_2 is the background dielectric constant of the semiconductor and d_{oi} is the transition dipole matrix element for the transition $o \rightarrow i$. The index o refers to the ground state (a state without any electron-hole pairs), while the index i refers to the one-electron-hole-pair state. The energy eigenvalue of the quantum-confined electron-hole transition is $\hbar\omega_i$, as given by Eq. (9.42), $\mathcal{E}_i = \hbar\omega_i$. Thus, $\hbar\omega_i$ explicitly depends on the size of the crystallites R. The homogeneous linewidth of the transition is γ_i. Equation (9.49) shows that the absorption spectrum of a single quantum dot consists of a series of Lorentzian peaks centered around the one-electron-hole-pair energies $\hbar\omega_i$. The inhomogeneous broadening may be taken into account by assuming that the particles have a size distribution given by f(R) around a mean value $\bar{R}$. Since $\alpha(\omega)|_R$ in Eq. (9.49) is the absorption coefficient for a given radius R, the average absorption is then

$$\alpha(\omega)\Big|_{\text{average}} = \int_0^\infty dR\, f(R)\, \alpha(\omega)|_R \,. \tag{9.50}$$

Using a Gaussian distribution around the mean radius, $\bar{R}$ = 20 Å, for f(R) and different Gaussian distribution widths, we calculate the results shown in Fig. 9.10 for CdS quantum dots. It is clear that the quantum-confined transitions broaden and merge to a continuous structure with increasing

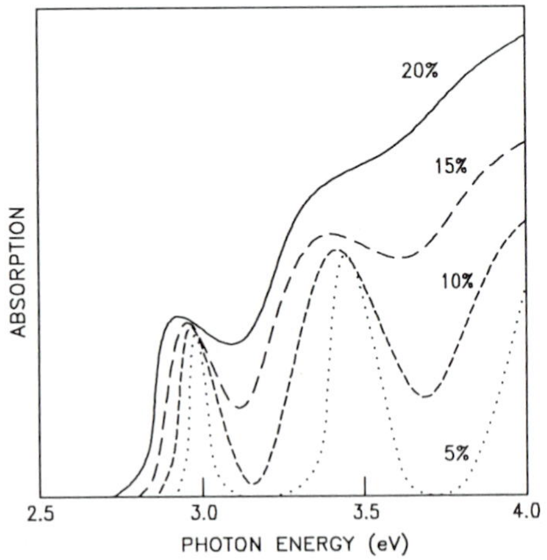

Figure 9.10. Linear absorption for CdS quantum dots with a Gaussian size distribution around a mean radius of 20 Å. The different curves correspond to different widths of the Gaussian distribution. For example, a 20% width corresponds to a 4 Å width for the Gaussian size distribution around a mean radius of 20 Å.

width of the size distribution. The spectrum shown by the full line closely resembles the observed spectrum in Fig. 9.9. This suggests that the samples in Fig. 9.9 have a size distribution of 15% to 20%.

In summary, quantum wires and quantum dots are materials with interesting optical and electronic properties. The challenge is to grow these quantum-confined structures with uniform size distribution. Such materials would lead to new physics and possibly revolutionary devices.

9-6. Problems

9.1. Show that the joint density of states in one dimension is given by Eq. (9.19).

9-7. References

1. L. Banyai, I. Galbraith, C. Ell, and H. Haug, Phys. Rev. B **36**, 6099 (1987).

2. M. Abramowitz and I. Stegun, *Handbook of Mathematical Functions* (Dover Publ., Washington, 1964).

3. T. Ogawa and T. Takagahara, Phys. Rev. B **43**, 14325 (1991).

4. D. Heitmann, H. Lage, M. Kohl, R. Cingolani, P. Grambow, and K. Ploog, in *Optical Excitons in Confined Systems*, Institute of Physics Conference Series 123 (Inst. of Physics, Bristol, 1992), pp. 109-118, and references therein.

Chapter X
ELECTRO-OPTICAL PROPERTIES OF SEMICONDUCTORS

In previous chapters we discussed the semiconductor absorption spectra in the absence of any external forces. We now consider the optical response when a static electric field is applied.

In 1958 Franz and Keldysh independently calculated the effect of such an electric field on the optical interband transitions in semiconductors, ignoring the effects of Coulomb interactions. They predicted that the applied field causes a broadening and low-energy shift of the band-edge absorption spectrum. The red shift of the band-edge results in an increase of optical absorption below the zero-field bandgap. Furthermore, these theories predict that the absorption above the bandgap develops some oscillatory structure. The electric field modifications of the optical absorption are commonly referred to as the Franz-Keldysh effect. An analysis of the field-induced absorption changes, including the electron-hole Coulomb attraction, shows that the oscillations above the bandgap are enhanced (for some material parameters) and that broadening splitting and a shift of exciton resonances toward lower energies occur. Changes of the exciton resonance by an electric field are referred to as the *dc-Stark effect.*

10-1. Franz-Keldysh Effect

We first consider the effect of an electric field in the absence of Coulomb interaction. The application of a uniform electric field across a bulk semiconductor leads to an acceleration of the electron motion through the lattice. The interaction Hamiltonian associated with the applied electric field is given by

$$H^e_{int} = -e\mathbf{E}_0 \cdot \mathbf{r}_e \ , \qquad (10.1)$$

where $\mathbf{E}_0$ is the applied field strength and the index e stands for electron. The corresponding interaction Hamiltonian for holes is identical to

Eq. (10.1), except that the minus sign is replaced by a plus sign. This perturbation Hamiltonian does not have the periodicity of the lattice because it represents a force that accelerates electrons in the direction of the applied field and, consequently, destroys the translational invariance of the Hamiltonian in the field direction. We note that, in general, the application of an electric field reduces the overall symmetry of the system.

The total Franz-Keldysh Hamiltonian for a noninteracting electron-hole pair in a uniform electric field is similar to that given in Chap. VI by Eq. (6.4), except for the addition of the electric-field-interaction term and the neglect of the Coulomb interaction; i.e.,

$$H = -\frac{\hbar^2}{2m_e}\nabla_e^2 - \frac{\hbar^2}{2m_h}\nabla_h^2 - e\mathbf{E}_0\cdot\mathbf{r}_e + e\mathbf{E}_0\cdot\mathbf{r}_h \; . \tag{10.2}$$

The Schrödinger equation for this Hamiltonian is

$$\left[-\frac{\hbar^2}{2m_e}\nabla_e^2 - \frac{\hbar^2}{2m_h}\nabla_h^2 - e\mathbf{E}_0\cdot(\mathbf{r}_e - \mathbf{r}_h)\right]\Phi(\mathbf{r}_e, \mathbf{r}_h) = \mathscr{E}\,\Phi(\mathbf{r}_e, \mathbf{r}_h) \; , \tag{10.3}$$

where $\Phi(\mathbf{r}_e, \mathbf{r}_h)$ and $\mathscr{E}$ are the eigenfunction and energy eigenvalue of the electron-hole pair, respectively. Similar to the procedure we adopted in Chap. VI, the wave function $\Phi(\mathbf{r}_e, \mathbf{r}_h)$ may be separated into the center-of-mass and relative coordinates, $\Phi(\mathbf{r}_e, \mathbf{r}_h) = g(\mathbf{R})\phi(\mathbf{r})$, with $\mathbf{r}$ and $\mathbf{R}$ given by Eqs. (6.5) and (6.6). Again, the center-of-mass equation is an equation for a free particle, not affected by the electric-field term, since the center of mass of an electron-hole pair does not have a net charge. However, the field term enters into the relative-coordinate equation,

$$\left[-\frac{\hbar^2}{2m_r}\nabla_r^2 - e\mathbf{E}_0\cdot\mathbf{r}\right]\phi_n(\mathbf{r}) = \mathscr{E}_n\,\phi_n(\mathbf{r}) \; , \tag{10.4}$$

where $\mathscr{E}_n$ represents the energy eigenvalues for the relative motion, and m_r is the reduced mass of the electron-hole pair. Assuming that the electric field is applied along the z axis, Eq. (10.4) becomes

$$\left(\frac{\hbar^2}{2m_r}\nabla_r^2 + eE_0 z + \mathscr{E}_n\right)\phi_n(\mathbf{r}) = 0 \; . \tag{10.5}$$

Solving this equation yields the energy eigenvalues $\mathcal{E}_n$ and eigenfunction $\phi_n(\mathbf{r})$. The absorption coefficient of the semiconductor in the presence of the electric field is then calculated by inserting the resulting expressions for $\mathcal{E}_n$ and $\phi_n(\mathbf{r})$ into Eq. (6.33) for the absorption coefficient,

$$\alpha(\omega) = \frac{8\pi^2\omega|d_{cv}|^2}{n_b c} \sum_n |\phi_n(\mathbf{r} = 0)|^2 \, \delta(\hbar\omega - \mathcal{E}_n) \, . \tag{10.6}$$

Here we only give the final results of such a calculation and discuss the physical interpretations in detail. The interested reader may find the details of the calculations in reviews such as Refs. 10.4 and 2.5. These calculations give the electroabsorption spectrum for direct-allowed transition as

$$\alpha(\omega) = \frac{2\pi\omega|d_{cv}|^2}{n_b c} \left(\frac{2m_r}{\hbar^2}\right)^{4/3} (eE_0)^{1/3} \left[-\mathcal{E}' \left(\mathrm{Ai}(\mathcal{E}')\right)^2 + \left(\mathrm{Ai}'(\mathcal{E}')\right)^2\right] , \tag{10.7}$$

where

$$\mathcal{E}' = \left(\frac{2m_r}{\hbar^2}\right)^{1/3} \left(\frac{1}{eE_0}\right)^{2/3} (E_g - \hbar\omega) \, , \tag{10.8}$$

and Ai is the Airy function, given by

$$\mathrm{Ai}(x) = \frac{1}{\pi} \int_0^\infty ds \cos\left(\frac{s^3}{3} + sx\right) \tag{10.9}$$

and

$$\mathrm{Ai}'(x) = d\mathrm{Ai}(x)/dx \, . \tag{10.10}$$

The Airy function has the property of decaying exponentially for positive arguments and has an oscillatory behavior for negative arguments. Therefore, for energies above the bandgap where $\hbar\omega > E_g$, the argument of the Airy function $\mathcal{E}'$ in Eq. (10.8) is negative, $\mathcal{E}' < 0$, and the absorption coefficient of Eq. (10.7) becomes an oscillatory function of energy and electric field. The amplitude of these oscillations decreases with an increase of $\hbar\omega$ above the bandgap. An increase in magnitude of the electric field also causes the oscillation amplitudes and the periodicity to increase. The period of the oscillations is related to the field E_0 and reduced mass m_r, like $(E_0^2/m_r)^{1/3}$. Figure 10.1 displays the calculated free-carrier absorption at zero field and a field of 10^5 V/cm for a bulk semiconductor, clearly showing the oscillatory structure for above-gap energies. For large photon energies, $\hbar\omega >> E_g$, the oscillations damp out and Eq. (10.7) reduces to free-carrier absorption at zero field, as discussed in Chap. V [Eq. (5.69)].

For photon energies below the gap, $\hbar\omega < E_g$, $\mathcal{E}'$ is positive and the Airy function decays exponentially. In fact, it can be verified that for direct-allowed transitions, when $\hbar\omega < E_g$, Eq. (10.7) becomes (see Prob. 10.1)

$$\boxed{\alpha_{FK}(\omega) = \frac{\omega|d_{cv}|^2 \left(\dfrac{2m_r}{\hbar^2}\right) eE_0}{4n_b c(E_g - \hbar\omega)} \exp\left[-\frac{4}{3eE_0}\left(\frac{2m_r}{\hbar^2}\right)^{1/2} (E_g - \hbar\omega)^{3/2}\right].}$$

free-carrier absorption in an electric field (10.11)

This result is often called the Franz-Keldysh electroabsorption. The exponent in Eq. (10.11) has a negative sign for $\hbar\omega < E_g$, describing a decaying function below the bandgap. Hence, the free-carrier absorption has an exponential tail in the presence of an electric field, as shown in

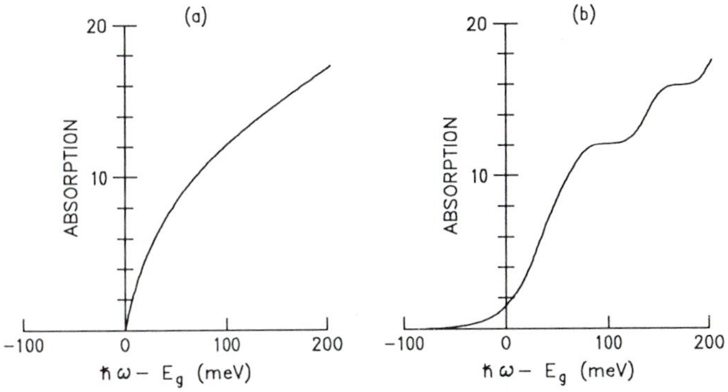

Figure 10.1. Calculated free-carrier absorption (a) at zero field and (b) with a field of $E_0 = 10^5$ V/cm (after Ref. 10.4).

Fig. 10.1(b). The extent of this tail depends on the magnitude of the field; it extends further for higher fields.

The presence of additional absorption below the zero-field bandgap may be understood as a field-induced tunneling of an electron from the valence band into the conduction band, which is assisted by the photon. This process is schematically shown in Fig. 10.2. The electric field tilts the electron states in space as a result of the additional energy term $\pm eE_0z$ for electrons and holes. The conduction states are tilted downward because the electron energy is reduced by a field energy of $-eE_0z$. The valence states are also tilted downward because the hole energy is increased by a field energy of $+eE_0z$. Figure 10.2(b) indicates that the tilt is in the same direction for both bands, keeping the energy separation equal to E_g. A photon with energy of $\hbar\omega$ less than the bandgap may promote the electron from the valence band to the conduction band if the electric field provides the energy deficiency, i.e., if $eE_0\Delta z = E_c - E_v - \hbar\omega$.

To summarize, the Franz-Keldysh effect is referred to as the effect of an external electric field on the band-edge absorption of semiconductors in the absence of Coulomb interaction. It leads to the appearance of absorption below the bandgap and to oscillatory absorption variations for energies above the gap.

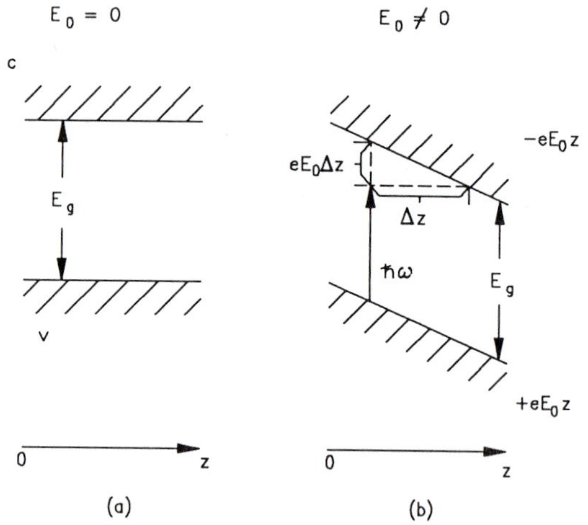

Figure 10.2. Effect of applying an electric field on the energy states in a semiconductor. The electric field (along z direction) tilts the states. In (a), $E_0 = 0$ and in (b), E_0 has a finite value.

10-2. DC-Stark Effect

We now improve our analysis of the bandgap electroabsorption by including the Coulomb interaction between the electrons and the holes in bulk semiconductors. The combined potential of the electron-hole pair with electric field is given by

$$H_{int} = -\frac{e^2}{\epsilon_0 r} - e\mathbf{E}_0 \cdot \mathbf{r} = -\frac{e^2}{\epsilon_0 r} - eE_0 z \,, \tag{10.12}$$

where $\mathbf{r} = \mathbf{r}_e - \mathbf{r}_h$. This potential is plotted in Fig. 10.3 for zero- and finite-field magnitudes. We see that the major effect of the applied field is the modification of the purely attractive Coulomb well into a potential barrier, where truly bound states no longer exist. The state with the lowest potential energy is clearly the one where electron and hole are infinitely separated. However, quasi-bound states still exist. These quasi-bound states are what is left of the excitons. The field now makes it energetically more favorable for the electron and hole to separate. This *field ionization* decreases the overall lifetime of the exciton, leading to a broadening of the resonances. Another effect of the electric field is the widening of the potential well, resulting in a *red shifting* of the excitonic lines. This shift, which is a second-order effect, is referred to as the dc-Stark shift. The shift increases quadratically with the field.

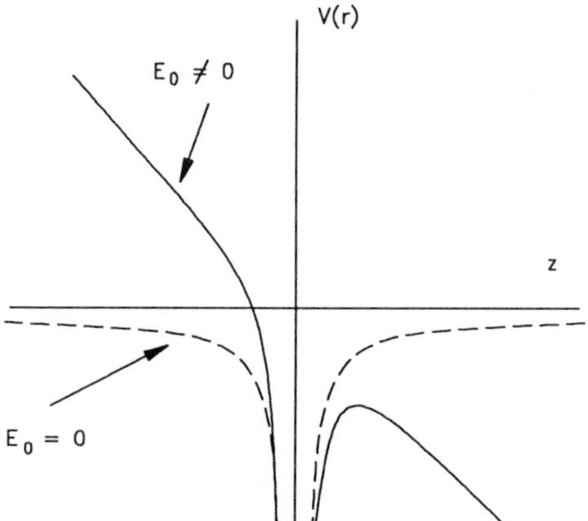

Figure 10.3. The dashed curve represents the electron-hole interaction in the absence of an external field. The solid curve gives this Coulomb interaction in the presence of the applied electric field (after Ref. 10.2).

To ionize the exciton into an unbound electron-hole pair, the electric field should provide an energy that is at least equal to the binding energy of the exciton across the exciton Bohr radius. That is, for $E_0 = E^0_{ion}$, we should have

$$ea_B E^0_{ion} = E_B \tag{10.13}$$

or

$$E^0_{ion} = \frac{E_B}{ea_B}, \tag{10.14}$$

where E_B and a_B are the binding energy and the Bohr radius of the exciton given by Eqs. (6.18) and (6.19), respectively. This ionization field is about 5×10^3 V/cm in GaAs.

The Schrödinger equation for this case may be obtained from Eq. (10.5) by replacing the potential term eE_0z by Eq. (10.12). That is,

$$\left[\frac{\hbar^2}{2m_r} \nabla_r^2 + eE_0z + \frac{e^2}{\epsilon_0 r} + \mathcal{E}_n\right] \phi_n(r) = 0\ , \tag{10.15}$$

where $\mathcal{E}_n = \hbar\omega - E_g$. The solution of this equation allows the evaluation of $\phi_n(r = 0)$, from which the absorption coefficient is then calculated as

$$\boxed{\alpha(\omega) = \alpha_{FK}(\omega) \left\{\Gamma\left(1 - \frac{1}{\sqrt{\mathcal{E}''}}\right) \exp\left[\frac{1}{\sqrt{\mathcal{E}''}}\, \ell n\left(\frac{8\, E_B\, \mathcal{E}''^{3/2}}{ea_B E_0}\right)\right]\right\}^2 ,}$$

electron-hole-pair absorption spectrum in an electric field (10.16)

where

$$\mathcal{E}'' = \frac{E_g - \hbar\omega}{E_B}, \tag{10.17}$$

and $\alpha_{FK}(\omega)$ is the Franz-Keldysh absorption coefficient given by Eq. (10.11). $\Gamma(1 - 1/\sqrt{\mathcal{E}''})$ is a Γ function defined as

$$\Gamma(x) = \int_0^\infty dy\; y^{x-1}\, e^{-y} \; . \tag{10.18}$$

Equation (10.16) shows the enhancement of the free-carrier Franz-Keldysh absorption due to the electron-hole Coulomb attraction. Depending on the detuning, $\mathcal{E}''$, and the field, E_0, this enhancement of the electroabsorption tail may be up to a factor of $\simeq 10^3$. For large detunings, $\alpha(\omega)$ approaches asymptotically the Franz-Keldysh spectrum.

The computed electroabsorption spectra for a 1s exciton (n = 1) at three field magnitudes is displayed in Fig. 10.4. As the electric field increases, the exciton shifts to lower energies and broadens. When the field strength is increased to the ionization value given by Eq. (10.14), the bound-exciton level is no longer distinguished and is broadened and mixed with continuum. In practice, the shift is relatively small and broadening usually dominates the spectra.

In addition to Stark shifting, the electric field may also result in Stark splitting of the degenerate excitonic levels. For instance, the n = 2 exciton corresponds to two different excitation states, p-like and s-like, which are normally degenerate. As we have seen in Chap. VI, in a crystal with inversion symmetry, the n = 2, p exciton is forbidden when the n = 2, s exciton is allowed. In such a case, under an applied electric field along z, a combination of s and p_z functions is formed, resulting in the appearance of a doublet in the absorption spectrum, as shown in Fig. 10.5. The shifting and broadening of the lines is also evident in this figure as the field increases.

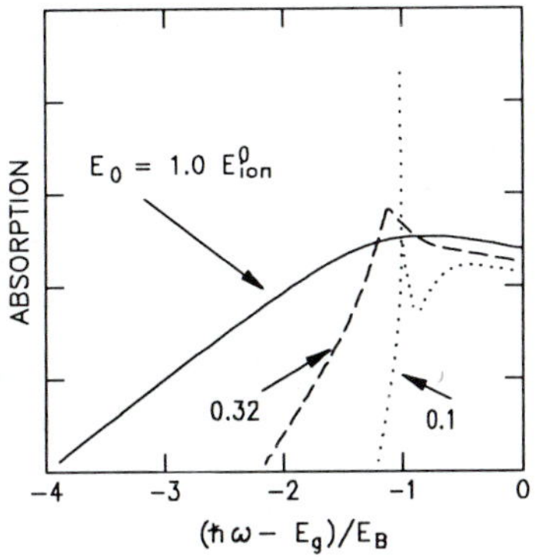

Figure 10.4. Calculated electroabsorption effect for n = 1 exciton for three field magnitudes, $E_0 = 0.1\ E^0_{ion}$, $0.32\ E^0_{ion}$, and $1.0\ E^0_{ion}$ (after Ref. 10.5).

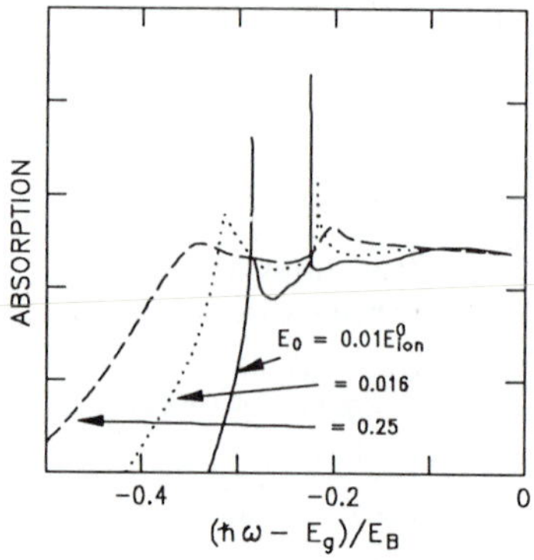

Figure 10.5. Calculated electroabsorption effect for n = 2 exciton for three field magnitudes, $E_0 = 0.01\ E^0_{ion}$, $0.016\ E^0_{ion}$, and $0.25\ E^0_{ion}$. The n = 1 exciton is omitted in the calculation. Stark splitting of the n = 2 exciton is evident in this figure (after Ref. 10.5).

The oscillatory behavior of the absorption for photon energies above the bandgap still remains when excitonic effects are included. Figure 10.6 displays the calculated electroabsorption for a temperature-broadened exciton and continuum states at several field magnitudes. The oscillations on the high-energy side of the exciton, as well as shifting and broadening of the resonance, are clearly displayed.

The electroabsorption spectra have been measured in many materials. Figure 10.7 displays the absorption coefficient for the example of PbI_2 with and without a field, and the absorption difference $\Delta\alpha$. These experimentally obtained electroabsorption spectra agree well with the

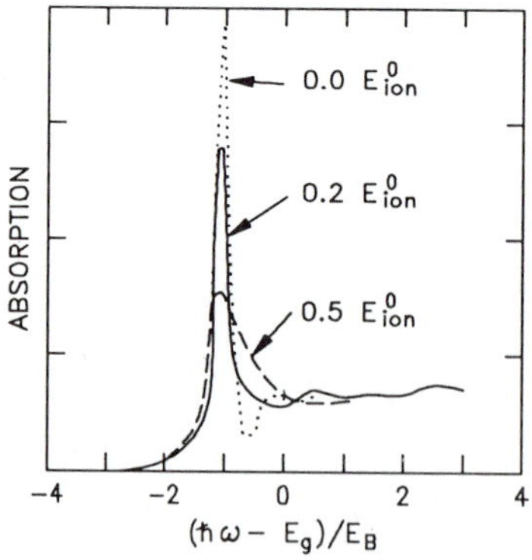

Figure 10.6. Calculated electroabsorption effect of a broadened exciton with various field magnitudes. The dotted line corresponds to no field and the dashed and full lines are for $E_0 = 0.2\ E^0_{ion}$ and $0.5\ E^0_{ion}$, respectively. The broadening and shifting of exciton and oscillations above bandgap are demonstrated (after Ref. 10.2).

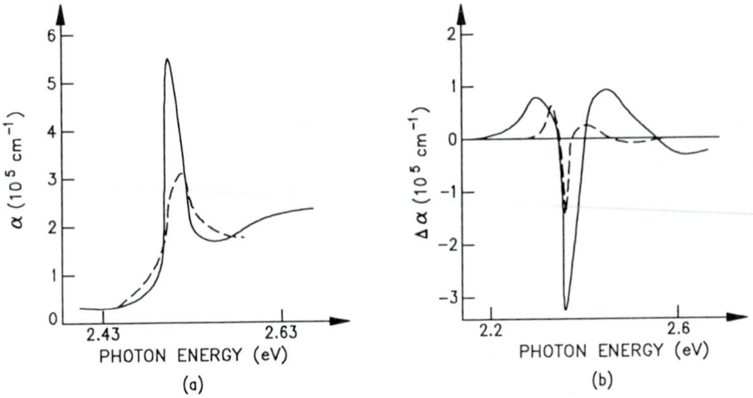

Figure 10.7. Measured electroabsorption of PbI_2 in the exciton region. (a) Absorption spectrum for no field (continuous curve) and for $E_0 = 2.85 \times 10^5$ V/cm (dashed curve). (b) Change in absorption coefficient at $E_0 = 2.85 \times 10^5$ V/cm (continuous curve) and $E_0 = 0.79 \times 10^5$ V/cm (dashed curve) (after Ref. 10.6).

theory that includes excitonic effects. Figure 10.8 shows the change in absorption spectra observed in PbI_2 with and without an electric field of 2.85×10^5 V/cm, and its comparison to excitonic electroabsorption and free-carrier electroabsorption theories.

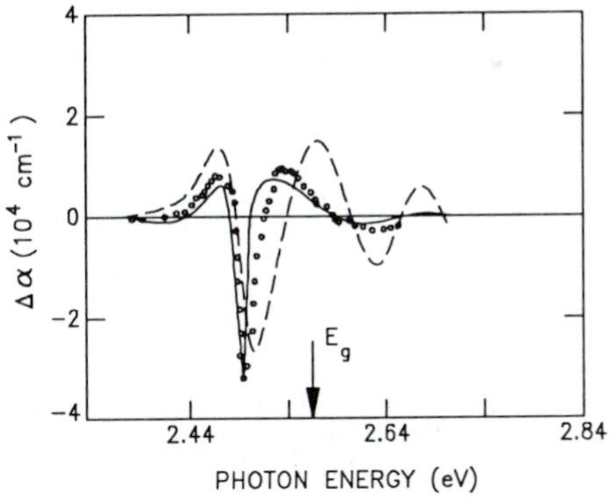

Figure 10.8. Comparison of the experimental and theoretical electroabsorption spectra in PbI_2 for a field magnitude of 2.85×10^5 V/cm. The dotted curve is the experimental data of Fig. 10.7(b). The theory of excitonic electroabsorption shown by the continuous curve is a much better fit to the experimental data, compared with the theory excluding Coulomb interaction (dashed curve) (after Ref. 10.2).

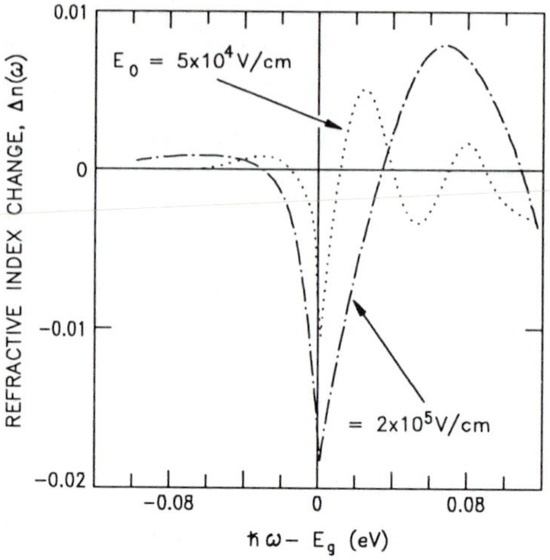

Figure 10.9. The calculated change in refractive index induced by an electric field of various magnitudes in GaAs, obtained by a Kramers-Kronig transformation of absorption changes (after Ref. 10.7).

Finally, we note that the electroabsorption changes, $\Delta\alpha(\omega)$, discussed so far also give rise to refractive index changes, $\Delta n(\omega)$, in the vicinity of the bandgap energy through Kramers-Kronig transformation. These index changes are typically a few percent when fields of magnitudes of approximately 10^5 V/cm are applied. Figure 10.9 shows the change in the refractive index of GaAs induced by various field strengths. The oscillatory structures also appear in the $\Delta n(\omega)$ spectra as one expects.

10-3. Electric Field Effects in Two Dimensions: Quantum-Confined Franz-Keldysh and Quantum-Confined Stark Effects

We treated the effect of an external electric field on the optical absorption of three-dimensional bulk semiconductors in the previous sections. Here we consider how the presence of an electric field modifies the optical properties of quasi two-dimensional GaAs-$Al_xGa_{1-x}As$ multiple quantum wells (MQWs). The external electric field may be applied parallel or normal to the quantum-well layers.

The electroabsorption of MQWs for fields parallel to the plane of the layers is qualitatively similar to that in bulk semiconductors as expected, since the motion of the carriers in the layers is not affected by confinement. The main effect of the field is the broadening of the exciton resonance. Figure 10.10 shows the measured electroabsorption spectra for

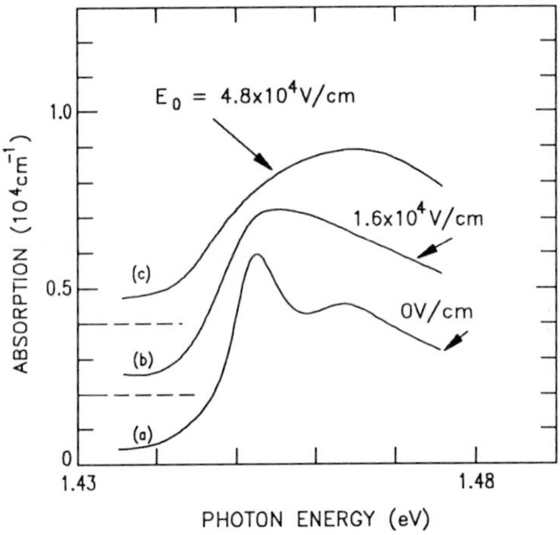

Figure 10.10. Electroabsorption for various fields applied parallel to the plane of the layers in a MQW sample, (a) 0 V/cm, (b) 1.6 × 10^4 V/cm and (c) 4.8 × 10^4 V/cm (after Ref. 10.8).

three field magnitudes that were applied parallel to plane of the layers of a GaAs-Al_xGa_{1-x}As MQW. The excitons in MQWs have larger binding energy than bulk and, consequently, are not as easily ionized. Otherwise, the general behavior of electroabsorption in this case is similar to that in bulk material.

The effect of fields perpendicular to the layers in MQWs is, however, qualitatively different from that in bulk materials. When the electron-hole Coulomb interaction is neglected in the treatment of the perpendicular field case, the resulting electroabsorption may be described as a quantum-confined Franz-Keldysh effect. With Coulomb interaction, the electro-absorption is described by a model referred to as the quantum-confined Stark effect. Of course, in practice, the Coulomb interaction cannot be neglected and the relevant model is the quantum-confined Stark effect.

Figure 10.11 shows the modified energy levels and wave functions for electrons and holes in an infinitely deep potential well as a result of the application of an electric field. The quantum wells are skewed by the field as shown in the figure. The field pushes the electrons and holes to the opposite walls of the well because of the opposite charges of the two particles, thereby modifying the overlap between the corresponding confined wave functions. The field ionization of the exciton, which is the dominant effect in three-dimensional bulk materials, is inhibited in two dimensions by the walls of the quantum well. Therefore, the exciton shifts and persists up to high field magnitudes. Figure 10.12 displays the observed large red shift of the exciton peak. The exciton persists for field strengths as high as 1.8 × 10^5 V/cm (20 V). The experiment was

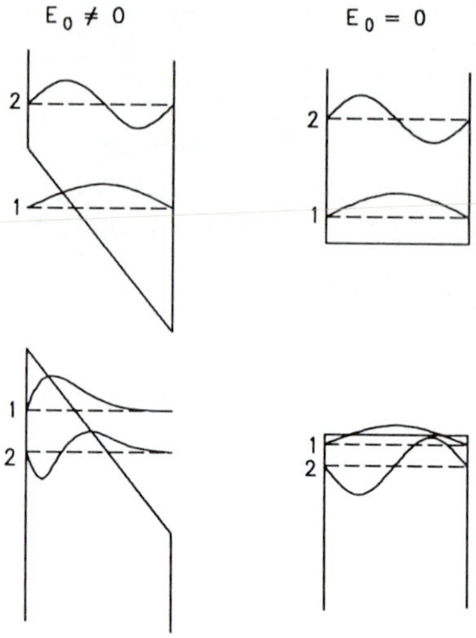

Figure 10.11. The effect of an external electric field on the energy levels, wave functions, and the bands of a quantum well.

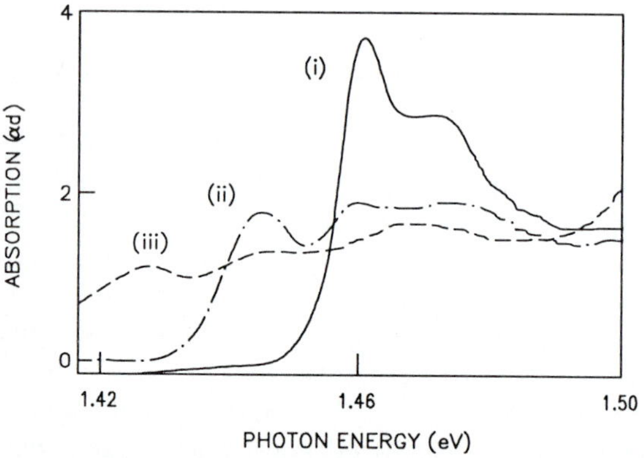

Figure 10.12. Electroabsorption spectra of a GaAs quantum well waveguide device as a function of applied field perpendicular to the plane of the layers. The optical field is polarized parallel to the layers and has magnitudes (i) 1.6×10^4 V/cm, (ii) 1.3×10^5 V/cm, and (iii) 1.8×10^5 V/cm (after Ref. 10.9).

performed using two 94-Å GaAs quantum wells centered in a superlattice for various field strengths applied perpendicular to the layers. The shift is approximately quadratic at low fields. The exciton is not destroyed by such high fields and remains resolved. This is in drastic contrast to Fig. 10.10 where the exciton had already ionized at a field of 4.8×10^4 V/cm.

Another important consequence of the application of an electric field is the lowering of the symmetry and, thus, the removal of the strict transition selection rules. As explained earlier, one such selection rule is that the allowed transitions are those between electron and hole confined states with $\Delta j = 0$. Transitions with $\Delta j \neq 0$ are forbidden in the absence of an external field (forbidden transitions). The presence of an electric field makes the transitions with $\Delta j \neq 0$ allowed because the electron and hole wave functions are no longer sinusoidal, and the overlap integral of Eq. (8.91) is generally not zero for $i \neq j$. Figure 10.13 displays the observed forbidden transitions and their enhancement with larger fields. In this figure, the first two subscripts refer to the electron and hole states, respectively. The letters h and ℓ in the subscript refer to the heavy- or light-hole states. Transitions that are allowed in the absence of the electric field are $\mathcal{E}_{11h}$, $\mathcal{E}_{11\ell}$, $\mathcal{E}_{22h}$, and $\mathcal{E}_{22\ell}$. Transitions $\mathcal{E}_{12h}$, $\mathcal{E}_{13h}$, $\mathcal{E}_{21h}$, and $\mathcal{E}_{21\ell}$ are forbidden transitions that have become allowed as a result of the electric-field application. The field magnitude is maximum for the top trace and is reduced for the lower two traces. As the field increases, the forbidden transitions become stronger, as expected.

An important consequence of the appearance of the forbidden transitions in the electroabsorption of the MQWs derives from sum rules for the interband transitions. One such sum rule states that the field-

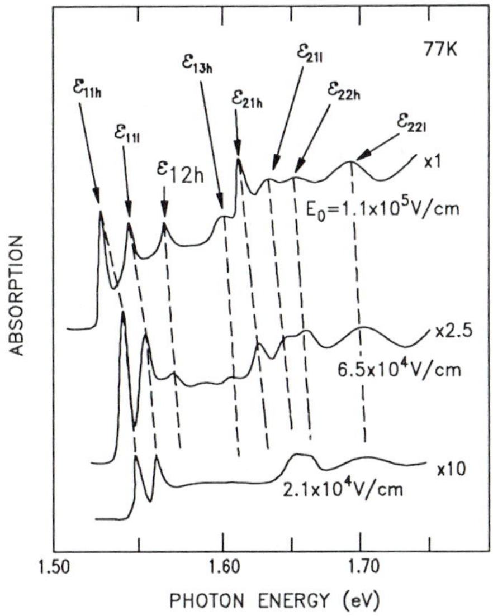

Figure 10.13. Photocurrent spectra for a 105-Å MQW at three field magnitudes of 2.1×10^4 V/cm (bottom trace), 6.5×10^4 V/cm (middle trace), and 1.1×10^5 V/cm (top trace). The photocurrent spectrum effectively gives the absorption spectrum of the sample (after Ref. 10.10).

induced absorption change over the entire spectrum must be zero. In other words, the increase of the absorption at higher photon energies is accompanied by a reduction of the absorption peak of the allowed transitions at lower photon energies as a result of the application of the field. Application of this sum rule to the electroabsorption spectra of Fig. 10.12 implies that the change in absorption [e.g., between spectra (i) and (iii) in Fig. 10.12, $\Delta\alpha = \alpha(i) - \alpha(iii)$] is zero if one inspects the entire spectrum, including photon energies much higher than those plotted in this figure. This sum rule has been tested experimentally and its validity has been demonstrated. There are other sum rules that will not be discussed here.

In summary, the electric-field-induced changes in the optical spectra near the band edge and exciton are significant and result in a rich variety of effects, including exciton Stark shifting, Stark splitting, broadening, and Franz-Keldysh oscillations. We will see in Chap. XVIII that such effects are extremely important in the realization of electro-optical devices.

10-4. Problems

10.1 Use the asymptotic expressions for the Airy function

$$\lim_{x\to\infty} \mathrm{Ai}(x) = \frac{1}{2\sqrt{\pi}x^{1/4}} e^{-\frac{2}{3}x^{3/2}} \left[1 - \frac{3c_1}{2x^{3/2}}\right], \qquad (10.19)$$

with $c_1 = 15/216$ and its derivative

$$\lim_{x\to\infty} \mathrm{Ai}'(x) = \frac{x^{1/4}}{2\sqrt{\pi}} e^{-\frac{2}{3}x^{3/2}} \left[1 + \frac{21c_1}{10x^{3/2}}\right], \qquad (10.20)$$

to derive the Franz-Keldysh absorption spectrum (10.11) from Eq. (10.7) using only the leading order contribution.

10-5. References

1. L. V. Keldysh, Soviet Physics, JETP 7, 788 (1958); W. Franz, Z. Naturforsch **139**, 484 (1958).

2. D. F. Blossey, Phys. Rev. B **3**, 1382 (1971).

3. K. Tharmalingam, Phys. Rev. **130**, 2204 (1963).

4. D. A. B. Miller, D. S. Chemla, and S. Schmitt-Rink, Phys. Rev B **33**, 6976 (1986).

5. D. F. Blossey, Phys. Rev. B **2**, 3976 (1970).

6. P. I. Perov, L. A. Avdeeva, and M. I. Elinson, Sov. Phys. Solid State **11**, 438 (1969).

7. B. O. Seraphin and N. Bottka, Phys. Rev. Lett. **6**, 134 (1965).

8. D. A. B. Miller, D. S. Chemla, T. C. Damen, A. C. Gossard, W. Wiegmann, T. H. Wood, and C. A. Burrus, Phys. Rev. B **32**, 1043 (1985).

9. J. S. Weiner, D. A. B. Miller, D. S. Chemla, T. C. Damen, C. A. Burrus, T. H. Wood, A. C. Gossard, and W. Wiegmann, Appl. Phys. Lett. **47**, 1148 (1985).

10. K. Yamanada, T. Fukunnaga, N. Tsukada, K. L. I. Kobayashi, and M. I. Shii, Appl. Phys. Lett. **48**, 840 (1986).

Chapter XI*
TWO-PHOTON ABSORPTION SPECTROSCOPY

Absorption spectroscopy, the study of the variation of the absorption coefficient with light frequency, is one of the most powerful techniques for obtaining information on the optical and electronic properties of semiconductors. In Chaps. V and VI it was shown that the linear absorption spectrum depends sensitively on the band structure of a material. However, the amount of information that can be obtained from conventional linear absorption spectroscopy is limited. A more powerful, but experimentally more difficult method is two-photon spectroscopy.

In this chapter we discuss the selection rules that have to be satisfied for two-photon absorption to take place. We then give some examples of two-photon absorption spectra measured in bulk semiconductors. The use of two-photon absorption in quantum wells to obtain information about the valence and conduction band offsets is included. Finally, we describe two-photon absorption in quantum dots, which demonstrates the energy shifts caused by the mixing of the valence-band states due to the quantum confinement.

11-1. Selection Rules for Two-Photon Spectroscopy

In the two-photon absorption technique the crystal is illuminated with two light sources, one intense source of fixed frequency ω_1 and a weaker source of tunable frequency ω_2. Both photons, $\hbar\omega_1$ and $\hbar\omega_2$, correspond to the transparency region of the crystal with $\hbar\omega_1$, $\hbar\omega_2 < E_g$, and are therefore not absorbed. However, there is a possibility of inducing an electronic transition from the valence to the conduction band by a process in which a pair of photons, $\hbar\omega_1 + \hbar\omega_2$, share their energy to complete the transition. The probability for such a process is small and could only be measured after intense laser sources were developed. Both photons, $\hbar\omega_1$ and $\hbar\omega_2$, need to be simultaneously present for the process to occur. Therefore, an induced absorption of beam ω_2 is measured only during illumination of the crystal with laser beam ω_1, as shown in Fig. 11.1.

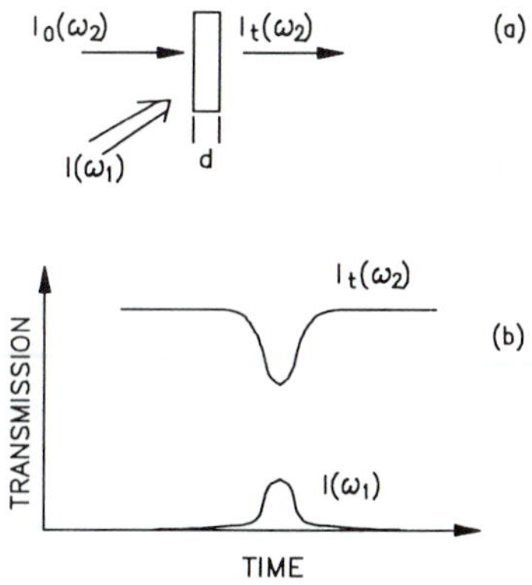

Figure 11.1. (a) Schematic diagram of a two-photon absorption experiment. The transmitted intensity at frequency ω_2 is monitored. (b) The lower curve shows the intensity of the ω_1 beam and the transmitted intensity of the ω_2 beam as a function of time. During the simultaneous presence of the two beams, two-photon absorption takes place, as indicated by the reduced transmission, $I_t(\omega_2)$.

Clearly, the process is nonlinear, with the absorption at frequency ω_1 being proportional to the light intensity at frequency ω_2.

As in Chap. VI, we start with a brief discussion of the general conservation rules in order to determine the conditions under which two-photon absorption is possible. The energy conservation law for a two-photon transition reads (see Fig. 11.2)

$$\hbar\omega_1 + \hbar\omega_2 = \mathcal{E}_n(K_c) \ . \tag{11.1}$$

Similarly, for the momentum conservation law, we have

$$\hbar\mathbf{q}_1 + \hbar\mathbf{q}_2 = \hbar\mathbf{K}_c \ . \tag{11.2}$$

In the two-photon transition, an exciton with energy $\mathcal{E}_n(K_c)$ and with center-of-mass momentum K_c is created. Since the photon wave vectors add vectorially, it is possible to explore the dispersion of the exciton state $\mathcal{E}_n(K_c)$ over a range

$$|q_2| - |q_1| \le K_c \le |q_1| + |q_2| \tag{11.3}$$

by changing the relative orientation of both beams, ω_1, q_1 and ω_2, q_2. We have seen that dipole-allowed excitonic states have a strong dispersion in

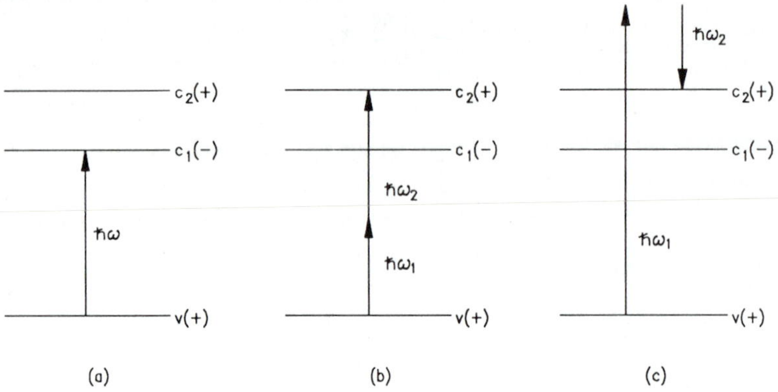

Figure 11.2. (a) Schematic representation of a one-photon absorption process. (b) Schematic representation of a two-photon absorption process starting from the valence band v to the conduction band c_2. (c) Two-photon process where $\hbar\omega_1$ is absorbed and $\hbar\omega_2$ is emitted.

the $\hbar K_c \sim \hbar q$ domain because of the polariton effect. Therefore, two-photon absorption allows the measurement of the exciton-polariton dispersion curve, as shown in Fig. 11.3. In this experiment, two laser beams with wave vectors $\mathbf{q}_1$ and $\mathbf{q}_2$ separated by an angle θ are used. The angle θ is varied and the transmission of one of the beams is measured. The data points in this figure correspond to the peak of the absorption when conditions given by Eqs. (11.1) through (11.3) are satisfied. Using the two-photon process depicted in Fig. 11.2(c), with simultaneous absorption of a photon, $\hbar\omega_1$, and emission of a photon, $\hbar\omega_2$, the phonon-polariton dispersion curve could be similarly measured.

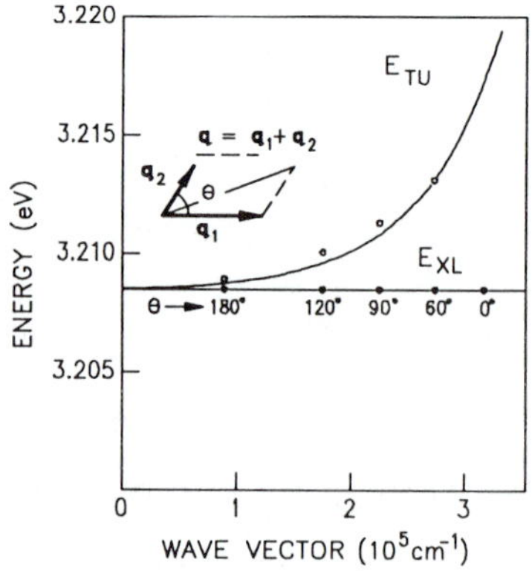

Figure 11.3. Polariton dispersion curve of the n = 1 exciton in CuCl as measured by two-photon absorption. E_{TU} = upper transverse polariton and E_{XL} = longitudinal exciton (after Ref. 11.1).

We now consider the angular momentum and parity conservation selection rules. In a crystal with inversion symmetry when parity is a good quantum number, one-photon transitions are dipole-allowed between states of opposite parity. Two-photon transitions are allowed between states with the same parity. This is so because each photon involved in the transition causes a parity change of one (see Chap. VI). The double change of parity restores the parity of the initial state. Therefore, one-photon allowed transitions are two-photon forbidden, and vice-versa (see Fig. 11.2).

In a noncentrosymmetric crystal, parity selection rules do not hold since parity is not a good quantum number. In such crystals, exciton states allowed in one-photon transition can also be observed in two-photon absorption. However, the number of possible excitonic states that can be reached in a two-photon transition is larger. Therefore, valuable additional information about the band structure of the crystal can be obtained from multiphoton spectroscopy.

From angular momentum conservation it is possible to understand which type of excitonic transitions are expected in a two-photon process. Using similar arguments as those presented in Chap. VI, we conclude that the total angular momentum transferred from the photons to the exciton is

$$\Delta \ell = \ell_1 - \ell_2 = 0 \quad \text{or} \quad \Delta \ell = \ell_1 + \ell_2 = 2 \,. \tag{11.4}$$

Therefore, the total angular momentum of the exciton is 0 or 2. As already discussed in Chap. VI, the exciton angular momentum can be split into an internal part and an envelope contribution,

$$\ell_{exc} = \ell_{int} + \ell_{env} \,. \tag{11.5}$$

In a centrosymmetric crystal with p-like valence band and s-like conduction band, the two-photon transition to a 1s exciton is not allowed because $\ell_{exc} = 1$ ($\ell_{int} = 1$, $\ell_{env} = 0$). However, the two-photon transition to a p-like exciton is possible since $\ell_{exc} = 2$ ($\ell_{int} = 1$, $\ell_{env} = 1$). Therefore, the two-photon absorption spectrum will start at the photon energy

$$\hbar\omega_2 + \hbar\omega_1 = E_g - E_B/4 \,, \tag{11.6}$$

corresponding to a p-like, n = 2 exciton.

On the other hand, if the valence band has s- or d-like character and the conduction band has s-like character, the two-photon transition is possible to the s-like, n = 1 exciton with total angular momentum $\ell = 0$ or 2, with $\ell_{int} = 0$ or 2 and $\ell_{env} = 0$. We have seen in Chap. VI that the one-photon absorption of such a crystal starts with the n = 2, p-like exciton. Therefore, a two-photon absorption spectrum has the characteristic of a first-class exciton spectrum when the one-photon spectrum is of the second-class type, and vice-versa. The precise selection rules for two-photon absorption in crystals are obtained from group theory, which is beyond the scope of this book. The interested reader is referred to Refs. 11.2-11.4.

The two-photon absorption coefficient $\alpha(\omega_2)$ depends on the polarization vectors $\mathbf{e}_1$ and $\mathbf{e}_2$ of the two incident beams with respect to the crystal axes. This information may be used to identify the symmetry of the exciton state accurately (see Ref. 11.5). The expected variation of the two-photon absorption has been calculated for all possible excitonic states corresponding to high symmetry points in the first Brillouin zone, such as the Γ point or X point, for the different crystal structures. The basic expression for the probability of a two-photon transition process is obtained from second-order time-dependent perturbation theory:

$$P = \frac{2\pi}{\hbar} \sum_{i,f} |W_{if}|^2 \, g(\mathcal{E}) \ . \tag{11.7}$$

Here $g(\mathcal{E})$ is the joint density of states that we have encountered in Chaps. V and VI for the three-dimensional case, and in Chaps. VIII and IX for the two-dimensional, one-dimensional, and zero-dimensional cases. In the dipole approximation, the composite transition matrix element W_{if} is:

$$W_{if} = \frac{e^2}{m^2c^2} A_1 A_2 \sum_{\ell} \left(\frac{\langle f|\mathbf{e}_1 \mathbf{p}|\ell\rangle\langle\ell|\mathbf{e}_2 \mathbf{p}|i\rangle}{\mathcal{E}_{\ell i} - \hbar\omega_2} + \frac{\langle f|\mathbf{e}_2 \mathbf{p}|\ell\rangle\langle\ell|\mathbf{e}_1 \mathbf{p}|i\rangle}{\mathcal{E}_{\ell i} - \hbar\omega_1} \right) . \tag{11.8}$$

A_1 and A_2 are the vector potentials at frequencies ω_1 and ω_2. Here $\mathcal{E}_{\ell i}$ is the energy difference between the initial state and an intermediate state. The summation runs over all states of the systems (all valence and conduction bands) called intermediate states.

It is worth examining one aspect of the two-photon matrix element in more detail. It is first noted that it contains energy denominators of the form

$$\Delta\mathcal{E}_1 = \mathcal{E}_{\ell i} - \hbar\omega_1 \quad \text{and} \quad \Delta\mathcal{E}_2 = \mathcal{E}_{\ell i} - \hbar\omega_2 \ . \tag{11.9}$$

Suppose that we consider a two-photon transition from the valence band v to the second highest conduction band c_2 and that we pick c_1, the lowest conduction band, as an intermediate state (see Fig. 11.4). The contribution of this particular path to the transition probability amplitude can be viewed as follows. The first photon, $\hbar\omega_1$, lifts the electron from v to c_1.

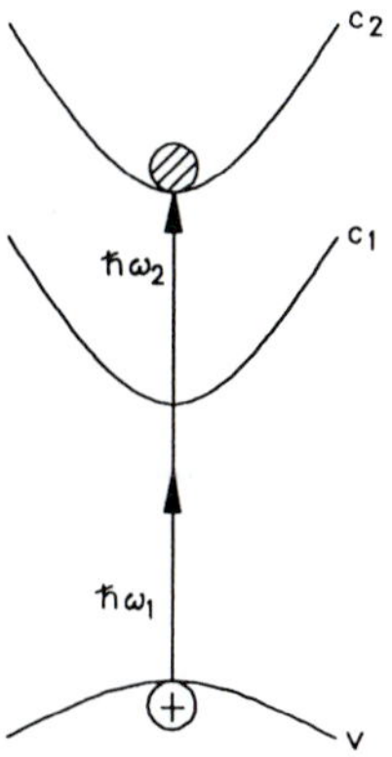

Figure 11.4. Two-photon absorption process with the first conduction band c_1 as the resonant intermediate state.

However, since energy conservation is not satisfied, the transition only occurs "virtually," during a time Δt, given by the uncertainty principle $\Delta t = \hbar/\Delta\mathcal{E}_1$. During the uncertainty time the presence of a photon, $\hbar\omega_2$, is required to complete the transition by inducing the electron transfer from c_1 to c_2. Both photons are absorbed after completion of the process, as required by the energy conservation law

$$\hbar(\omega_1 + \omega_2) = \mathcal{E}_{c_2} - \mathcal{E}_v \, . \tag{11.10}$$

We note that the transition amplitude depends on the individual photon energies for a given total photon energy and constant photon flux. In particular, the transition probability diverges as the energy of one of the incident photons approaches a resonance in the system (for example, when $\hbar\omega_1 = \mathcal{E}_{c_1} - \mathcal{E}_v$ in Fig. 11.4). Furthermore, we can see that the matrix element has the property of a tensor depending on terms like $(\mathbf{e}_1 p)(\mathbf{e}_2 p)$, as mentioned already.

An exact calculation of the two-photon transition probability requires the knowledge of the matrix elements connecting the initial and final states to all other states of the system, including excitonic effects. Therefore, approximations are introduced that often involve dropping most of the terms in the summation. For instance, one useful approximation consists of reducing the crystal band structure to the initial valence band and the final conduction band. This is referred to as the two-band model (see Ref. 11.6).

As we mentioned before, a two-photon spectrum is experimentally obtained by measuring the change of transmission of a probe source ω_2 induced by illumination of the crystal with an intense short pulse of light at frequency ω_1 as a function of ω_2. The induced attenuation of the probe beam, ω_2, can be written as (see Fig. 11.1)

$$\frac{dI(\omega_2)}{dz} = -\beta I(\omega_1)\, I(\omega_2) \ , \tag{11.11}$$

or

$$\frac{dI(\omega_2)}{I(\omega_2)} = -\beta I(\omega_1)\, dz \ , \tag{11.12}$$

or

$$I(\omega_2, t) = I_0(\omega_2)\, e^{-\beta I(\omega_1, t)\, d} \ . \tag{11.13}$$

Here β is the two-photon absorption coefficient, $I_0(\omega_2)$ is the incident probe intensity, and d is the sample thickness. In deriving Eq. (11.13) we ignored the depletion (space dependence) of $I(\omega_1)$. The exponent in Eq. (11.13) is sometimes written in terms of the photon flux $F(\omega_1)$ at frequency ω_1 (the number of photons crossing a unit area per unit time) with $\hbar\omega_1 F(\omega_1) = I(\omega_1)$. Sometimes the two-photon cross-section per unit cell is introduced, $\delta' = \beta\hbar\omega_1/N$. Typical values of δ' in a semiconductor are $\delta' \sim 10^{-49}$ cm^4 sec., corresponding to an induced absorption, $\beta I(\omega_1) \sim 1$ cm^{-1}, for $I(\omega_1) \sim 10^7$ W/cm^2. Thus, induced absorption signals are rather weak and require the precise time-resolved measurement of I_0, $I_t(\omega_2)$, and $I(\omega_1)$ at time $\Delta t = 0$.

A simpler experimental method is two-photon excitation spectroscopy. Here the sum of the two photon energies again adds up to the transition energy; however, the luminescence resulting from the creation of electron-hole pairs by two-photon absorption is measured instead of the transmission of one beam. The amount of luminescence resulting from the creation of electron-hole pairs is measured as a function of the frequency of the incident tunable laser source, keeping its intensity constant. If the luminescence intensity is directly proportional to the density of electron-hole pairs, the magnitude of the luminescence signals gives the value of the two-photon absorption coefficient. The two-photon excitation method may be quite sensitive as a result of the extremely low background signal as opposed to the transmission measurement. It amounts to detecting a luminescence signal at frequency $\hbar\omega \simeq E_g$, while rejecting incident laser photons at $\hbar\omega_i \sim E_g/2$. However, caution must be exercised since this method does not discriminate between signals from two-photon absorption and from two-step transitions, with real impurity levels in the forbidden gap acting as real intermediate states. Also, it is more difficult to obtain the two-photon transition cross-section because it requires a knowledge of the luminescence efficiency (the probability of one excited electron-hole pair to emit a photon).

11-2. Examples of Two-Photon Absorption Spectra in Bulk Semiconductors

Figure 11.5 shows the one- and two-photon absorption spectra in Cu_2O. Because Cu_2O has an inversion symmetry, one expects different exciton states in each case. The one-photon absorption spectrum is a good example of second-class transitions, starting with the n = 2 term with p envelope. As discussed before, s and d excitons are allowed in a two-photon transition and are indeed clearly observed. The comparison between both spectra reveals the large splitting between 2p and 2s excitons.

Figure 11.6 shows how two-photon spectroscopy can bring new information, not available from one-photon spectroscopy. In CuCl there is no inversion symmetry so that the parity selection rule does not apply. In the one-photon absorption spectrum, excitons are observed below the bandgap at $E_g \simeq 3.5$ eV, but no other structure is seen at higher photon energies [see Fig. 11.6(a)]. In the two-photon spectrum the same excitons are observed. However, in addition, an exciton transition is clearly seen at

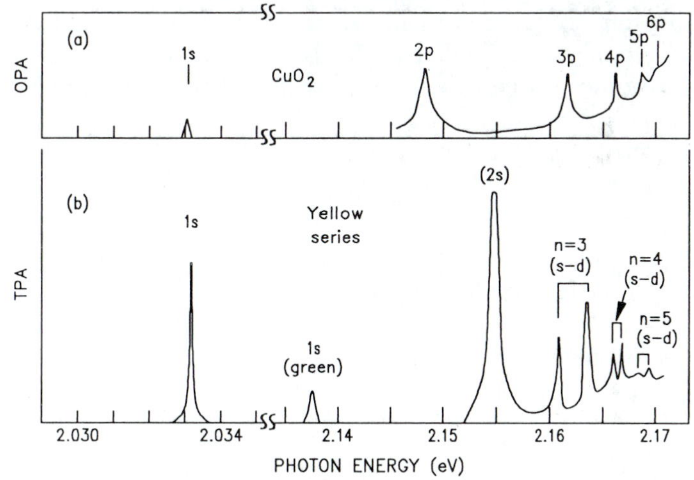

Figure 11.5. (a) One-photon absorption (OPA) and (b) two-photon absorption spectra of Cu_2O at T = 2 K. The different exciton lines of the yellow series are clearly seen. The s-d lines denote exciton states with s and d characters in the envelope function (after Ref. 11.7).

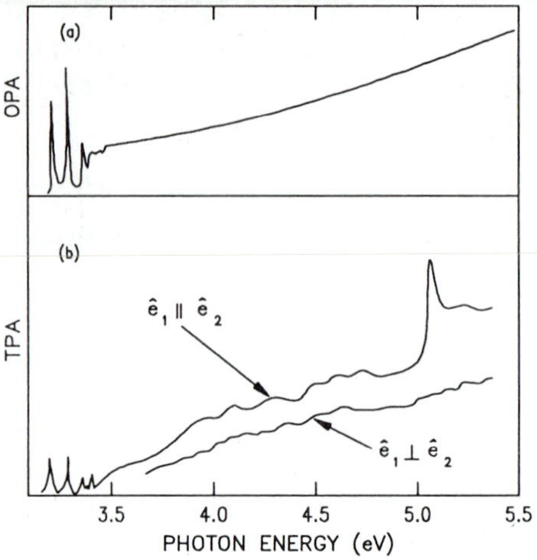

Figure 11.6. One-photon absorption (OPA) and two-photon absorption of CuCl. In || configuration, both polarization vectors $\mathbf{e}_1$ and $\mathbf{e}_2$ are parallel to [100] direction; in $\perp$configuration, $\mathbf{e}_1$ is in [100] and $\mathbf{e}_2$ is in [010] direction. The line at 5.08 eV seen in two-photon absorption (TPA) corresponds to an exciton from a deeper valence band (after Ref. 11.8).

$\hbar\omega_1 + \hbar\omega_2 = 5.08$ eV, but only for parallel polarization vectors $\mathbf{e}_1 \parallel \mathbf{e}_2$. This exciton can be assigned to transitions starting from another valence band, lying 1.79 eV below the upper valence band.

Two-photon spectroscopy can be extended to three-photon processes. In a centrosymmetric crystal, the dipole operator with negative parity is applied three times, so that the final state must have negative parity. Thus, three-photon spectroscopy reaches the final states that are accessible in linear, one-photon spectroscopy. However, additional excitonic states with other symmetries are also accessible. An example of three-photon spectroscopy is shown in Fig. 11.7(a).

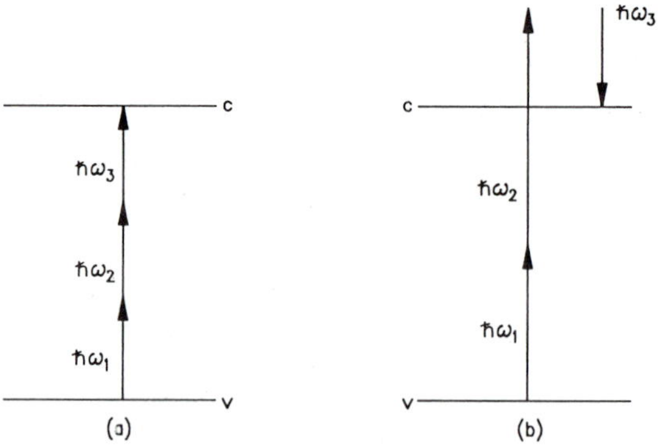

Figure 11.7. Three-photon transition; (a) three photons are absorbed; (b) two photons are absorbed and one photon is emitted.

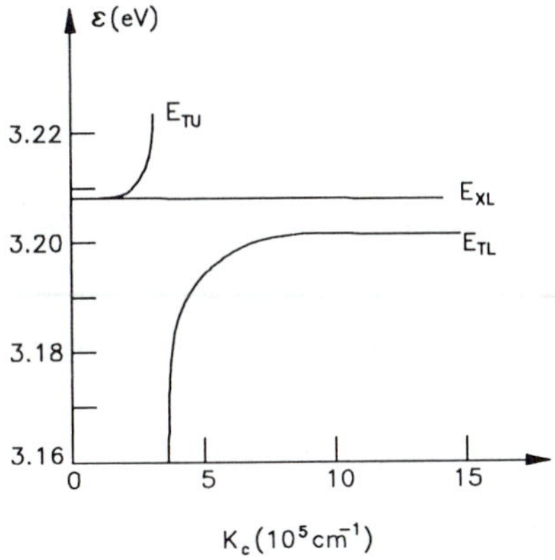

Figure 11.8. The lower polariton (E_{TL}) and the longitudinal exciton (E_{XL}) dispersion curves measured by two-photon Raman scattering in CuCl at 4 K (after Ref. 11.9). The data points for the upper polariton (X_{TU}) are the same as those in Fig. 11.3, measured by two-photon absorption (after Ref. 11.1).

A variant of the three-photon absorption process is one in which two photons are absorbed, with simultaneous induced emission of a third photon [see Fig. 11.7(b)]. This allows probing of a polariton curve in a larger region of k space. The dispersion curve of the polariton measured in this way is shown in Fig. 11.8.

11-3. Quantum-Well Structures

We now examine the kind of information that two-photon absorption spectroscopy can bring when applied to the case of quantum-well structures. We recall that a single quantum well consists of a very thin layer of a semiconductor with a thickness of the order of 10 nm or less, imbedded between two thicker layers of a semiconductor of the same crystal lattice constant but larger bandgap energy (see Chap. VIII). The exciton states in the quantum well have nearly two-dimensional character, due to the confinement of electrons and holes in the thin layer. Similar to Eq. (8.81), the exciton wave function in a quantum well at $K_c = 0$ can be written as

$$\psi(r) = \phi_n^{xy}(r_{xy})\, u_c u_v \zeta_{ci} \zeta_{vj} \,, \tag{11.14}$$

where $\phi_n^{xy}(r_{xy})$ is the exciton envelope function in the xy plane parallel to the layer surface, n is the principal quantum number for the hydrogenic envelope function (n = 1 for s-like function, n = 2 for s- and p-like function, etc.), $u_c(r)$ and $u_v(r)$ are the Bloch functions for the electron and the hole which are the atomic functions at the origin of the bands, and $\zeta_{ci}(z)$ and $\zeta_{vj}(z)$ are the quantized electron and hole wave functions in the z direction normal to the layer. The quantum numbers labeling the quantized electron and hole subbands are represented by i and j. We also recall that the transition selection rule $\Delta j = 0$ applies between electron and hole subbands.

By applying similar arguments as in Sec. 11-1, we can identify the transitions that are two-photon allowed. It is necessary to first specify the polarization configuration for the incident beams. If both beams propagate along the z axis, then the polarization vectors always lie in the xy plane. Since the dipole operator is directed along the incident polarization vector (see Chap. V), it only affects the part of the wave function lying in the xy plane. In the two-band approximation, therefore, we only need to consider transition matrix elements of the type $\langle u_{ci}|ep|u_{vi}\rangle$ for the first step in the transition, and $\langle \phi_n|ep|\phi_n\rangle$ for the second step. The first step corresponds to an interband transition with the selection rule $\Delta j = 0$ applying, and the second step corresponds to a transition from an exciton with s-like to p-like envelope function. The situation does not differ from that of the three-dimensional case. Assuming a material with inversion symmetry in the xy plane, we retrieve the case discussed in Sec. 11-1. Transitions that display n = 1 excitons with s-like envelope function $\phi(xy)$ in one-photon absorption will start with the n = 2 exciton with p-like envelope function in the xy plane in two-photon absorption, and vice-versa. An example is shown in Fig. 11.9. From the energy difference between the 1s exciton observed in one-photon absorption and the 2p exciton seen in two-photon absorption, it is possible to obtain the exciton Rydberg energy (the 1s exciton binding energy) of the quantum well.

We now consider the case when the polarization vector lies along the z axis. This is obtained, for instance, by illuminating a sample sideways containing a quantum-well structure embedded in a waveguide geometry in order to increase the coupling between the incident radiation and the sample (see Fig. 11.10). When the polarization vector is parallel to z, it operates on the part of the wave function that depends on z. As before, we have to apply the dipole operator twice, remembering that in each step a change of parity must take place, as shown schematically in Fig. 11.11. The first transition promotes the electron from valence subband hi to conduction subband ej with the same subband quantum number because of selection rule $\Delta j = 0$ and $\Delta \ell = 1$. In the second step, the transition occurs between valence or conduction subbands of different parity, with a

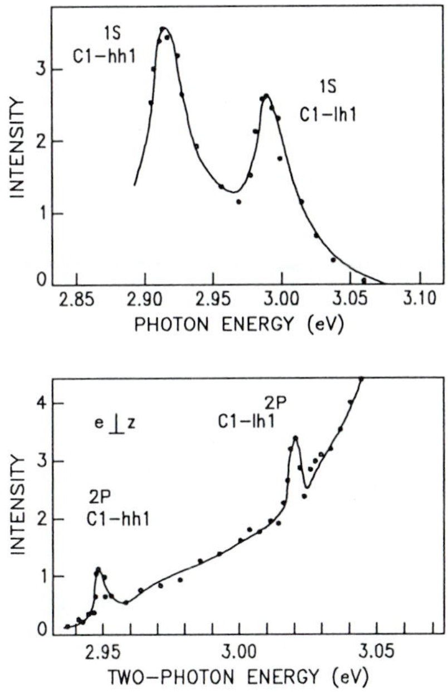

Fig. 11.9. (a) One- and (b) two-photon absorption spectra in ZnSe/ZnS quantum wells with 20-Å well and barrier sizes at 2 K. In Fig. 11.9(a) the two peaks are the heavy- and light-hole excitons. Note the energy scale difference between the two parts of the figure (after Ref. 11.10).

transition matrix element of the form $\langle \zeta_{ej} | \mathbf{e}p | \zeta_{ej'} \rangle$ or $\langle \zeta_{hi} | \mathbf{e}p | \zeta_{hi'} \rangle$. This type of transition is allowed between functions of different parity, i.e., between even and odd states with $\Delta j = 1,3$. Therefore, the two-photon selection rule $\Delta j \neq 0$ applies in quantum wells if the polarization vector is parallel to z. Note that if one includes excitonic effects, the excitonic envelope function in the final state should be s-like to satisfy the angular momentum selection rule, which requires $\Delta \ell = 0$ or 2.

By comparing the results obtained in one- and two-photon absorption with the $\mathbf{e} \parallel z$ configuration, one directly obtains the energy difference between successive subband states either in the conduction or in the valence band. This type of information is useful since it can clarify the band offset in the quantum-well structure. Indeed, if the electron (hole) mass is known, it is possible to determine the band offset ΔE_c (ΔE_v) using the procedure described in Chap. VIII. As a consistency check, the values obtained must satisfy the condition

$$\Delta E_c + \Delta E_v = E_g^A - E_g^B \,, \tag{11.15}$$

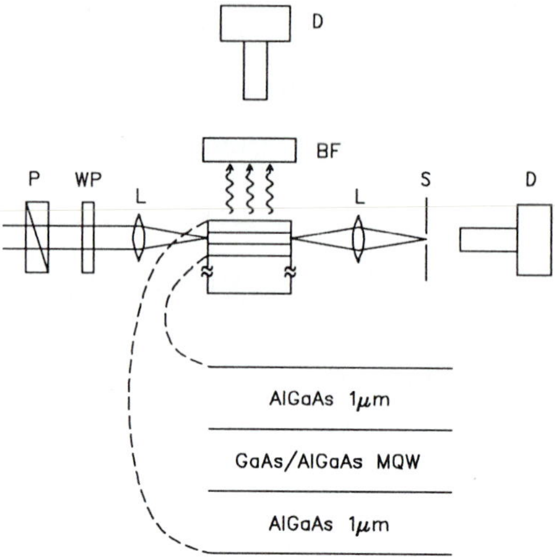

Fig. 11.10. Experimental setup for the waveguide sample geometry. P, polarizer; WP, zero-order half-wave plate; L, lens; S, slit; D, detector; and BF, 10-nm bandpass filter. The transmission band of the filter coincides with the sample luminescence (after Ref. 11.11).

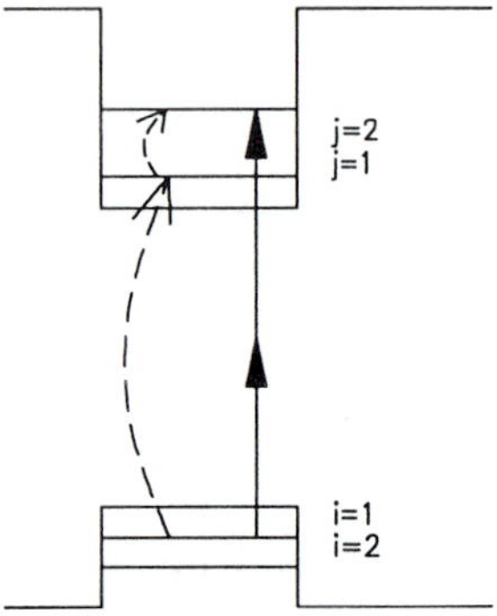

Fig. 11.11. Schematic diagram showing a two-photon transition for **e** || **z**. The dotted lines show a transition path. In the first step, an interband transition occurs with $\Delta j = 0$. The second step completes the two-photon transition to a final state with $\Delta j = 1$.

where E_g^A and E_g^B are the bulk bandgap values of material A and B, respectively, forming the wells and barriers. An example is shown in Fig. 11.12 for the case of a GaAs quantum well.

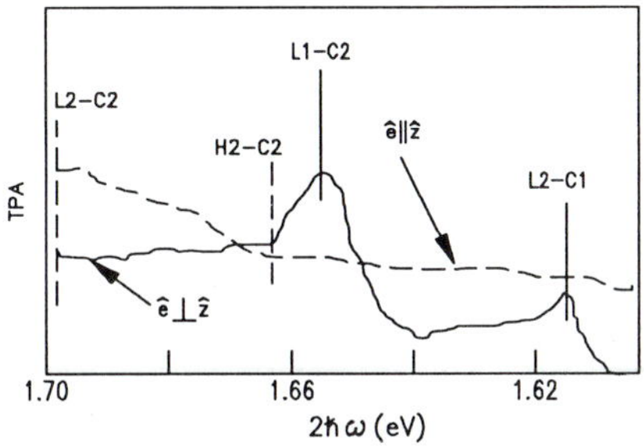

Fig. 11.12. Two-photon absorption spectra at 5 K of GaAs/AlGaAs quantum wells with 100Å/150Å well and barrier thickness for the **e** ∥ **z** (solid line) and **e** ⊥ **z** (dashed line) configurations. Vertical lines indicate the positions of the exciton features observed in two-photon (solid line) and one-photon (dashed line) absorption. The symbols like L1 - C2 indicate transition from the light-hole exciton with quantum number i = 1 to the electron sublevel with quantum number j = 2 (after Ref. 11.11).

11-4. Quantum Dots

Two-photon spectroscopy also yields valuable information about the energy states in semiconductor quantum dots. We recall from our discussion in Chap. IX that the energy levels in quantum dots under idealized conditions can be labeled with the quantum numbers n,ℓ,m (see Sec. 9-4). As shown in Fig. 9.6, dipole-allowed one-photon transitions are possible only if the quantum numbers of the electron and hole are the same, $\Delta\ell = 0$. Hence, only the $e_{1s}h_{1s}$, $e_{1p}h_{1p}$ ... states can be observed through one-photon spectroscopy.

As discussed in the previous sections of this chapter, two-photon absorption leads to an additional angular momentum change of $\Delta\ell = \pm 1$. Hence, in quantum dots the transitions to the states $e_{1s}h_{1p}$, $e_{1p}h_{1s}$, ... are two-photon allowed (see Fig. 11.13). Taking this model at face value, one can use Eq. (9.42) to calculate the one- and two-photon absorption spectra for CdS quantum dots, as shown in Fig. 11.14. Effective masses appropriate for the A band in bulk wurzite CdS, $m_e \simeq 0.2\ m_0$ and $m_h \simeq 1.35\ m_0$ ($m_\perp = 0.7\ m_0$, $m_\parallel = 5\ m_0$, and thus, $m_h = \sqrt[3]{(m_\perp^2 m_\parallel)} = 1.35\ m_0$), have been used. A phenomenological broadening parameter has been introduced to account for the homogeneous linewidth and the inhomogeneous contribution from the distribution of dot sizes. The energy differences between the peaks observed in one- and two-photon absorption (see Fig. 11.13) are due to the different allowed transitions. These energy differences are related to the ratio of electron and hole effective masses by

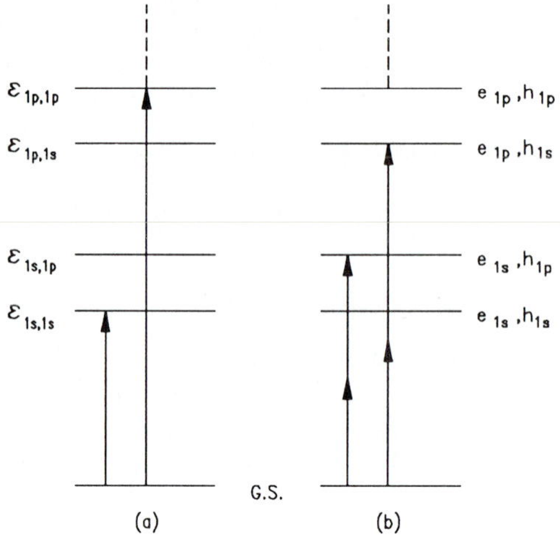

Figure 11.13. Schematic representation of (a) one-photon and (b) two-photon excitation in a semiconductor quantum dot.

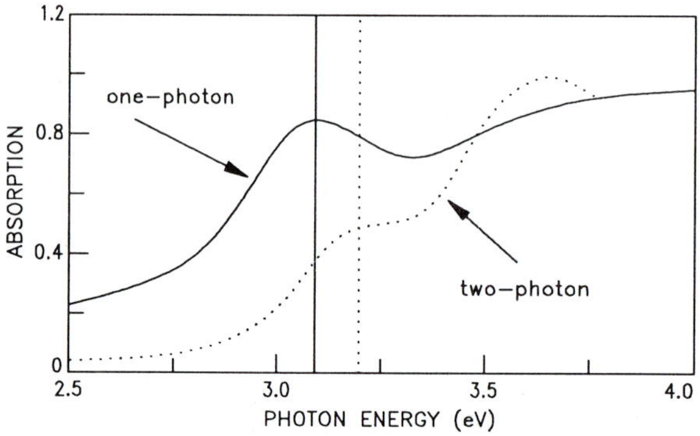

Figure 11.14. Theoretical calculation of one- and two-photon absorption based on a parabolic band model (after Ref. 11.12).

$$\frac{\mathcal{E}_{(1p_h - 1s_e)} - \mathcal{E}_{(1s_h - 1s_e)}}{\mathcal{E}_{(1s_h - 1p_e)} - \mathcal{E}_{(1s_h - 1s_e)}} = \frac{\mathcal{E}_{(1p_h - 1p_e)} - \mathcal{E}_{(1s_h - 1p_e)}}{\mathcal{E}_{(1p_h - 1p_e)} - \mathcal{E}_{(1p_h - 1s_e)}} = \frac{m_e}{m_h}, \tag{11.16}$$

where $\mathcal{E}$ is the energy of the subscripted transition. Using the masses for CdS, the first two-photon resonances are expected to be situated between the first two one-photon transitions, as seen in Fig. 11.14. The splitting between the one- and two-photon absorption peaks is a result of the nondegeneracy of the valence-band states.

To check the predictions of this simple calculation, experimental one- and two-photon absorption spectra for CdS quantum dots in glass were compared. In contrast to the simple model calculations of Fig. 11.14, the experiment shows (see Fig. 11.15) a nearly degenerate one- and two-photon resonance.

These experimental results could be reproduced by calculations that include the mixing of the different valence bands due to the confinement potential of the quantum dot. Even though the details of these calculations are beyond the scope of this book, it is intuitively clear that it is likely that the quantum confinement changes the allowed energy eigenvalues. From our discussion of the properties of semiconductor materials in Chap. VII, we recall that the band structure of the different materials is strongly influenced by the symmetry properties. Now in a quantum dot, the finite size and the resulting confinement potential clearly lower the symmetry of the structure in comparison to the corresponding bulk material. This symmetry reduction leads to band structure modifications, which, for the case of CdS quantum dots, explains the near degeneracy of the one- and two-photon transition.

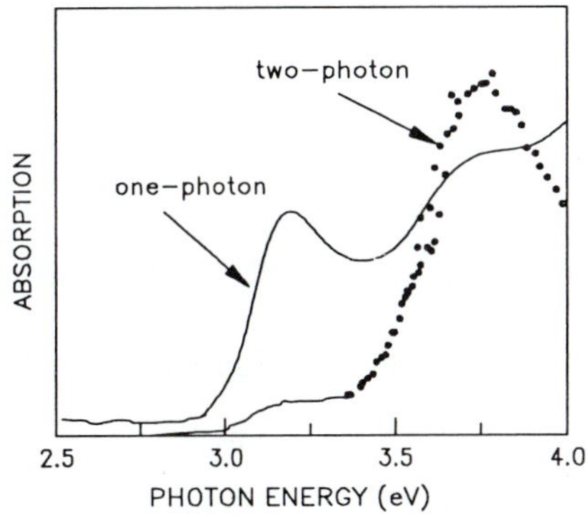

Figure 11.15. Experimental results of one-photon (solid line) and two-photon (dots) absorption spectra for a CdS quantum dot sample with average radius of $R = 0.9 \pm 0.7$ nm. No observable differences are seen in the transition energies between the one- and two-photon spectra (after Ref. 11.12).

11-5. Problems

11.1. What are the units of the two-photon absorption coefficient β and the photon flux $F(\omega)$?

11.2. Write down an equation similar to Eq. (11.11) for the case where the two-photon absorption occurs for identical photons; i.e., both photons are absorbed from the same beam (a one-beam experiment). Solve the resulting differential equation for the transmitted light.

11.3. Prove Eq. (11.16).

11-6. References

1. D. Fröhlich, E. Mohler, and P. Wiesner, Phys. Rev. Lett. **26**, 554 (1971).

2. T. R. Bader and A. Gold, Phys. Rev. **171**, 997 (1968).

3. M. Inoue and Y. Toyozawa, Phys. Soc. Jpn. **20**, 363 (1965).

4. A. Pasquarello and A. Quattropani, Phys. Rev. B **43**, 3837 (1991); see also, M. M. Denisov and V. P. Makarov J. Phys. C **5**, 2651 (1972).

5. D. Fröhlich, B. Staginnus, and E. Schönherr, Phys. Rev. Lett. **19**, 1032 (1967).

6. G. D. Mahan, Phys. Rev. B **170**, 825 (1968).

7. D. Fröhlich, R. Kenklies, Ch. Uihlein, and C. Schwab, Phys. Rev. Lett. **43**, 1260 (1979); see also, Phys. Rev. B **23**, 2731 (1981).

8. D. Fröhlich and M. Volkenandt, Solid State Commun. **43**, 189 (1982).

9. B. Hönerlage, A. Bivas, and Vu Duy Phach, Phys. Rev. Lett. **41**, 49 (1978).

10. F. Minami, K. Yoshida, J. Gregus, K. Inoue, H. Fujiyasu, in *Optics of Excitons in Confined Systems*, edited by A. D'Andrea, R. Del Sole, R. Girlanda, and A. Quattropani (IOP Publishing Ltd., 1991), pp. 249-253.

11. K. Tai, A. Mysyrowicz, R. J. Fisher, R. E. Slusher, and A. Y. Cho, Phys. Rev. Lett. **62**, 1784 (1989).

12. K. I. Kang, B. P. McGinnis, Sandalphon, Y. Z. Hu, S. W. Koch, N. Peyghambarian, A. Mysyrowicz, L. C. Liu, and S. H. Risbud, Phys. Rev. B **45**, 3465 (1992).

Chapter XII*
BIEXCITONS, ELECTRON-HOLE LIQUID, AND PLASMA

It was shown in Chap. VI that light absorption by a semiconductor can be described in terms of generation of electron-hole pairs. Electron-hole pairs initially created by photons with energy $\hbar\omega$, larger than the exciton ground-state energy $\mathcal{E}_1$ by an excess amount $\Delta\mathcal{E}$, given by $\Delta\mathcal{E} = \hbar\omega - \mathcal{E}_1$, rapidly lose their excess energy by emission of optical and acoustic phonons until they reach the lowest exciton level, n = 1. We have also seen that each excitonic level, n = 1, 2, ..., forms a band of energy by itself, having an energy dispersion [see Eq. (6.20)]

$$\mathcal{E}_1 = E_g - E_B + \frac{\hbar^2 K_c^2}{2M} \qquad \text{for } n = 1$$

$$\mathcal{E}_2 = E_g - \frac{E_B}{4} + \frac{\hbar^2 K_c^2}{2M} \qquad \text{for } n = 2, \text{ etc. ,} \tag{12.1}$$

where K_c and M are the exciton center-of-mass wave vector and mass, respectively. The last term in Eq. (12.1) expresses the kinetic energy of the freely propagating exciton. Thus, after fast energy relaxation (the process of dissipation of excess energy) and before recombination (the process of annihilation of an electron-hole pair either by emission of photons or through a nonradiative process), the ensemble of excitons forms a gas similar to a gas of free hydrogen atoms. As the exciton density increases as a result of increased excitation intensity, the excitons can bind to form excitonic molecules, also called biexcitons, a process similar to the case of hydrogen molecules. For even higher densities, larger exciton complexes may also form, leading to electron-hole droplets, or even an electron-hole liquid (EHL) phase. In this chapter we discuss optical properties of semiconductors associated with these forms of excitonic complexes.

12-1. Biexcitons or Excitonic Molecules

The formation of a biexciton due to the coupling of a pair of excitons and its analogy to a hydrogen molecule have been experimentally established. Evidence for the existence of such biexciton states in semiconductors was first seen in CuCl (see Ref. 12.1), and later in a large number of other materials. Each excitonic molecule is similar to a hydrogen molecule with a ground state where two holes of opposite spins share two electrons, also with opposite spins. The energy of a biexciton is given by

$$\mathcal{E}_{xx} = 2\mathcal{E}_x^{(0)} - E_{Bxx} + \frac{\hbar^2 K_{xx}^2}{2M_{xx}} = \mathcal{E}_{xx}^{(0)} + \frac{\hbar^2 K_{xx}^2}{2M_{xx}}, \tag{12.2}$$

where $2\mathcal{E}_x^{(0)} = 2(E_g - E_B)$ is the internal energy of two noninteracting excitons at zero wave vectors, $\mathcal{E}_{xx}^{(0)}$ is the biexciton energy at zero wave vector, E_{Bxx} is the biexciton binding energy, and K_{xx} and M_{xx} are the center-of-mass wave vector and mass of the biexciton, respectively. The biexciton mass is twice the exciton mass, $M_{xx} = 2M$.

There are fundamental differences, however, between biexcitons and real hydrogen molecules. First of all, excitons and biexcitons have a finite lifetime. Furthermore, due to the small mass of excitons, typically of the order of the free electron or less (i.e., three orders of magnitude less than real hydrogen atoms), quantum effects are very important for biexcitons. The biexciton zero point energy is comparable to the molecular binding energy itself. In the case of a simple band structure, such as CuCl, at most, one molecular excited level is stable (the first rotational level), in contrast to hydrogen molecules with a rich manifold of excited rotational and vibrational levels.

The binding energy of the biexciton depends on the ratio of effective masses, $\sigma = m_e/m_h$, which are listed in Table 12.1 for selected materials. Results of a variational calculation are shown in Fig. 12.1. The limiting case $\sigma \to 0$ corresponds to the hydrogen molecule, with a well-established experimental value. No experimental verification of the other limit, $\sigma = 1$ (molecule of positronium), exists yet. However, calculations predict that the biexciton binding energy is smaller by one order of magnitude than the binding energy of the hydrogen molecule.

Table 12.1. Exciton and biexciton parameters of some direct-gap materials (after Ref. 12.3).

Material	Point-group	$\sigma = m_e/m_h$	$\mathcal{E}_x$ [eV] exper.	E_B [meV] exper.	$\mathcal{E}_{xx}$ [eV] exper.	E_{Bxx} [meV] exper.
CuCl	T_d	0.25	3.200	190	6.3720	28
CuBr	T_d	0.2	2.9633	108	5.9103	20.5
CdS	C_{6v}	0.14	2.552	28	5.0983	5.7
ZnO	C_{6v}	0.47	3.3750	60.8	6.7310	19
ZnSe	T_d	0.23	2.8025	21	5.6018	2.2
ZnTe	T_d	0.16	2.381	11	4.7605	1.5

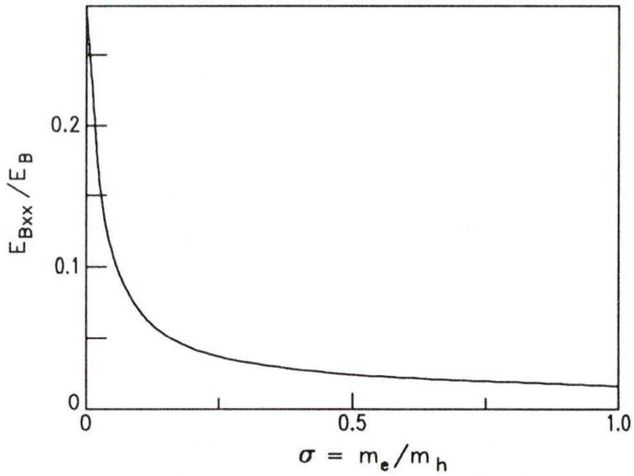

Figure 12.1. Binding energy, E_{Bxx}, of the biexcitons normalized to the exciton binding energy E_B, as a function of the electron-hole-mass ratio, $\sigma = m_e/m_h$ (after Ref. 12.2).

12-2. Biexciton Two-Photon Absorption

Biexcitons may be created by two-photon absorption, thereby giving rise to a strong resonance in the two-photon absorption spectrum. There are two reasons for this strong two-photon absorption process. First, there is an enhancement of the transition rate due to the proximity of an individual photon energy with the exciton energy, as can be seen in Fig.

12.2(a). The detuning between the energy of the exciton and one of the two photons needed for the two-photon absorption process, $\Delta\mathcal{E} = \mathcal{E}_{\mathbf{x}}^{(0)} - \hbar\omega = E_{Bxx}/2$, appears in one of the denominators of the general expression describing two-photon transitions [see Eq. (11.8)]. This leads to an enhancement of the transition rate with the exciton state acting as a nearly resonant intermediate state. A second reason is related to the particular type of transitions involved. Instead of the usual two-photon process, in which one electron is promoted from the valence to the conduction band by interacting with two photons, here we excite two electrons and two holes to form a biexciton. The intermediate state corresponds to one virtual electron-hole pair. To complete the transition, a second electron-hole pair must be created within a distance of the order of a_B. This geometric factor brings a second enhancement to the transition probability. Both effects combine to give an enhancement of the two-photon absorption, typically 5-6 orders of magnitude larger than a usual two-photon transition toward an exciton (see Ref. 12.2). An example of a two-photon absorption to the biexciton state of CuCl is shown in Fig. 12.2(b).

By using two independent laser beams so that one photon from each beam is absorbed in the two-photon absorption process, it is possible to obtain information on the symmetry of the biexciton state by varying the polarization vector of each beam. For instance, if both laser beams have the same circular polarization, there is no allowed two-photon absorption to the biexciton state, while two-photon absorption is allowed if the two beams are oppositely circularly polarized. This confirms that the created stable biexciton state has two electrons with opposite spin orientation. To understand this, we note that the angular momentum carried by each

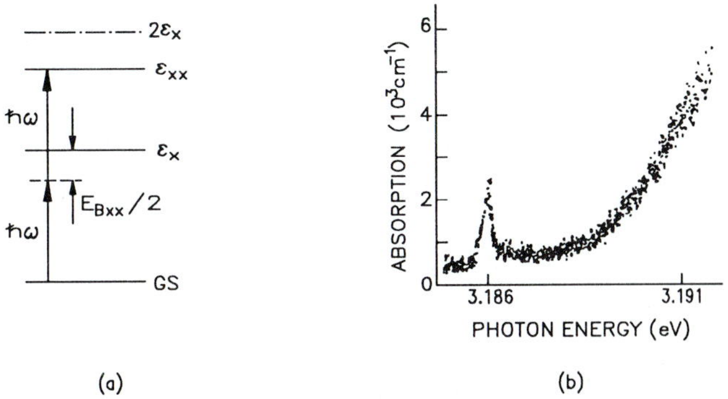

Figure 12.2. (a) Schematic representation of exciton and biexciton energy levels and the two-photon absorption process to the biexciton state. (b) The biexciton two-photon absorption observed in a 4-μm thin film of CuCl grown on NaCl (after Ref. 12.4).

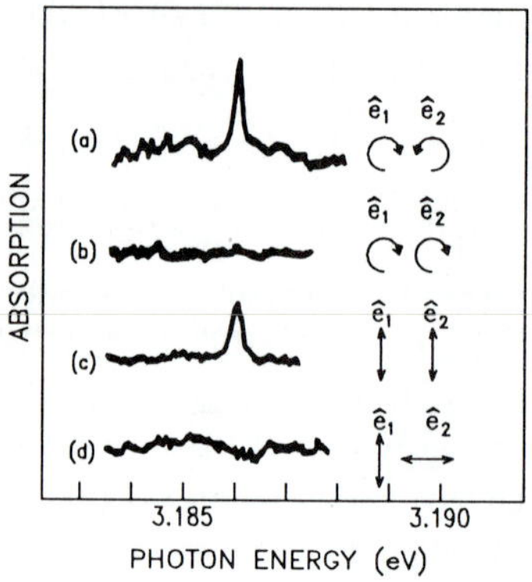

Figure 12.3. Two-photon absorption due to biexcitons in CuCl at 5 K for different polarization vectors of the incident beams. The transmission of beam 1 with polarization vector $\mathbf{e}_1$ is measured in the presence of beam 2 with polarization $\mathbf{e}_2$. The disappearance of the two-photon signal for configurations presented in (b) and (d) allows the determination of the symmetry of the biexciton state (after Ref. 12.5).

absorbed photon is transferred to the spin of the electrons in the biexciton. Since the photon angular momenta are opposite, the spins of the electrons are also opposite. Therefore, the biexciton is a paramolecule, similar to the ground state of the hydrogen molecule. The dependence of the biexciton two-photon absorption with laser polarization is shown in Fig. 12.3.

12-3. Biexciton Luminescence

We now examine how the presence of biexcitons in a crystal can be detected by luminescence. An elementary step in the radiative decay process of a biexciton is shown in Fig. 12.4. A biexciton disintegrates into a free exciton and a photon. The energy and momentum conservation laws governing this decay process read

$$\mathcal{E}_{xx}(K_{xx}) = \hbar\omega + \mathcal{E}_x(K_c) \tag{12.3}$$

or

$$\mathcal{E}_{xx}^{(0)} + \frac{\hbar^2 K_{xx}^2}{4M} = \hbar\omega + \mathcal{E}_x^{(0)} + \frac{\hbar^2 K_c^2}{2M} . \tag{12.4}$$

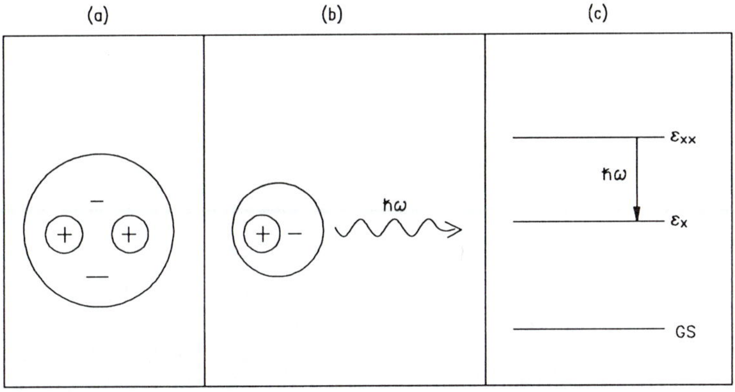

Figure 12.4. Elementary process leading to biexciton luminescence; (a) a biexciton before radiative recombination and (b) the state of the system after radiative recombination. (c) Schematic representation of the energy levels involved in the optical transition leading to biexciton luminescence.

The momentum conservation is

$$K_{xx} = K_c + q \,, \tag{12.5}$$

where q is the emitted photon wave vector.

The biexciton luminescence lineshape can be expressed as

$$I(\hbar\omega) = C\int d\mathcal{E} \mid W_{if} \mid^2 N_{xx}\, \delta[\hbar\omega - (\mathcal{E}_{xx} - \mathcal{E}_x)]\, \delta(K_{xx} - K_c - \mathbf{q}), \tag{12.6}$$

where W_{if} is the matrix element for the transition between the biexciton and exciton branch, N_{xx} is the biexciton energy distribution, C is a constant which includes the collection efficiency to the detector, and the δ functions represent the energy and momentum conservations. In Eq. (12.6) we have assumed that $q << K_{xx}, K_c$ so that $K_{xx} \simeq K_c$ and only vertical transitions are considered.

The distribution function of biexcitons N_{xx} may be calculated using Eq. (6.53) with the distribution function $f(\mathcal{E})$ being given by the Bose-Einstein distribution [Eq. (6.55)], which can be approximated at low densities, $|\mu/k_B T|>1$, by the classical Maxwell-Boltzmann distribution and the density of states similar to Eq. (6.54). Thus,

$$N_{xx} \sim \mathcal{E}_{xx}^{1/2} \exp(-\mathcal{E}_{xx}/k_B T) \ . \tag{12.7}$$

For a constant transition matrix element, independent of wave vector, Eq. (12.6) reduces to the simple form

$$I(\hbar\omega) \propto (\mathcal{E}_{xx}^{(0)} - \mathcal{E}_{x}^{(0)} - \hbar\omega)^{1/2} \exp(-\mathcal{E}_{xx}^{(0)} - \mathcal{E}_{x}^{(0)} - \hbar\omega)/k_B T \ . \tag{12.8}$$

Thus, the biexciton emission line takes the form of an inverted Maxwell-Boltzmann shape. Photons of lower energies correspond to decaying biexcitons with larger kinetic energy (see Fig. 12.5).

In reality, the situation is more complex because of the longitudinal-transverse splitting of the exciton and because of the polariton dispersion of the transverse exciton at small wave vectors, $K_c \sim q$, as we have seen in Chap. VI. Therefore, there are two final exciton branches for the transition, the transverse and longitudinal branches, as shown in Fig. 12.5. Two corresponding emission bands appear for the decay of a biexciton, called M_L and M_T. The emission from the M_T branch deviates from the inverted Maxwell-Boltzmann lineshape near $K_c = 0$ because of the strong dispersion due to the polariton effect (see Chap. VI). Figure 12.6 shows a measured emission due to biexcitons in CuCl, together with a calculated lineshape, where these effects have been taken into account.

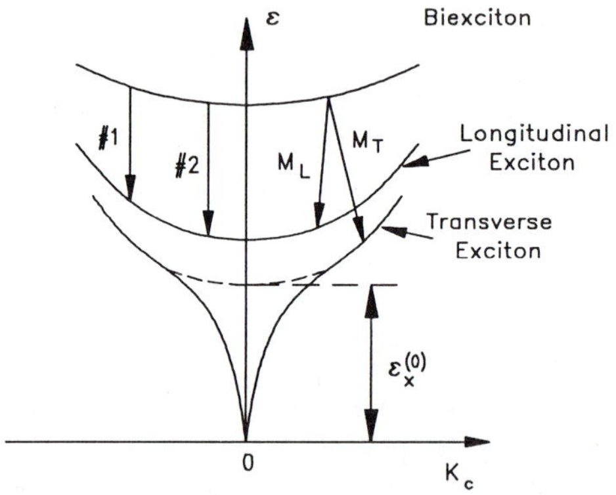

Figure 12.5. Exciton and biexciton dispersions. The biexcitons can decay to either the transverse or longitudinal exciton branches, giving rise to emitted photons at energies M_T or M_L, respectively. The smaller curvature of the biexciton dispersion curve is due to the heavier biexciton mass. Consequently, biexcitons with larger wave vectors emit photons of lower energy in the radiative decay process (the energy of photon # 2 is larger than the energy of photon # 1).

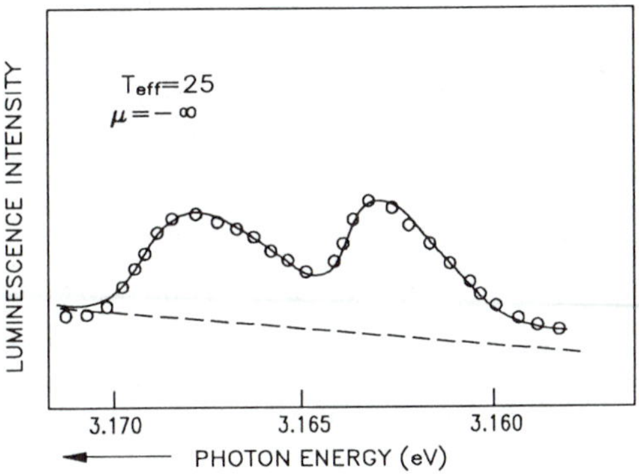

Figure 12.6. Luminescence spectrum of biexcitons created at $K_{xx} = 0$ at T = 25 K. Open circles are fits to the emission spectrum using Eq. (12.8) (after Ref. 12.6).

12-4. Biexciton Gain and Lasing

An interesting property of the biexciton emission is the possibility of large optical gain, in contrast to the direct free exciton decay where substantial optical gain is not possible because of reabsorption. One method of measuring optical gain is shown in Fig. 12.7. The excitation beam is focused along a line of length ℓ on the sample surface. The emission intensity along the focus axis is

$$I = I_s \frac{e^{g\ell} - 1}{g\ell} , \tag{12.9}$$

where I_s is the spontaneous emission intensity (see Prob. 12.2). A plot of I/I_s as a function of excited length ℓ yields the optical gain.

A more direct method measures the amplification of a weak test beam through an excited volume of the sample of thickness d. We have

$$I(\hbar\omega) = I_0(\hbar\omega)e^{gd} . \tag{12.10}$$

Optical gain in excess of 10^4 cm^{-1} has been obtained at the biexciton emission peak using both methods. Figure 12.8 shows the gain observed using the direct method.

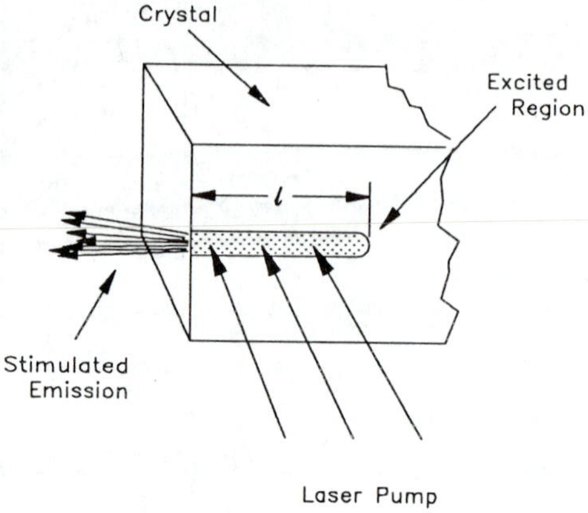

Figure 12.7. One of the geometries to measure optical gain in semiconductors (after Ref. 12.7).

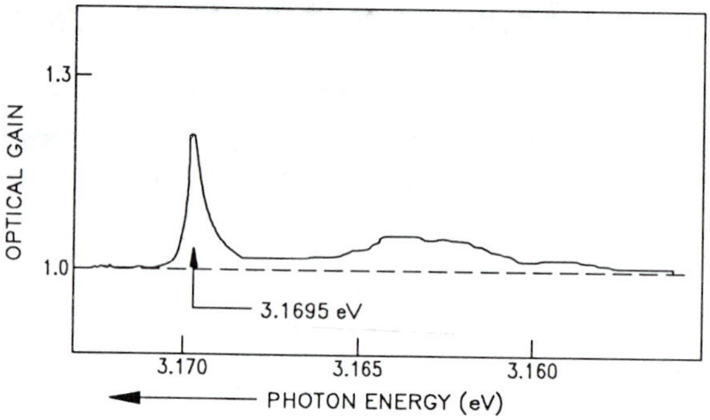

Figure 12.8. Optical gain of a broadband probe beam propagating in the backward direction relative to a resonant pump beam with intensity $I_0 \sim 1\ MW/cm^2$ in a CuCl film of thickness 2 μm at 5 K. The dashed line corresponds to a gain of unity (no net amplification) (after Ref. 12.6).

12-5. Bose-Einstein Condensation of Biexcitons

As we discussed earlier, excitons and biexcitons with integer spin quantum numbers obey the Bose-Einstein statistics, in contrast to electrons and holes, which follow the Fermi-Dirac statistics. To be more precise, excitons or biexcitons can be considered Bose particles at densities such that the average distance between them is large compared to their radius, corresponding to the condition $na_B \ll 1$. According to the Pauli exclusion principle, no more than one identical Fermion may occupy a quantum state. However, more than one Boson may accumulate in one quantum state, as can be seen by examining Eq. (12.11), which gives the occupation probability of state of energy $\mathcal{E}$,

$$f_{BE}(\mathcal{E}) = \frac{1}{\exp(\mathcal{E} - \mu)/k_B T - 1} . \tag{12.11}$$

It is important to realize that the chemical potential μ associated with a gas of Bosons must always be negative. Otherwise, one would obtain a negative value for the number of particles in the vicinity of $\mathcal{E} = 0$, a physically unrealistic situation, as can be seen by inspecting Eq. (12.11) (see Prob. 12.5). We now evaluate the total number of Bose particles, given by the density of available energy states multiplied by the probability of occupation of each state, integrated over all possible energy states of the system,

$$n_t = \int_0^\infty d\mathcal{E}\, g_{BE}(\mathcal{E})\, f_{BE}(\mathcal{E}) . \tag{12.12}$$

The density of states $g_{BE}(\mathcal{E})$ for Bose particles is similar to the one we derived for Fermions in Eq. (2.89) as

$$g_{BE}(\mathcal{E}) = \frac{p}{4\pi^2} \left(\frac{2m}{\hbar^2}\right)^{3/2} \mathcal{E}^{1/2} , \tag{12.13}$$

except that the spin degeneracy factor of 2 for Fermions is replaced by p with p = 1 for spin 0 particles, p = 3 for spin 1 particles, etc. Substituting Eq. (12.13) into Eq. (12.12) gives

$$n_t = \frac{p}{4\pi^2} \left(\frac{2m}{\hbar^2}\right)^{3/2} \int_0^\infty d\mathcal{E} \frac{\sqrt{\mathcal{E}}}{\exp[(\mathcal{E} - \mu)/k_B T] - 1} . \tag{12.14}$$

The maximum value for n_t is given when the chemical potential μ approaches zero from negative values:

$$n_c(T) = \frac{p}{4\pi^2}\left(\frac{2m}{\hbar^2}\right)^{3/2}\int_0^\infty d\mathcal{E}\,\frac{\sqrt{\mathcal{E}}}{\exp(\mathcal{E}/k_BT) - 1}\,. \tag{12.15}$$

The integral in Eq. (12.15) has a finite value of $\simeq 2.3(k_BT)^{3/2}$ (see Prob. 12.6), yielding

$$n_c(T) = 6.4 \times 10^{15}\, p\left(\frac{m}{m_0}\right)^{3/2} T^{3/2}\,, \tag{12.16}$$

where we normalized the particle mass m to the free electron mass m_0 and substituted for the constants. In Eq. (12.16), T is in degrees Kelvin and n_c is in cm^{-3}. The fact that the integral Eq. (12.14) remains finite, even for the maximum allowed value of μ, leads to the following problem: What happens if more particles are added to the system at a fixed temperature once the value $\mu = 0$ is reached, or equivalently, what happens if the temperature is decreased for a fixed number of particles? The solution to this problem has been given by Einstein. All additional particles collect into the ground state $\mathcal{E} = 0$. Therefore, the occupation number n_0 of the ground state may become very large. For instance, if the total number of particles is twice the critical value n_c, then as many particles occupy the ground state as all other available states. This phase transition, called Bose-Einstein condensation, may be viewed as a condensation in momentum or **k** space, in contrast to gas-liquid condensation which occurs in real space. Thus, for particle densities above $n_c(t)$, we may write

$$n_t = n_0 + n_c(T)\,, \tag{12.17}$$

indicating that for increasing density, the ground-state population n_0 begins to increase when the particle density exceeds $n_c(T)$. A critical temperature T_c may also be defined, below which Bose-Einstein condensation occurs when the particle density is above n_c. The temperature-density relationship is given by (see Prob. 12.3)

$$\frac{n_0}{n_t} = 1 - \left(\frac{T}{T_c}\right)^{3/2}, \tag{12.18}$$

with

$$T_c = \frac{3.31\,\hbar^2}{mk_B}\left(\frac{n_t}{p}\right)^{2/3} = 2.9 \times 10^{-11}\left(\frac{m_0}{m}\right)\left(\frac{n_t}{p}\right)^{2/3} . \qquad (12.19)$$

Excitons and biexcitons offer an interesting system for verifying this fundamental prediction of quantum mechanics. The density of excitonic particles may be easily changed by varying the light intensity. Also, because of the small effective mass, typically on the order of or less than the free electron mass, the critical temperature is well above the transition temperature expected from a gas of real atoms, as may be seen from Eq. (12.19). Evidence for Bose-Einstein condensation may be obtained from the appearance of a sharp emission line above a certain threshold intensity and below a threshold temperature, corresponding to the decay of the particles from the condensed state. In other words, at low densities the distribution of the kinetic energy of the particles should obey the classical Maxwell-Boltzmann distribution, resulting in a broad Maxwellian luminescence lineshape, but at high densities, above n_c, the Bose-Einstein phase transition should occur, resulting in a sharp emission line. Such experiments have been performed for biexcitons in CuCl, as shown in Fig. 12.9. The biexcitons were generated by direct two-photon absorption using a single laser beam tuned to half the biexciton energy, and the luminescence was detected at low temperatures. Figure 12.9(a) shows the luminescence spectra obtained for various pump intensities at a fixed temperature of 20 K. At low intensities [the bottom spectrum of Fig. 12.9(a)] the generated biexciton density is small, and the biexcitons act like a dilute gas following the classical Maxwell-Boltzmann distribution. Therefore, the emission is characterized by two broad lines with Maxwellian lineshapes, corresponding to the decay of thermalized biexcitons to the longitudinal and transverse excitons. As the pump intensity is increased, the emission lineshape changes and sharp lines appear on the high-energy sides of the broad Maxwellians. The sharp lines only appear above a critical intensity, corresponding to a critical biexciton density resulting from the decay of that portion of population of biexcitons that are Bose condensed, giving evidence for Bose-Einstein condensation in this system. This evidence is further strengthened by the temperature dependence of the luminescence for a biexciton density above n_c, as shown in Fig. 12.9(b). The sharp lines again appear below a critical temperature, as expected for Bose-Einstein condensation. This evidence was backed up by a series of other measurements that we will not elaborate on here. There is also experimental evidence that a gas of free excitons in Cu_2O displays Bose-Einstein statistics and condensation.

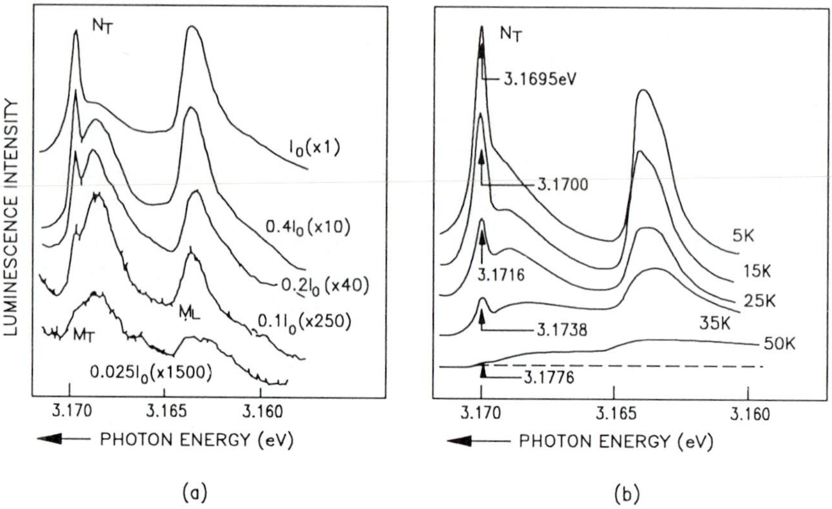

Figure 12.9. (a) The biexciton luminescence as a function of emitted photon energy for various incident pump intensities ($I_0 \sim 1\ \mathrm{MW/cm^2}$) at $T = 20$ K. The sharp emission line on the high-energy side of the broad Maxwellians appears only above a critical intensity at this constant temperature. (b) The temperature dependence of the biexciton luminescence for a fixed pump intensity of $\simeq 1\ \mathrm{MW/cm^2}$, showing the sharp luminescence lines, appears only below a critical temperature (after Ref. 12.6).

12-6. Electron-Hole Liquid

In 1967 L. V. Keldysh pointed out that excitons in semiconductors can undergo a phase transition to an EHL phase similar to the condensation of water vapor into liquid drops. This EHL phase, which has been observed in several semiconductors, has a metallic character, like a molten metal. It is characterized by a well-defined interface with its surroundings, a surface tension, and a definite equilibrium density of electron-hole pairs, n_L (see Fig. 12.10). Since the equilibrium density n_L is fixed, an increase of crystal excitation merely increases the size of the crystal volume occupied by the liquid. This volume is given by the relation

$$V_L = \frac{N_L}{n_L}, \tag{12.20}$$

where N_L is the total number of electron-hole pairs inside the liquid.

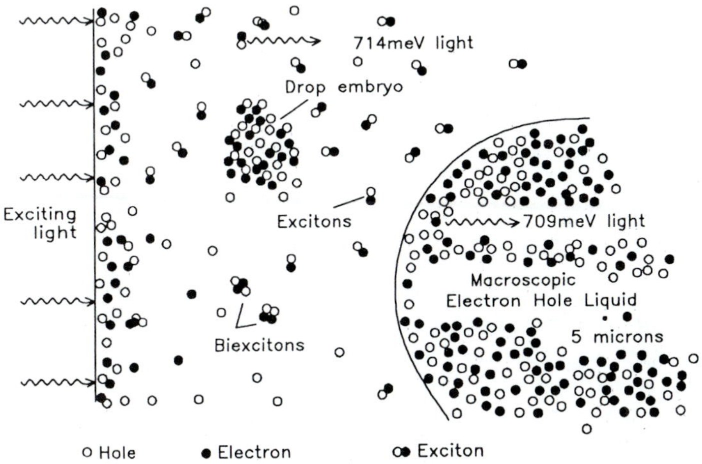

Figure 12.10. Schematic representation of the liquid phase of excitons in a semiconductor like Ge (after Ref. 12.9).

The existence of the EHL is determined by the condition

$$E_L > \frac{E_{Bxx}}{2} + E_B \,, \tag{12.21}$$

stating that the binding energy per electron-hole pair E_L of the N_L electron-hole pairs in the liquid droplet must be larger than the sum of the exciton binding energy and half of the total biexciton binding energy.

If the condition expressed by Eq. (12.21) is not satisfied, it is energetically more favorable for the system to remain in the form of a gas of biexcitons or excitons. Generally, E_L - $(E_{Bxx}/2 + E_B)$ is small, typically a few meV, so that the EHL forms only at very low temperatures. It represents a rather unique phase, with pronounced Fermi quantum statistics for both electron and hole distributions.

To evaluate E_L, one must calculate the minimum energy required to add an electron-hole pair inside the liquid and compare this value to the energy required to form a biexciton. Since the density n_L of electron-hole pairs is high, we first introduce a dimensionless parameter r_s, which gives the average distance between particles normalized to the exciton Bohr radius,

$$r_s = \left(\frac{3}{4\pi n_L}\right)^{1/3} \frac{1}{a_B} . \tag{12.22}$$

The average energy, or the ground-state energy, of the electron-hole pair inside the liquid is

$$\mathcal{E}(n_L) = (E_g - E_L) = \mathcal{E}_{kin} + \mathcal{E}_{xc} , \tag{12.23}$$

where $n_L(E_g - E_L)$ is the total excitation energy in the liquid phase, $\mathcal{E}_{kin}$ is the kinetic energy comprising the sum of the kinetic energies of electrons and holes, and $\mathcal{E}_{xc}$ is the sum of exchange and correlation energies between the particles in the liquid. As the exchange energy, one denotes that part of the total energy that results from the Fermi nature (Fermi-Dirac statistics) of the electrons and holes. For these particles the Pauli exclusion principle states that each quantum state can be occupied by no more than one Fermion. This principle keeps equal particles (particles with same quantum numbers and same charge) away from each other, leading to a reduction of their repulsive interaction. The correlation energy is due to the Coulomb interaction of the charge carriers, also introducing correlations for particles in different quantum states. Hence, exchange effects occur for Fermions with equal quantum numbers, whereas correlation effects influence all charged carriers.

$\mathcal{E}_{kin}$ is always positive and, therefore, competes against liquid formation. We calculate $\mathcal{E}_{kin}$ by considering the kinetic energy of electrons filling the bottom of the conduction band and holes occupying the top of the valence band. To simplify the calculation, we will assume that we deal with a simple direct bandgap semiconductor at T = 0. We have [using Eq. (2.82)]

$$n_L = n_e = n_h = 2\int \frac{dk}{(2\pi)^3} = \frac{1}{\pi^2}\int_0^{k_f} dk\, 4\pi k^2 = \frac{k_f^3}{3\pi^2} , \tag{12.24}$$

where the factor of 2 has been included to take the electron spin into account. The Fermi wave number for electrons or holes is k_f and may be written in terms of r_s as

$$k_f\, a_B = (3\pi^2 n_L)^{1/3}\, a_B = \left(\frac{9\pi}{4}\right)^{1/3} \frac{1}{r_s} = \frac{1.92}{r_s} , \tag{12.25}$$

where we have used Eqs. (12.24) and (12.22). The average kinetic energy at T = 0 may be written as

$$\mathcal{E}_{kin} = \frac{\int_0^{\mu} d\mathcal{E}\ \mathcal{E}\ g(\mathcal{E})}{\int_0^{\mu} d\mathcal{E}\ g(\mathcal{E})}, \tag{12.26}$$

where $g(\mathcal{E})$ is the density of states and $\mu = \mu_e + \mu_h$ is the total chemical potential (see Fig. 12.11). The denominator of Eq. (12.26) describes the total number of states. Substituting for $g(\mathcal{E})$ from Eq. (2.89) in Eq. (12.26), we get

$$\mathcal{E}_{kin} = \frac{\int_0^{\mu} d\mathcal{E}\ \mathcal{E}^{3/2}}{\int_0^{\mu} d\mathcal{E}\ \mathcal{E}^{1/2}} = \frac{5}{3}\mu = \frac{5}{3}(\mu_e + \mu_h) = \frac{5}{3}\left(\frac{\hbar^2 k_f^2}{2m_e} + \frac{\hbar^2 k_f^2}{2m_h}\right). \tag{12.27}$$

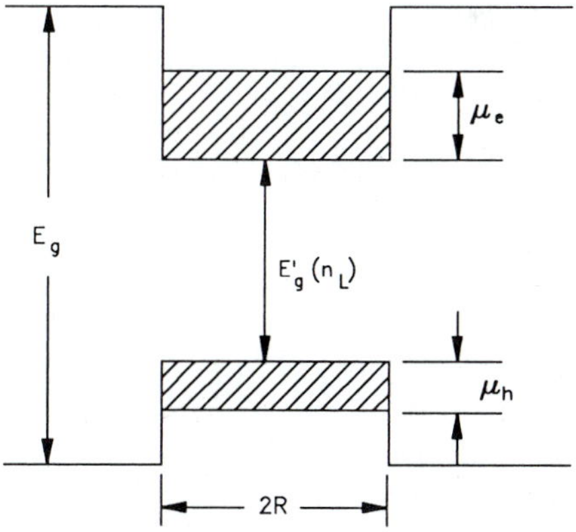

Figure 12.11. Schematic energy representation of an EHL drop of radius R. Inside the liquid of density n_L, the energy gap takes its renormalized value E'_g. At $T = 0$, electrons occupy the conduction band up to the electron chemical potential μ_e, and the holes up to μ_h.

Substituting for k_f^2 from Eq. (12.25) and using Eq. (6.18), we get

$$\mathcal{E}_{kin} = \frac{2.21}{r_s^2} E_B . \tag{12.28}$$

$\mathcal{E}_{xc}$ is more difficult to calculate because it requires a many-body treatment of the carrier-carrier interaction in the plasma. The result of many-body calculations yields that $\mathcal{E}_{xc}$ is negative and strongly dependent on n_L. Therefore, $\mathcal{E}_{xc}$ reduces the minimum energy required to create a new electron-hole pair, leading to a reduction of the bandgap inside the liquid compared with E_g. This reduced bandgap energy, E'_g, is often called the renormalized gap with $E'_g - E_g = \mathcal{E}_{xc}$ (see Fig. 12.11). The concepts of exchange and correlation effects, screening, and bandgap renormalization will be discussed in more detail in Chap. XIII.

The calculated variation of $\mathcal{E}_{xc}$ or E'_g with r_s is shown in Fig. 12.12 (see Ref. 12.10). The data points are obtained from experiments performed in Ge and Si. In Fig. 12.13 the sum $\mathcal{E}_{kin} + \mathcal{E}_{xc}$ is plotted versus r_s for the case of a simple band structure. Although $\mathcal{E}(n_L) = \mathcal{E}_{kin} + \mathcal{E}_{xc}$ goes through a minimum around $r_s = 2$, the value at the minimum lies above $E_g - E_B$, the n = 1 exciton energy. This indicates that liquid formation does not occur in a semiconductor with a simple band structure. However, this treatment neglects the interaction between electrons or holes and the lattice. Inclusion of the electron- (hole) phonon interaction in polar materials may allow liquid formation, even in a simple direct-gap material.

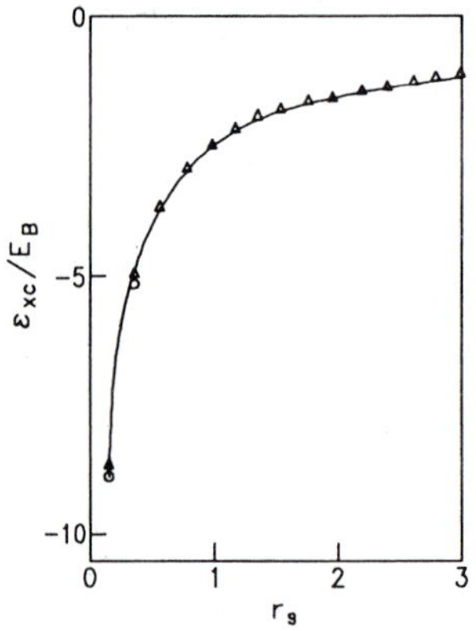

Figure 12.12. The reduction of the energy gap as a result of Coulomb correlation and exchange effects as a function of r_s. The data points are those for Ge and Si (after Ref. 12.10).

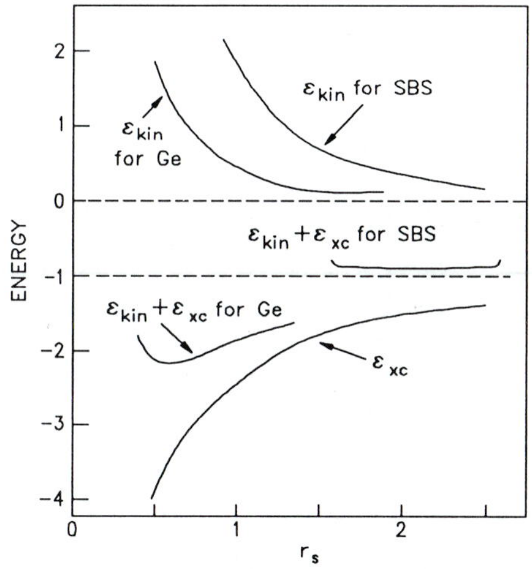

Figure 12.13. The normalized kinetic energies (normalized to the exciton binding energy) $\mathcal{E}_{kin}$ for a semiconductor with a simple band structure and for Ge. The normalized exchange energy $\mathcal{E}_{xc}$ is the same for both cases and is also plotted. The sum of $\mathcal{E}_{kin}$ + $\mathcal{E}_{xc}$ is shown for a simple band structure and Ge as well (after Ref. 12.11). The horizontal line at an energy of -1 corresponds to the exciton energy, E_g - E_B.

A different situation occurs in a material with a complex band structure, such as an indirect bandgap material. As we have seen in Chap. V, when the extrema of the valence or conduction band are not both at the Γ point of the Brillouin zone, there are several identical minima of the conduction band (or maxima of the valence band) corresponding to equivalent points in the first Brillouin zone. In the general case of such a complex band structure, it can be shown that the kinetic energy is expressed as

$$\mathcal{E}_{kin} = \frac{2.21}{r_s^2}\left(\frac{m_r}{\nu_e^{2/3}\, m_{de}} + \frac{m_r}{\nu_h^{2/3}\, m_{dh}}\right), \tag{12.29}$$

where ν_e and ν_h are the number of equivalent valleys in the conduction band and valence band. The density-of-states masses of electrons and holes, m_{de} and m_{dh}, are given by

$$m_{de} = (m_{\|e}\, m_{\perp e}^2)^{1/3} \qquad m_{dh} = (m_{\|h}\, m_{\perp h}^2)^{1/3}, \tag{12.30}$$

and m_r is the reduced mass,

$$\frac{1}{m_r} = \frac{1}{m_{oe}} + \frac{1}{m_{oh}} , \tag{12.31}$$

where $1/m_{oe} = 1/3\ (1/m_{\| e} + 2/m_{\perp e})$ is the optical mass of the electron, and similarly for the hole. For such complex band structures, the kinetic energy is reduced in the liquid, since the electrons or holes are equally distributed in the equivalent valleys with a concomitant decrease of the kinetic energy in each valley. Similarly, a band degeneracy or mass anisotropy decreases the positive contribution of $\mathcal{E}(n_L)$ through a reduction of $\mathcal{E}_{kin}$. As a consequence, the EHL is stable in indirect gap semiconductors such as Ge or Si. The effect of band complexity is shown for Ge in Fig. 12.13. Now the average energy for an electron-hole pair reaches a minimum, below the free exciton energy, for a value $r_s \leq 1$, and the EHL becomes stable. The density of electron-hole pairs in the liquid state is 2.3×10^{18} cm^{-3} for Si and 2.3×10^{17} cm^{-3} for Ge.

12-7. Luminescence of Electron-Hole Liquid

The presence of drops of EHL in a semiconductor can be detected by its characteristic emission. The luminescence lineshape due to the radiative recombination of electron-hole pairs inside the liquid can be obtained simply in an indirect semiconductor by noting that the optical transition is similar to the band-to-band absorption in an indirect-gap material, except for the fact that the populations are inverted near the extrema of the valence and conduction bands. The EHL luminescence lineshape is expressed as (see Prob. 12.7)

$$I(\hbar\omega) \sim \int_0^{\Delta\mathcal{E}} d\mathcal{E}_1\, f_e\, f_h\, \mathcal{E}_1^{1/2}\, (\hbar\omega \pm \hbar\Omega - E_g^{ind} - \mathcal{E}_1)^{1/2} , \tag{12.32}$$

with $\Delta\mathcal{E} = \mathcal{E}_1 + \mathcal{E}_2 = \hbar\omega \pm \hbar\Omega - E_g^{ind}$, and $\hbar\Omega$ is the energy of the participating optical phonon [see Fig. 12.14(a)]. An example of the luminescence due to the EHL is shown schematically in Fig. 12.14(b). The measured EHL luminescence spectra of Ge is shown in Fig. 12.15.

Except for very high excitation levels, the liquid is in the form of small drops with a typical size of radius < 10 μm, limited by the lifetime of the electron-hole pairs inside the drops. A spectacular increase of the drop size has been obtained by applying a contact force with a rounded end of a rod of radius R to a face of a crystal. The effect of the contact stress is to

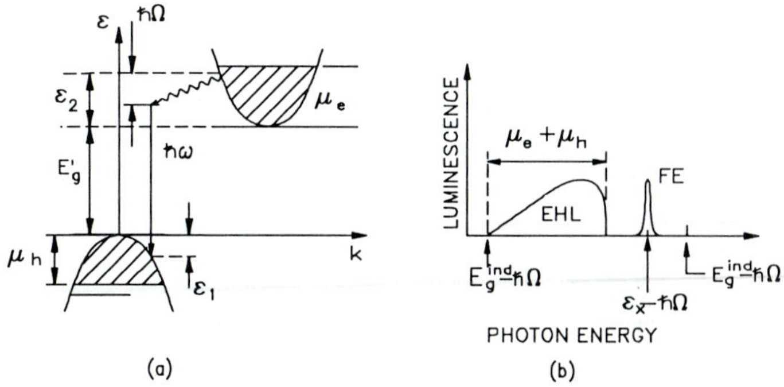

Figure 12.14. (a) Energy diagram showing the electron transition in the recombination of an electron-hole pair inside a liquid in an indirect-gap semiconductor. (b) The expected luminescence lineshape of the EHL and the free exciton.

produce a nonuniform strain inside the crystal, as represented schematically in Fig. 12.16. In turn, the strain reduces the bandgap energy, resulting in the formation of a potential well inside the crystal at a distance of approximately 0.5 R from the contact surface. Electron-hole pairs are attracted toward the bottom of the well where they form a single large

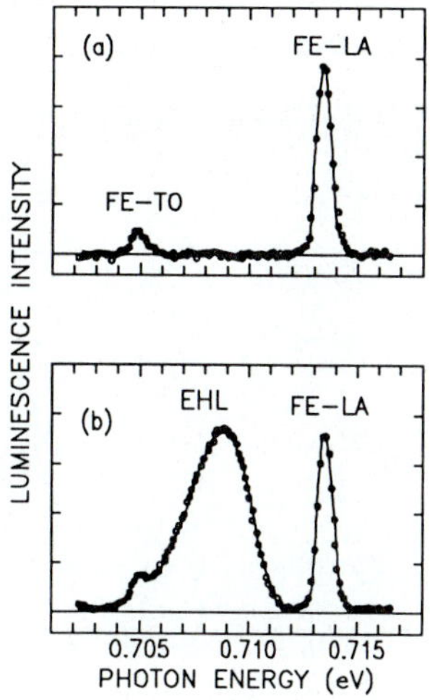

Figure 12.15. (a) Luminescence spectrum of free exciton (FE) in ultrapure Ge at 2 K surface excited below the threshold for EHL formation. The peaks correspond to LA and TO phonon-assisted exciton luminescence lines, referred to as the phonon replicas. (b) Same as for (a) except for a 2.2 times higher excitation power. The peak labeled EHL is the luminescence of the EHL assisted by LA phonon, and the small bump on the low-energy side of the EHL line is the TO phonon replica for FE (after Ref. 12.12).

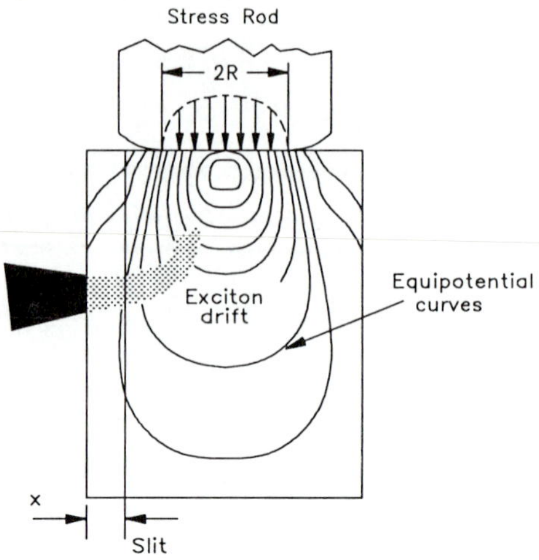

Figure 12.16. A nonuniform strain is applied to a silicon crystal by pressure from a contact stress rod. The strain narrows the bandgap energy. Theoretical contours of constant bandgap are plotted. Excitons move in the force field $\mathbf{F} = -\nabla E_g$ toward the potential minimum (after Ref. 12.13).

drop of radius, as large as 300 μm. The image of a large electron-hole drop can be obtained by focusing the EHL luminescence onto a detector sensitive to the emitted radiation (see Fig. 12.17).

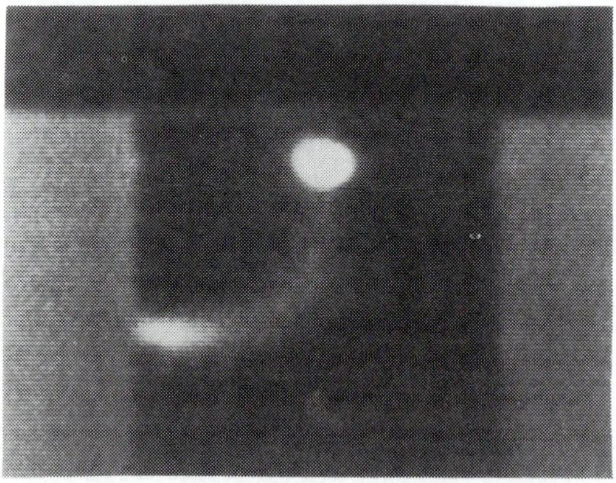

Figure 12.17. Luminescence from excitons that recombine while drifting toward the potential minimum in a crystal of pure silicon at 10 K. The excitons are generated at the left crystal surface by light from a continuous argon laser. They are drawn into the crystal by a nonuniform strain using the technique shown in Fig. 12.16. Most excitons are eventually trapped in a potential well (bright spot) created by the strain maximum beneath the contact area (entire dark region at top) (after Ref. 12.13).

12-8. Thermodynamics of the Electron-Hole Liquid

As is the case for usual liquids, it is possible to draw a phase diagram for the EHL. Figure 12.18 shows a temperature-density diagram, characterized by a hatched region. In this region, the system of n carriers decomposes spontaneously into a low-density vapor phase of free excitons coexisting with a high-density liquid phase of free electron-hole pairs. The critical point defining the temperature above which no phase separation occurs obeys the empirical rule

$$k_B T_c \simeq 0.1 \, E_B \, . \tag{12.33}$$

Strictly speaking, the electron-hole system in a semiconductor is not a thermodynamic system because of the finite lifetime of the carriers. The system is not "closed" in the thermodynamic sense. We speak about an "open" system since a flow of energy (optical excitation) is needed to maintain a constant density. The phase transitions under such conditions are nonequilibrium transitions that require methods of analysis beyond ordinary equilibrium thermodynamics. A more detailed discussion of the nonequilibrium thermodynamics of EHLs and of the kinetics of droplet formation is given in Ref. 12.16.

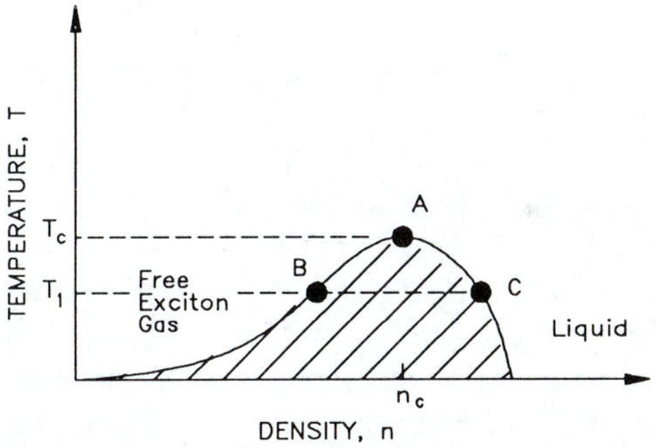

Figure 12.18. Schematic phase diagram of temperature versus density of electron-hole pairs in a semiconductor on a logarithmic-density scale. In the hatched region the system decomposes into a low-density gas phase of excitons (B) and a high-density liquid phase (C). Point A is the critical point, with associated temperature T_c and density n_c. Above T_c there is no liquid phase (see Refs. 12.15 and 12.16).

12-9. Electron-Hole Plasma

Even in the absence of an EHL, the insulating phase of the excitonic gas becomes unstable at high densities and transforms into a conducting phase of unbound electron-hole pairs. At high densities, the nature of the conducting state is similar to that inside the EHL. This configuration of unbound electron-hole pairs is referred to as the "electron-hole plasma." However, the plasma is not a thermodynamic phase. In a plasma state the entire excited volume is filled by electron-hole pairs. The density is variable, determined by the degree of excitation of the crystal. The plasma phase also occurs in simple direct-gap semiconductors, provided the excitation intensity is strong enough. The gradual change from exciton gas to electron-hole plasma is often called the "Mott transition" or "exciton ionization." At low temperatures, it occurs at densities such that

$$n\, a_B{}^3 \simeq 1 \; . \tag{12.34}$$

In an electron-hole plasma we have no excitons, and hence, no excitonic absorption or luminescence features. The occurrence of a Mott transition is, therefore, accompanied by a drastic change of the optical absorption of the excited volume, with the disappearance of exciton absorption lines. By contrast, in the case of liquid formation, exciton absorption vanishes only in the volume occupied by the drops of the EHL.

The plasma may also be recognized by its characteristic luminescence. The recombination process is the same as for the EHL. Again, one expects a broad emission line starting at the value of the renormalized energy gap E'_g with a spectral width given by the sum of the chemical potentials of the electrons and holes [see Fig. 12.14(b)]. However, in contrast to the EHL, for which an increase of excitation intensity merely increases the emission intensity without affecting its spectral shape, the spectral lineshape of the plasma emission changes with the intensity of excitation, indicating the change of the plasma density, as shown in Fig. 12.19.

The appearance of the plasma luminescence is accompanied by the disappearance of the exciton absorption line, as shown in Fig. 12.20. The exciton bleaches as a result of screening of the Coulomb potential responsible for the binding of the electron and the hole. This screening effect will be discussed in more detail in Chap. XIII.

It is possible to obtain information on the exciton saturation dynamics by applying the time-resolved luminescence detection technique. The luminescence spectra shown in Fig. 12.19 are time-integrated; i.e., the photons emitted at various times are not differentiated. A time-resolved detection of the luminescence reveals the appearance of the plasma emission only at early times, followed by the emission of the biexcitons once the plasma density has decreased. Figure 12.21 shows the time-

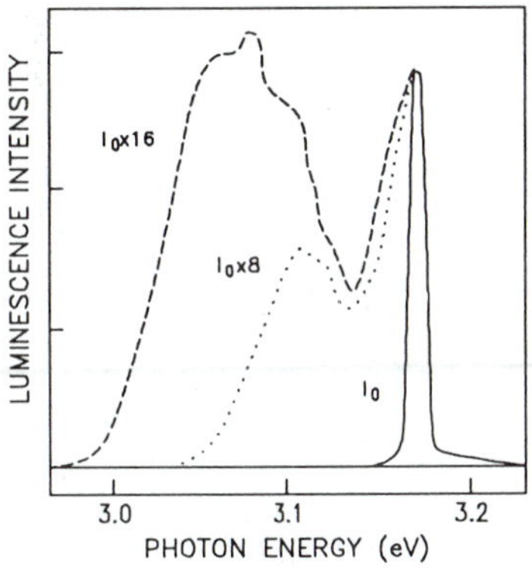

Figure 12.19. Time-integrated luminescence of CuCl at 13 K for different excitation intensities and an excitation energy of $\simeq$ 3.87 eV. The line at ~ 3.162 eV corresponds to the biexciton luminescence; the broadband around 3.1 eV is the result of electron-hole plasma recombination (after Ref. 12.17).

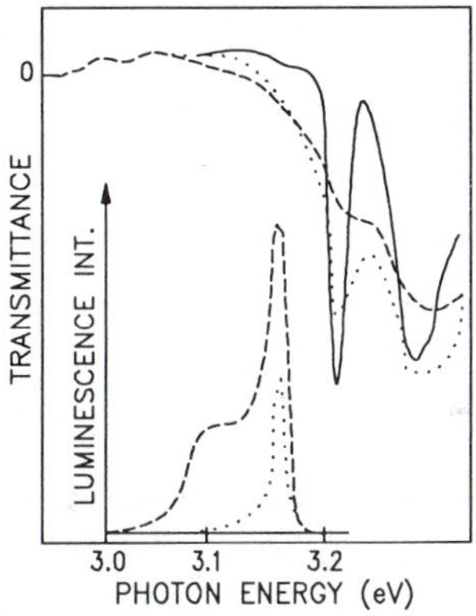

Figure 12.20. Time-integrated luminescence and the corresponding absorption spectra of CuCl at low temperatures, showing that the occurrence of the electron-hole plasma luminescence is accompanied by bleaching of the exciton absorption peak (after Ref. 12.17).

resolved luminescence of CuCl at different time delays after excitation with pump photons with energies above the bandgap. The luminescence from various time slots shows that at early times (2 ps) the luminescence only originates from the plasma. However, after 8 ps, when the plasma has decreased below the value fixed by the Mott criterion, only the biexciton luminescence remains. This indicates that a reverse Mott transition has taken place, with the conducting plasma phase being transformed into an insulating phase of biexcitons. The dashed curve in Fig. 12.21 shows the time-integrated luminescence again for comparison.

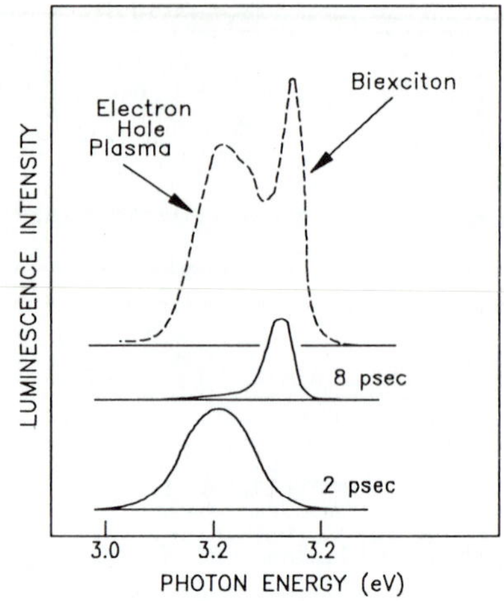

Figure 12.21. Time-resolved luminescence of CuCl excited above the bandgap with subpicosecond pulses. The data show that 2 ps after the excitation the luminescence primarily results from the electron-hole plasma, while after 8 ps only the biexciton luminescence remains. The dashed curve is the time-integrated luminescence, also shown in Fig. 12.19 (after Ref. 12.18).

12-10. Problems

12.1. Describe an experimental technique to measure the biexciton binding energy.

12.2. For the experimental configuration of Fig. 12.7, show that the emission intensity is given by Eq. (12.9).

12.3. Starting from Eq. (12.17), prove Eqs. (12.18) and (12.19).

12.4. Compute the critical density and temperature for Bose-Einstein condensation for exciton molecules in CdS using $m_e = .235\ m_0$ and the material parameters from Table 12.1.

12.5. (a) Show that the Bose-Einstein distribution must always have a negative chemical potential.

(b) What is the mean occupation number when $\mu = k_B T$ or $0.1\ k_B T$?

(c) What is the corresponding occupation number for a Fermi distribution?

(d) What is the occupation number of a Fermi distribution when $\mu = 0$? Can μ become positive?

12.6. Show that the integral of Eq. (12.14) is finite for $\mu = 0$. Find this value.

12.7. Noting that the luminescence intensity is proportional to the total number of electron-hole pairs in the system, derive the luminescence lineshape of an EHL drop, as expressed in Eq. (12.32). Verify that the lower and upper boundaries of the luminescence occur at photon energies of $E_g^{ind} - \hbar\Omega$ and $E_g^{ind} - \hbar\Omega + \mu_e + \mu_h$.

12-11. References

1. A. Mysyrowicz, J. B. Grun, A. Bivas, R. Levy, and S. Nikitine, Phys. Lett. A **26**, 615 (1968).

2. O. Akimoto and E. Hanamura, J. Phys. Soc. Jpn. **33**, 1537 (1972); see also, E. Hanamura, Solid State Commun. **12**, 951 (1973).

3. C. Klingshirn and H. Haug, Phys. Rep. **70**, 315 (1981).

4. N. Peyghambarian, G. R. Olbright, D. A. Weinberger, H. M. Gibbs, and B. D. Fluegel, J. Lumin. **35**, 241 (1986).

5. L. L. Chase, N. Peyghambarian, G. Grynberg, and A. Mysyrowicz, Opt. Commun. **28**, 129 (1979).

6. N. Peyghambarian, L. L. Chase, and A. Mysyrowicz, Phys. Rev. **27**, 2325 (1983).

7. J. M. Hvam, J. Appl. Phys. **49**, 3124 (1978); see also, K. L. Shaklee and R. F. Leheny, Appl. Phys. Lett. **18**, 475 (1971).

8. L. V. Keldysh, in *Proceedings of the 19th International Conference on Physics of Semiconductors* (Moscow, 1968) p. 1303; see also, M. Combescot, P. Nozierer, J. Phys. C **5**, 2369 (1972); and for a review see, T. M. Rice, J. Hensel, T. Phillips, and G. A. Thomas, Solid State Physics **32**, 1 (1977).

9. C. D. Jeffries, Science **189**, 955 (1975).

10. P. Vashishta and R. K. Kalia, Phys. Rev. B **25**, 6492 (1982).

11. C. Benoit a la Guillaume, in *Collective Excitations in Solids*, edited by B. D. Bartolo (Plenum Press, New York, 1981) p. 633.

12. R. M. Westervelt, in *Electron-hole Droplets in Semiconductors*, edited by C. D. Jeffries and L. V. Keldysh (North Holland, Amsterdam, 1983), pp. 187-266.

13. J. P. Wolfe, Phys. Today **35**, 46 (1982).

14. G. A. Thomas, T. M. Rice, and J. C. Hensel, Phys. Rev. Lett. **33**, 219 (1974).

15. J. Shah, M. Combescot, and A. H. Dayem, Phys. Rev. Lett. **38**, 1497 (1977).

16. S. W. Koch, "Dynamics of First-Order Phase Transitions in Equilibrium and Nonequilibrium Systems," Springer Lecture Notes in Physics 207 (Springer, Berlin, 1984).

17. D. Hulin, A. Mysyrowicz, A. Migus, and A. Antonetti, J. Lumin. **30**, 290 (1985).

18. D. Hulin, J. Etchepare, A. Antonetti, L. L. Chase, G. Grillon, A. Migus, and A. Mysyrowicz, Appl. Phys. Lett. **45**, 993 (1984).

Chapter XIII
SEMICONDUCTOR OPTICAL NONLINEARITIES

So far in our discussion of the optical absorption and refraction, we have assumed that these material parameters do not depend on the intensity of the optical fields. However, this assumption is correct only for sufficiently low excitation intensities, that is, in the linear regime. For higher excitation intensities, the optical material properties may be modified, and the response changes as the excitation density is varied. Such excitation-dependent optical properties are commonly called *optical nonlinearities*. It turns out that under sufficiently strong excitation the optical properties of basically all semiconductors exhibit nonlinear characteristics. They may give rise to effects such as excitation-dependent absorption and refraction, nonlinear wave mixing, optical bistability, or other optical instabilities. Some of these effects have attracted considerable attention, since they may be useful for applications in optical switching devices, optical logic gates, or even optical computing.

We first discuss different nonlinear effects, beginning with nonlinearities that are associated with real carrier generation. This is the case when an intense incident light beam has its wavelength tuned to an absorption region of the crystal. Nonlinear mechanisms such as bandfilling, bandgap renormalization, and thermal effects are introduced. The phenomenon of plasma screening is analyzed in some detail. Since the full theoretical analysis of most carrier nonlinearities is beyond the scope of this book, we present simple arguments to underline the main physical mechanisms. This is done with figures for representative materials and a discussion of the basic nonlinear effects. Next we discuss carrier-related optical nonlinear effects in quantum wells and quantum dots.

Examples of transient nonlinearities are summarized, and a brief theoretical treatment is given. We then consider the case when the intense incident light is tuned to a transparency region of the medium. The optical Stark effect is discussed as a representative example of such a nonlinearity. Finally, for completeness we introduce the third-order optical susceptibility $\chi^{(3)}$ and discuss its limited validity for the case of semiconductors.

13-1. Classification of Optical Nonlinearities

As a rough categorization, optical nonlinearities in semiconductors may be classified into two categories: quasi-equilibrium and transient or coherent nonlinearities.

Quasi-equilibrium nonlinearities are associated with real carrier generation. They occur after relatively long time scales when the created electrons and holes have had sufficient time to interact among themselves and establish Fermi-Dirac distributions within each band. The precise mechanism of carrier generation is not important, since memory of the excitation process is lost. Carrier generation can occur by resonant laser excitation in the exciton resonance or above the bandgap, or by carrier generation through injection pumping (see Chap. XVIII). The resulting nonlinearities are called *quasi-equilibrium* nonlinearities, since the carriers responsible for the nonlinear material response are in "local" equilibrium among themselves. This means that after their creation, the carriers in the different bands have had sufficient time to interact and establish a plasma temperature and a chemical potential, which are the two thermodynamic quantities needed to define a Fermi distribution function for a given density. In quasi-equilibrium the dynamics of the electron-hole density, and not the excitation dynamics such as laser pulse duration, etc., determines the dynamics of the nonlinearity.

Quasi-equilibrium conditions have to be distinguished from the full thermodynamic equilibrium. Quasi-equilibrium refers to thermal equilibrium within each band, but the semiconductor as a whole is out of thermal equilibrium. A true thermodynamic equilibrium would imply an empty conduction band and a full valence band, i.e., the crystal ground state. The quasi-equilibrium Fermi-Dirac distribution for electrons and holes within their respective bands is given by

$$f_{e,h} = \frac{1}{\exp[(\mathcal{E}_{e,h} - \mu_{e,h})/k_B T] + 1} , \tag{13.1}$$

where T is the quasi-equilibrium temperature (the effective temperature of the carriers), $\mathcal{E}_e$ and $\mathcal{E}_h$ are the energies, and μ_e and μ_h are the quasi-chemical potentials of the electrons and holes, respectively. The quasi-chemical potentials are determined from the particle number in each band, as in Eq. (2.72). In an intrinsic, i.e., undoped perfect semiconductor, there are an equal number of electrons and holes, $n_e = n_h$. However, the respective chemical potentials are not equal, $\mu_e \neq \mu_h$, since the effective masses of electrons and holes are generally different. We also note that the temperature of the carriers may be significantly different from the lattice temperature. During their limited lifetime, carriers may not have sufficient time to establish thermal equilibrium with the lattice, even if the multiple particle-particle interaction allows a definition of carrier temperature. In this case, one speaks of an *effective carrier temperature*, different from lattice temperature.

Quasi-equilibrium is typically realized on a time scale of several hundred femtoseconds (1 femtosecond = 10^{-15} sec) up to a few picoseconds after excitation, so that the carriers have ample time for collisions, driving the electrons and holes into the quasi-equilibrium state. On the other hand, *transient* or *coherent nonlinearities* occur on sufficiently short time scales when the coherent light-matter coupling is not relaxed, either through carrier-carrier or through carrier-lattice interactions. The time dynamics of these nonlinearities basically follows the exciting laser pulse time profile for very fast laser pulses of the order of a few tens of femtoseconds. Experimentally, the exciting laser frequency is often tuned below the exciton resonance in the transparency region of the semiconductor so that real carrier generation is minimized. In this case, only coherent nonlinearities are left. However, resonant excitation also produces a coherent nonlinearity component that coexists with the carrier nonlinearity. The time response of this resonantly induced coherent nonlinearity is determined by the dephasing time of the electronic resonance, which is much shorter than the carrier lifetime.

In the subsequent sections of this chapter we discuss mechanisms for transient and quasi-equilibrium nonlinearities. We start with a discussion of the carrier density-dependent effects, which lead to phenomena such as plasma screening and state filling.

13-2. Plasma Screening

We assume a semiconductor where a finite density of electrons (holes) has been excited in the conduction (valence) band. One of the most important consequences of the interaction among these excited carriers is that they screen the Coulomb interaction among the carriers in the same band as well as between the electrons and holes in different bands. Here screening simply refers to the reduction of the range of the Coulomb potential of a charge in the presence of other charges. This is a source of optical nonlinearities, since (as we saw in Chap. VI) the existence of an exciton resonance depends on the attractive Coulomb interaction between an electron and a hole. Through a density-dependent modification of this attractive potential we obtain a density-dependent modification of the absorption and, since carrier density depends on the light intensity, the material response is optically nonlinear.

To investigate the phenomenon of plasma screening in its pure form, let us first treat the case of a one-component plasma of electrons, which we assume has a constant density n_0. We consider the effect of placing a test charge into the plasma. The electrons in the plasma will redistribute as a result of the repulsive interaction with the inserted charge, and a density change $\Delta n(\mathbf{r})$ will be induced. This redistribution occurs in such a way as to minimize the total energy of the system and may be expressed as the effect of decreasing (screening) the effective potential energy of the test charge.

Consider a local test charge -e at position $\mathbf{r}$ denoted as $n_t(\mathbf{r})$ inside a plasma with mean density n_0. The test charge may be expressed as a delta function by

$$n_t(\mathbf{r}) = -e\delta(\mathbf{r}) \ . \tag{13.2}$$

The test charge induces a change in the charge-density distribution, $-e\Delta n(\mathbf{r})$, within the plasma as a result of the repulsion of the electrons in the plasma from the neighborhood of the test charge. The effective electrostatic potential resulting from the test charge is the sum of the direct potential Φ_t and the induced potential Φ_{ind},

$$\Phi_{eff}(\mathbf{r}) = \Phi_t(\mathbf{r}) + \Phi_{ind}(\mathbf{r}) \ . \tag{13.3}$$

The potential is related to the charge density by Poisson's equation as

$$\nabla^2\Phi(\mathbf{r}) = -4\pi\rho(\mathbf{r}) \ , \tag{13.4}$$

where $\rho(\mathbf{r}) = -en(\mathbf{r})$ is the charge density and $n(\mathbf{r})$ is the total electron number density,

$$n(\mathbf{r}) = n_0 + \Delta n(\mathbf{r}) \ . \tag{13.5}$$

Using Eq. (13.4) and noting that the induced charge density is $-e\Delta n(\mathbf{r})$, we may write

$$\nabla^2\Phi_{ind}(\mathbf{r}) = 4\pi e\Delta n(\mathbf{r}) \ , \tag{13.6}$$

$$\nabla^2\Phi_t = 4\pi e\delta(\mathbf{r}) \ . \tag{13.7}$$

Therefore, taking ∇^2 from both sides of Eq. (13.3) and substituting from Eqs. (13.6) and (13.7), we get

$$\nabla^2 \Phi_{eff}(\mathbf{r}) = 4\pi e[\delta(\mathbf{r}) + \Delta n(\mathbf{r})] \ . \tag{13.8}$$

Introducing the potential energy $V_{eff}(\mathbf{r}) = -e\Phi_{eff}(\mathbf{r})$, Eq. (13.8) becomes

$$\nabla^2 V_{eff}(\mathbf{r}) = -4\pi e^2[\delta(\mathbf{r}) + \Delta n(\mathbf{r})] . \tag{13.9}$$

We may substitute for $V_{eff}(\mathbf{r})$, $\delta(\mathbf{r})$, and $\Delta n(\mathbf{r})$ their Fourier transforms in Eq. (13.9), to get

$$\nabla^2 \int d\mathbf{k}\, e^{i\mathbf{k}\cdot\mathbf{r}}\, V_{eff}(\mathbf{k}) = -4\pi e^2 \int d\mathbf{k}\, e^{i\mathbf{k}\cdot\mathbf{r}} - 4\pi e^2 \int d\mathbf{k}\, e^{i\mathbf{k}\cdot\mathbf{r}}\, \Delta n(\mathbf{k}) . \tag{13.10}$$

The first integral on the right-hand side of Eq. (13.10) is just the delta function $\delta(\mathbf{r})$. On the left-hand side the derivative is with respect to $\mathbf{r}$, and the only term inside the integral that is $\mathbf{r}$ dependent is $\exp(i\mathbf{k}\cdot\mathbf{r})$. Thus, $\nabla^2\exp(i\mathbf{k}\cdot\mathbf{r})$ gives $-k^2\exp(i\mathbf{k}\cdot\mathbf{r})$ in the integrand. For Eq. (13.10) to hold, the integrands on both sides need to be the same and we get

$$k^2 V_{eff}(\mathbf{k}) = 4\pi e^2[1 + \Delta n(\mathbf{k})] . \tag{13.11}$$

Assuming a plasma in quasi-equilibrium, the electron density is related to the chemical potential through Eq. (2.72). Due to the presence of the test charge, the chemical potential is changed from $\mu(n_0)$ to $\mu[n_0 + \Delta n(\mathbf{r})]$. Since the plasma is in quasi-equilibrium, the change in the chemical potential has to be compensated by V_{eff}, so

$$\mu(n_0) - \mu[n_0 + \Delta n(\mathbf{r})] = V_{eff}(\mathbf{r}) . \tag{13.12}$$

We treat the situation such that the test charge is a small perturbation of the total electron system, so $n_0 >> \Delta n(\mathbf{r})$. Hence, we may expand $\mu[n_0 + \Delta n(\mathbf{r})]$ in a Taylor series, keeping only the linear terms

$$\mu[n_0 + \Delta n(\mathbf{r})] \simeq \mu(n_0) + \frac{\partial\mu}{\partial n}\Big|_{n = n_0} \Delta n(\mathbf{r}) \tag{13.13}$$

or

$$\mu[n_0 + \Delta n(\mathbf{r})] - \mu(n_0) = \frac{\partial\mu}{\partial n}\Big|_{n = n_0} \Delta n(\mathbf{r}) . \tag{13.14}$$

Substituting into Eq. (13.12), we get

$$V_{eff}(\mathbf{r}) = -\frac{\partial\mu}{\partial n}\Big|_{n = n_0} \Delta n(\mathbf{r}) \tag{13.15}$$

or

$$\Delta n(\mathbf{r}) = \frac{V_{eff}(\mathbf{r})}{-\partial\mu/\partial n|_{n = n_0}} . \qquad (13.16)$$

The Fourier transform of Eq. (13.16) merely replaces the variable **r** with its conjugate transform variable **k**. Thus, Eq. (13.16) becomes

$$\Delta n(\mathbf{k}) = -\frac{\partial n(\mathbf{k})}{\partial\mu} V_{eff}(\mathbf{k}) . \qquad (13.17)$$

Inserting Eq. (13.17) into Eq. (13.11) yields

$$\left[k^2 + 4\pi e^2 \frac{\partial n(\mathbf{k})}{\partial\mu}\right] V_{eff}(\mathbf{k}) = 4\pi e^2 . \qquad (13.18)$$

If we now define the *inverse-screening length* or the *screening wave number* as

$$\kappa = \sqrt{4\pi e^2\, \partial n/\partial\mu} , \qquad (13.19)$$

we find the effective or screened potential,

$$V_{eff}(\mathbf{k}) \equiv V_s(\mathbf{k}) = \frac{4\pi e^2}{k^2 + \kappa^2} . \qquad (13.20)$$

The Fourier transform of Eq. (13.20) yields the screened Coulomb potential (Yukawa potential) in real space as

$$V_s(\mathbf{r}) = \int \frac{dk}{(2\pi)^3}\, e^{i\mathbf{k}\cdot\mathbf{r}}\, \frac{4\pi e^2}{(k^2 + \kappa^2)} , \qquad (13.21)$$

which gives (see problem 13.1)

$$V_s(\mathbf{r}) = \frac{e^2}{r}\, e^{-\kappa r} . \qquad (13.22)$$

Such a screened potential was first introduced by Yukawa for interactions inside the nucleus of atoms. Therefore, it is often called the Yukawa potential. The screened Coulomb potential $e^2\, e^{-\kappa r}/r$ is plotted in Fig. 13.1 together with the unscreened potential e^2/r. The comparison shows that

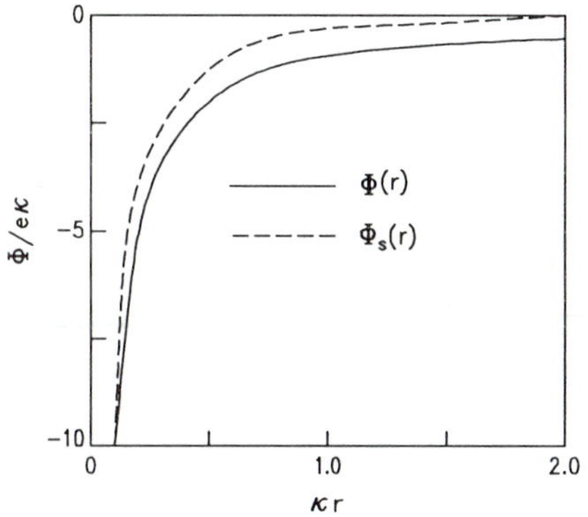

Figure 13.1. Comparison of the screened (dashed line) and unscreened (full line) Coulomb potential for a three-dimensional system of a point charge as seen by another point charge in an electron plasma.

the long-ranged, unscreened Coulomb potential is screened to a distance $1/\kappa$.

For the example of a nondegenerate electron plasma, where the Fermi distribution function can be approximated by the Maxwell-Boltzmann distribution function [Eq. (2.97)], the screening wave number (13.19) can be evaluated analytically. From Eq. (2.96) we have

$$\frac{\partial \mu}{\partial n} = \frac{k_B T}{n}, \tag{13.23}$$

and Eq. (13.19) becomes

$$\kappa = \sqrt{\frac{4\pi e^2 n}{k_B T}}, \tag{13.24}$$

which is the *Debye-Hückel* screening wave number. Equation (13.24) shows that the screening wave number in three-dimensional systems increases like $n^{1/2}$ with increasing plasma density, reducing the effective range of the Coulomb interaction potential.

It is important to emphasize that the aforementioned treatment of the screened Coulomb potential was classical. The quantum theory of plasma screening in the random phase approximation (RPA) yields the Lindhard formula, describing the frequency and wave vector dependent dielectric function $\epsilon(\omega, q)$ as

$$\epsilon(\omega, q) = \lim_{\delta \to 0} \left[1 - \frac{4\pi e^2}{\epsilon_0 V q^2} \sum_{\mathbf{k}} \frac{f_{\mathbf{k}-\mathbf{q}} - f_{\mathbf{k}}}{\hbar\omega + i\delta + \mathcal{E}(\mathbf{k} - \mathbf{q}) - \mathcal{E}(\mathbf{k})} \right],$$

Lindhard (RPA) dielectric function (13.25)

with δ being the broadening factor. The frequency dependence of $\epsilon(\omega, q)$ is referred to as temporal dispersion, while the wave vector dependence of the dielectric function is called the spatial dispersion.

From the Lindhard-formula one can derive several important limiting cases, including the Debye-Hückel result discussed above. As another illustrative example, we discuss the static limit, $\omega \to 0$, where Eq. (13.25) simplifies to

$$\epsilon(\omega = 0, q) = 1 - \frac{4\pi e^2}{\epsilon_0 V q^2} \sum_{\mathbf{k}} \frac{f_{\mathbf{k}-\mathbf{q}} - f_{\mathbf{k}}}{\mathcal{E}(\mathbf{k} - \mathbf{q}) - \mathcal{E}(\mathbf{k})} . \tag{13.26}$$

Expanding the difference between the Fermi functions for small momentum values q, we obtain

$$\begin{aligned} f_{\mathbf{k}-\mathbf{q}} - f_{\mathbf{k}} &= f_{\mathbf{k}} - \mathbf{q}\cdot\nabla_{\mathbf{k}} f + \cdots - f_{\mathbf{k}} \\ &\simeq f_{\mathbf{k}} - q \frac{\partial f}{\partial k} \cos(\gamma) + \cdots - f_{\mathbf{k}} \simeq -q \frac{\partial f}{\partial k} \cos(\gamma) , \end{aligned} \tag{13.27}$$

where γ is the angle between $\mathbf{q}$ and $\mathbf{k}$. Similarly, for the energy difference we obtain

$$\begin{aligned} \mathcal{E}(\mathbf{k} - \mathbf{q}) - \mathcal{E}(\mathbf{k}) &= \mathcal{E}(\mathbf{k}) - q \frac{\partial \mathcal{E}(\mathbf{k})}{\partial k} \cos(\gamma) + \ldots - \mathcal{E}(\mathbf{k}) \\ &\simeq -q \frac{\partial \mathcal{E}(\mathbf{k})}{\partial k} \cos(\gamma) = -\frac{\hbar^2}{m} q\, k \cos(\gamma) . \end{aligned} \tag{13.28}$$

Inserting Eqs. (13.27) and (13.28) into Eq. (13.26) yields

$$\epsilon(\omega = 0, q) = 1 - \frac{4\pi e^2}{\epsilon_0 V q^2} \sum_{\mathbf{k}} \frac{1}{k} \frac{\partial f}{\partial k}$$

$$= 1 - \frac{4\pi e^2}{\epsilon_0 V q^2} \frac{m}{\hbar^2} \frac{V}{(2\pi)^3} 8\pi \int_0^\infty dk\, k^2 \frac{1}{k} \frac{\partial f}{\partial k}, \tag{13.29}$$

where we used the three-dimensional conversion from k summation to integration and included a factor 2 from the spin summation. In the Thomas-Fermi limit, degenerate electron distributions (i.e., step-like Fermi functions) are assumed

$$f_{\mathbf{k}} = \theta(k_F - k), \tag{13.30}$$

where k_F is the Fermi wave number. For this case the integral in Eq. (13.29) is easily evaluated by partial integration,

$$\int_0^\infty dk\, k \frac{\partial f}{\partial k} = kf\Big|_0^\infty - \int_0^\infty dk\, f = -\int_0^{k_F} dk = -k_F . \tag{13.31}$$

Inserting Eq. (13.31) into Eq. (13.29) we obtain [(see problem (13.3)]

$$\epsilon(q, 0) = 1 + \frac{\kappa^2}{q^2}, \tag{13.32}$$

where

$$\boxed{\kappa = \sqrt{6\pi e^2 n/\mathcal{E}_F} = \sqrt{\frac{12\pi e^2 m_e}{\hbar^2 (3\pi^2)^{2/3}}}\, n^{1/6} .}$$

Thomas-Fermi screening wave number (13.33)

Equation (13.33) gives the *Thomas-Fermi screening wave number*. This approximation of screening is suitable for low temperatures where the electronic distribution is degenerate.

It should be noted that Thomas-Fermi screening is often a good approximation at metallic densities. In semiconductors, however, screening described by the Debye-Hückel model is a more reasonable approximation when the temperature is not too low. The holes in semiconductors almost always follow the Debye-Hückel screening because of their larger mass, which causes the distribution to be nondegenerate.

13-3. Exciton Ionization

Screening of the attractive electron-hole Coulomb interaction leads to one of the largest optical nonlinearities in semiconductors. The observed effect is the excitation-dependent bleaching of the exciton resonance (Mott transition) and of the Coulomb enhancement of the interband absorption. As shown in Chap. VI, in an unexcited semiconductor the exciton resonance exists as basically an isolated peak below the semiconductor bandgap in the optical absorption spectrum because of the electron-hole binding via the attractive Coulomb interaction. Such a bound state in a three-dimensional system can exist only if the attractive interaction is sufficiently strong. In other words, by weakening the electron-hole Coulomb attraction, the exciton resonance disappears from the spectrum.

As we discussed in the previous section, the strength of the Coulomb potential of a single charge is reduced by screening due to the presence of other charges. For a finite density of carriers, the bare Coulomb potential is replaced by the screened potential, Eq. (13.20), which is weaker than the unscreened one. As a consequence, the attractive Coulomb interaction between and electron and the oppositely charged hole is reduced. Hence, when a large number of electron-hole pairs is created in a semiconductor (e.g., by high-intensity optical excitation), the Coulomb potential is screened, causing the excitons to ionize. Exciton ionization means that they no longer exist as a bound state.

As an illustration, Fig. 13.2(a) shows a typical (good sample quality) low-temperature absorption spectrum of a medium-gap bulk semiconductor like GaAs near the band edge for two carrier densities. The spectrum labeled 1 is plotted for a small carrier density, while curve 2 represents the absorption when the carrier density is above the Mott density. The main features of the spectrum at low carrier density are the narrow excitonic lines (only $n = 1$ and $n = 2$ fundamental lines are spectrally resolved for the chosen broadening) on the low-energy side of the absorption edge. The energies are plotted relative to the bandgap; thus, the band-to-band absorption edge is located at position 0.0 on the horizontal axis. The frequency spectrum describing the continuum states extends to higher energies beyond the excitonic resonances, but these states are almost unchanged under the present conditions. Curve 2 exhibits a modified absorption spectrum due to the pump-generated electron-hole pairs. The excitonic resonances have disappeared as a result of screening of the Coulomb interaction by the electrons and holes.

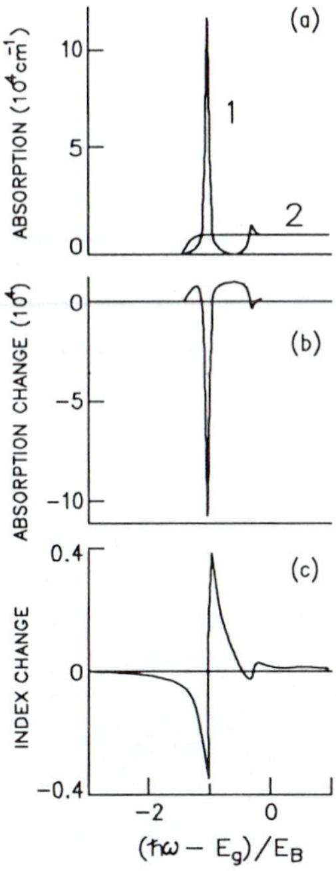

Figure 13.2. Excitonic optical nonlinearity in a semiconductor such as GaAs. (a) The absorption spectra at a low (curve 1) and high (curve 2) carrier density. (b) The change in the absorption coefficient, $\Delta\alpha$. (c) The change in the index of refraction, obtained by a Kramers-Kronig transformation of $\Delta\alpha$. The horizontal axes in Fig. 13.2(a), (b), and (c) have the same scale.

The change in absorption coefficient, $\Delta\alpha = \alpha(n) - \alpha_0$, can be obtained by direct subtraction of curves 1 and 2 in Fig. 13.2(a). The result is plotted in Fig. 13.2(b). A Kramers-Kronig transformation of $\Delta\alpha$ gives the nonlinear change in the index of refraction, $\Delta n(\omega)$,

$$\Delta n(\omega) = \frac{c}{\pi} \, \Pr \int_0^\infty \frac{\Delta\alpha(\omega')}{\omega'^2 - \omega^2} \, d\omega' , \tag{13.34}$$

where Pr stands for the principal value of the integral. Equation (13.34) is derived from Eq. (3.58) assuming

$$|\epsilon''| << |\epsilon'| , \tag{13.35}$$

which is generally true for semiconductors. In this case, Eq. (3.26) can be approximated by

$$\boxed{n(\omega) \simeq \sqrt{\epsilon'(\omega)} \ .}$$

refractive index (13.36)

Writing $\epsilon'(\omega) = \epsilon_0' + \Delta\epsilon'(\omega)$, where $|\Delta\epsilon'| << |\epsilon_0'|$, we get

$$n(\omega) \simeq n_0(\omega) + \Delta n(\omega) , \tag{13.37}$$

with $n_0(\omega) = \sqrt{\epsilon_0'}$ as the background refractive index and $\Delta n(\omega) = \Delta\epsilon_0'/2n_0(\omega)$. [Note that these quantities should not be confused with the mean density n_0 and the induced density $\Delta n(\mathbf{r})$ in Sec. 13-2.] For the absorption, Eq. (3.28), we write

$$\boxed{\alpha(\omega) = \frac{\omega}{cn(\omega)} \epsilon''(\omega) \simeq \alpha_0(\omega) + \Delta\alpha(\omega) ,}$$

absorption coefficient (13.38)

where $\alpha_0(\omega) = \omega\epsilon_0''(\omega)/cn_0(\omega)$ and $\Delta\alpha(\omega) = \omega\Delta\epsilon''(\omega)/cn_0(\omega)$. Using Eq. (3.41), we see that the refractive index change can be written as

$$\Delta n(\omega) = \frac{2\pi}{n_0(\omega)} \Delta\chi'(\omega) , \tag{13.39}$$

and the absorption change is

$$\Delta\alpha = \frac{4\pi\omega}{cn_0(\omega)} \Delta\chi''(\omega) \ . \tag{13.40}$$

Now we write the Kramers-Kronig relation (3.58) twice, once for $\chi'(\omega, n_1)$ and once for $\chi'(\omega, n_2)$, where n_1 and n_2 are constant parameters that affect the optical properties of the medium. n_1 and n_2 typically represent two different carrier densities, but could also correspond to two electron temperatures in the case of thermal nonlinearities. Taking the difference and using

$$\Delta\chi(\omega) \equiv \chi(\omega, n_1) - \chi(\omega, n_2) , \tag{13.41}$$

we get

$$\Delta\chi'(\omega) = \frac{2}{\pi} \, \mathrm{Pr} \int_0^\infty d\omega' \, \frac{\omega'}{\omega'^2 - \omega^2} \, \Delta\chi''(\omega') , \tag{13.42}$$

which is identical to Eq. (13.34) when we replace $\Delta\chi'$ and $\Delta\chi''$ by $\Delta n(\omega)$ and $\Delta\alpha$ using Eqs. (13.39) and (13.40), respectively. It should be noted at this point that the validity of Eq. (13.34) depends on the fact that the parameters n_1 and n_2 in Eq. (13.41) are in fact frequency and time independent. For very slow temporal variations of n, Eq. (13.34) may be regarded as approximately correct, but for fast changes, as in pump-probe experiments with ultrashort pulses, a Kramers-Kronig relation is not applicable.

Using Eq. (13.34) to compute the index change related to the absorption change of Fig. 13.2(b), we obtain the result plotted in Fig. 13.2(c). As can be noted from Fig. 13.2(c), $\Delta n(\omega)$ is negative on the low-energy side of the exciton and positive on the high-energy side. A negative index change, $\Delta n(\omega)$, means that the index of refraction is decreased at higher pump intensities. In other words, the larger pump intensity, which gives a larger carrier density [curve 2 of Fig. 13.2(a)], causes the index of refraction to become smaller on the low-energy side and larger on the high-energy side of the exciton resonance. The laser-induced negative index change is referred to as a *self-defocusing* optical nonlinearity. The positive $\Delta n(\omega)$ on the high-energy side of the exciton corresponds to a *self-focusing* optical nonlinearity. The reason for such terminology is that the semiconductor medium behaves like a lens when a laser beam with a Gaussian spatial profile is incident on it. For example, on the blue side of the exciton line with $\Delta n(\omega) > 0$, the medium acts as a positive lens. A laser beam with a Gaussian shape has an intensity profile such that the intensity in the center is larger than the sides. The larger intensity in the center of the beam induces a larger index of refraction in the semiconductor [$\Delta n(\omega) > 0$] compared with the sides of the beam. The optical pathlength for the beam [$n(\omega)d$ where $n(\omega)$ is the index and d is the thickness] in the center is, thus, larger than the sides. The material effectively behaves like a positive lens for the Gaussian beam. (As is well known, the optical pathlength is larger in the center of a positive lens.)

The optical nonlinearity that was just described as a result of the bleaching of the excitonic resonances usually has a large magnitude and a relatively rapid response time. This nonlinearity is often dominant in the spectral vicinity of the exciton in larger-gap semiconductors.

13-4. Bandfilling

As discussed in earlier chapters, the electrons and holes as Fermions can occupy each quantum state only once according to the Pauli exclusion principle. Hence each **k** state in a semiconductor band can be occupied twice, once with a spin-up and once with a spin-down carrier. An occupied state is no longer available as a final state in an optical absorption process. Due to the principle of energy minimization, the carriers in quasi-equilibrium occupy the available states from the bottom of the band, so that the energetically lowest states are occupied first. This results in filling the states near the bottom of the conduction band by electrons and the top of the valence band by holes. The schematics of the bandfilling process is shown in Fig. 13.3. The simultaneous availability of an electron and a hole state with momentum k is determined by the so-called bandfilling factor,

$$A(\mathcal{E}) = 1 - f_e(\mathcal{E}) - f_h(\mathcal{E}) \; , \tag{13.43}$$

where f_e and f_h are the Fermi functions for the electrons and holes, respectively. The bandfilling factor arises from the consideration that for a valence-to-conduction-band ($v \rightarrow c$) transition to occur, the valence-band state has to be filled by an electron ($\propto f_v$) and the conduction-band state has to be empty ($\propto 1 - f_c$), so the transition rate is proportional to the product,

$$A_{v\rightarrow c} = f_v(1 - f_c) \; . \tag{13.44}$$

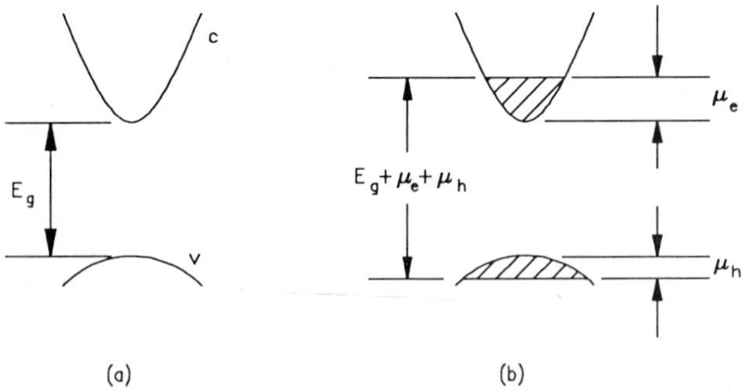

Figure 13.3. Schematic representation of optical nonlinearity due to the bandfilling effect in a semiconductor. (a) The semiconductor band structure in the absence of electron-hole pairs. (b) The bands when a large electron-hole density is generated, causing filling of the bands near their extrema. Here the temperature T is assumed to be zero.

Furthermore, to obtain the net transition rate, one has to subtract the inverse process (i.e., the conduction-to-valence-band-transition rate (c → v) that describes carrier recombination)

$$A_{c\rightarrow v} = f_c(1 - f_v) , \tag{13.45}$$

since it leads to a reduction of the bandfilling. Subtracting Eq. (13.45) from (13.44) and using $f_v = 1 - f_h$ yields Eq. (13.43).

As discussed in Sec. 2-8, the Fermi functions can only vary between 0 and 1, where 0 indicates vanishing occupation probability and 1 indicates that the state is occupied. When both f_e and f_h are equal to zero (i.e., for the unexcited semiconductor), A becomes unity and all states are available for optical absorption. For higher carrier densities, f_e and f_h are larger than zero, and for sufficiently high densities, there exists an energy for which the sum of both Fermi functions equals unity, making A = 0. This occurs exactly for the energy

$$\mathcal{E} = \mu_e + \mu_h \equiv \mu , \tag{13.46}$$

i.e., at the sum of the quasi-chemical potentials of electrons and holes (see Prob. 13.4). The vanishing of the bandfilling factor indicates that the semiconductor is transparent at this frequency; the optical absorption has been "bleached" to zero. For even higher carrier densities, $f_e + f_h > 1$ for $\mathcal{E} < \mu$. In that case, $A(\mathcal{E}) < 0$ and the optical absorption becomes negative. Negative absorption means amplification, as can be seen from Beer-Lambert's law [Eq. (3.30)],

$$I(x) = I_0 \, e^{-\alpha x} . \tag{13.47}$$

For $\alpha < 0$, the transmitted intensity is higher than the input intensity. This *optical gain* is the basis of optical amplification and semiconductor lasers, as will be discussed in Chap. XVII.

For densities that are not too high (i.e., when optical gain does not yet exist), the filling of the bands results in a gradual bleaching of the absorption in the vicinity of the bandgap and appears as if the onset of absorption has shifted to higher energies. Figure 13.4(a) displays this bandfilling effect on the absorption spectrum of a narrow-gap semiconductor like InSb. Again, curve 1 is plotted for a low carrier density and curve 2 for a higher carrier density. The bleaching of the band-edge absorption, shown in curve 2, manifests itself as an apparent blue shift of the absorption edge.

The change in the absorption coefficient, $\Delta\alpha$, is shown in Fig. 13.4(b), and its Kramers-Kronig transformation, $\Delta n(\omega)$, is in Fig. 13.4(c). The nonlinear index is negative at frequencies below the absorption edge, similar in sign to the exciton bleaching nonlinearity.

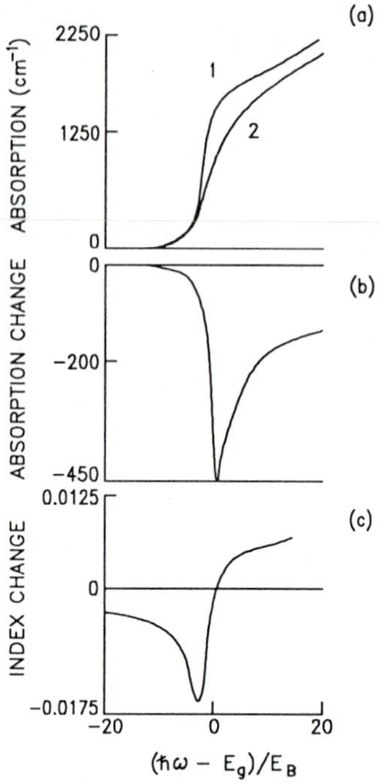

Figure 13.4. Bandfilling nonlinearity in a narrow-gap semiconductor. (a) The absorption spectra at low temperatures for low (curve 1) and high (curve 2) carrier densities. (b) The change in the absorption coefficient, $\Delta\alpha$. (c) The change in the refractive index, $\Delta n(\omega)$, obtained from a Kramers-Kronig transformation of $\Delta\alpha$ of Fig. 13.4(b).

13-5. Bandgap Renormalization*

Another consequence of the Coulomb interaction among the laser-excited electronic excitations is the effect of the bandgap shrinkage or bandgap renormalization; i.e., the energies of the electrons and holes in their respective bands are reduced. This energy reduction is a consequence of the exchange effect for particles with equal spin, and Coulomb correlation effect for all particles. The exchange effect is caused by the Pauli exclusion principle. The probability that two Fermions with identical quantum numbers are at the same point in real space is zero. For increasing separation between the particles, the probability slowly approaches unity. Hence, the Pauli exclusion leads to a reduction of the probability that equally charged particles come close to each other, and this in turn reduces the repulsive (i.e., positive Coulomb energy) contribution. This situation for particles with equal spins is often described by the presence of an "exchange hole," where each Fermion is surrounded by a region where the probability for the existence of another identical Fermion is very small. Correspondingly, equally charged Fermions with different quantum numbers (e.g., electrons with different spins) avoid each other because of the Coulomb repulsion. As in the case of the exchange hole, this "Coulomb hole" also leads to a decrease of the overall energy.

A good approximation for the bandgap reduction is

$$\boxed{\delta E_g = \sum_{q \neq 0} [V_s(q) - V(q)] - \sum_{q \neq 0} V_s(q)\,[f_e(q) + f_h(q)]\,,}$$

electronic bandgap reduction (13.48)

where the first and second terms are called the "Coulomb-hole" and "screened-exchange" contributions, respectively. Here $V(q)$ and $V_s(q)$ are the Fourier transform of the unscreened and screened Coulomb potentials, respectively. The renormalized bandgap is then

$$E_g' = E_g + \delta E_g \,, \tag{13.49}$$

with $\delta E_g < 0$ [the first term in Eq. (13.48) has a negative sign since the screened Coulomb potential is smaller than the bare Coulomb potential, $V_s(q) < V(q)$]. Figure 13.5 shows the bandgap shift as a function of the dimensionless distance r_s between two carriers,

$$r_s \equiv \left(\frac{1}{na_0^3} \frac{3}{4\pi} \right)^{1/3} , \tag{13.50}$$

We see that the screened exchange and the Coulomb-hole contributions both increase with increasing carrier density (decreasing particle separation). For low carrier densities, the dominating contribution comes from the Coulomb-hole term, whereas at elevated densities, both terms are equally important. An often useful approximation for the bandgap reduction has been derived in Ref. 13.1.

$$\frac{\delta E_g}{E_B} = E_{xc} + n \frac{\partial E_{xc}}{\partial n} \,, \tag{13.51}$$

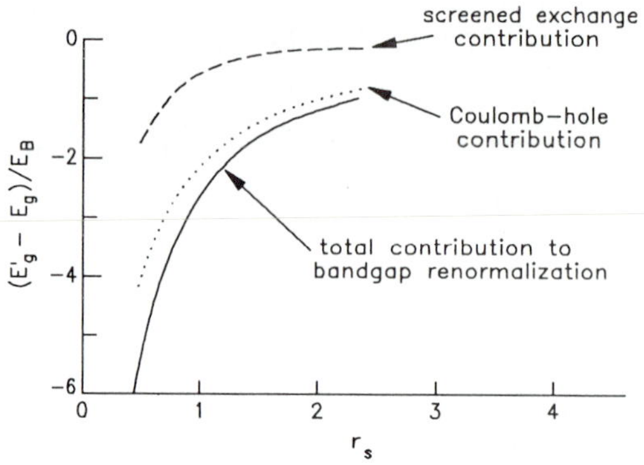

Figure 13.5. Renormalized semiconductor bandgap as a function of dimensionless carrier distance r_s for T = 300 K and material parameters appropriate for bulk GaAs. The shift of the bandgap energy normalized to the exciton binding energy is plotted. The long-dashed line is the screened-exchange contribution, the short-dashed line is the Coulomb-hole contribution, and the full line is the sum of both, respectively.

where E_B is the exciton Rydberg energy (see Chap. VI) and

$$E_{xc} = \frac{4.8316 + 5.0879\, r_s}{0.0152 + 3.0426\, r_s + r_s^2} \; . \tag{13.52}$$

Bandgap reduction leads to a monotonous red shift of the onset of the continuum absorption in semiconductors. At the Mott density the bandgap has shifted one exciton Rydberg energy below the zero-density bandgap E_g. For even higher densities, bandgap renormalization may cause increasing absorption in the spectral region below the exciton resonance. Whether such an increasing absorption is visible in an experimental spectrum depends on the magnitude of the bandgap renormalization versus the increasing chemical potential due to the bandfilling effect discussed in Sec. 13-4. If the chemical potential, $\mu = \mu_e + \mu_h$, is greater than E_g', the absorption becomes negative for $E_g' < \hbar\omega < \mu$ and optical gain occurs in that regime. The density-dependent renormalized bandgap can be extracted from measurements of optical gain spectra by monitoring the onset of gain.

An example of increasing absorption due to bandgap renormalization calculated for low-temperature CdS is shown in Fig. 13.6. The absorption spectra are plotted in Fig. 13.6 at low excitation densities (curve 1) and under higher excitation (curve 2). It is clearly shown in curve 2 that the absorption has increased in the spectral region just below the exciton

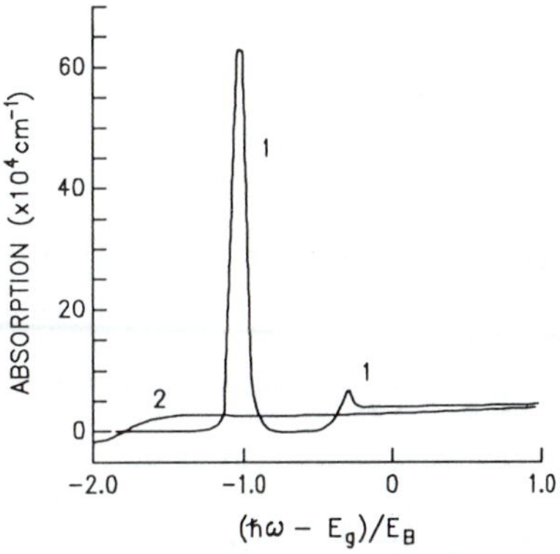

Figure 13.6. Computed semiconductor absorption spectra for the parameters of bulk CdS at T = 30 K. Curves 1 and 2 are the absorption coefficients for low and high carrier densities, respectively. The increased absorption on the low-energy side of the lowest exciton resonance is due to bandgap reduction. A region of optical gain (negative absorption) is also visible in curve 2.

resonance. This increase is a direct consequence of the bandgap reduction, which in this case has caused the band edge to shift below the exciton resonance. The excitons are completely ionized at the carrier densities needed to observe bandgap reduction, such as that in Fig. 13.6. The change in the absorption coefficient as a result of the bandgap renormalization also gives rise to an index change.

13-6. Thermal Nonlinearities

Some of the laser-excited electron-hole pairs recombine nonradiatively, giving their energy to the crystal through phonon emission, which corresponds to lattice heating. In most materials the temperature change causes a shift of the bandgap. There is no general law expressing the thermal shift of the gap with temperature. In most crystals, heating of the sample leads to a bandgap reduction, with a corresponding red shift of the onset of absorption. There are exceptions, however, such as CuCl for which heating leads to an increase of the bandgap. In many crystals the variation of the bandgap has been determined empirically. For instance, in bulk GaAs, a good fit to the measured thermal bandgap reduction is given by

$$E_g(T) = 1522 \text{ meV} - \frac{0.58T^2}{T + 226\text{K}} \frac{\text{meV}}{\text{K}} . \tag{13.53}$$

Figure 13.7 displays the effect of laser heating on the optical absorption and the index of refraction. Curve 1 is plotted for a low laser intensity, where the heating by the laser is negligible. For a larger laser intensity (curve 2), the generated heat has caused the band edge to shift toward the red. The change in the absorption spectrum is shown in Fig. 13.7(b). The shift of the absorption spectrum toward lower energies gives rise to an index change which is shown in Fig. 13.7(c). As can be seen from this figure, in this case a thermally induced index change is positive below the band edge. Due to the ease in obtaining thermal nonlinearities and their existence at room temperature, researchers have extensively used them, mainly for proof-of-principle experiments. However, for many practical applications thermal nonlinearities are not desirable, notably because of the slow response time, determined by the

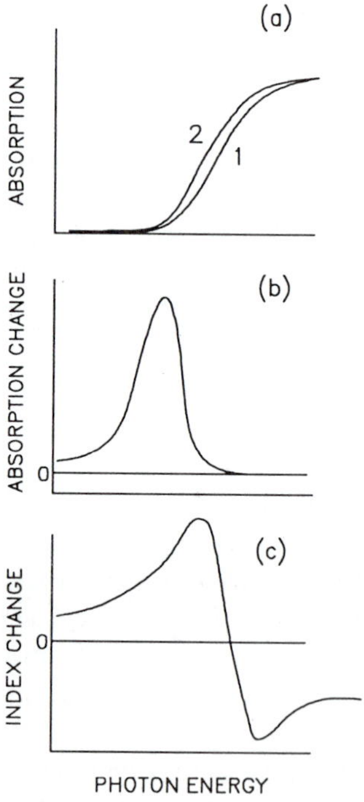

Figure 13.7. Thermal optical nonlinearity in semiconductors. (a) Schematics of the absorption spectra in the vicinity of the bandgap for low (curve 1) and high (curve 2) laser intensities. (b) The change in the absorption coefficient. (c) The change in the refractive index.

time necessary to cool the crystal. This slow response helps distinguish thermally induced nonlinearities from optical nonlinearities of electronic origin.

13-7. Theory of Nonlinearities: Semiconductor Bloch Equations*

So far our discussion of nonlinear mechanisms has been mostly descriptive and schematic. Even though a detailed discussion of the many-body theory needed to analyze electronic semiconductor nonlinearities is beyond the scope of this book, we outline the basic steps in such a treatment of carrier-dependent nonlinearities in this section.

The interactions between the laser-excited electrons and holes lead to a generalization of the Wannier equation (which is often called the Bethe-Salpeter equation) or the semiconductor Bloch equation for the interband polarization P(k) (see Sec. 15-4 for the full semiconductor Bloch equations). One form of this equation is (see Ref. 2.5)

$$\left[\hbar\omega - E_g{}' - \mathcal{E}_e(k) - \mathcal{E}_h(k) + i\delta\right] P(\mathbf{k}) =$$

$$- A(k)\left[d_{cv} E + \sum_{\mathbf{k}'} V_s(|\mathbf{k} - \mathbf{k}'|) P_{\mathbf{k}'}\right], \tag{13.54}$$

where δ describes the broadening of the resonances, A(k) is the bandfilling factor of Eq. (13.43), E is the applied light field, $\mathcal{E}_e$ and $\mathcal{E}_h$ are the electron and hole energies, $E_g{}'$ is the renormalized bandgap, d_{cv} is the valence-band-to-conduction-band dipole matrix element, V_s is the screened Coulomb potential, and $P(\mathbf{k})$ is the interband polarization of the material. The solution of Eq. (13.54) yields P(k), from which optical susceptibility χ_k can be obtained,

$$\chi_k = \frac{P(k)}{E}, \tag{13.55}$$

which in turn is related to the dielectric function via

$$\epsilon = 1 + 4\pi \sum_{\mathbf{k}} d_{cv}^{*} \chi_{\mathbf{k}} \, . \tag{13.56}$$

Under quasi-equilibrium conditions, this dielectric function depends parametrically on the electron-hole-pair density n.

In Eq. (13.54) we recognize the optical nonlinearities discussed in the earlier sections of this chapter. The term E_g' contains the bandgap renormalization, V_s contains the screening of the Coulomb potential, and the prefactor A(k) contains the bandfilling nonlinearity. Actually, a more detailed analysis of Eq. (13.54) shows that the usual bandfilling nonlinearity results from the term $\propto$ A(k)E on the right-hand side of Eq. (13.54). As we see, however, A(k) also multiplies the Coulomb term,

$$\sum_{\mathbf{k}'} A(k)\, V_s(|\mathbf{k} - \mathbf{k}'|) P_{\mathbf{k}'} \, . \tag{13.57}$$

Since A(k) is always smaller than unity, the multiplication of V by A leads to an additional reduction of the electron-hole Coulomb attraction. This reduction is also a consequence of the Pauli exclusion principle, which in this case assures that only those states can contribute to the formation of bound electron-hole pairs that are not already occupied by other electrons or holes. Thus, the presence of electrons and holes reduces the effective electron-hole attraction, not only through the screening, but also through Pauli blocking. Both effects are always simultaneously present, but their relative importance changes with the dimension of the system. The general influence of A(k) on the optical response is also known as *phase-space-filling* nonlinearity. The density-dependent modifications caused by the blocking term A(k) in Eq. (13.57) are similar to those caused by screening, since both mechanisms lead to a reduced electron-hole attraction. It turns out that for bulk semiconductors at densities that are not too low, screening is usually a stronger effect than blocking. However, in semiconductor quantum wells, quite the opposite is true, as we will discuss in more detail in the next section.

A considerable simplification of the solution of Eq. (13.54) is obtained for bulk semiconductors by ignoring the blocking factor of Eq. (13.57), i.e., by replacing

$$\sum_{\mathbf{k}'} A(k)\, V_s(|\mathbf{k} - \mathbf{k}'|) P_{\mathbf{k}'} \rightarrow \sum_{\mathbf{k}'} V_s(|\mathbf{k} - \mathbf{k}'|) P_{\mathbf{k}'} \,. \tag{13.58}$$

Under this assumption, an often useful approximation to the nonlinear optical absorption of bulk semiconductors may be obtained by a partly phenomenological generalization of the Elliott formula [Eq. (6.33)], as

$$\alpha(\omega) = \frac{8\pi^2\omega\, |d_{cv}|^2}{n_0(\omega)c} \tanh\left[\frac{\hbar\omega - \mu(N)}{2k_B T}\right]$$

$$\times \sum_n |\phi_n(r = 0)|^2\, \delta_\Gamma(\hbar\omega - \mathcal{E}_n - E_g') \,, \tag{13.59}$$

where $\delta_\Gamma(x)$ is a broadened delta function, and $\phi_n(\mathbf{r})$ and $\mathcal{E}_n$ are the electron-hole pair wave functions and energy eigenvalues, respectively. They are the solutions of the modified Schrödinger equation (Wannier equation) in which the bare Coulomb potential is replaced by the screened Coulomb potential. Equation (13.59) can be evaluated for various semiconductors. This plasma theory gives good agreement with experimental results for bulk semiconductors, as shown in Chap. XII.

13-8. Optical Nonlinearities of Quantum Wells and Quantum Dots*

The dominant contributions to the optical nonlinearities in semiconductor quantum wells are somewhat different from those of bulk materials. Most importantly, Coulomb screening is weaker in quantum wells than in bulk. Intuitively, one can understand this by noting that screening is a spatial rearrangement of the electrons and holes to "intercept" the field lines between the interacting carriers. As shown schematically in Fig. 13.8, a significant part of the electric field lines in a quantum well are outside the layer in which the carriers can move. Therefore, the possibilities for the carriers to modify the interaction between the electron and hole are very restricted, and the effect of screening is reduced in comparison to bulk materials. However, Eq. (13.57) shows that the Coulomb attraction is also weakened by Pauli blocking. Since the **k** space obviously has one dimension less in two-dimensional systems than in three-dimensional ones, it takes fewer carriers to fill the available states. Hence, phase-space filling in two dimensions is more efficient than in three dimensions, whereas plasma screening is weaker.

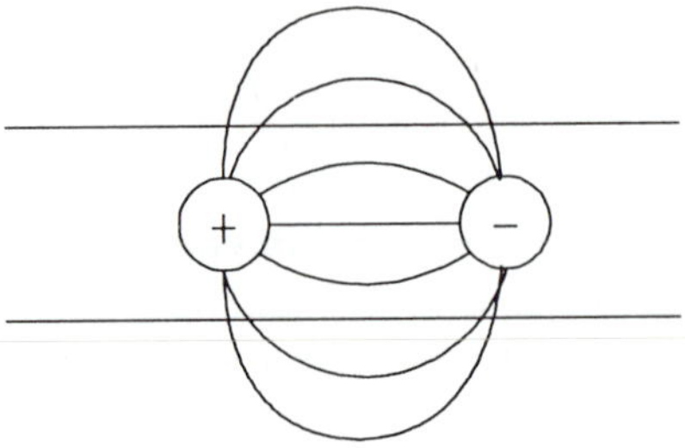

Figure 13.8. Schematic drawing of the field lines between an electron and a hole in a quantum well.

We note that even though the contributions of the different physical processes causing the changes in the optical spectra near the band edge of bulk and multiple quantum well systems are somewhat different, the overall features of the nonlinearities are quite similar.

As discussed in Chap. X, the linear absorption spectrum of quantum dots consists of a series of lines corresponding to transitions between quantum-confined electron-hole states. The origin of optical nonlinearity in small quantum dots is mainly the state-filling effect, resulting in bleaching of the quantum-confined transitions. Screening of the Coulomb interaction by excited carriers is not important in quantum dots. This may be understood by noting that screening occurs in bulk semiconductors by carrier excitation because the excited electrons and holes can easily undergo intraband transitions, since a large number of states are available for such transitions. Such transitions lead to a rearrangement of carriers in real space, resulting in the weakening of the Coulomb potential by screening. In quantum dots, however, the energy states that electrons or holes may occupy are quantized with a relatively large energy separation. Therefore, transitions between those states require a significant amount of energy transfer and are very unlikely to occur as a consequence of Coulomb effects alone. Consequently, plasma screening in small quantum dots is basically absent. However, the Coulomb interactions between the charged particles is still important and leads to some other interesting nonlinear effects (such as energy changes for two electron-hole-pair states) that will be discussed in Chap. XIV. Here we only stress that since the Pauli exclusion principle prevents occupation of an electronic state by more than one identical particle, the quantum-confined transitions undergo absorption saturation if we try to excite additional charged particles into the quantum-confined states. This process of state filling is mainly responsible for quantum dot optical nonlinearities.

13-9. Transient Nonlinearities: Optical Stark Effect

In previous sections of this chapter we discussed the fact that quasi-equilibrium nonlinearities are controlled by real carrier generation as a result of resonant laser excitation. Thus, the response time of such nonlinearities is dictated by the carrier lifetime, which is typically in the nanosecond range in most direct-gap semiconductors. However, semiconductors also exhibit transient or coherent nonlinearities with response times that either follow the excitation laser pulse time profile or are limited by the dephasing time [the dephasing time is the inverse polarization damping, δ^{-1}, in Eq. (13.54)] of the electronic resonance. Transient nonlinearities are best described in off-resonance pumping situations, where the excitation laser frequency is tuned below the exciton resonance. However, coherent nonlinearities may also be detected at the early stage of resonant pumping using laser pulses of very short duration. This topic will be further treated in Chap. XV. Here we only treat the optical Stark effect as an example of nonresonant pumping nonlinearity.

As we pointed out in Chap. VI, ideal semiconductors are transparent for photon frequencies below the n = 1 exciton energy. A laser pulse tuned in this transparency region is not absorbed and, thus, real carrier creation is not expected. Nevertheless, the presence of the intense below-bandgap laser changes the optical properties of the semiconductor, inducing optical nonlinearity during the presence of the strong optical field. The incident light causes nonresonant, often called "virtual" carrier excitation. The uncertainty principle states that the energy of an electron is only specified within an accuracy of the order of $\Delta\mathcal{E} = \hbar/\Delta t$. Thus, loosely speaking, electrons are present in the conduction band during a time determined by the energy offset, $\Delta\mathcal{E} = \mathcal{E}_x - \hbar\omega$.

In the optical Stark effect the *virtual* excitation of electron-hole pairs leads to a high-energy shift of the exciton absorption line. However, this excitonic shift only lasts as long as the intense nonresonant pump field is present. Associated with the excitonic blue shift, there is a transient index change, $\Delta n(\omega)$. The correct theoretical description of the optical Stark effect requires a numerical solution of the semiconductor Bloch equations described in Sec. 15-4. We give a simple physical picture here, treating the case of a two-level system. Despite its simplicity, this description exhibits features that are useful to analyze the exciton Stark effect. We assume that the laser has a large detuning from exciton, $\Delta\omega$, expressed as

$$\Delta\omega = \omega_x - \omega \, , \tag{13.60}$$

where $\mathcal{E}_x = \hbar\omega_x$ is the exciton energy and $\hbar\omega$ is the energy of the laser beam. The coupling of the photon and the material leads to a blue shift of the resonance with a magnitude given by

$$\delta\omega = \sqrt{(\Delta\omega)^2 + 4(d_{cv}E/\hbar)^2} - \Delta\omega \,, \qquad (13.61)$$

where d_{cv} is the dipole matrix element between the valence and conduction bands, and E is the electric field amplitude of the pump laser. For large detunings, $\Delta\omega >> 2\, d_{cv}E/\hbar$, Eq. (13.61) reduces to

$$\delta\omega = \Delta\omega \left[1 + 4 \left(\frac{d_{cv}E/\hbar}{\Delta\omega} \right)^2 \right]^{1/2} - \Delta\omega$$

$$\simeq \Delta\omega \left[1 + 2 \left(\frac{d_{cv}E}{\hbar\Delta\omega} \right)^2 \right] - \Delta\omega \simeq 2\, \frac{|d_{cv}E|^2}{\hbar^2\Delta\omega} \,, \qquad (13.62)$$

stating that at large detunings the resonance shift is directly proportional to the electromagnetic field intensity and inversely proportional to the laser detuning from the resonance.

13-10. $\chi^{(3)}$ Formalism

It is customary in traditional nonlinear optics to describe nonlinear effects in transparent materials such as liquids (like CS_2) with the concept of a field-dependent nonlinear optical susceptibility. In this scheme the relation

$$\mathbf{P} = \chi\, \mathbf{E} \qquad (13.63)$$

is expanded in successive higher-order terms of the radiation fields,

$$\mathbf{P} = \chi^{(1)}\mathbf{E} + \chi^{(2)}:\mathbf{E}:\mathbf{E} + \chi^{(3)}:\mathbf{E}:\mathbf{E}:\mathbf{E} \,, \qquad (13.64)$$

where the symbolic multiplication, ":" , indicates proper tensor multiplication, since the susceptibility coefficients $\chi^{(1)}$, $\chi^{(2)}$, $\chi^{(3)}$, ... are tensors. The first term in the expansion, Eq. (13.64), is linear in the electric field. The first-order complex susceptibility $\chi^{(1)}$ describes the linear response of the medium to the incident electric field, giving the linear absorption and refractive index. The second term in the right-hand side of Eq. (13.64) describes second-order nonlinearities. It leads to phenomena such as second-harmonic generation or sum or difference-frequency generation and optical rectification. In sum-frequency generation, two of the incident photons with frequency ω mix to generate a third photon of frequency, $\omega + \omega = 2\omega$ (or in general for two different incident frequencies, ω_1 and ω_2, the generated wave has a frequency of $\omega_3 = \omega_1 + \omega_2$). The second-order susceptibility $\chi^{(2)}$ is nonvanishing only in media without inversion symmetry (see Prob. 13.5). The third term in the expansion represents third-order nonlinearities with $\chi^{(3)}$ being the third-order nonlinear susceptibility. Using this term, phenomena such as four-wave mixing (FWM), phase conjugation (PC), third-harmonic generation (THG), and intensity-dependent nonlinearity, such as Kerr-type nonlinearity, etc., may be treated. The FWM processes will be described in Chap. XII. In THG, three waves, each having a frequency ω, mix to generate a fourth wave of frequency 3ω. In treating the intensity-dependent nonlinearity with the help of $\chi^{(3)}$, it is customary to consider excitation below the resonance. In an isotropic medium (where $\chi^{(2)}$ is zero), only one independent element of the $\chi^{(3)}$ tensor exists, so that $\chi^{(3)}$ becomes a scalar (see Prob. 13.6). The absorption is also ignored when the excitation frequency is far below resonance. It is then found that the refractive index may be written as

$$\begin{aligned} n(\omega) &= n_0(\omega) + \Delta n(\omega) \\ &= n_0(\omega) + n_2\, I\,, \end{aligned} \tag{13.65}$$

where $n_0(\omega)$ is the linear index of refraction, $\Delta n(\omega)$ is the change in the index, and I is the light intensity. The coefficient n_2 is called the Kerr-coefficient and is related to $\chi^{(3)}$ by

$$n_2 = \frac{1}{3}\left(\frac{4\pi}{n_0(\omega)}\right)^2 \chi^{(3)}\,. \tag{13.66}$$

Here n_2 is in units of (cm²/kW) and $\chi^{(3)}$ has esu units (see Ref. 13.6). Therefore, the index change $\Delta n(\omega)$ is linearly dependent on intensity. The magnitude of n_2, which is intensity-independent, has been employed as a measure of the magnitude of the nonlinearity.

Even though the use of the $\chi^{(3)}$ formalism in atomic systems and in liquids has been successful, its application in treating semiconductors is limited. In particular, this formalism is not well adapted to the description of resonant situations. As we described for the case at the beginning of this chapter, quasi-equilibrium semiconductor optical nonlinearities depend on the light intensity only indirectly through carrier density n.

Furthermore, it is the carrier density that changes the semiconductor absorption and refraction, and not the light intensity directly.

13-11. Two-Photon Absorption Nonlinearity for Optical Limiting

Until now we have dealt primarily with nonlinear mechanisms such as exciton ionization and bandfilling, where increasing the incident laser intensity results in a reduction of the absorption. There are also mechanisms where the light absorption increases as the laser intensity is increased. One such mechanism is the bandgap renormalization that we discussed in Sec. 13-5, where for frequencies below the bandgap, the increase in laser intensity results in an increase in absorption as the band edge shifts to lower energies. Another mechanism that also leads to intensity-dependent absorption is the two-photon-absorption process that we discussed in Chap. XI. Here we consider the effect of two-photon absorption on the transmittance of an intense beam. Thus, in this process two photons of the same frequency are absorbed in order to complete the absorption. For photon energies in the vicinity of half the bandgap energy, $\hbar\omega \simeq E_g/2$, two-photon absorption occurs as schematically shown in Fig. 13.9. The crystal becomes electronically excited through the transfer of an electron from the valence band to the conduction band by the simultaneous absorption of two photons, $\hbar\omega_1$ and $\hbar\omega_2$. Each individual photon energy is less than the required transition energy and, therefore, falls in the transparent region of the crystal. However, the sum of the two photons, $\hbar(\omega_1 + \omega_2)$, satisfies the energy conservation law for the transition.

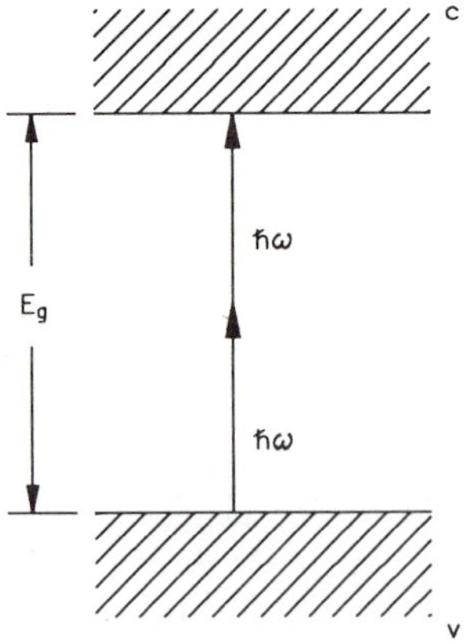

Figure 13.9. Schematic representation of two-photon absorption in a semiconductor.

For two-photon absorption, the variation of the light intensity inside the semiconductor as it propagates through it may be written as

$$\frac{dI}{dz} = -\beta I^2 , \tag{13.67}$$

where β is the two-photon absorption coefficient, which is assumed to be a positive quantity, and z is in the direction of laser propagation. Equation (13.67) specifies that the change in laser intensity is proportional to the square of the light intensity, as expected for a two-photon process. In writing Eq. (13.67), we ignored possible one-photon absorption that may also occur simultaneously at $\hbar\omega$. In general, Eq. (13.67), including one-photon absorption, is written as

$$\frac{dI}{dz} = -\alpha I - \beta I^2 . \tag{13.68}$$

Equations (13.68) or (13.67) may be easily solved. Note that when $\beta = 0$ the solution of Eq. (13.68) gives the Beer-Lambert law, Eq. (3.30); i.e.,

$$I(z) = I_0 e^{-\alpha z} , \tag{13.69}$$

where I_0 is the incident laser intensity. Equation (13.68) may be simply solved when $\alpha = 0$ with a solution given by

$$\int_{I_0}^{I(z)} \frac{dI}{I^2} = \int_0^z -\beta dz , \tag{13.70}$$

where I(z) is the light intensity at distance z inside the semiconductor. Integration on both sides of Eq. (13.70) gives

$$I(z) = \frac{I_0}{1 + \beta z I_0} , \tag{13.71}$$

indicating that for a fixed I_0, I(z) decreases as z increases. The behavior of I(z) as a function of intensity at a fixed position z is plotted in Fig. 13.10. I(z) is limited to a value of $1/\beta z$ as the I_0 becomes very large. This limiting behavior of two-photon absorption has been experimentally observed, as shown in Fig. 13.11, and may be potentially useful as an optical limiter in some applications.

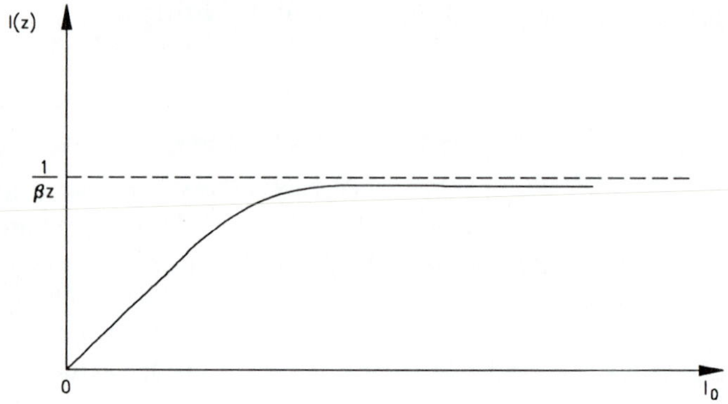

Figure 13.10. The limiting behavior of two-photon absorption.

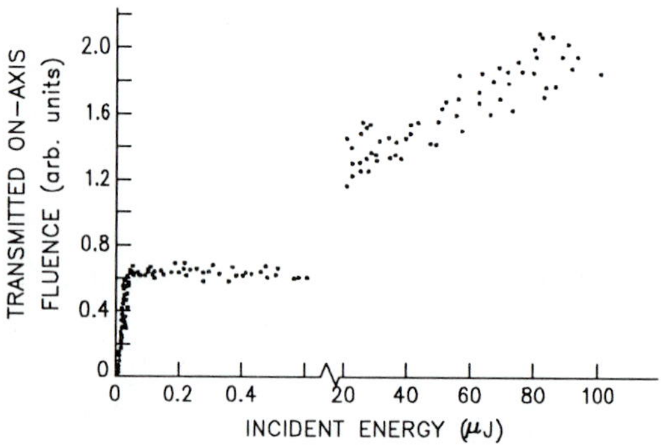

Figure 13.11. Experimentally measured input-output characteristic of a 32-mm-thick ZnSe. The transmitted beam was imaged onto a pinhole, and the detector was placed behind the pinhole (after Ref. 13.7).

13-12. Problems

13.1. (a) Show that the Fourier transform of the Coulomb potential e/r is $4\pi e/k^2$.

(b) Show that the Fourier transform of the screened Coulomb potential $(e/r)e^{-\kappa r}$ is $4\pi e/(\kappa^2 + k^2)$.

13.2. Show that the following equation holds:

$$\delta(\mathbf{r}) = \int \frac{d\mathbf{k}}{(2\pi)^3}\, e^{i\mathbf{k}\cdot\mathbf{r}} . \tag{13.72}$$

13.3. Insert the result $\int_0^\infty dkk\, \frac{\partial f}{\partial k} = -k_F$ into Eq. (13.29) and show that the inverse Thomas-Fermi screening length is

$$\kappa = \sqrt{6\pi e^2 n/\mathcal{E}_F} = \sqrt{\frac{12\pi e^2 m_e}{\hbar^2(3\pi^2)^{2/3}}}\, n^{1/6}. \tag{13.73}$$

Hint: Compute the expression for the particle density n at T = 0 to express $\mathcal{E}_F = \hbar^2 k_F{}^2/2m$ and k_F in terms of n.

13.4. Show that the bandfilling factor

$$A(\mathcal{E}) = 1 - f_e(\mathcal{E}) - f_h(\mathcal{E}) \tag{13.74}$$

is zero at the energy $\hbar\omega = E_g + \mu_e + \mu_h$.

13.5. Show that $\chi^{(2)}$ is zero in a medium with inversion symmetry.

13.6. Show that in an isotropic medium, $\chi^{(3)}$ reduces to a scalar.

13.7. Derive the law expressing the transmission of a beam of incident intensity I_0 as a function of sample thickness taking both linear and two-photon absorption into account.

13-13. References

1. P. Vashista and R. K. Kalia, Phys. Rev. B **25**, 6412 (1982).

2. S. Schmitt-Rink, D. A. B. Miller, and D. J. Chemla, Phys. Rev. B **35**, 8113 (1990).

3. J. M. Hvam and C. Dornfeld, in *Optical Switching in Low-Dimensional Systems*, edited by H. Haug and L. Banyai (Academic Press, New York, 1989) pp. 233-241.

4. A. Mysyrowicz, D. Hulin, A. Antonetti, A. Migus, W. T. Masselink, and H. Morkoc, Phys. Rev. Lett. **56**, 2748 (1986).

5. N. Peyghambarian, S. W. Koch, M. Lindberg, B. Fluegel, and M. Joffre, Phys. Rev. Lett. **62**, 1185 (1989).

6. H. M. Gibbs, *Optical Bistability: Controlling Light with Light* (Academic Press, Orlando, 1985).

7. E. W. Stryland, Y. Y. Wu, D. J. Hagan, M. J. Soileau, and K. Mansour, J. Opt. Soc. Am. B **5**, 1980 (1988).

Chapter XIV
MEASUREMENT TECHNIQUES OF OPTICAL NONLINEARITIES

A variety of experimental methods have been employed to study mechanisms responsible for optical nonlinearities in the absorption spectra and to measure the nonlinear refractive indices in semiconductors. Pump-probe spectroscopy, nonlinear interferometry, beam-distortion measurements, four-wave mixing, and phase conjugation are among these techniques. In this chapter we discuss some of the most commonly used experimental methods and show some representative results.

14-1. Pump-Probe Spectroscopy

Consider the experimental arrangement shown in Fig. 14.1. Two laser beams are focused on the same spot of a semiconductor sample. The laser labeled "pump" has a relatively large intensity, denoted as $I_x(t)$, and is usually tuned to an energy of $\hbar\omega_x$, within the absorption region of the semiconductor. The pump is absorbed, resulting in the generation of a density $n(t,r)$ of electron-hole pairs. In its simplest form, this carrier generation may be described by a rate equation such as

$$\boxed{\frac{\partial n}{\partial t} = \frac{\alpha[\omega_x, n(t,r)]\, I_x(t)}{\hbar\omega_x} - \frac{n}{\tau} - \nabla D \nabla n ,} \tag{14.1}$$

electron-hole-pair density rate equation

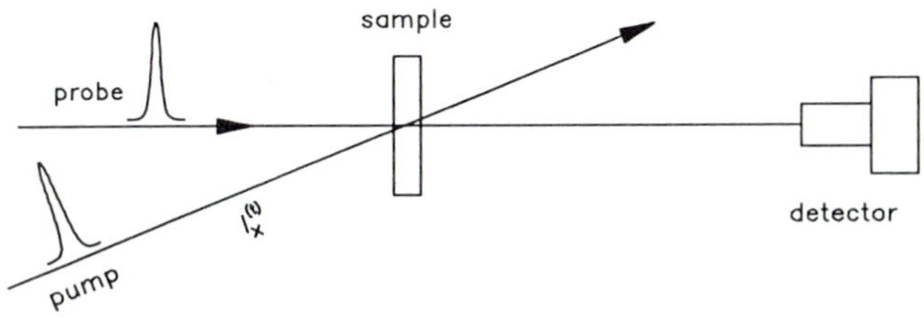

Figure 14.1. Schematics of pump-probe laser spectroscopy.

where $\alpha(\omega_x, n)$ is the carrier density-dependent semiconductor absorption coefficient at the pump frequency, τ is the carrier lifetime, and D is the electron-hole-pair diffusion coefficient, which is generally a function of carrier density. The first term on the right-hand side of Eq. (14.1) describes the generation rate of electron-hole pairs, while the last two terms specify the carrier decay rates, either by recombination or spatial diffusion, respectively. In Eq. (14.1) we assume linear nonradiative decay n/τ, and for simplicity ignore higher-order nonlinear mechanisms such as Auger recombination or radiative decay. One often expresses the diffusion coefficient D in the form of the diffusion length ℓ_d, which is defined as

$$\ell_D = \sqrt{D\tau} \,. \tag{14.2}$$

Typically, the diffusion length is a few micrometers in high-quality semiconductor materials. For the case of spatially homogeneous excitation, the last term in Eq. (14.1) is zero because $\nabla n = 0$, reducing this equation to

$$\frac{dn}{dt} = \frac{\alpha[\omega_x, n(t)]\, I_x(t)}{\hbar\omega_x} - \frac{n}{\tau} \,. \tag{14.3}$$

The solution to this first-order differential equation is obtained as (see problem 14.1)

$$n(t) = \int_0^t dt'\, e^{(t' - t)/\tau}\, \frac{I_x(t')\, \alpha[\omega_x, n(t')]}{\hbar\omega_x} \,, \tag{14.4}$$

showing that the density of excited electron-hole pairs n(t) depends directly on the integral over the intensity of the pump laser. Note, however, that in general, n(t) is not directly proportional to I_x, since the absorption at the excitation frequency usually changes during the excitation. If this absorption change at the pump frequency is negligible, and for steady-state conditions where $dn/dt = 0$, Eq. (14.3) yields

$$n = \frac{\alpha(\omega_x)\, I_x\, \tau}{\hbar\omega_x} \,. \tag{14.5}$$

More generally, however, the carrier density follows the integral over the pulse multiplied by the changing absorption, as shown by Eq. (14.4). In any case, varying the pump intensity causes the number of excited electron-hole pairs to change. The interaction processes among these laser-excited electrons and holes lead to density-dependent changes of the optical semiconductor properties, i.e., to the optical nonlinearities discussed in the previous chapter.

In pump-probe spectroscopy, a second light beam, referred to as the "probe" beam, is used to monitor the changes in the optical properties caused by the pump. The probe intensity is very small, not inducing any changes by itself. The transmission spectrum of the probe beam detected in the presence of the pump beam is compared to the spectrum without pump beam, giving the frequency-dependent absorption of the sample for different intensities of the pump. For this purpose it is convenient to have a probe beam that is either spectrally broad or easily tunable in wavelength, making it possible to monitor the entire band-edge absorption region. Often, the probe beam is also a laser beam, and both pump and probe laser originate from a common laser source. For instance, the pump laser is a dye laser or a solid-state laser with a fixed or adjustable wavelength. The probe laser is another tunable-dye or solid-state laser, or it consists of the broad luminescence of a dye. Care must be taken to prevent pump laser radiation from reaching the detector. This can be done by polarization and angle discrimination.

To minimize problems associated with carrier diffusion (i.e., to achieve spatially homogeneous excitation conditions), the pump beam is usually focused to a relatively large diameter (> 50 μm), and the probe transmission is monitored only around the center of the pump beam. The electron-hole pair density is nearly homogeneous in this small region, and spatial variations may be ignored in a first approximation. In any case, the correct determination of the carrier density for a given excitation intensity is a difficult task, requiring special care for a truly quantitative analysis of experiments.

The pump-probe technique described above may be employed to construct the nonlinear refractive index, as well as to determine the physical origin of the optical nonlinearities. Without pump, the probe measures the linear absorption spectra, already discussed in Chap. VI. The changes in the optical properties are monitored through the analysis of the probe absorption changes. The absorption changes caused by the pump-generated electron-hole pairs lead to changes in the index of refraction of the material via the Kramers-Kronig transformation (see Sec. 13-3). It should be noted that absorption changes should be measured in a large frequency range for the Kramers-Kronig transformation to be accurate. Examples of pump-probe measurements of the nonlinearities in several bulk semiconductors, quantum wells, and quantum dots are given in the following sections.

14-2. An Example of Bulk Semiconductor Nonlinearities*

Here we describe the nonlinear behavior of ZnSe as an example of a bulk semiconductor. The nonlinear absorptive changes in laser-excited bulk ZnSe, which are measured by the pump-probe scheme, are reproduced in Fig. 14.2. Figure 14.2(a) shows the measured ZnSe absorption spectra as functions of probe wavelength at T = 150 K when the pump is absent, and also with the pump present. The spectrum labeled 0, which is taken in the absence of the pump beam, exhibits a pronounced exciton peak, together with the Coulomb-enhanced continuum states at higher energies. In the presence of the pump pulse [curve 1 of Fig. 14.2(a)], the exciton bleaches as a result of Coulomb screening. Simultaneously, a small broadening on both sides of the exciton appears. The change in the index of refraction $\Delta n(\omega)$ is obtained by making a

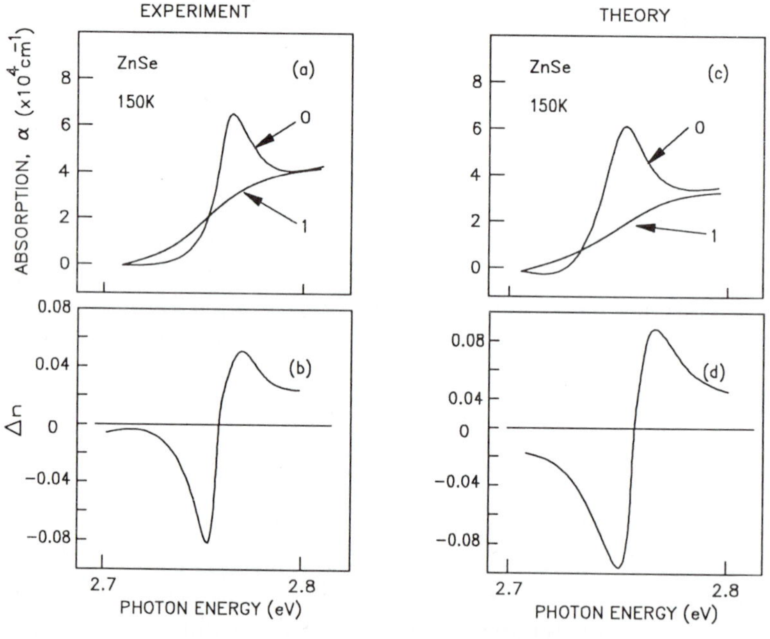

Figure 14.2. Experimental and theoretical spectra of nonlinear absorption and nonlinear refractive index for a 0.55-μm ZnSe thin film at T = 150 K. (a) Experimental absorption spectra for: (0) linear (no pump), (1) with a pump of intensity 46.5 kW/cm^2. (b) The nonlinear refractive index change corresponding to the measured absorption spectra of (a). (c) Calculated absorption spectra for different electron-hole-pair densities n: (0) 1×10^{15} cm^{-3}, (1) 3×10^{17} cm^{-3}. (d) Calculated nonlinear refractive index change (Ref. 14.1).

Kramers-Kronig transformation, Eq. (13.34), of the measured change in absorption coefficient $\Delta\alpha$, as shown in Fig. 14.2(b).

The experimental data are analyzed using the plasma theory, Eq. (13.48). Examples of the results are shown in Fig. 14.2(c). The low-density absorption curve (0) is the spectrum of the unexcited crystal. The 1s-exciton peak vanishes at higher electron-hole (plasma) densities because of the strong screening. The detailed analysis of the various nonlinear mechanisms included in the theory show that the main mechanism for the optical nonlinearity in this case is the screening. Bandfilling and bandgap renormalization are not very important under the present, relatively low-excitation conditions. Figure 14.2(d) displays the calculated refractive index as a function of photon energy. The overall agreement with the experimental data is good.

14-3. An Example of Quantum Well Nonlinearities*

The nonlinear absorption and refraction has been measured for a variety of different GaAs/AlGaAs multiple quantum well samples. As an example, Fig. 14.3 shows the linear and nonlinear absorption spectra of the 76-Å sample discussed in Sec. 8-6 and Fig. 8.20. The absorption changes in Fig. 14.3 have been obtained for a pump with a fixed energy of $\simeq 1.519$ eV, i.e., well above the lowest heavy- and light-hole exciton resonances. We see that these exciton resonances bleach when the pump is present, as a consequence of screening and phase-space filling caused by the optically generated electron-hole pairs. For higher excitation intensities

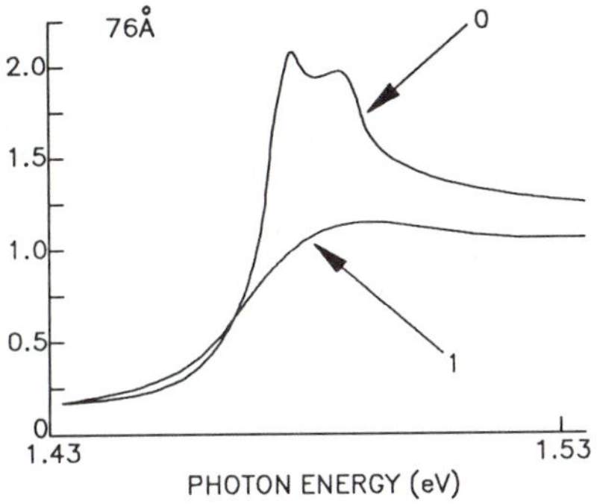

Figure 14.3. Nonlinear absorption spectra for a GaAs 76-Å multiple quantum well. The linear spectrum (0) shows the two exciton structures associated with the heavy and light holes (see also Fig. 8.20). The spectrum labeled 1 is the probe absorption in the presence of the pump (from Ref. 14.2).

(not shown in Fig. 14.3), the near-bandgap absorption can be completely bleached, and even optical gain (negative absorption) regions develop.

14-4. An Example of Quantum Dot Nonlinearities*

Here we only show an example of measured optical nonlinearities for quantum dots in glass. As briefly mentioned in Sec. 13-8, the main source of optical nonlinearities in the quantum dots is blocking of the transitions by state filling (Pauli exclusion principle). An example of the experimentally measured absorption change for resonant excitation into the energetically lowest quantum-confined transition is shown in Fig. 14.4. The top spectrum in Fig. 14.4 displays the linear absorption of a CdS-doped glass quantum-dot sample at T = 10 K. This spectrum has two peaks at energies of approximately 2.95 eV and 3.26 eV. The low-energy peak originates from the transition between the 1s-hole and 1s-electron states. The high-energy peak is due to the transition between the 1p-hole and 1p-electron states. These transitions shift to higher energies as the crystallite sizes are reduced, demonstrating clear evidence for quantum confinement [see Eq. (9.39)].

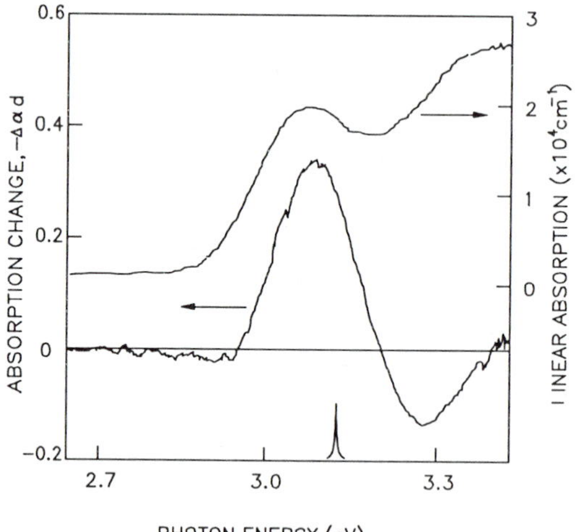

Figure 14.4. The top curve is the linear absorption spectrum of a sample of CdS quantum dots in glass. The bottom trace shows the energetic position of the pump pulse. The intermediate trace is the change in the linear absorption spectrum as a result of excitation by the pump pulse. d is the thickness of the sample (Ref. 14.1).

To investigate the optical nonlinearity of these transitions, a pump-probe technique was employed. The photon energy of the pump pulse was tuned inside the energetically lowest transition. The narrow peak at the bottom of Fig. 14.4 displays the spectral position of the pump pulse. A photon from the pump excites the first electron-hole pair in the dot. A broadband probe pulse detects the changes made in the absorption spectra as a result of excitation by the pump pulse. Probe absorption without previous excitation by the pump is equivalent to the generation of one electron-hole pair in the unexcited quantum dot. If, however, a pump photon has been absorbed previously, the probe beam generates a second electron-hole pair in the presence of the pump-generated pair. These situations are schematically shown in Fig. 14.5 for the two cases: When only the pump or only the probe is present, one pair exists in the dot [Fig. 14.5(a)], but when both pump and probe pulses are present, two electron-hole pairs are inside the dot [Fig. 14.5(b)].

The lower spectrum in Fig. 14.4 shows the measured changes in the absorption, $-\Delta\alpha d = d\,[\alpha_{probe}$ (without pump) - α_{probe} (with pump)]. In such a spectrum, a positive peak corresponds to bleaching and a negative peak indicates an induced absorption, i.e., an increase in the probe absorption as a consequence of the presence of the pump-generated electron-hole pair. The nonlinear spectrum in Fig. 14.4 shows a positive peak around the $1s_h$-$1s_e$ transition, centered around the pump, and a negative peak on the high-energy side, centered around 3.178 eV. The

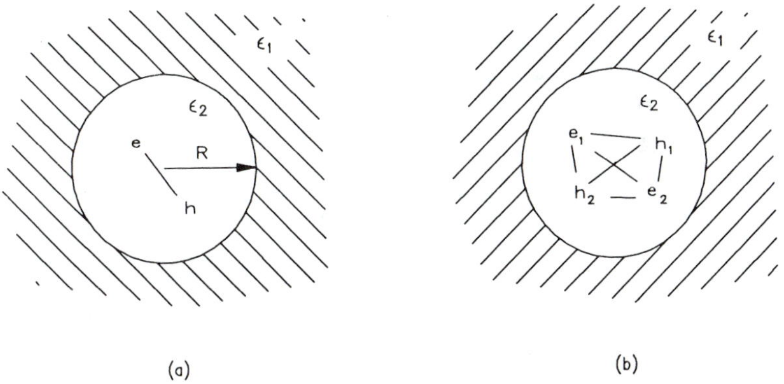

Figure 14.5. (a) One electron-hole pair inside a semiconductor quantum dot of dielectric constant ϵ_2 embedded in a glass matrix of dielectric constant ϵ_1. (b) Two electron-hole pairs in the quantum dot. The Coulomb interaction is attractive between the electrons and holes; it is repulsive between electrons and electrons and between holes and holes.

bleaching of the $1s_h$-$1s_e$ transition is the result of state filling. The generation of one electron-hole pair by the pump causes saturation of the one-pair transition (bleaching of the linear absorption peak).

The origin of the induced absorption feature (the negative peak in the lower spectrum of Fig. 14.4) is assigned to the generation of two-pair states in the quantum dot. The pairs are created by absorption of one pump and one probe photon. The theoretical analysis of the linear and third-order nonlinear optical properties has to include the one- and two-electron-hole-pair states (see Ref. 14.3). These investigations consistently lead to the conclusion that Coulomb effects are important even for the smallest quantum dots. Using numerical matrix diagonalization techniques, the energies and wave functions were obtained for the one- and two-pair ground states and for all the excited pair states. With these wave functions, the various dipole matrix elements were evaluated for transitions between the ground state and the one- and two-electron-hole-pair states. The changes in the absorption, $-\Delta\alpha d$, were then calculated, as shown in Fig. 14.6. Like the experiment, the theory also exhibits a decreasing absorption feature (bleaching) around the lowest quantum-confined transition and an increasing absorption feature on the high-energy side.

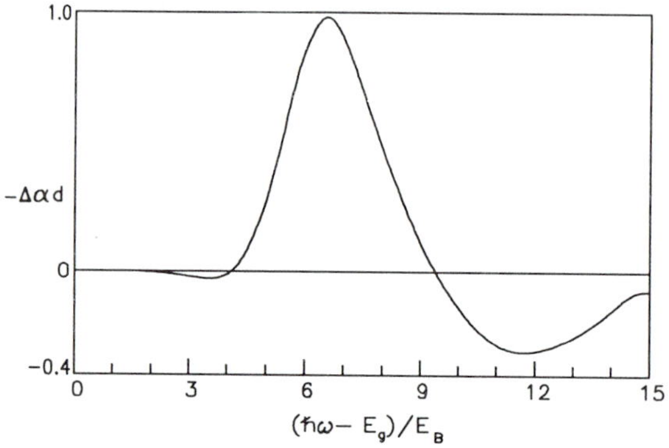

Figure 14.6. Computed absorption changes for a semiconductor quantum dot of radius $R = a_B$ (where a_B is the exciton Bohr radius), assuming a pump and probe geometry and pumping into the energetically lowest one-pair state (Ref. 14.3).

14-5. An Example of Transient Nonlinearity*

As an example of transient nonlinearities, we describe some measurements of the optical Stark effect (OSE) in GaAs-AlGaAs multiple quantum wells. As discussed earlier, OSE occurs for photon energies below the bandgap, in the transparency region of the semiconductor. Pump-probe spectroscopy with very short duration laser pulses, in the femtosecond time regime, may be employed to observe the effect and time resolve its response.

Figure 14.7 exhibits the results of the OSE measurements in a GaAs multiple quantum well at low temperatures for pumping below the exciton resonance at $\hbar\omega_{\text{pump}} \simeq 1.56$ eV. The solid curve represents the linear absorption spectrum of the sample in the absence of the pump pulse. It clearly shows the heavy- and light-hole excitons. The dotted curve corresponds to the probe absorption in the presence of the pump pulse (when pump and probe have almost complete time overlap). The dotted spectrum is shifted to high energies as a result of pump excitation. This blue shift lasts as long as the pump is present at the sample. In addition to the shift, the pump has also resulted in partial bleaching of the exciton resonance. Most of the exciton Stark shift and its bleaching recover after the pump pulse exits the sample. This transient shift and bleaching are the result of the OSE.

The OSE also gives rise to a nonlinear index of refraction. Since the OSE is a transient process, the generated index change is also transient with a rapid response time. Figure 14.8 shows the numerically evaluated absorption and index changes. The semiconductor Bloch equations (15.15)-(15.17) were numerically solved to obtain the induced polarization,

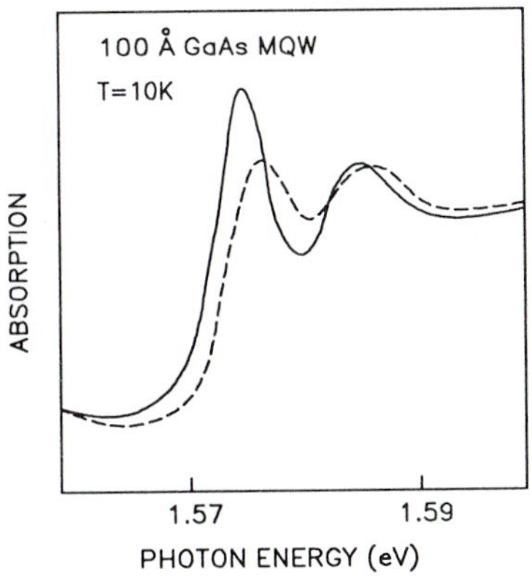

Figure 14.7. The absorption spectra of a GaAs-AlGaAs multiple quantum well at T = 10 K (after Ref. 14.4) with a pump (dashed spectrum) and without a pump (solid curve).

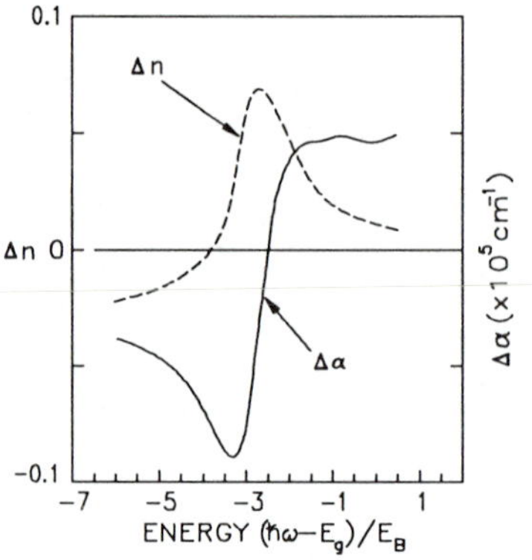

Figure 14.8. Calculated transient nonlinear index change $\Delta n(\omega)$ resulting from the OSE in a GaAs-AlGaAs multiple quantum well sample (dashed curve). The absorption change $\Delta\alpha$ is also shown (solid curve). Both $\Delta\alpha$ and $\Delta n(\omega)$ were numerically evaluated for a complete time overlap between the pump and probe pulses using the semiconductor Bloch equations (after Ref. 14.4).

whose real part yields the nonlinear refractive index. The solid curve shows the absorption change $\Delta\alpha$, and the dashed curve represents the refractive index change $\Delta n(\omega)$. This nonlinear index change has a substantial magnitude, and may be used to demonstrate high-speed nonlinear devices. It is noted that the Kramers-Kronig relation, Eq. (13.34), between $\Delta\alpha$ and $\Delta n(\omega)$ is no longer valid in this case because of the dynamic response. Applicability of the Kramers-Kronig requires that the system response is quasi-stationary, so that the optical susceptibility is $\chi(\omega) = P(\omega)/E(\omega)$ [Eq. (3.38)].

14-6. Nonlinear Interferometry

A relatively direct method of measuring nonlinear index changes is nonlinear interferometry. In this technique an interferometer containing the semiconductor material is constructed. The interferometer may be a Michelson interferometer with the nonlinear material in one of the arms, or a Fabry-Perot with the semiconductor inserted between the two partially reflecting mirrors. A low-intensity laser beam is first directed through the interferometer, and the resulting fringe pattern is detected. By applying a strong beam in the arm containing the semiconductor, or by increasing the laser intensity, the index of refraction of the material changes, causing a phase change that shifts the fringe pattern. From the measured fringe shift, the nonlinear phase shift and the nonlinear index of refraction may be obtained. For example, Fig. 14.9 schematically shows the fringe pattern in a Fabry-Perot interferometer at low intensity (dashed line), and the shifted pattern at high intensity (solid line). The separation between two successive fringes is referred to as the free spectral range (FSR) of the Fabry-Perot, and it corresponds to a 2π phase change. The laser-induced phase change $\Delta\phi$, corresponding to the fringe shift $\Delta\lambda$, is given by

$$\Delta\phi = \frac{2\pi\Delta\lambda}{\mathrm{FSR}} . \tag{14.6}$$

This phase change during a round trip in the Fabry-Perot is related to the index change $\Delta n(\omega)$ by

$$\Delta\phi = 2\Delta n(\omega) \left(\frac{2\pi}{\lambda}\right) d , \tag{14.7}$$

where d is the thickness of the semiconductor, which is assumed to completely fill the gap between the two mirrors of the Fabry-Perot. The index change is then

$$\Delta n(\omega) = \frac{\lambda\Delta\lambda}{2d(\mathrm{FSR})} . \tag{14.8}$$

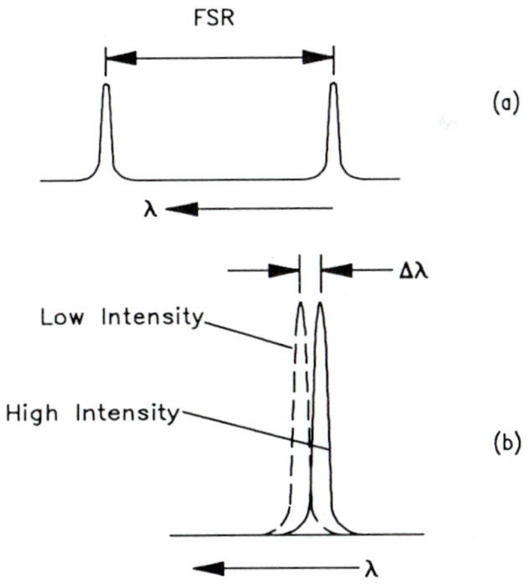

Figure 14.9. (a) Schematic representation of the fringe pattern in a nonlinear Fabry-Perot interferometer. (b) Shift of one of the Fabry-Perot peaks. The dashed curve (solid curve) represents the Fabry-Perot peak at low intensities (high intensities).

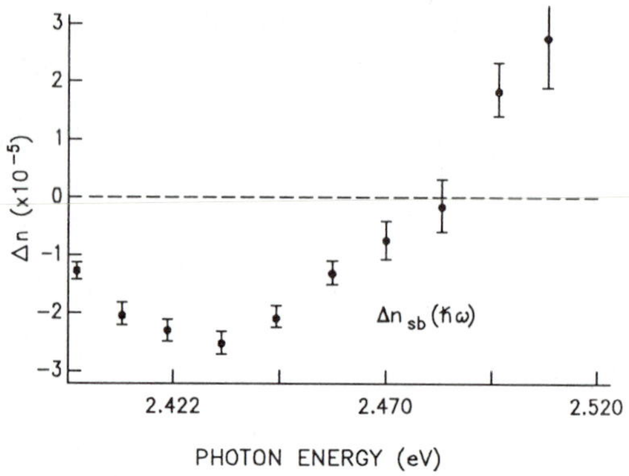

Figure 14.10. Measured nonlinear index of a semiconductor material embedded in a glass matrix by using a Twymann-Green interferometer (after Ref. 14.5).

Similar expressions may be found for the nonlinear index in other types of interferometers. As an example, Fig. 14.10 shows the measured nonlinear index in a semiconductor embedded in glass matrix at room temperature. The origin of the nonlinearity in this sample was found to be bandfilling. Consequently, the index change was expected to have a negative sign below the bandgap and a positive sign above the gap. As shown in Fig. 14.10, such behavior is indeed observed. Thus, nonlinear interferometry not only provides an accurate measurement of the size of the nonlinearity, but also gives its sign, contributing to the identification of the origin of the nonlinear mechanism.

14-7. Beam-Distortion Technique

We pointed out in the previous chapter that the nonlinearity of an optical material such as a semiconductor makes the material act like a lens, leading to self-focusing or self-defocusing of the incident laser beam. If the nonlinear index change is negative, the nonlinear semiconductor behaves like a negative diverging lens with defocusing property, while for positive index changes, the material acts like a positive converging lens with focusing behavior. Therefore, the spatial profile of the incident laser beam becomes distorted by the optical nonlinearities. Such spatial beam distortion provides another technique to measure the nonlinear index of refraction. Several ways exist to measure the spatial beam distortion, and consequently, the nonlinear refraction. Here we describe only a simple single-beam technique, referred to as a Z-scan, that allows the measurement of both the sign and magnitude of the nonlinear index change (see Ref. 14.9 for more details).

The experimental arrangement for the Z-scan technique is shown in Fig. 14.11. A laser beam with a Gaussian spatial beam profile is directed toward the sample after passing through a tight focusing lens L. The amount of light transmitted through the sample is detected by detector D_2 through a finite aperture, while the incident laser intensity is detected by D_1. No additional lens is used after the sample, so that the far-field profile of the transmitted intensity may be measured by D_2. The position of the sample z with respect to the focal plane of the lens is varied (this is why the technique is called Z-scan), and at each position the laser transmission intensity is measured. The transmitted intensity through the aperture as a function of position z is related to the nonlinear index. This can be simply understood by the following example. Let us assume that the semiconductor has a negative nonlinear index of refraction and is very thin, being regarded as a thin negative lens. Initially the sample is put far away from the focal plane of the lens on the -z side and we scan it toward +z direction. The laser intensity inside the sample is low and, therefore, the sample does not experience any appreciable index change. It behaves like a thin plate, which does not perturb the beam profile. Thus, the transmission through the aperture is low, since the beam diverges and the aperture is not at the focus of the lens. As the sample is moved toward the focal plane of the lens, the laser intensity (power per unit area) at the sample increases, causing a nonlinear index change. The sample, therefore, behaves like a negative lens and self-defocusing of the beam occurs. The combination of self-defocusing by the sample and focusing of the lens L tends to collimate the beam as long as the sample is in the region of negative z. The collimated beam can go through the aperture (as its waist has been narrowed) and results in the enhancement of the transmittance. When the sample is scanned to positions with $z > 0$, where it is located after the focal plane of the lens, the combined effect of self-defocusing by

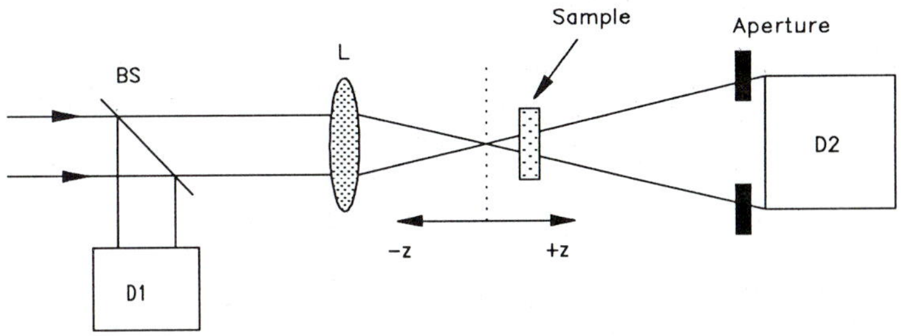

Figure 14.11. Experimental arrangement for the Z-scan technique. The position of the sample z with respect to the focal plane of the lens is varied, and the transmittance is measured (see Ref. 14.9).

the sample and lens action leads to the increased divergence of the beam. Therefore, for $z > 0$, the transmittance is low. The high transmittance for $z < 0$ and low transmittance for $z > 0$ suggest that at $z = 0$ the transmittance should go through zero. The solid curve in Fig. 14.12 schematically shows such behavior for a sample with a negative index change. The normalized transmission shows a peak for the -z position and a valley for the +z position, indicative of a negative nonlinear index in the material. Using similar arguments, it can be shown that for a material with positive nonlinearity, the positions of the peak and valley are reversed, as displayed by the dashed curve in Fig. 14.12. In this case, the peak of the transmittance occurs at the +z side, while the valley is at the -z side.

Therefore, the locations of the transmittance peak and valley in the Z-scan experiment immediately reveal the sign of the nonlinearity. The magnitude of the index change may be obtained by analyzing the transmittance of the Z-scan. Here we only quote the results without going through the detailed derivations. The interested reader may consult Ref. 14.9 for a detailed analysis. The magnitude of the nonlinear index change $\Delta n_0(\omega)$ (the on-axis value at the focus, $z = 0$) gives rise to a phase change of $\Delta\Phi_0 = (2\pi/\lambda)\, \Delta n_0(\omega)\, d$, where d is the effective length of the sample and λ is the laser wavelength. The phase change and, consequently, the index change are obtained from the measurable quantity ΔT_{pv}, which is defined as the difference between the transmittances at the peak and valley, $\Delta T_{pv} = T_p - T_v$, where T_p (T_v) is the peak (valley) transmittance. It has been shown that

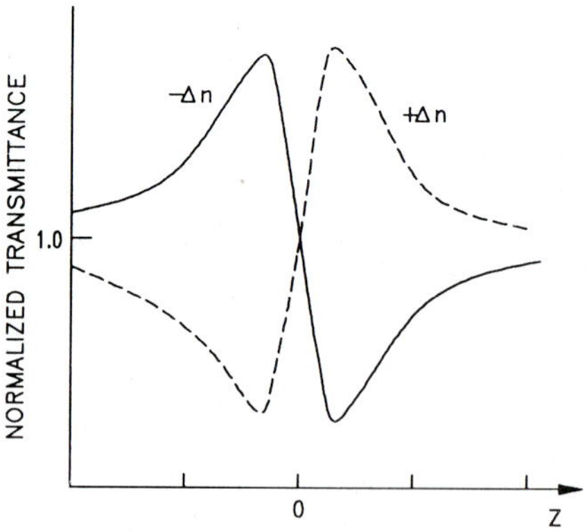

Figure 14.12. Schematic behavior of the transmittance in a Z-scan measurement for negative (solid curve) and positive (dashed curve) index changes.

$$\Delta T_{pv} \simeq 0.405 \,|\, \Delta\Phi_0 \,|\, \text{ for } |\, \Delta\Phi_0 \,| \leq \pi \,. \tag{14.9}$$

Therefore, the magnitude and sign of the nonlinearity may be determined from the magnitude and locations of the peak and valley of the Z-scan.

It should be noted that the above discussion only applies to purely refractive nonlinearities where one assumes that no simultaneous absorption changes occur. When absorptive nonlinearities are also present, the Z-scan profile no longer behaves as simply as Fig. 14.12, and the peak or the valley may be suppressed. For example, for semiconductors near the bandgap, where exciton bleaching, bandgap renormalization, and other types of nonlinear mechanisms are present, such Z-scans may not be easily interpreted. However, for far-off-resonance excitation in semiconductors, this technique is useful and attractive because of its simplicity.

14-8. Four-Wave Mixing

Another common experimental technique is based on the interference between several beams inside the nonlinear medium. In this "wave-mixing spectroscopy" the material is excited with two or more beams, which may or may not have different frequencies and are incident under specified angles. These beams generate interference patterns inside the medium called gratings (see Fig. 14.13). Depending on the number of different beams involved in such a wave-mixing process, the technique is known as three-wave mixing, four-wave mixing, etc. If the frequencies of the beams are identical, the mixing process is called "degenerate"; otherwise it is nondegenerate or partially degenerate.

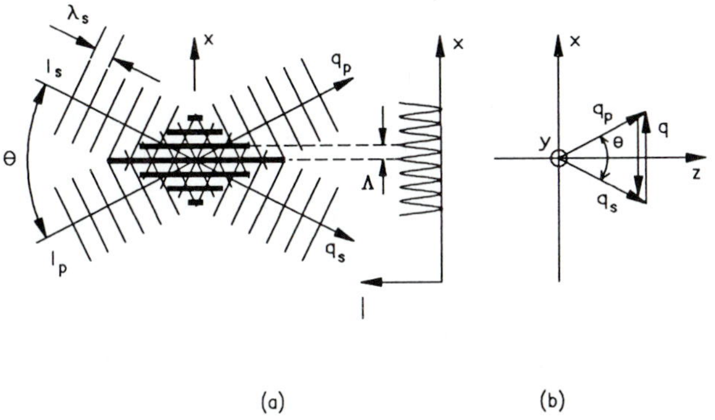

Figure 14.13. Grating produced by the interference of two light waves. (a) Experimental geometry and (b) vectorial representation (see Ref. 14.17).

The schematic of the standard degenerate four-wave-mixing (DFWM) geometry is shown in Fig. 14.14(a). Two copropagating laser beams of equal frequency, and with electric field and wave vectors $\mathbf{E}_p, \mathbf{E}_s$ and $\mathbf{q}_p, \mathbf{q}_s$, respectively, are incident on the material at an angle θ. The two beams interfere on the material, forming a spatial fringe pattern with intensity maxima and minima. In the space regions where the intensity is maximum, the nonlinearity of the medium is maximum. For instance, if the incident laser frequency is tuned to an exciton resonance, a high density of carriers is generated, leading to the effects described in Chap. XI. Obviously, the nonlinearities are small in the regions of intensity minima. Therefore, a nonlinear grating is created because the electron-hole concentration is spatially modulated between high and low values. A third beam, which may or may not be coincident with the two beams $\mathbf{q}_s$ and $\mathbf{q}_p$, is diffracted by this grating along certain directions.

A mathematical description of the grating produced by the interference of the two degenerate laser beams may be obtained by analyzing the geometry depicted in Fig. 14.13. The wave vectors of the incident beams

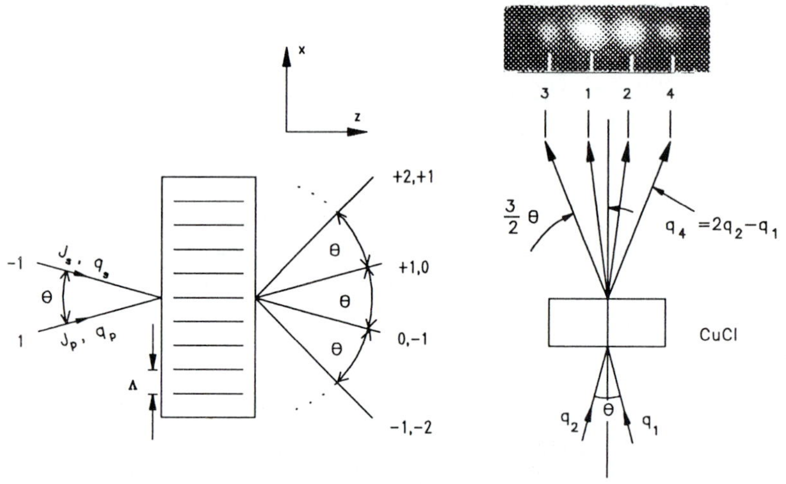

Figure 14.14. (a) Schematics of DFWM spectroscopy. I_p and I_s are the intensities of the input beams, x is the transverse direction, and z is the propagation direction, respectively. The medium length in propagation direction is d. The interference grating inside the medium is depicted by parallel lines, and the resulting scattering orders are denoted by ±1, ±3. (b) Experimentally measured four-wave-mixing signals in the directions of $2\mathbf{q}_1 - \mathbf{q}_2$ and $2\mathbf{q}_2 - \mathbf{q}_1$ on both sides of the transmitted beams, with wave vectors $\mathbf{q}_1$ and $\mathbf{q}_2$ for a CuCl sample (after Ref. 14.24).

$\mathbf{q}_p$ and $\mathbf{q}_s$ are assumed to be in the xz plane, making an angle of $\theta/2$ with respect to the z axis and having components

$$\mathbf{q}_p = q_z \, \mathbf{e}_z + q_x \, \mathbf{e}_x \tag{14.10a}$$

and

$$\mathbf{q}_s = q_z \, \mathbf{e}_z - q_x \, \mathbf{e}_x \, . \tag{14.10b}$$

The interference of the two laser beams takes place in the xy plane of the sample. We refer to this xy plane as the interference plane for the remaining part of this section. We also assume that the incident fields have Gaussian transverse profiles, varying like $\exp(-x^2 + y^2/W^2)$ in the x and y directions, respectively, with W being the spot size or the radius of the beam.

The general form of the electric-field amplitudes may be written as

$$E_p = E_p^0 \, e^{-\sigma^2} \, e^{i\mathbf{q}_p \cdot \mathbf{r}} \tag{14.11}$$

and

$$E_s = E_s^0 \, e^{-\sigma^2} \, e^{i\mathbf{q}_s \cdot \mathbf{r}} \, , \tag{14.12}$$

where the time variations of the field amplitudes are included in E_p^0 and E_s^0; $\sigma^2 = (x^2 + y^2)/W^2$ and $\mathbf{r} = x \, \mathbf{e}_x + y \, \mathbf{e}_y + z \, \mathbf{e}_z$. The field amplitudes in the interference plane are obtained by writing $\mathbf{q}_p$ and $\mathbf{q}_s$ in terms of their components from Eq. (14.10) and noting that the fields are evaluated at $z = 0$ in the interference plane. We get

$$E_p = E_p^0 \, e^{-\sigma^2} \, e^{iq_x x} \tag{14.13}$$

and

$$E_s = E_s^0 \, e^{-\sigma^2} \, e^{-iq_x x} \, . \tag{14.14}$$

The total field in the interference region is the sum of the two fields as

$$E = E_p + E_s = e^{-\sigma^2} \, [E_p^0 \, e^{iq_x x} + E_s^0 \, e^{-iq_x x}] \, . \tag{14.15}$$

The intensity in the interference region is given by

$$I = \frac{cn_b}{8\pi} |E|^2$$

$$= \frac{cn_b}{8\pi} e^{-2\sigma^2} [|E_p^0|^2 + |E_s^0|^2 + 2|E_p^0||E_s^0|\cos(2q_x x)] , \qquad (14.16)$$

where n_b is the refractive index of the material (the background refractive index). In writing Eq. (14.16) we ignored a possible phase difference between E_p^0 and E_s^0 which would give a simple additive term in the argument of the cosine function. The presence of the periodic cosine function explicitly indicates that the intensity I in the interference region is modulated, creating an intensity fringe pattern in the x direction.

Usually, the *grating* vector $\mathbf{q}$ is defined as

$$\mathbf{q} = \pm(\mathbf{q}_p - \mathbf{q}_s) , \qquad (14.17)$$

as graphically shown in Fig. 14.13(b). The magnitude of the grating vector may be determined using the geometry of Fig. 14.13(b) and noting that $q_p = q_s$, as

$$q^2 = q_p^2 + q_s^2 - 2p_p q_s \cos\theta = 4q_p^2 \sin^2(\theta/2) \qquad (14.18)$$

or

$$q = 2q_p \sin(\theta/2) . \qquad (14.19)$$

The examination of Fig. 14.13(b) indicates that vector $\mathbf{q}$ is along x direction and $q_x = q_p \sin(\theta/2)$. Thus, Eq. (14.19) can also be written as

$$q = 2q_x . \qquad (14.20)$$

Substituting for q_x in terms of q in Eq. (14.16) yields

$$I = \frac{cn_b}{8\pi} e^{-2\sigma^2} [|E_p^0| + |E_s^0|^2 + 2|E_p^0||E_s^0|\cos q x] . \qquad (14.21)$$

The spatial period Λ of the intensity fringe pattern from Eq. (14.21) is

$$\Lambda = 2\pi/q . \qquad (14.22)$$

Expressing q in terms of the incident field wavelength using Eq. (14.19) gives

$$q = 2\left(\frac{2\pi}{\lambda_p}\right)\sin(\theta/2)$$

$$= \frac{4\pi}{\lambda_p}\sin(\theta/2)\ . \tag{14.23}$$

The spatial period of the grating may now be written in terms of λ_p by the substitution of q in Eq. (14.22); we get

$$\boxed{\Lambda = \frac{\lambda_p}{2\sin(\theta/2)}\ .}$$

spatial period of grating (14.24)

For small angles of incidence, $\sin(\theta/2) \simeq \theta/2$, and Eq. (14.24) becomes

$$\Lambda \simeq \frac{\lambda_p}{\theta}\ . \tag{14.25}$$

Therefore, the grating wave vector specifies the direction and period of the fringe pattern. The grating period may be changed by varying the incident angle θ. However, Λ cannot be larger than the diameter of the incident laser beam.

We now examine the way a third probe beam is scattered by the formed grating. To simplify, we will only consider the case where the probe beam is coincident with either incident pump beam. This corresponds to the "self diffraction" case, also called two-beam interference. If the angle θ between the beams is sufficiently small, the input beams can be scattered by the induced refractive index and absorption gratings to produce several angular scattering orders at the output, as shown schematically in Fig. 14.14(a). This is the *Raman-Nath regime*. One example of such a four-wave-mixing process is displayed in Fig. 14.14(b) for the case of CuCl, where the diffracted beams can be clearly seen on each side of the undiffracted transmitted beams. The light frequency ω is close to the exciton or half the biexciton resonance. In contrast, for larger angles only one scattering order occurs, giving rise to *Bragg scattering*. For semiconductors in which the excited carriers are sufficiently mobile to migrate significant distances before recombining, one usually operates in the Raman-Nath regime, since the density grating period exceeds the diffusion length of the excitation.

In the Bragg scattering regime, which is also referred to as the *thick grating* regime, usually no more than one scattering order can be detected. This is due to the fact that the phase-matching condition should be satisfied for scattering to take place. The phase-matching condition is equivalent to the momentum conservation requirement in scattering processes. In general, the phase-matching condition is satisfied when the grating vector is equal to the difference between the scattered (diffracted) and probe wave vectors. Consider a situation where the grating is created by two beams with wave vectors $\mathbf{q}_p$ and $\mathbf{q}_s$ and the grating vector is $\mathbf{q} = \pm(\mathbf{q}_p - \mathbf{q}_s)$. A probe photon with wave vector $\mathbf{q}_c$ can be diffracted from this grating only in the phase-matching direction. If $\mathbf{q}_D$ denotes the wave vector of the diffracted photon, this condition using the above recipe reads

$$\mathbf{q} = \mathbf{q}_D - \mathbf{q}_c \tag{14.26}$$

or

$$\mathbf{q}_D - \mathbf{q}_c = \pm(\mathbf{q}_p - \mathbf{q}_s) \; . \tag{14.27}$$

In the thick grating geometry of Fig. 14.15(a), the probe photon is one of the incident photons that creates the grating; i.e., $\mathbf{q}_c = \mathbf{q}_p$ or $\mathbf{q}_s$. For the self-diffraction case, where the probe photon has wave vector $\mathbf{q}_c = \mathbf{q}_p$, Eq. (14.27) yields

$$\mathbf{q}_D - \mathbf{q}_p = \pm(\mathbf{q}_p - \mathbf{q}_s) \tag{14.28}$$

or

$$\mathbf{q}_D = 2\mathbf{q}_p - \mathbf{q}_s \text{ or } \mathbf{q}_D = \mathbf{q}_s \; . \tag{14.29}$$

Condition $\mathbf{q}_D = \mathbf{q}_s$ corresponds to a scattering of photons from beam $\mathbf{q}_p$ in the direction of beam $\mathbf{q}_s$. Condition $\mathbf{q}_D = 2\mathbf{q}_p - \mathbf{q}_s$ is graphically shown in Fig. 14.15(b). Similarly, if the probe photon is $\mathbf{q}_c = \mathbf{q}_s$, then the phase-matching condition of Eq. (14.27) gives $\mathbf{q}_D = 2\mathbf{q}_s - \mathbf{q}_p$ or $\mathbf{q}_D = \mathbf{q}_p$, as shown in Fig. 14.15(c).

In contrast to the Bragg regime, in the Raman-Nath regime or the *thin grating* regime, the strict phase-matching condition is somewhat relaxed and many scattering orders may be observed. For the thin medium geometry of Fig. 14.14, one has the familiar diffraction law given by

$$\Lambda \sin \phi_m = m\lambda_c \; , \quad m = 0, \pm 1, \pm 2, \ldots \; , \tag{14.30}$$

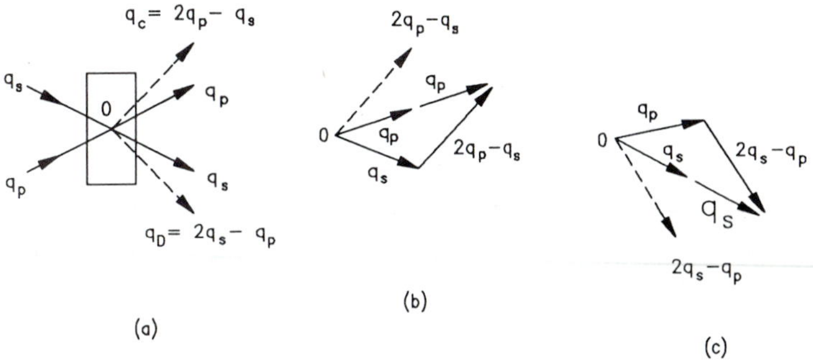

Figure 14.15. (a) The thick grating or the Bragg regime geometry showing that one scattering order is present either at $\mathbf{q}_D = 2\mathbf{q}_p - \mathbf{q}_s$ or at $\mathbf{q}_D = 2\mathbf{q}_s - \mathbf{q}_p$, where a $\mathbf{q}_p$ or a $\mathbf{q}_s$ photon is scattered, respectively; (b) and (c) The phase-matching or momentum conservation requirement for the scatterings of part (a). The dashed lines here are the reconstructed scattered photon wave vectors from origin.

where we assumed that the probe photon with wavelength λ_c has a small incidence angle in agreement with the Raman-Nath regime. The angle of mth diffraction order is ϕ_m, Λ is the spatial period of the grating given by Eq. (11.24), and m can take any positive or negative integer values. For the self-diffraction geometry of Fig. 14.14, the probe photon wavelength λ_c is again equal to either of the incident beam photons with $\lambda_c = \lambda_p$ or $\lambda_c = \lambda_s$. Equation (14.30) indicates that several scattering orders are present in this regime. The equivalent vectorial form of Eq. (14.30) is the wave vector conservation, similar to Eq. (14.26), given by

$$m\mathbf{q} = \mathbf{q}_D - \mathbf{q}_c \ , \quad m = 0, \pm 1, \pm 2, \ldots \tag{14.31}$$

or

$$m(\mathbf{q}_p - \mathbf{q}_s) = \mathbf{q}_D - \mathbf{q}_c \ , \tag{14.32}$$

specifying that $\mathbf{q}_D - \mathbf{q}_c$ can be equal to integer multiples of the grating vector $\mathbf{q}$. The ± sign is eliminated in Eq. (14.32), since m can take both negative and positive values. With the geometry of Fig. 14.14, with $\mathbf{q}_c = \mathbf{q}_p$, Eq. (14.32) gives

$$\begin{aligned} \mathbf{q}_D &= \mathbf{q}_p + m(\mathbf{q}_p - \mathbf{q}_s) \\ &= \mathbf{q}_p, \mathbf{q}_s, 2\mathbf{q}_p - \mathbf{q}_s, 2\mathbf{q}_s - \mathbf{q}_p, 3\mathbf{q}_p - 2\mathbf{q}_s, 3\mathbf{q}_s - 2\mathbf{q}_p \ , \ldots \ , \end{aligned} \tag{14.33}$$

corresponding to different m values, as shown schematically by different scattered beams in Fig. 14.14.

In addition to conservation of momentum, conservation of energy should also obey in nonlinear wave mixing experiments. The general form of the conservation of energy is

$$\omega_D - \omega_c = \omega_p - \omega_s \; , \tag{14.34}$$

where ω_p and ω_s are the frequencies of incident waves, while ω_c and ω_D are the frequencies of the probe and diffracted beams, respectively. For DFWM process, $\omega_p = \omega_s = \omega_c = \omega_0$, thus, $\omega_D = \omega_0$, indicating that the diffracted wave has the same frequency as the incident waves.

In the Raman-Nath approximation, the field transmitted through the sample of length d can be written as (see Ref. 14.25)

$$E = E_0 \, e^{iq_0 \Delta n(\omega,n)d \, - \, \alpha(\omega,n)d/2} \; , \tag{14.35}$$

where $q_0 = \omega/c$, and $\Delta n(\omega, n)$ denotes the frequency and density-dependent refractive-index change whose physical origins were discussed in more detail in Chap. XI. In writing Eq. (14.35), we used the fact that the field experiences amplitude and phase changes after transmission through the medium. The amplitude changes follow the Beer-Lambert law [see Eq. (3.30) where $I \sim E^2 \sim \exp(-\alpha d)$ and, consequently, $E \sim \exp(-\alpha d/2)$]; the phase change is given by $\exp(i\, q_0\, \Delta n(\omega)\, d)$. Equation (14.35) shows how the transmitted field depends on the material excitation through the nonlinear absorption $\alpha(\omega, n)$ and refractive index change $\Delta n(\omega)$. The spatial variations of the transmitted field of Eq. (14.35) give rise to different Raman-Nath scattering orders of Fig. 14.14. Since the directly transmitted beams that correspond to scattering orders ± 1 exist even in the linear regime, with excitation independent optical properties of the medium, the strength of the higher-order scattering contributions is a measure for the nonlinearity of the material. In the most simplified approximation, the area under the third-order peak is a measure for the third-order nonlinearity, i.e., the part of the nonlinearity that is directly proportional to the incident intensity. More generally, however, such an assignment is reasonably correct only for sufficiently low intensities. In all other cases one has to use the analysis based on the equations outlined in this section in combination with a proper theory for the material nonlinearities. From the experimental point of view, DFWM is one of the simpler techniques to obtain an order-of-magnitude estimate of the material nonlinearities.

14-9. Optical Phase Conjugation*

Consider an optical wave propagating in the z direction,

$$\mathbf{E}(\mathbf{r},t) = \frac{1}{2}\left[\mathbf{E}_0(x,y)\, e^{i(\omega t - q_0 z + \phi_0)} + \text{c.c.}\right] . \tag{14.36}$$

Here $\mathbf{E}_0$ is the slowly varying complex amplitude of the field, $\mathbf{q}_0$ is the optical wave vector, and $\phi_0(x,y)$ is the phase of the wave that is assumed to be a nearly plane wave. The properties of the wave after reflection by an ordinary mirror are well known and can be found in textbooks of classical optics.

Now consider a phase-conjugate mirror. By definition, such a mirror transforms the incident wave into a reflected wave, $\mathbf{E}_r(r,t)$, propagating in the reverse direction and having the same temporal variation but a reversed spatial phase,

$$\mathbf{E}_r(\mathbf{r},t) = \frac{1}{2}\left[\mathbf{E}_0^*(x,y)\, e^{i(\omega t + q_0 z - \phi_0)} + \text{c.c.}\right] . \tag{14.37}$$

Formally, this amounts to an operation of time inversion performed on the incident wave (leaving the spatial part unchanged), as can be immediately seen by explicitly writing the complete expression

$$\mathbf{E}_r(\mathbf{r},t) = \frac{1}{2}\left[\mathbf{E}_0^*(x,y)\, e^{i(\omega t + q_0 z - \phi_0)} + \mathbf{E}_0(x,y) e^{-i(\omega t + q_0 z - \phi_0)}\right]$$

$$= \mathbf{E}(\mathbf{r},-t) . \tag{14.38}$$

Thus, the reflected wave $\mathbf{E}_r(\mathbf{r},t)$ behaves as if it retraces the path of the original wave back in time, like a movie played backwards. Many applications derive from this intriguing property, leading to an automatic compensation of distortions of the phase front experienced by the primary wave on its way to the conjugate mirror. Figure 14.16 illustrates this property with a simple example.

A versatile method to fabricate a phase-conjugate mirror resulting from four-wave mixing has been proposed by Hellwarth (Ref. 14.23) and is now the most commonly used in applications. It exploits a special configuration of DFWM, shown schematically in Fig. 14.17. Here three incident beams with common frequency ω interact via the nonlinearity of the medium to generate a fourth coherent wave called the signal wave. Two of the incident waves (the pump wave labeled 2 and 3 here) are counterpropagating, while the third wave (the probe wave labeled 4) overlaps the pump waves in the nonlinear medium at an arbitrary angle.

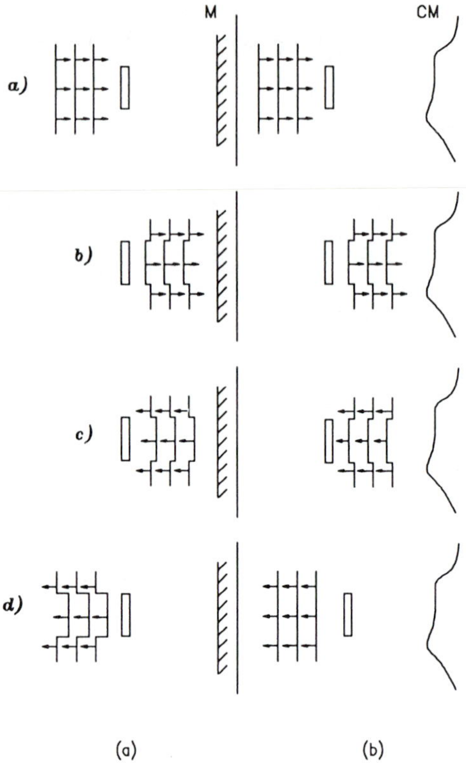

Figure 14.16. (a) Behavior of a plane wave front. (*a*) Incident on a piece of glass (aberrator). (*b*) After transmission through the aberrator. (*c*) After reflection on an ordinary mirror (M). (*d*) On the way back after transmission through the aberrator. (b) Same as in (a) except that reflection occurs on a phase conjugate mirror (CM).

Because of energy and momentum conservation, the generated signal wave 1 will have the same frequency ω as the incident beams and a propagation vector opposite to that of the probe beam. The nonlinear medium activated by the two counterpropagating pump beams of frequency ω behaves like a conjugate mirror. The signal wave is generated by a polarization proportional to the complex conjugate of the probe field and possesses the properties of phase conjugation with respect to the incident probe wave.

Interpretation of the physical phenomena taking place in the crystal may be obtained by noting that the probe-field amplitude interferes with each of the pump fields to create a standing wave pattern, as seen in Sec. 14-8. This in turn leads to a static-volume grating of the refractive

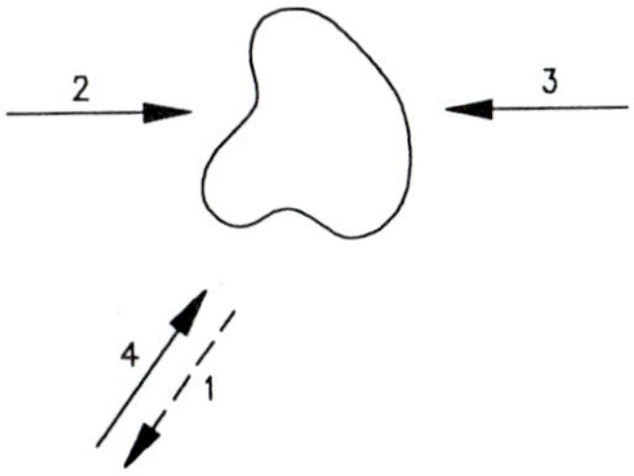

Figure 14.17. Geometric configuration for fully DFWM. Probe beam 4 interacts in the nonlinear medium with pump beams 2 and 3 to generate a signal beam 1. All beams have a common frequency ω.

index (if the incident waves fall in a transparent region of the medium), or a population grating (if the incident waves are tuned in an absorption region of the crystal). The second pump beam is diffracted at the Bragg angle in the direction opposite to the incident probe beam. This interpretation, in terms of real-time holographic writing and reading process, explains how the reflectivity of the conjugate mirror, $R_c = I_1/I_4$, can exceed 100% (where I_4 and I_1 are the probe and signal intensities). As illustrated in Fig. 14.18, two gratings are formed with spacings $\Lambda_1 = \lambda/2\sin\theta/2$ and $\Lambda_2 = \lambda/2\sin(\pi - \theta)/2$, where θ is the angle between the probe beam 4 and the pump beam 2, and λ is the wavelength inside the medium.

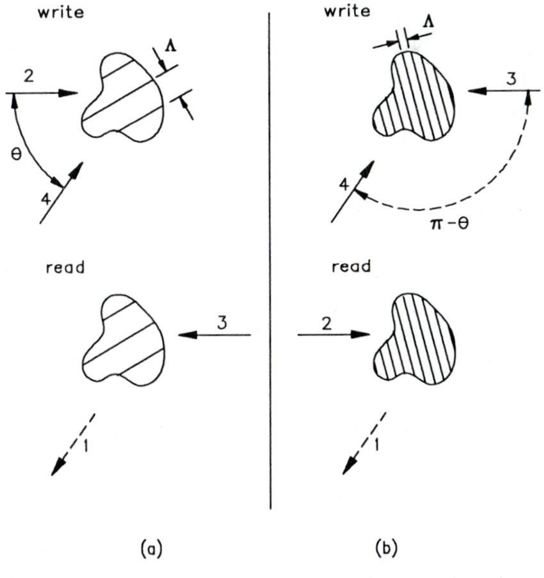

Figure 14.18. Holographic analogy for optical phase conjugation via four-wave mixing. (a) Large-period grating. (b) Small-period grating.

There also exists another small contribution to the phase conjugate signal (see Fig. 14.19) from simultaneous absorption of two photons (one from each pump beam), giving a two-photon coherence at frequency 2ω, followed by the emission of two photons (one probe, one signal). This process should not be confused with that resulting from the standing wave pattern formed by the interference of the two incident pump beams, with period $\lambda/2$. Such a stationary index grating is indeed formed in the medium, giving rise to a diffracted signal wave, but not in the direction $\mathbf{q}_1 = -\mathbf{q}_4$, and is therefore of no concern here.

This two-photon contribution is usually negligible compared to those originating from the grating and can only be observed under special conditions of strong enhancement due to a two-photon resonance. However, it has some unique properties. Since it results from a spatially uniform two-photon excitation of the medium, it leads to a phase-conjugate reflectivity that is independent of the incidence angle θ of the probe beam. Also, it is free from effects associated with a motion of the excitation in the volume. Furthermore, it can be observed with a probe polarization vector orthogonal to both pump-beam-polarization vectors, in contrast with the conjugate signal originating from the grating, which requires a nonvanishing component of the probe vector along one of the pump-polarization vectors.

An example of measured optical phase conjugation in a semiconductor is shown in Fig. 14.20. The experiments were performed on two samples of Ge, an intrinsic and a doped sample. The phase-conjugation reflectivities rise as the incident pump intensity increases, such that total reflection with $R \sim 1$ is approached at higher input intensities of $\simeq 100$ MW/cm^2.

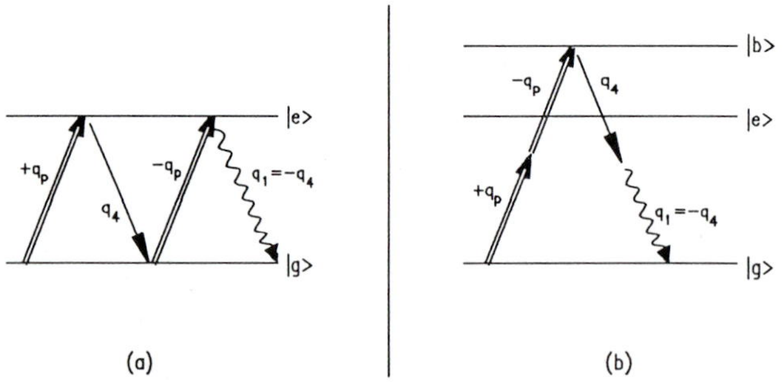

Figure 14.19. Schematic energy diagram for resonant processes leading to optical phase conjugation. (a) One-photon resonant process described in terms of simultaneous writing and reading of a population volume grating. (b) Two-photon resonant process corresponding to a spatially uniform coherent excitation of the system at frequency 2ω. In this figure we use the notation $\mathbf{q}_p = \mathbf{q}_2$ and $-\mathbf{q}_p = \mathbf{q}_3$.

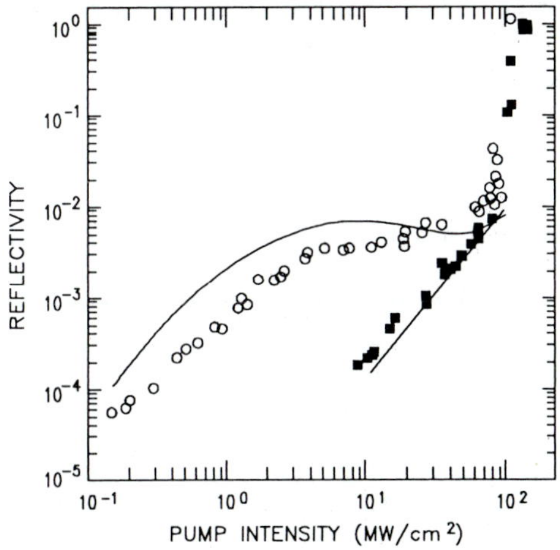

Figure 14.20. Optical phase conjugation measured in a 3-mm sample of intrinsic (squares) and impurity-doped (circles) Ge. Total reflection (R ~ 1) is reached for high input intensities (after Ref. 14.26).

14-10. Problems

14.1. Obtain the formal solution of the carrier density rate equation (14.3). Hint: Use the substitution $n(t) = n'(t)\, e^{-t/\tau}$.

14.2. Show that the angle between different scattering orders of Fig. 14.14 is the same as the incident angle θ in the Raman-Nath regime.

14.3. Show that the vectorial form of the grating equation $\Lambda \sin \theta_m = m\lambda_c$ is equivalent to $m\mathbf{q} = \mathbf{q}_D - \mathbf{q}_c$.

14-11. References

1. N. Peyghambarian, S. H. Park, S. W. Koch, A. Jeffery, J. E. Potts, and H. Cheng, Appl. Phys. Lett. **52**, 182 (1988).

2. S. H. Park, J. F. Morhange, A. D. Jeffery, R. A. Morgan, A. Chavez-Pirson, H. M. Gibbs, S. W. Koch, N. Peyghambarian, M. Derstine, A. C. Gossard, J. H. English, and W. Wiegmann, Appl. Phys. Lett. **52**, 1201 (1988).

3. Y. Z. Hu, S. W. Koch, M. Lindberg, N. Peyghambarian, E. L. Pollock, and F. F. Abraham, Phys. Rev. Lett. **64**, 1185 (1990); see also, Y. Z. Hu, M. Lindberg, and S. W. Koch, Phys. Rev. B **42**, 1713 (1990).

4. S. G. Lee, P. Harten, J. P. Sokoloff, R. Jin, B. Fluegel, K. E. Meissner, C. L. Chuang, R. Binder, S. W. Koch, G. Khitrova, H. M. Gibbs, N. Peyghambarian, J. N. Polky, and G. A. Pubanz, Phys. Rev. B **43**, 1719 (1991).

5. G. R. Olbright and N. Peyghambarian, Appl. Phys. Lett. **48**, 1184 (1986).

6. Y. H. Lee, A. Chavez-Pirson, B. K. Rhee, H. M. Gibbs, A. C. Gossard, and W. Wiegmann, Appl. Phys. Lett. **49**, 1505 (1986).

7. M. C. Downer and C. V. Shank, Phys. Rev. Lett. **56**, 761 (1986).

8. D. A. B. Miller, C. T. Seaton, M. E. Price, and D. S. Smith, Phys. Rev. Lett. **47**, 197 (1981).

9. M. Sheik-Bahae, A. A. Said, and E. W. Van Stryland, Opt. Lett. **14**, 955 (1989); see also, M. Sheik-Bahae, A. A. Said, T. H. Wei, D. J. Hagen, and E. W. Van Stryland, IEEE J. Quantum Electron. **26**, 760 (1990).

10. M. J. Weber, D. Milan, and W. L. Smith, Opt. Engin. **17**, 463 (1978).

11. M. J. Moran, C. Y. She, and R. L. Carman, IEEE J. Quantum Electron. **QE-11**, 259 (1975).

12. S. R. Friberg and P. W. Smith, IEEE J. Quantum Electron. **QE-23**, 2089 (1987).

13. R. Adair, L. L. Chase, and S. A. Payne, J. Opt. Soc. Am. B **4**, 875 (1987).

14. A. Owyoung, IEEE J. Quantum Electron. **QE-9**, 1064 (1973).

15. W. E. Williams, M. J. Soileau, and E. W. Van Stryland, Opt. Commun. **50**, 256 (1984).

16. G. R. Olbright, N. Peyghambarian, S. W. Koch, and L. Banyai, Opt. Lett. **12**, 413 (1987).

17. H. J. Eichler, P. Günter, and D. W. Pohl, *Laser-Induced Dynamic Gratings*, Springer Series in Optical Sciences, Vol. 50 (Springer-Verlag, Berlin, 1986).

18. D. Weaire, B. S. Wherrett, D. A. B. Miller, and S. D. Smith, Opt. Lett. **4**, 331 (1979).

19. J. R. Hill, G. Parry, and A. Miller, Opt. Commun. **43**, 151 (1982).

20. T. F. Boggess, S. C. Moss, I. W. Boyd, and A. L. Smirl, in *Ultrafast Phenomena IV*, edited by D. H. Huston and K. B. Eisenthal (Springer-Verlag, New York, 1984) pp. 202-204.

21. A. Maruani and D. S. Chemla, Phys. Rev. B **23**, 841 (1981).

22. J. M. Hvam and C. Dörnfeld, J. Phys. C (Paris) **2**, 205 (1988); J. M. Hvam, I. Balsev, and B. Hönerlage, Europhys. Lett. **4**, 839 (1987).

23. R. W. Hellwarth, J. Opt. Soc. Am. **67**, 1 (1977); see also, the *Special Issue on Nonlinear Optical Phase Conjugation*, Opt. Engin. **21** (1982).

24. Y. Masumoto and S. Shionoya, J. Phys. Soc. Jpn. **49**, 2236 (1980); Solid State Commun. **38**, 865 (1981).

25. D. Richardson, E. M. Wright, and S. W. Koch, Phys. Rev. A **41**, 1620 (1990); see also, D. Richardson, B. P. McGinnis, E. M. Wright, N. Peyghambarian, and S. W. Koch, Phys. Rev. A **44**, 628 (1991).

26. D. E. Watkins, C. R. Phipps, Jr., and W. W. Rigard, J. Opt. Soc. Am. **73**, 624 (1983).

Chapter XV*
FEMTOSECOND SPECTROSCOPY

Until recently, time-resolved optical studies of electron-hole interband transitions in semiconductors were restricted to incoherent processes because of the rapid scattering among the elementary excitations. This situation has been changed in the last few years through the development of techniques to produce ultrashort pulses with durations of only a few femtoseconds (1 fs = 10^{-15} s). Spectroscopy with these pulses provides the possibility to investigate materials in the very early stages after the excitation, when coherent properties are still at least partially present.

In this chapter we give an introduction to the relatively new field of femtosecond laser spectroscopy. First we describe two examples of experimental techniques to generate femtosecond pulses, and then we discuss applications to the study of ultrafast carrier dynamics in semiconductors (spectral-hole burning), exciton bleaching, the light-induced shift of the exciton resonance, the optical Stark effect, and the time-resolved photon echo.

15-1. Femtosecond Pulse Generation

In this section we qualitatively describe the principle of femtosecond pulse generation. It is beyond the scope of this book to give a detailed and accurate description of the state-of-the-art in a fast evolving field. The interested reader is referred to Ref. 15.1 and the proceedings of the topical meetings on the subject, such as the conferences on ultrafast phenomena.

We consider a broadband light source with inverted population, emitting in a random fashion into a large number of radiation modes. Such a light amplifier (e.g., a dye such as Rhodamine 6G (Rh 6G) pumped by an Argon laser) will emit a chaotic radiation field due to the incoherent superposition of the radiation modes. This chaotic field contains spikes of intensity corresponding to the fortuitous additive superposition of a large number of modes. The typical duration of these spikes is of the order of the inverse spectral bandwidth of the source, $1/\delta\omega$. This is schematically represented in Fig. 15.1(a).

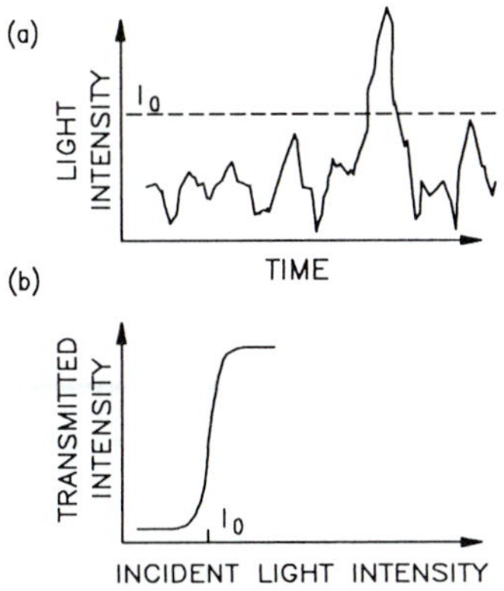

Figure 15.1. (a) Schematic representation of a chaotic radiation field due to the superposition of many radiation modes without phase relationship. The dotted horizontal line shows the discriminative effect of a saturable absorber with critical intensity I_0. Photons with intensities above I_0 are transmitted through the saturable absorber, and those below I_0 get absorbed. (b) Schematic representation of the transmission of a saturable absorber with critical intensity I_0.

Now consider the action of a fast saturable absorber with fast recovery time, opaque to low-input light intensities but transparent to high intensities, as shown schematically in Fig. 15.1(b). If both the light emitter and saturable absorber are put in a cavity capable of accommodating many radiation modes, pulse shaping will take place inside the cavity. The randomly generated initial spike of short duration is transmitted through the saturable absorber, then amplified when crossing the optical amplifier, whereas the rest of the emission is attenuated. As a result, after several round trips in the cavity, a stable mode of operation is achieved, whereby a train of short pulses, with a duration of the order of $1/\delta\omega$, and a separation between pulses equal to the round-trip time in the cavity are emitted by the oscillator. Figure 15.2 shows a widely used example of such an oscillator called a colliding-pulse modelocked (CPM) dye laser. It takes the form of a ring cavity, containing two counterpropagating pulses that simultaneously cross the saturable absorber and are mutually effective for the saturation process. Note the presence of a system of prisms inside the ring cavity. The purpose of the prisms is to compensate for the group velocity dispersion due to the other optical elements of the cavity. Group velocity dispersion is detrimental to the stability of a short pulse, since it introduces a phase delay between the different radiation modes and, therefore, it tends to destroy (after a few round trips) the positive interference at the origin of the short spike.

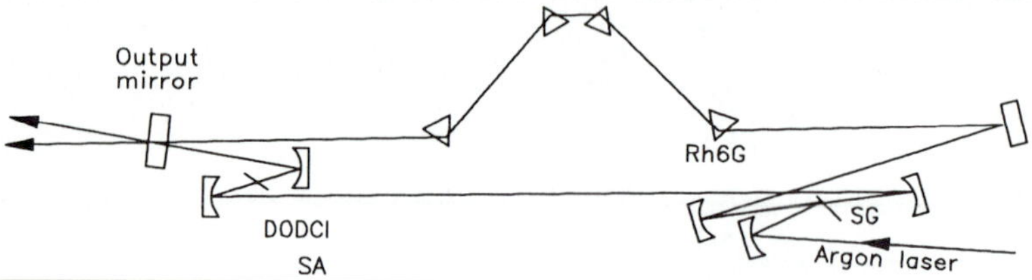

Figure 15.2. Colliding-pulse modelocked (CPM) dye laser cavity emitting femtosecond pulses. The amplifying medium (the saturable gain, SG) consists of a thin jet of Rh 6G dye. The saturable absorber (SA) consists of DODCI dye. The amplifier cell is pumped with an Argon laser.

Figure 15.3 shows how a pair of prisms can compensate for a normal group velocity dispersion by delaying the path of rays that have longer wavelength with respect to those having a shorter wavelength. Lasers emitting a train of short pulses are called modelocked lasers. To understand more formally why locking of many modes to a common phase leads to short pulses, we consider the total field from N + 1 modes, m = 0, 1, 2, 3, ... N, with frequencies $\omega_0 + m\Delta\omega$,

$$E(t) = a_0 \sum_{m=0}^{N} \exp[i(\omega_0 + m\Delta\omega)\, t - \Phi_m] \,, \tag{15.1}$$

where a_0 is a constant. We use the relation

$$\sum_{0}^{N} x^n = \frac{1 - x^N}{1 - x} \,. \tag{15.2}$$

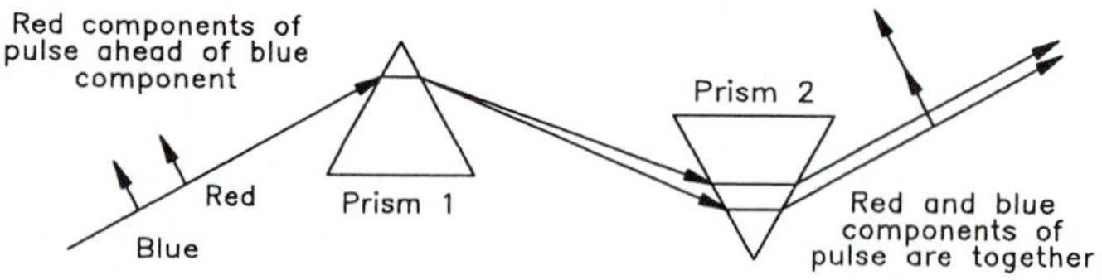

Figure 15.3. A pair of prisms compensates for normal group velocity dispersion, which leads to the delay of red components of a pulse with respect to its blue components.

Since the phase Φ_m is the same for all the modes, it can be taken out of the summation, and we obtain

$$E(t) = a_0 \exp(i\omega_0 t - \Phi) \frac{1-\exp(iN\Delta\omega t)}{1-\exp(i\Delta\omega t)} . \tag{15.3}$$

The intensity is given by

$$I(t) = \frac{c\sqrt{\mathscr{E}_0}}{8\pi} |E(t)|^2 = \frac{c\sqrt{\mathscr{E}_0}}{8\pi} EE^* \tag{15.4}$$

or

$$I(t) = a_0^2 \frac{\sin^2 \frac{N}{2}\Delta\omega t}{\sin^2 \frac{\Delta\omega t}{2}} , \tag{15.5}$$

where use has been made of the relation

$$1 - \cos 2x = 2 \sin^2 x . \tag{15.6}$$

Equation (15.5) may be plotted for different values of N. It is found that the coherent superposition of N modes leads to pulses of duration $\delta t = 2\pi/N\Delta\omega$ with a repetition rate of $T = L/c$, where c/L is the frequency interval between modes, L is the cavity length, and c is the speed of light (see Fig. 15.4).

Recently, a new type of modelocked laser, called the Ti:sapphire laser, has been developed. It has the advantage over CPM dye lasers that the femtosecond pulse wavelength can be continuously tuned over a broad frequency range, between 670 nm and 1000 nm, and that the output power is higher by one or two orders of magnitude. The Ti:sapphire laser does not contain a saturable absorber but makes use of the nonlinear change of refractive index of the amplifier to lock together the phase of many modes. A schematic description of such a laser is shown in Fig. 15.5. These short pulses can be amplified with relatively modest means to an energy per pulse exceeding 100 mJ. When tightly focused, this translates into light intensities beyond 10^{18} W/cm^2.

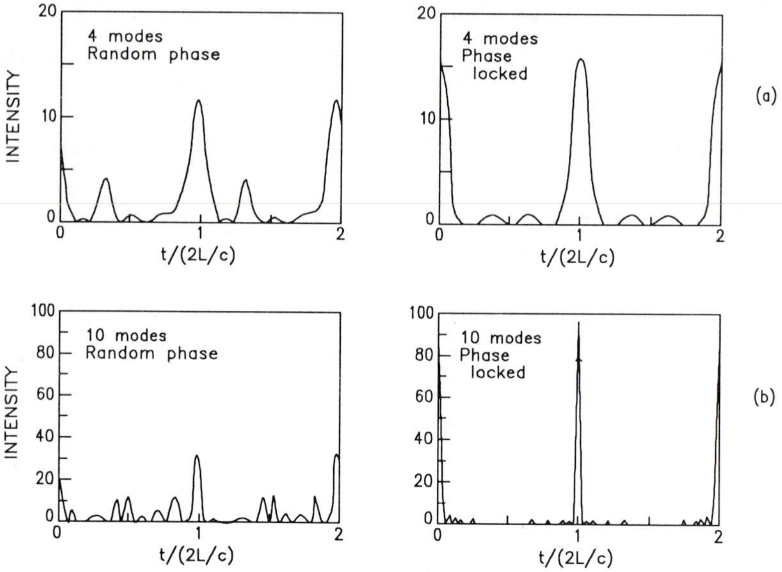

Figure 15.4. The superposition of (a) 4 and (b) 10 equally spaced radiation modes when the phase of the modes is random (left) and when the phase is locked (right).

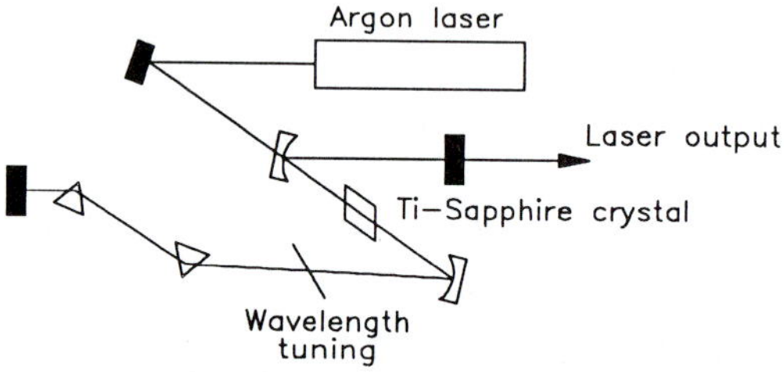

Figure 15.5. Schematic representation of a Ti:sapphire laser.

15-2. Pulse-Duration Measurement

One key issue in femtosecond spectroscopy is the measurement of the pulse duration. Since there is no electronic detector capable of responding on a subpicosecond time scale, one must resort to nonlinear optical techniques. The basic idea is to convert a measurement in the time domain into a measurement in the space domain by increasing the path of part of the optical pulse with respect to the rest of the pulse, and to convert this path delay into an electric signal directly proportional to the path delay.

Since the speed of light is 30 μm/100 fs in free space, it is easy to obtain an accuracy down to the femtosecond time scale by controlling the delay length with submicron accuracy. In principle, any optical nonlinearity with instantaneous on and off response time can be used to convert the time measurement into a length measurement. The most commonly used method is the autocorrelation technique based on second-harmonic generation. As mentioned in Chap. XIII, a transparent medium with no inversion symmetry generates light at the harmonic frequency 2ω if it is irradiated with intense light at frequency ω. It is important to realize that this harmonic generation occurs only during the presence of the radiation at ω. Light at frequency 2ω will be emitted preferentially in certain directions that satisfy the phase-matching conditions $n(\omega) = n(2\omega)$. The phase-matching conditions translate the fact that both beams ω and 2ω propagate at the same phase velocity, so that energy transfer always occurs constructively from beam ω to beam 2ω. If the phase-matching condition is not satisfied, the harmonic generated at a given location x in the crystal will interfere destructively with the harmonic generated at location $x + \delta$, where δ is the coherence length [defined as $\delta = 2\pi/\Delta q$, where $\Delta q = q(2\omega) - 2q(\omega)$], with the net result that very little harmonic light will emerge from the crystal. To satisfy the phase-matching conditions, one selects a birefringent crystal having two different refractive indices, the ordinary and extraordinary index for orthogonal polarization directions, and chooses the incident light propagation vector such that the ordinary index at ω and extraordinary index at 2ω are equal. The principle of such an autocorrelator is shown in Fig. 15.6. The optical pulse to be measured is divided into two pulses of equal intensity. After delaying one pulse with respect to the other, both pulses are recombined in a nonlinear crystal transparent at both frequencies ω and 2ω, usually a KDP crystal that is oriented for phase matching. Second-harmonic light originating from the two beams is created, not only in the direction of the two incident beams,

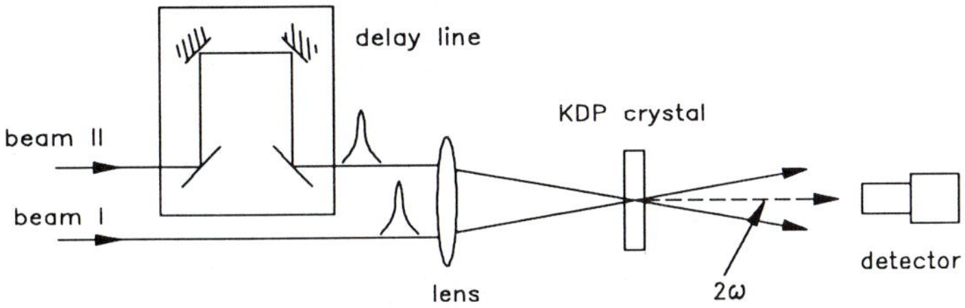

Figure 15.6. Schematic setup of a background-free autocorrelation technique for measuring pulse duration. Beams I and II are the replicas of the same pulse that are focused on a second-harmonic crystal like KDP with a variable time delay. The second-harmonic signal is shown by the dashed line.

but also in the direction along the bisector of the two incident beams. This second-harmonic light resulting from the time overlap between both pulses is detected with a conventional photodetector. Its intensity is proportional to the autocorrelation function,

$$I(t_p) = \int_{-\infty}^{+\infty} I(t)\, I(t - t_p)dt\,, \tag{15.7}$$

where t_p is the time delay between the two pulses. Note that the autocorrelation function is an even function of time and, therefore, cannot yield the true temporal shape of a pulse with asymmetric temporal shape, giving only the pulse duration. An example of an autocorrelation trace from a 55-fs pulse emitted by a CPM dye laser is shown in Fig. 15.7.

For very short pulses, below 100 fs, it is better to choose an interferometric autocorrelator. Here both beams are adjusted to propagate collinearly inside the KDP crystal after the delay line (see Fig. 15.8). The optical arrangement corresponds to a Michelson interferometer. Since the two beams can interfere, the signal recorded by the detector displays fringes corresponding to successive constructive and destructive interferences, according to the time delay between both pulses. These interferences appear only when both beams overlap in time inside the nonlinear crystal. To obtain the pulse duration, it is sufficient to count the number of fringes between the full width at half maximum of the autocorrelation trace, since each fringe corresponds to a wavelength optical path delay. An example of such an autocorrelation trace for a pulse of 16-fs duration is shown in Fig. 15.9.

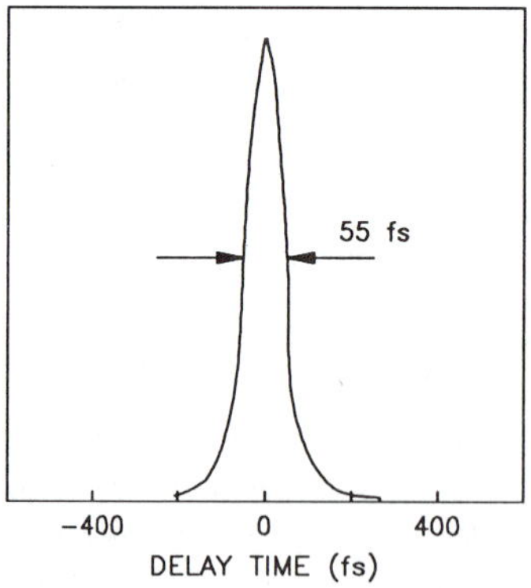

Figure 15.7. Autocorrelation trace of a 55-fs pulse emitted by a CPM dye-laser cavity.

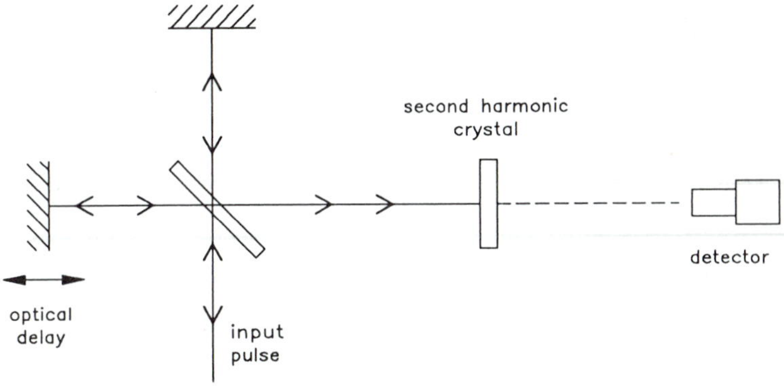

Figure 15.8. Schematic setup for an interferometer autocorrelator. The two pulses arrive collinearly at the second-harmonic crystal.

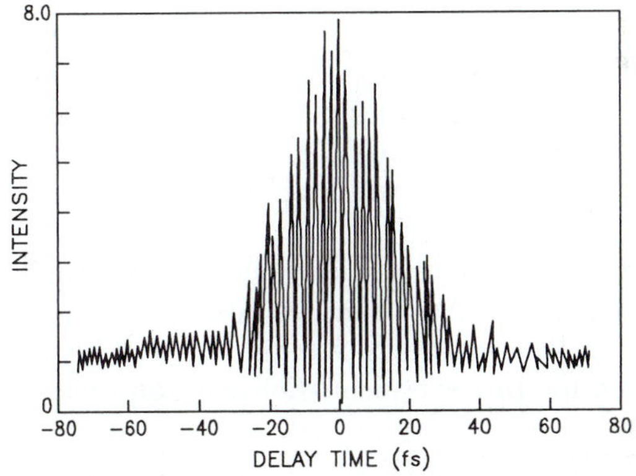

Figure 15.9. Interferometric trace of an autocorrelation corresponding to an optical pulse of 16-fs duration.

15-3. Femtosecond Pump-Probe Spectroscopy

The pump-probe measurement technique was described in Chap. XIV. The same technique is employed in order to measure the dynamics of excited states of semiconductors and other materials in the sub-picosecond domain. Femtosecond pump pulses induce changes in the transmittance of the material of interest. Broadband femtosecond probe pulses are used to detect those changes (see Fig. 15.10). To obtain a broad bandwidth probe pulse, a fraction of the amplified femtosecond pulse is focused on a transparent dielectric medium, such as a jet of ethylene glycol. The dielectric medium radiates a collimated broadband emission of the same

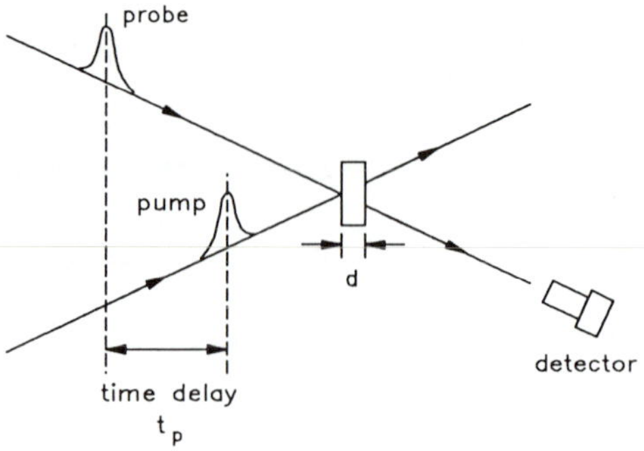

Figure 15.10. Schematics of femtosecond spectroscopy. Pump pulse and probe pulse are delayed by t_p with respect to each other. The transmission of the probe is measured with and without the presence of the pump to obtain the DTS.

duration as the incident pulse. The linearly polarized pump and probe pulses are usually orthogonally polarized to facilitate isolation of the probe pulse from the pump pulse. A delay line with variable length optical path difference between pump and probe path is used to generate a variety of pump-probe time delays, t_p. The pump and probe beams are then focused to the same spot on the sample.

So-called differential transmission spectra (DTS) often yield sensitive and reliable information on the semiconductor transmission spectra. The schematics of DTS measurements are shown in Fig. 15.10. DTS is the difference between the probe transmission with and without the pump,

$$\mathrm{DTS} = \frac{|E_p(\omega)|^2_{\text{pump on}} - |E_p(\omega)|^2_{\text{pump off}}}{|E_p(\omega)|^2_{\text{pump off}}} \, . \qquad (15.8)$$

Here $E_p(\omega)$ denotes the amplitude of the probe field after going through the sample. The DTS described by Eq. (15.8) depends implicitly on the time delay t_p between pump and probe. In the experiments, the DTS is measured at various time delays. One obtains transmittance changes with an accuracy of a few parts in 10^4 with a resolution better than 50 fs.

To illustrate the basic information contained in DTS, let us assume that the pump-induced changes in the probe transmission are small at early time delays. Then we can write

$$E_p(\omega)|_{\text{pump on}} = E_p(\omega)|_{\text{pump off}} + \Delta E_p(\omega)|_{\text{pump on}} \, . \qquad (15.9)$$

Inserting Eq. (15.9) into (15.8) and neglecting the quadratic terms in ΔE_p,

$$\text{DTS} \simeq \frac{2\,\text{Re}\,[E_p{}^*(\omega)|_{\text{pump off}}\ \Delta E_p(\omega)|_{\text{pump on}}]}{|E_p(\omega)|^2{}_{\text{pump off}}}\ . \tag{15.10}$$

This relation suggests that the DTS can be viewed as the *interference* between the unperturbed transmitted probe field and the pump-induced change in the transmitted probe field. Under certain coherent conditions, this interference can give rise to oscillatory features in the DTS, as we will show below.

Alternatively, if coherent features are absent, we can write

$$|E_p(\omega)|^2{}_{\text{pump off}} = |E_0|^2\, e^{-\alpha d}\ , \tag{15.11}$$

where E_0 is the amplitude of the incident field, and α is the linear absorption coefficient. Furthermore,

$$|E_p(\omega)|^2{}_{\text{pump on}} = |E_0|^2\, e^{-(\alpha+\Delta\alpha)d}\ , \tag{15.12}$$

where $-\Delta\alpha$ are the pump-induced absorption changes. Inserting Eqs. (15.11) and (15.12) into (15.8) yields

$$\text{DTS} = e^{-\Delta\alpha d}\ . \tag{15.13}$$

If we again assume that the pump-induced changes are small, Eq. (15.13) becomes

$$\text{DTS} \simeq -\ \Delta\alpha d\ , \tag{15.14}$$

showing that for incoherent processes the DTS is directly proportional to the absorption changes generated by the pump pulse. Generally, however, one has to analyze the full Eq. (15.8) to include both coherent and incoherent effects.

15-4. Semiconductor Bloch Equations

In order to theoretically analyze ultrafast semiconductor nonlinearities, we have to use the dynamic many-body equations for the semiconductor response. In the simplest form, these equations can be written in the form of *semiconductor Bloch equations*, similar to the optical Bloch equations in atomic physics. The semiconductor Bloch equations are (Ref. 2.5)

$$\frac{\partial P(\mathbf{k})}{\partial t} = -\, i[\mathcal{E}_e(\mathbf{k}) + \mathcal{E}_h(\mathbf{k})]\, P(\mathbf{k})/\hbar$$

$$-\, i\,(n_e(\mathbf{k}) + n_h(\mathbf{k}) - 1)\, \omega_R(\mathbf{k}) + \left.\frac{\partial P(\mathbf{k})}{\partial t}\right|_{col}, \qquad (15.15)$$

$$\frac{\partial n_e(\mathbf{k})}{\partial t} = -\, 2\, \mathrm{Im}\,(\, \omega_R(\mathbf{k})\, P^*(\mathbf{k})\,) + \left.\frac{\partial n_e(\mathbf{k})}{\partial t}\right|_{col}, \qquad (15.16)$$

$$\frac{\partial n_h(\mathbf{k})}{\partial t} = -\, 2\, \mathrm{Im}\,(\, \omega_R(\mathbf{k})\, P^*(\mathbf{k})\,) + \left.\frac{\partial n_h(\mathbf{k})}{\partial t}\right|_{col}. \qquad (15.17)$$

semiconductor Bloch equations

Here $n_e(\mathbf{k})$ and $n_h(\mathbf{k})$ are the nonequilibrium electron and hole distributions, and $P(\mathbf{k})$ is the interband polarization. The effective electron ($i = e$) or hole ($i = h$) energies are $\mathcal{E}_i(\mathbf{k})$, i.e., the band energies plus the exchange and correlation energy renormalizations (see Chap. XIII). Furthermore, we introduced the *generalized Rabi frequency*,

$$\omega_R(\mathbf{k}) = \frac{1}{\hbar}\left[d_{cv} E + \sum_{\mathbf{q} \neq \mathbf{k}} V(|\mathbf{k} - \mathbf{q}|)\, P(\mathbf{q}) \right],$$

generalized Rabi frequency (15.18)

which is the sum of the Rabi frequency of the applied field $d_{cv}\mathcal{E}/\hbar$ and the internal dipole field $\sum_{\mathbf{q} \neq \mathbf{k}} V(|\mathbf{k} - \mathbf{q}|)P(\mathbf{q})/\hbar$. Here V is the Coulomb interaction potential. The appearance of the effective Rabi frequency instead of the bare Rabi frequency of the applied field emphasizes the fact that the electron-hole pairs in the semiconductor react to an effective field made up of applied field plus dipole field of all other generated electron-hole excitations.

The collision terms $\left.\frac{\partial}{\partial t}\,..\right|_{col}$ in the semiconductor Bloch equations describe the various electron and hole scattering and polarization decay processes, such as carrier-carrier scattering. Without going into the details, we just mention at this point that in a systematic theory, the collision terms also include the effects of screening of the Coulomb potential.

As limiting cases, the semiconductor Bloch equations reproduce the atomic Bloch equations if we set the Coulomb potential to zero everywhere in the equations. Furthermore, one can verify that the linearized Eq. (15.15) is nothing but the Fourier-transform of the exciton Wannier equation (6.16) (see Prob. 15.4). The semiconductor Bloch equations contain the nonlinearities discussed in Chap. XIII. For example, the energies contain the bandgap renormalization, and the factor $(n_e(\mathbf{k}) + n_h(\mathbf{k}) - 1)$ expresses the Fermionic phase-space filling effects, leading, for example, to the bandfilling nonlinearities. The numerical solution of the semiconductor Bloch equations for $E(t) = E_{pump}(t) + E_{probe}(t)$ allows the analysis of femtosecond pump-probe experiments. Some examples will be discussed in the following sections.

15-5. Spectral Hole Burning and Exciton Bleaching

Ultrafast carrier dynamics is best analyzed by photoexciting electrons and holes initially in a nonequilibrium distribution with a femtosecond-duration pump pulse, and subsequently monitoring their relaxation with a femtosecond probe continuum. As shown in Fig. 15.11, the pump laser is tuned above the semiconductor band edge, and the probe laser measures the spectral changes around the pump. Electron and hole population distributions can be monitored by observing absorption changes near the exciton resonance and in the spectral vicinity of the excitation pulse (Refs. 15.2-15.4).

The dynamic evolution of laser-excited band-to-band transitions can be roughly divided into three stages. First, in the collision-free or coherent regime, different k states react independently of each other to the exciting light. The pump pulse generates a high density of electron-hole pairs, which just start to interact among themselves and with phonons. These carriers are, hence, in highly nonequilibrium states. The light field introduces a coherent coupling between these valence-band and conduction-band states, driving an oscillating polarization in the matter. The semiconductor electrons oscillate between the respective states in the valence and conduction bands (Rabi flopping), not unlike the coherent electronic excitations in two-level-atom spectroscopy.

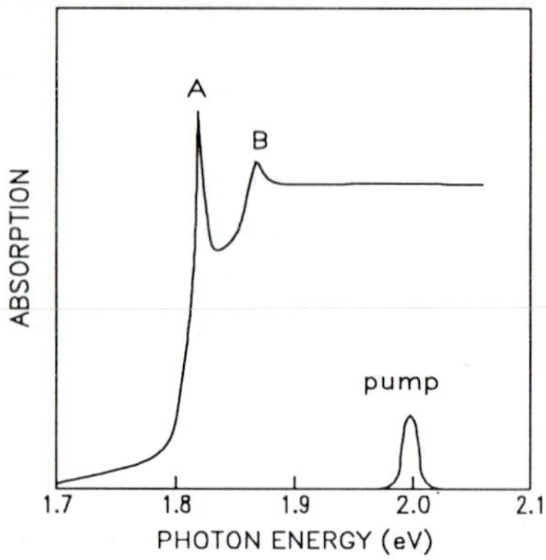

Figure 15.11. Typical spectral-hole-burning experiment. The pump pulse is tuned above the semiconductor bandgap, and the probe pulse monitors the transmission changes in the spectral region around and below the pump.

In the second stage of the band-to-band transition dynamics, phase relaxation has occurred and the electrons and holes are described by nonthermal distributions, which exist locally in k space. This is evidenced by the spectral hole around the center frequency of the excitation pulse. Electrons and holes are subject to the respective intraband Coulomb interaction and to the interaction with phonons. Both interaction mechanisms redistribute the electronic excitations toward the extrema of their respective bands. The optical excitation process acts as a local source of new electron-hole pairs. When the bands are filled up to the resonant k states, the transition is bleached and no more light is absorbed. Finally, after many collisions have occurred, quasi-thermal equilibrium is reached, where electron-hole pairs are created by the pump beam with the same rate as they recombine radiatively or nonradiatively. The optical nonlinearities in this quasi-equilibrium regime have been discussed in Chap. XIII.

These dynamic evaluation processes may be observed by femtosecond time-resolved pump-probe measurements. Typically, differential transmission is employed for such experiments, since some of the signals are too small to be observable in direct transmission. A schematic representation of such a measurement is shown in Fig. 15.12. The solid curve in Fig. 15.12(a) displays the linear absorption spectrum without the presence of the pump. The dashed curve shows the absorption of the sample in the presence of the pump. The pump spectrum is also shown. The exciton bleaches as a result of screening and blocking effects, as represented by the reduction of the exciton absorption strength. In addition, the absorption strength is reduced in the spectral vicinity of the

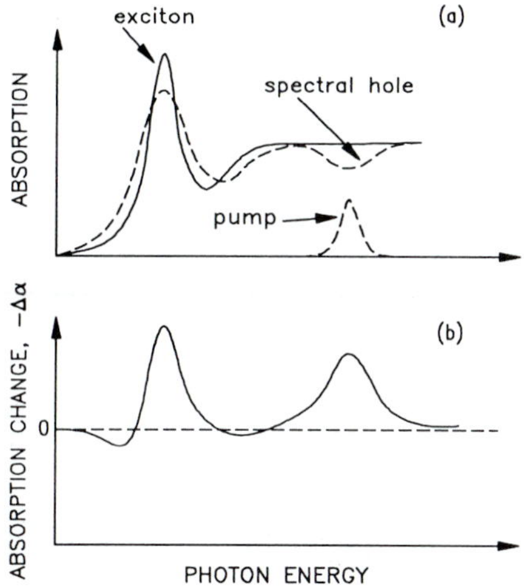

Figure 15.12. Schematic representation of a femtosecond pump-probe measurement in a semiconductor. (a) The linear absorption (solid curve) and the pump-induced absorption (dashed curve) are plotted. (b) The difference spectrum of the curves of part (a); i.e., $-\Delta\alpha = \alpha$ (without pump), $-\alpha$ (with pump).

pump. The state-filling effect is responsible for this spectral hole. The experiment usually measures the difference between the two curves of Fig. 15.12(a), as represented in Fig. 15.12(b). The exciton bleaching and the spectral hole appear as positive peaks. The broadening of exciton and bandgap renormalization increase the absorption below the exciton, as represented by the negative signals in the differential spectrum.

Figure 15.11 shows an example of the linear absorption spectrum of a bulk CdSe thin film at 10 K. The two excitonic peaks labeled A and B originate from the heavy-hole and light-hole valence bands split by crystal field interaction. The pump pulse has a 70-fs duration and the laser photon energy is at 1.99 eV, on the high-energy side of the excitons. Absorption changes, $-\Delta\alpha$, following excitation by the pump pulse, are measured as a function of time delay between the pump and probe pulses. Figure 15.13(a) shows the measured differential spectra in 50-fs intervals. The 0-fs spectrum corresponds to a complete time overlap of the pump and probe pulses, while the spectrum labeled 50 fs shows the absorption changes observed when the peak of the pump pulse precedes the peak of the probe by 50 fs. The 0-fs and 50-fs spectra show the presence of a nonthermal spectral hole, as indicated by the dotted area, on the high-energy side of the A and B exciton bleaching. As the nonthermal distribution thermalizes by carrier-carrier and carrier-phonon scattering events, the spectral hole washes out and only the bleached exciton remains.

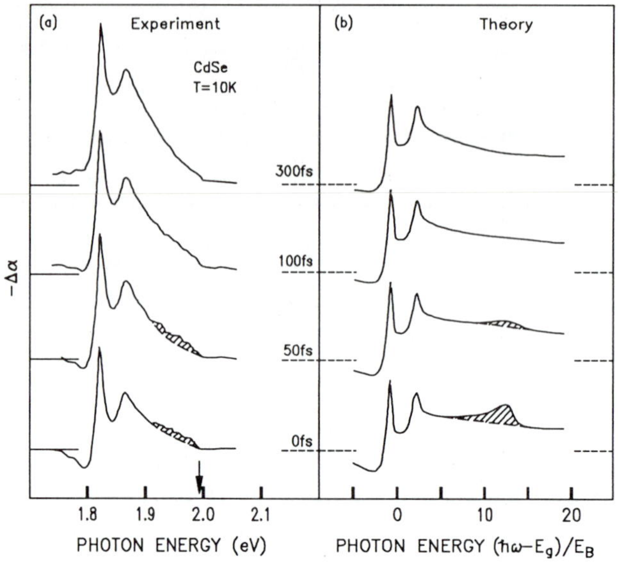

Figure 15.13. (a) The measured change in the absorption observed under the 1.99 eV excitation at 10 K. The time delay of each spectrum is shown in the figure. The dashed area shows the nonthermalized distribution of hot carriers, i.e., the spectral hole. (b) Calculated pump-induced absorption for various time delays between the pump and probe pulses (after Ref. 15.4).

After 300 fs the spectral hole has completely disappeared [see Fig. 15.13(a)].

To analyze these experiments, the semiconductor Bloch equations (15.15)-(15.17) have to be solved numerically, including carrier-carrier scattering and carrier-phonon scattering in the collision terms. For the present situation, these scattering terms have two main effects: the thermalization of a nonthermal carrier distribution and the decay of the phase coherence of the optical polarization (dephasing). The results of such a theory are displayed in Fig. 15.13(b). The 0-fs spectrum in Fig. 15.13(b) clearly displays the spectral dip at the pump position. Simultaneously, the two excitons are completely bleached. The spectral dip has a tail extending to low energies with respect to the pump energy, in good agreement with the experiment. Both experiment and theory show that the spectral hole rapidly washes out and the excitons stay bleached for later times, indicating very large carrier-carrier scattering rates.

15-6. Optical Stark Effect

As we pointed out in Chap. XIII, the optical Stark effect is a well-known phenomenon in atomic systems, whereby an atomic transition energy is shifted through irradiation of the medium with an optical beam with photon energy detuned from the transition energy. In semiconductors the optical Stark effect can be observed if the excitation pulse is tuned energetically below the exciton resonance by an amount ΔE, and the modifications of the exciton spectrum are monitored by the probe pulse (Refs. 13.4 and 15.5-15.7). As an example, Fig. 15.14 shows the transmission spectra of a GaAs-AlGaAs multiple quantum well sample at 15 K under pumping with femtosecond pulses with photon energy tuned below the exciton. For negative time delays, the transmission is monitored when the pump pulse arrives before the probe, representing the unexcited sample. Thus, the two transmission dips in the solid curve (-2-ps time delay) of Fig. 15.14 correspond to the heavy-hole and light-hole exciton absorption peaks in the absence of the pump. When the pump and probe overlap in time (0-ps dotted trace), the excitons shift to higher energies and partially bleach. After 1.2 ps (the dashed curve), the shift and bleaching have almost entirely disappeared. This shift and recovery of the exciton is evidence of the optical or ac Stark effect.

The experimental observations of the exciton blue shift, bleaching, and nearly complete recovery can be explained theoretically by solving the semiconductor Bloch equations (15.15)-(15.17). An example is shown in Fig. 15.15. The ultrafast bleaching and recovery of the exciton resonance

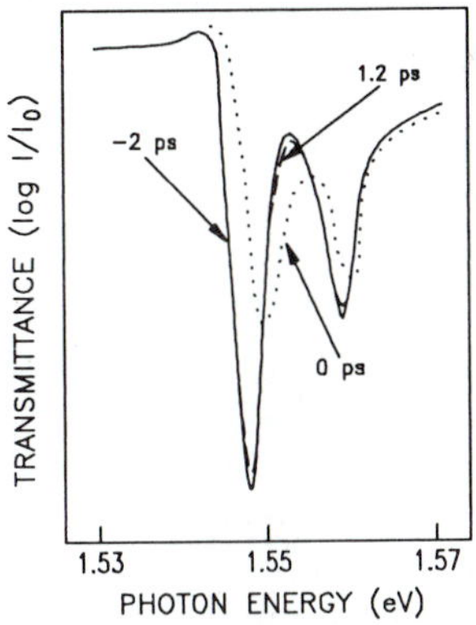

Figure 15.14. Subpicosecond time-resolved transmission at 15 K of a GaAs-AlGaAs multiple quantum well sample with well and barrier thickness of 100 Å, recorded at different delays from a nonresonant intense optical pump pulse. The spectral position of the pump pulse is also shown in the figure (after Ref. 13.4).

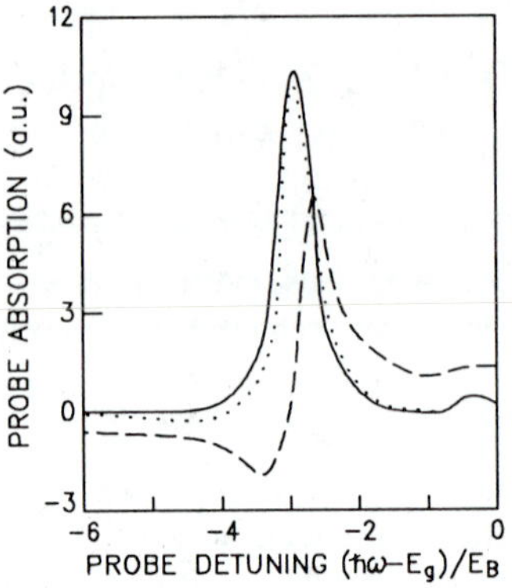

Figure 15.15. Computed probe absorption spectra for a 50-Å GaAs quantum well vs probe detuning without pump (solid line) and for various pump-probe time delays: $t_p = 0$ (dashed curve), $t_p = 120$ fs (dotted curve). The probe detuning is plotted in units of the bulk exciton Rydberg energy (4.2 meV in GaAs) with respect to the unrenormalized (zero-excitation) bandgap. The pump detuning is 9 E_B below the exciton (after Ref. 15.6).

is due to the process of ultrafast adiabatic following. The coherently excited carriers reduce the exciton oscillator strength via many-body effects, causing exciton bleaching. The time dependence of the light-induced shift follows the pump-pulse amplitude, even though the characteristic medium response times are much longer than the pulse duration.

15-7. Coherent Oscillations

At early times, just before the arrival of the pump pulse, interesting coherent effects can be observed. Figure 15.16 shows a series of probe differential transmission spectra obtained in a 0.5-μm-thick GaAs sample at 15 K, when the pump-pulse photon energy is well above the bandgap energy. The curves are 100 fs apart and centered around the exciton resonance. For early delays there are symmetric oscillations around the exciton. The amplitude of the oscillations increases with decreasing time delay between the pulses, and the frequency of the oscillations varies proportional to the time delay. With increasing time, the oscillations gradually disappear, and for zero time delay, the central peak assumes the spectral shape of the exciton, characteristic of the bleaching of the exciton resonance.

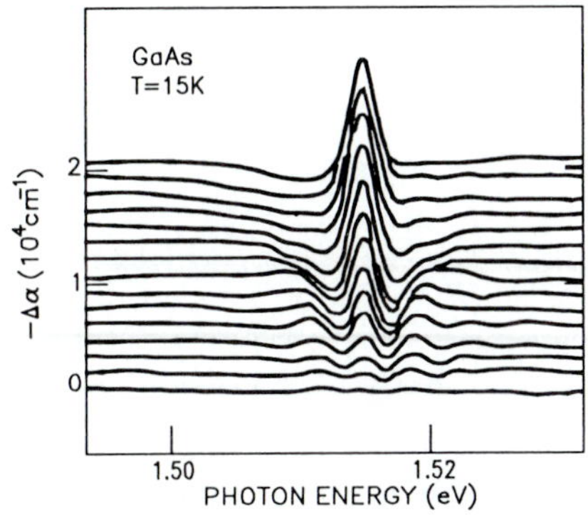

Figure 15.16. Measured DTS around the exciton resonance in a 0.5-μm molecular-beam-epitaxy-grown layer of GaAs at 15 K, for pumping above band. The pump pulse is centered around 620 nm with 60-fs duration. The spectra are taken 100 fs apart (after Ref. 15.9).

The origin of the oscillations is a transient grating coherently generated by the pump and probe pulses together. The generation of this grating does not necessarily require that pump and probe pulses overlap, but it is sufficient to have overlap of the effective field (15.18). This means that rather than requiring temporal pulse overlap, the occurrence of coherent effects requires only the temporal overlap of pulse-induced polarizations. In femtosecond experiments, the induced polarization may exist for much longer times than the pulse itself, since the polarization decay is determined by the dephasing time, which may be as long as a few picoseconds.

The semiconductor Bloch equations (15.16) and (15.17) show that the polarization leads to carrier generation, just as the applied field. Hence, the interference of the probe-induced polarization with the pump pulse leads to the generation of a population, which exhibits the interference structure. As discussed in Sec. 14-8, such an interference is called a population grating. This grating leads to the scattering of part of the pump pulse into the probe direction, and this is measured by the detector; see the discussion in Sec. 15-3 and Eq. (15.10).

In order to compute the differential transmission spectrum, one has to solve the semiconductor Bloch equations for the case of time-delayed pump and probe pulses. For the situation in Fig. 15.16, where the semiconductor is excited very high into the band and no spectral overlap between the excitation pulse and the exciton resonance exists, the dominant interaction between the exciton and the pump pulse is via the many-body effects caused by the electron-hole excitations. This case can be modeled simply assuming that the bleaching is due to a time-dependent dipole damping, i.e., by letting

$$\left.\frac{\partial P(\mathbf{k})}{\partial t}\right|_{col} \rightarrow -\gamma(t)\, P(\mathbf{k})\ , \tag{15.19}$$

in Eq. (15.15) where

$$\gamma(t) = \gamma_0 + \delta\gamma(t)\ . \tag{15.20}$$

For this case we obtain the differential transmission changes for negative time delays t_p (probe before pump) as

$$\mathrm{DTS} \propto \mathrm{Re}\left[\frac{1}{\gamma_0 - i\Delta}\, e^{(-i\Delta+\gamma_0)t_p}\right]\ , \tag{15.21}$$

where Δ is the detuning from the exciton resonance. Equation (15.21) describes a damped sinusoidal oscillation with the argument Δt_p. The evaluation of Eq. (15.21) is shown in Fig. 15.17, demonstrating good agreement with the experimental results in Fig. 15.16. The oscillation period gets longer as the pump-probe delay approaches zero.

Oscillations are not only observed in the spectral region around the exciton when the pump is high above the bandgap. They also appear in the frequency region around the pump when pumping above the bandgap, and around the exciton when pumping in the transparency region below the exciton.

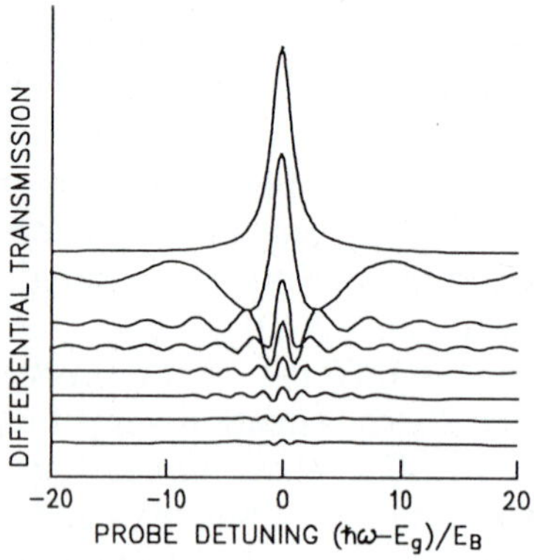

Figure 15.17 Differential transmission spectra calculated in the vicinity of the exciton resonance ω_x. The pump is high into the band with $\omega_{pump} >> \omega_x$. The curves are in 50-fs intervals with the bottom trace for -350-fs and the top trace for 0-fs time delays (after Ref. 15.9).

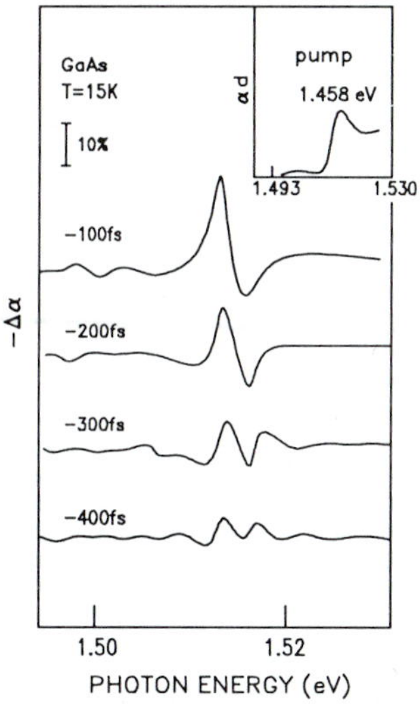

Figure 15.18. Measured DTS in the spectral region of the exciton for a 0.5-μm-thick bulk GaAs sample excited below the exciton resonance. The sample temperature is 15 K, and the pump pulse is centered around ≃ 1.49 eV with 200-fs duration. Delay times are -400, -300, and -200 fs (after Ref. 15.9).

As an example, Fig. 15.18 shows the differential transmission for pumping below the exciton in a thin CdS platelet at 150 K. We observe coherent transmission oscillations, which in this case develop into a dispersive feature characteristic of the blue shifting of the exciton resonance. The oscillations are now centered around the exciton with increasing amplitude as the time delay approaches zero. However, unlike the case for resonant pumping, the shape of the oscillations is asymmetric, always with the largest positive peak on the side toward the pump and the largest negative peak away from the pump. The oscillations later develop into the dispersive-looking feature, which shows the Stark shift of the exciton resonance. For more details see Ref. 15.9.

15-8. Photon Echo

In optical systems, which are inhomogeneously broadened, the phenomenon of photon echo occurs. Inhomogeneous broadening means that different transitions, such as those between the different **k** states in a semiconductor, with different energies $\mathcal{E}(\mathrm{k})$, contribute to an optical transition. If one excites such a system with a pulse at time $t = -t_p$, and then with another pulse at time $t = 0$, the system responds with an echo pulse at the time $t = +t_p$. The physical origin of this phenomenon is as follows. The first pulse (partially) excites a coherent superposition of many states of the system. Each state develops in time according to its transition energy. The instantaneous state is proportional to $\exp(i\mathcal{E}(k)t/\hbar)$; i.e., the different transitions are out of phase. The second pulse (partially) reverses this temporal evolution, which has progressed for the time period t_p between the pulses. Hence, at the time t_p after the second pulse, the states are (partially) back in phase, generating the echo pulse.

The measurement of photon echoes is a special example of the more general class of time-resolved multiwave mixing experiments, sketched in Fig. 15.19. The first pulse, E_1, is incident at time $t = -t_p$, and the second pulse, E_2, peaks at $t = 0$. The echo is observed most easily in the direction $2\mathbf{k}_2 - \mathbf{k}_1$, the so-called four-wave-mixing or photon-echo direction, where the momentum conservation law is satisfied. Such photon echoes are observed, for example, for femtosecond excitation of the semiconductor continuum states, since these states are intrinsically inhomogeneously broadened [the transition energy $\mathcal{E}(k)$ is different for different k].

Figure 15.20 shows an example of the numerical evaluations of the full semiconductor Bloch equations for the photon echo configuration. The time-resolved signal in the photon echo direction is calculated for the case of a CdSe sample, excited above the bandgap. Two pulses of 100-fs duration (four-wave harmonic mixing) in the geometry of Fig. 15.19 are used with 400-fs delay between the pulses. The arrows in Fig. 15.20 indicate the positions of the peak of the two incident pulses in time with one pulse arriving at the sample at $t = -400$ fs and the second at $t = 0$. Figure 15.20 shows that a photon echo signal is generated at $t = 400$ fs.

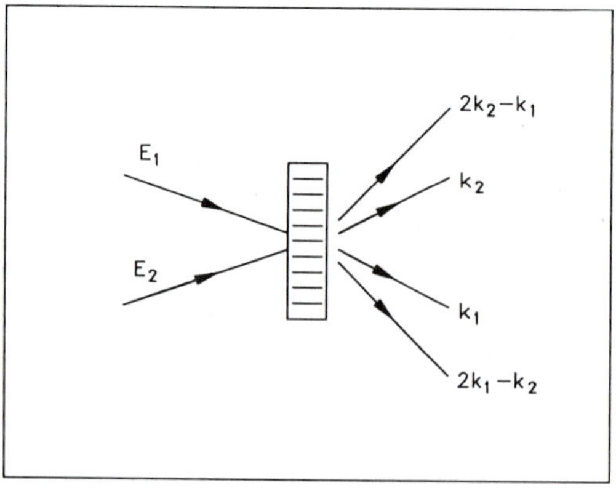

Figure 15.19. Schematic plot of a four-wave-mixing experiment. The photon echo signal is measured in the direction $2\mathbf{k}_2 - \mathbf{k}_1$.

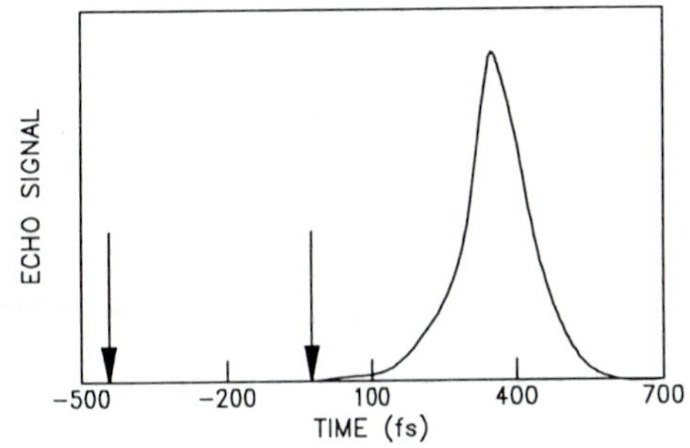

Figure 15.20. Time-resolved signal in photon echo direction for excitation above the bandgap of a CdSe sample. The dephasing time is 200 fs, the time delay is -400 fs, and the pulse four-wave harmonic mixing is 100 fs for both pulses. The peak value of the dipole coupling energy of the second pulse is $d_{cv} E_2 = 0.1\ E_B$, where the exciton binding energy $E_B = 16$ meV in CdSe and the peak amplitude of the first pulse is $d_{cv} E_1 = 0.1\ E_B$. The peak of the pulses in time are marked by arrows (after Ref. 15.10).

The echo signal occurs as a result of the intrinsic inhomogeneous broadening of the electron-hole continuum states.

15-9. Problems

15.1. Assuming Gaussian pulses with frequency bandwidth $\delta\nu$ and temporal width δt, calculate $\delta\nu \times \delta t$ for transform-limited pulses.

15.2. The autocorrelation technique measures the convoluted pulse duration. That is, if the autocorrelation trace has a four-wave harmonic mixing of δt, the true pulse duration would be $p\delta t$, with p being a constant. Calculate the constant p for Gaussian pulses.

15.3. Plot Eq. (15.5) for N = 20.

15.4. Show that the linearized Eq. (15.15),

$$\frac{\partial P_{\mathbf{k}}}{\partial t} = -\,i\,\frac{\hbar^2 k^2}{2m_r} P_{\mathbf{k}} + i \sum_{\mathbf{q} \neq \mathbf{k}} V_{|\mathbf{k}-\mathbf{q}|} P_{\mathbf{q}} \,, \tag{15.22}$$

is the Fourier-transform of the Wannier equation (6.16).

15-10. References

1. For a description of various techniques for femtosecond pulse generation, see, e.g., Special Issue of IEEE J. Quantum Electron., October 1992.

2. J. L. Oudar, D. Hulin, A. Migus, A. Antonetti, and F. Alexandre, Phys. Rev. Lett. **55**, 2074 (1985).

3. W. H. Knox, C. Hirlimann, D. A. B. Miller, J. Shah, D. S. Chemla, and C. V. Shank, Phys. Rev. Lett. **56**, 1191 (1986).

4. B. Fluegel, A. Paul, K. Meissner, R. Binder, S. W. Koch, N. Peyghambarian, F. Sasaki, T. Mishina, and Y. Masumoto, Solid State Commun. **83**, 17 (1992).

5. D. Fröhlich, A. Nöhte, and K. Reimann, Phys. Rev. Lett. **55**, 1335 (1985).

6. R. Binder, S. W. Koch, M. Lindberg, N. Peyghambarian, and W. Schäfer, Phys. Rev. Lett. **65**, 899 (1990).

7. A. Von Lehmen, D. S. Chemla, J. E. Zucker, and J. P. Heritage, Opt. Lett. **11**, 609 (1986).

8. B. Fluegel, N. Peyghambarian, M. Lindberg, S. W. Koch, M. Joffre, D. Hulin, A. Migus, and A. Antonetti, Phys. Rev. Lett. **59**, 2588 (1987).

9. S. W. Koch, N. Peyghambarian, and M. Lindberg, J. Phys. C **21**, 5229 (1988), review article.

10. M. Lindberg, R. Binder, and S. W. Koch, Phys. Rev. A **45**, 1865 (1992); see also, S. W. Koch, A. Knorr, R. Binder, and M. Lindberg, Phys. Status Solidi B **173**, 177 (1992).

Chapter XVI
ALL-OPTICAL NONLINEAR DEVICES

The optical nonlinearities discussed in Chaps. XIII and XIV may be used to demonstrate optical switches, gates, and bistable elements under suitable conditions. For most of these devices, the nonlinearity of the material is combined with a feedback mechanism for their operation (see Refs. 16.1-16.5). The most common device is a nonlinear etalon consisting of a nonlinear medium between two partially reflecting mirrors. Nonlinear waveguide devices have also been studied. In a waveguide structure, the laser beam (or the wave) travels parallel to the plane of the structure. This is in contrast to the nonlinear etalons where the wave propagates normal to the surface of the structure.

In this chapter we review various applications of the optical nonlinearities, both in etalons as well as in guided-wave structures.

16-1. Optical Bistability

Before we review the application possibilities of semiconductor materials as fast switches or nonlinear devices, we briefly discuss the basic features of the nonlinear transmission characteristics of a semiconductor etalon. Here the best known effect is the optical bistability or optical hysteresis. A device is said to be optically *bistable* if two stable output states for the transmitted light intensity exist for the same value of the input intensity over some range of input values. One further expects the nonlinear device to respond by changing abruptly from one value of the output to the other. For example, in Fig. 16.1(a) the intensity transmitted through a bistable device is plotted as a function of incident intensity. For an increasing incident intensity I_0, the transmission of the device remains low until I_0 is higher than some critical value $I_\uparrow$. The transmission then remains high even if I_0 is decreased below $I_\downarrow$ until another critical value $I_\downarrow (<I_\uparrow)$ is reached, and the device jumps to the low (off) state. This type of hysteresis is the signature of optical bistability. The actually realized transmission value depends on the excitation history. The system is in a different state depending on whether the input intensity is increased from zero or decreased from a sufficiently high level.

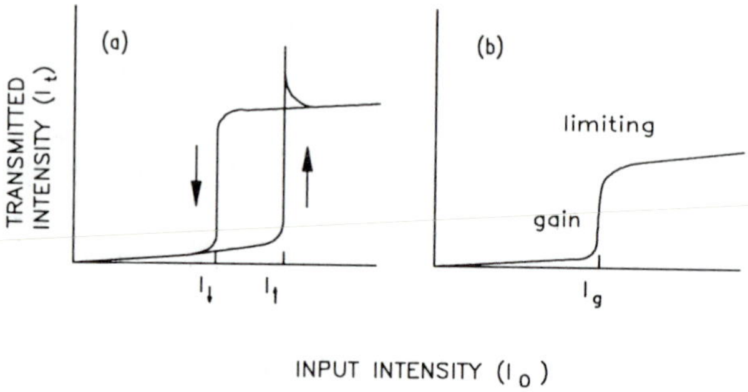

Figure 16.1. Transmission of a typical bistable optical device under conditions of (a) bistability (memory) and (b) high ac gain (optical transistor, discriminator, or limiter).

Optical bistable devices can perform a number of logic functions, such as optical memory, optical transistor, optical discriminator, optical limiter, optical oscillator, optical gate, etc. Figure 16.1(a) demonstrates how such a device can serve as an optical memory element, since either one of two output states can be maintained by an input beam of intensity between the two critical values $I_\uparrow$ and $I_\downarrow$. Under slightly modified operating conditions, the same device can be used as an optical transistor, as shown in Fig. 16.1(b). For intensities close to I_g, a small change in the input produces a large change in the output. The device in Fig. 16.1(b) can also be used as an optical limiter, since for input intensities above I_g, increasing the input does not change the output. Finally, such a system can serve as an optical discriminator; input intensities above I_g are transmitted with little attenuation, while those below I_g are highly attenuated.

Optical bistability requires a nonlinear medium with a feedback. Depending on whether the feedback is provided electronically or optically, the bistability is hybrid or all-optical, respectively. In a hybrid bistable device, the output voltage of a detector that senses the transmitted intensity is fed back to the device, which is usually an electro-optic modulator. On the other hand, in an all-optical device, the feedback is either intrinsic or provided by external mirrors. All-optical bistability can be absorptive or purely dispersive.

In the case of *absorptive bistability*, a saturable medium is confined between two partially reflecting mirrors. The laser frequency is tuned to an absorption resonance of the material. The Fabry-Perot cavity is also tuned to the absorption resonance. At low intensities, absorption in the medium is high, and transmission of the bistable etalon is low. With increasing intensity, the absorption decreases because of saturation and the intracavity intensity increases, which causes the absorption to decrease further. This positive feedback continues until the transmission jumps discontinuously to a high value, corresponding to the bleaching of the resonance at a certain switch-up intensity $I_\uparrow$. As this intensity is further increased, transmission does not increase appreciably because the transition is now fully saturated. If the intensity is decreased, the transmission stays high even below $I_\uparrow$ because the intracavity intensity is still high due to the storage property of the Fabry-Perot cavity. Decreasing intensity beyond $I_\downarrow$ causes the absorption in the cavity to set in and the transmission to go down.

Dispersive bistability uses a material with excitation-dependent refractive index inside a Fabry-Perot etalon. The laser frequency is initially detuned from the Fabry-Perot cavity peak frequency. By increasing the laser intensity, the Fabry-Perot peak frequency shifts toward the laser frequency, since the index of refraction of the material changes with laser irradiance. Switch-on from low to high transmitted intensity occurs when the Fabry-Perot peak sweeps through the laser frequency.

The operation of optical bistability may be simply described by a graphical method. Consider a nonlinear material with index of refraction $n(\omega) = n_0(\omega) + \Delta n(\omega)$ inside a Fabry-Perot etalon. The transmission of the etalon is determined by the equation

$$\mathcal{T} = \frac{I_t}{I_0} = \frac{(1 - R)^2 \, e^{-\alpha d}}{(1 - R\, e^{-\alpha d})^2 + 4Re^{-\alpha d} \sin^2(\phi_0 + \phi_n)} , \tag{16.1}$$

where I_0 and I_t are the incident and transmitted light intensities, respectively. The reflection coefficient of the mirrors is denoted by R. In many practical applications, these mirrors are actually the end faces of the semiconductor crystal itself, and R is just the natural reflectivity. R can be increased through additional high reflectivity coatings evaporated onto these surfaces. The quantity ϕ_0 is the linear round-trip phase shift of the light in the resonator for a weak light intensity, d is the length of the medium, and $\alpha = \alpha(\omega)$ and $\phi_n = \omega \Delta n(\omega) d/c$ are the carrier-density-dependent absorption and phase changes, respectively. The Fabry-Perot transmission depends on the carrier density n through α and $\Delta n(\omega)$. The carrier density in turn is determined by the intensity I_{in} of the light inside the resonator through a rate equation of the form [see Eq. (14.1)]

$$\frac{dn}{dt} = -\frac{n}{\tau} + \frac{\alpha(\omega, n)}{\hbar\omega} I_{in} . \tag{16.2}$$

The total round-trip phase shift $\phi = \phi_0 + \phi_n$ is carrier-density-dependent through the density dependence of $\Delta n(\omega)$. The carrier density and light intensity are related through Eq. (16.2). Note that in steady state (dn/dt = 0), Eq. (16.2) reads $n = \alpha I_{in} \tau/\hbar\omega$, so that carrier density n and laser intensity I_{in} are linearly dependent as long as the density dependence of α is ignored. To discuss the simplest possible case, we assume that the refractive index change $\Delta n(\omega)$ is linearly dependent on the carrier density and, therefore, also on laser intensity inside the cavity:

$$\phi = \phi_0 + \phi_n = \phi_0 + \phi_2 I_t . \qquad (16.3)$$

Here we have taken $\phi_n = \omega \Delta n(\omega) d/c = \phi_2 I_t$, noting that the laser intensity inside the cavity and the transmitted intensity are proportional. Using I_t from Eq. (16.3), the transmission of the Fabry-Perot may also be written as

$$\mathcal{T} = \frac{I_t}{I_0} = \frac{\phi - \phi_0}{\phi_2 I_0} . \qquad (16.4)$$

The steady-state solution is reached when both Eqs. (16.1) and (16.4) are satisfied. These two equations are plotted in Fig. 16.2(a) as a function of the nonlinear phase shift, $\phi - \phi_0 = \phi_n$. The solution of Eq. (16.1) is given by the typical Fabry-Perot transmission peaks. Equation (16.4) as a function of $\phi - \phi_0$ is a straight line with a slope that is inversely proportional to incident intensity I_0. For input intensities below that of line A in Fig. 16.2(a), the two curves intersect at only one point. This point is a solution of the two equations and represents the lower state of the system. In a plot of I_t versus I_0, as shown in Fig. 16.2(b), these states

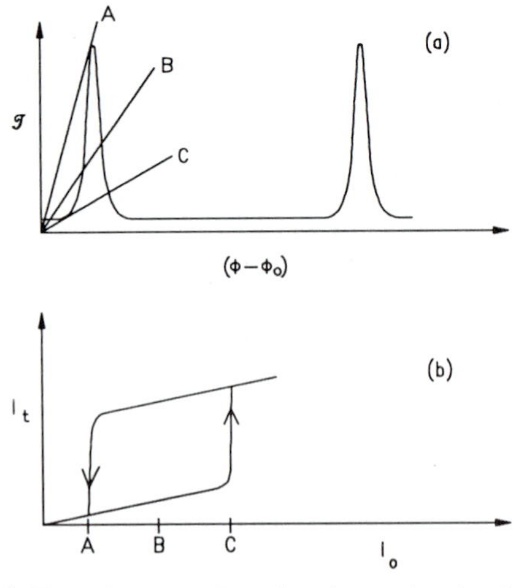

Figure 16.2. (a) The solution of Eqs. (16.1) and (16.4). (b) Transmitted versus input intensity graph. The laser intensities denoted by A, B, and C correspond to the straight lines in Fig. 16.2(a).

are represented by the curve between the origin and point A. As the input intensity increases, the slope of the straight line decreases. For intensities between A and C (such as line B), the two curves intersect at three points, showing that there are three output intensities for the same input. Analysis of the dynamics of the system indicates that in this region the system is metastable (see Ref. 16.1). Switching to the higher state occurs when the input intensity reaches C. At that intensity, again there is only one intersection between the two curves which gives the higher state. Figure 16.2(b) shows the switch-on when intensity reaches point C. If one reduces the intensity with the system being in the higher state, switching down does not occur until the intensity is lowered to point A. As before, for intensities between A and C, the system is unstable and stays in the higher state. At point A there is again only one intersection between the two curves.

The numerical solution of optical bistability may be obtained by solving the coupled Eqs. (16.1) and (16.2). The incident intensity I_0 may vary in time, as in the case of an excitation pulse, but the temporal variation has to be slow on the time scale of the resonator round-trip time for steady state operation. Examples of the results for the case of GaAs are shown in Fig. 16.3(a)-(c) for different temporal widths of the exciting pulse and in Fig. 16.3(d) for the steady state. The curves show well-developed hysteresis loops whose size increases if the temporal pulse width decreases. This is the well-known effect of dynamical hysteresis.

Increasing-absorption optical bistability is based on an *increase* of the absorption at the operating wavelength with an increase of laser intensity. The increase of absorption can be the result of a shift of a resonance or a band edge by thermal effects, or by the creation of an electron-hole plasma. The laser is initially tuned to the low-absorption region of the spectrum (on the low-frequency side of the band edge) to give high transmission. With an intensity increase, the band edge shifts toward the operating wavelength, which gives rise to an increase in absorption and a further shift of the band edge. This intrinsic positive feedback is sufficient for switching to a low-transmission state, and no external feedback is required. Increasing absorption bistability is characterized by a backward loop in the transmitted-versus-input intensity ($I_{\downarrow} > I_{\uparrow}$).

Optical bistability of electronic origin has been observed in semiconductor materials, such as bulk GaAs and GaAs MQWs, InSb, CdS, CuCl, etc. Semiconductor etalons have the attractive property of providing large optical nonlinearity in very short ($\simeq 1$-μm) lengths. A short length provides a faster cavity-buildup time, causing a fast switching time if all other time constants are short enough. It also permits tighter focusing before beam walk-off losses become significant, thus reducing input powers and switching energies. The flat parallel faces of the

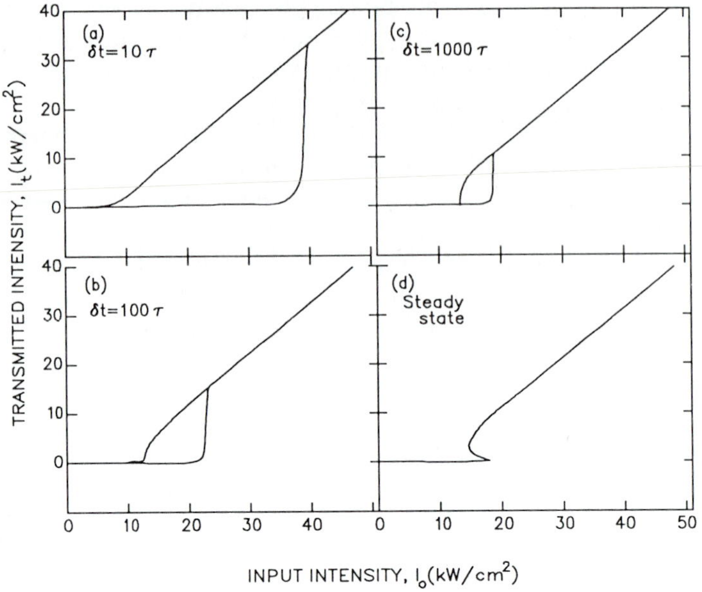

Figure 16.3. Transmitted intensity versus input intensity computed for a $\simeq 2$-μm GaAs etalon at room temperature for the excitation energy $\hbar\omega = 1.4032$ eV, well below the exciton resonance at 1.420 eV. Parts (a)-(c) are obtained assuming pulsed excitation with a triangular pulse of full width 10 τ (a), 100 τ (b), and 1000 τ (c). Part (d) shows the steady-state results, exhibiting the typical s-shaped characteristics. Note that the branch with negative slope is unstable. The frequency detuning of the laser from the Fabry-Perot frequency is -14.2 meV. τ is the carrier lifetime, and the mirror reflectivity R = 0.9 (after Ref. 16.6).

semiconductor are high reflectivity dielectrically coated with multiple layers of dielectrics of high reflectivity, forming a Fabry-Perot etalon. Figure 16.4 shows an example of optical bistability at room temperature measured for a GaAs MQW etalon.

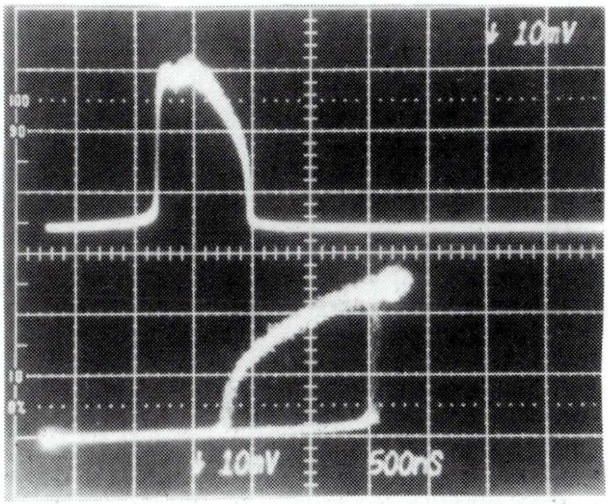

Figure 16.4. Room-temperature optical bistability in a GaAs-AlGaAs multiple quantum well (MQW) etalon. The top trace is the transmission of the device (output) versus time. The bottom trace is the output versus input, showing hysteresis. Input is a triangular pulse (after Ref. 16.1).

16-2. Nonlinear Optical Logic Gates

In addition to the optical bistable devices operating with a single beam, operation of nonlinear etalons using two light beams, with the additional flexibility of two wavelengths, has demonstrated various optical logic operations such as AND, OR, and NOR. These logic functions are often characterized by "truth tables." These tables list the output state S of the gate (0 for low, 1 for high) in the presence (1) or absence (0) of the input beams A and B. Examples for such truth tables are:

A	B	S
0	0	0
0	1	0
1	0	0
1	1	1

for an AND gate, and

A	B	S
0	0	0
0	1	1
1	0	1
1	1	1

for an OR gate. These tables show that the state S is high (1) only if beams A *and* B are high for the AND gate, and that S is high if A *or* B is high in an OR gate. The realization of such logic gates in an optical system is schematically shown in Fig. 16.5.

For the experimental realization of these optical logic gates, one exploits the optically induced shift of the Fabry-Perot transmission peak in response to input pulses. The nonlinear medium must be such that the absorption of one input pulse changes the refractive index at the wavelength of the other pulse sufficiently so that the spectral shift of the transmission peak of the etalon is about one instrument width. For example, if the wavelength is initially tuned to a transmission peak of the etalon, a NOR-gate operation results if the control pulse is able to detune the etalon from the probe wavelength. Figure 16.6 shows a demonstration

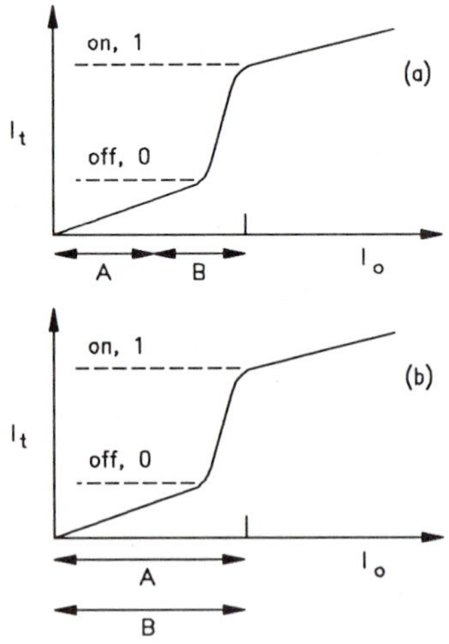

Figure 16.5. (a) Operation of an AND gate. Both beams of intensity A and B have to be present for the output to be high (on, 1). If no beam or only one beam is present, the output is low (off, 0). (b) OR gate; either A or B is sufficient to switch to high output.

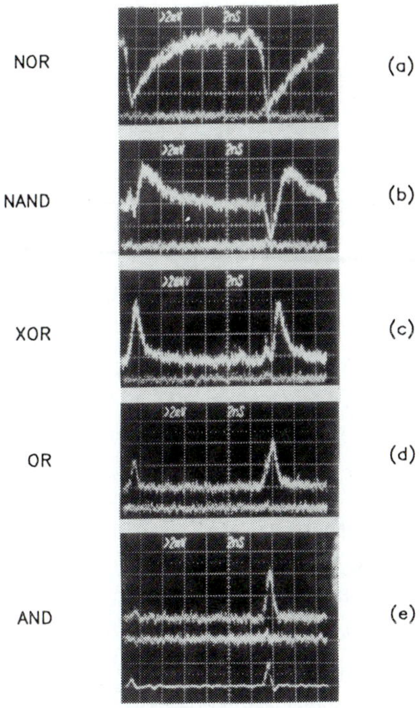

Figure 16.6. Logic operations NOR, NAND, XOR, OR, and AND in a GaAs etalon (after Ref. 16.8).

of logic NOR, NAND, XOR, OR, and AND gates using a GaAs etalon. The bottom trace in Fig. 16.6 displays the presence of input pulses at the gate in time. The first peak shows the appearance of one input pulse, and the presence of the second peak with twice the amplitude shows that two input pulses are applied. The AND gate of Fig. 16.6 does not switch until both pulses are applied. The straight line under each logic operation curve shows the zero line.

The switching speeds of the gate depend on the carrier lifetime when carrier excitation is the origin of the nonlinearity. For example, in Fig. 16.6 the recovery time of the NOR gate is slower than the switch-on time, since the carrier lifetime is in the nanosecond range.

A faster switching speed may be obtained by the use of the optical Stark effect. If the input-pulse frequency is tuned below the exciton, into the transparency region of the material, then the electromagnetic field of the laser may be used to switch the etalon. The advantage of this switching, exploiting the optical Stark effect, is that the carrier lifetime is no longer a limitation, since carrier generation is minimized and the system response directly follows the pulse (ultrafast adiabatic following) (see

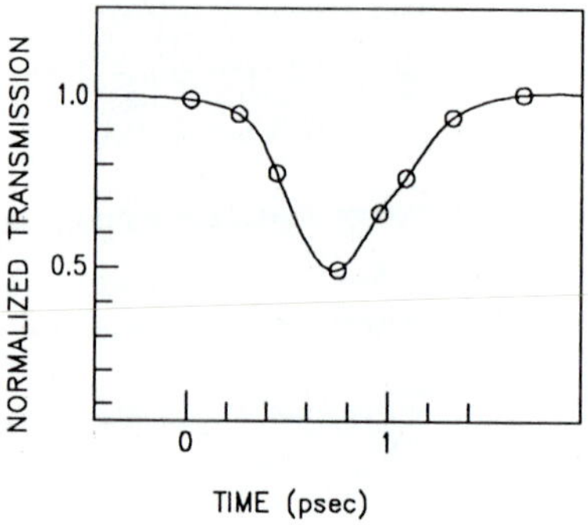

Figure 16.7. Picosecond switching of a GaAs logic gate. The optical Stark effect has been employed to obtain these switching speeds (after Ref. 16.9).

Chap. XV). The disadvantages of such high-speed devices are the higher input intensities required to realize them. A 1-ps recovery using the optical Stark effect has been demonstrated as shown in Fig. 16.7, which shows NOR-gate switch-on and switch-off times of $\simeq$ 1 ps.

16-3. Gain and Cascading in Nonlinear Etalons*

Some of the important requirements for optical bistable devices or nonlinear switches are cascadability, gain, fan-out capability, and good contrast. Cascadability refers to the operation by which the output pulse of the first device serves as the input to the next device. By fan-out one means the division of the output pulse of a device into N parallel output pulses, with each output pulse serving as an input pulse in the next stage of the system. These requirements are essential, because in order to cascade, signal levels have to be restored after each operation to drive the next state. For a reasonable fan-out, the output of one logic gate has to be used to switch several (at least two) logic gates. The contrast should be high enough so that signal-to-noise ratio can be kept high throughout the system.

For a nonlinear etalon operating in a bistable or thresholding mode (Fig. 16.8), the gain and the contrast can be defined in terms of the power of the light pulses as follows. Suppose a weak signal beam is amplified through a nonlinear etalon by a strong holding beam. We define the device gain as

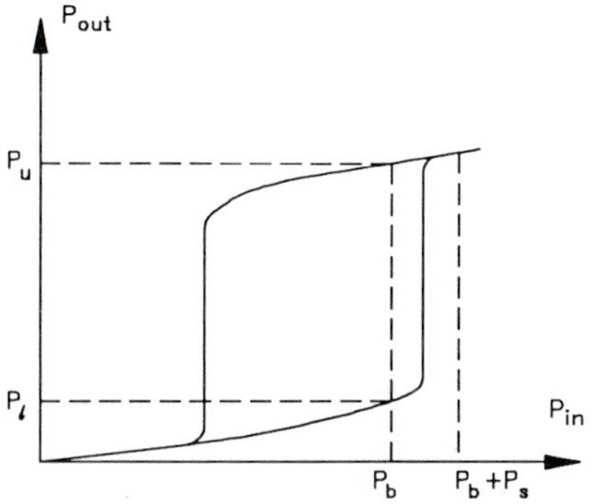

Figure 16.8. A bistable device as a latching logic gate. P_b is the bias power, P_s is the signal power, and P_u and P_ℓ are the upper- and lower-state output powers, respectively (after Ref. 16.6).

$$G_d = \frac{P_u - P_\ell}{P_s}, \tag{16.5}$$

where P_u is the output power when the device is in the upper bistable state, P_ℓ is the output in the lower state, and P_s is the signal-beam power, as shown in Fig. 16.8. This definition of G_d is suitable for the steady-state operation of a device, i.e., when the input pulse length is much longer than the device response time. Note that G_d is limited by practical considerations such as the stability of the laser amplitude and frequency, the etalon temperature stability, etc. Theoretically, G_d can approach infinity by decreasing P_s, but the switching time may increase.

The contrast of a logic gate is defined as

$$C_d = \frac{P_u}{P_\ell}, \tag{16.6}$$

representing the signal-to-noise ratio of a device. Thus, a large contrast is generally desired, especially for such applications as the nonlinear decision making in an optical associative memory. For an etalon operating in the transient mode, where the pulse length is comparable to the device response time, we can still define the device gain and the contrast in a similar way, with the powers in the above expressions replaced by energies.

The minimum gain required for cascading is $G_d > 1$. Generally, for a gate to have a fan-out of N, $G_d > N$ is needed if the output of the first gate is to be able to switch N succeeding gates by itself. This requirement can be reduced if the outputs of several gates are used to switch one gate (fan-in). The contrast must also be high enough that $P_u/N > P_s$, but with P_ℓ/N being too small to switch the device. This requirement in effect determines G_d. The number N of secondary states needed depends upon the architecture.

As an example, Fig. 16.9 shows the experimental setup for cascading with two etalons working in transmission as OR-gates where the transmitted signal of the first etalon (E1) is used to switch the second one (E2). A typical result of the cascading experiment is plotted in Fig. 16.10. Note that the results shown here are for the cases where the devices are operated in the bistable mode. It is also possible to adjust the experimental parameters (such as the etalon detuning) to make the bistable loop extremely narrow to approach the case of thresholding operation. The largest gain observed was 4, where a 0.25 mW change of the output signal of the first stage induced a change of 1 mW in the output from the second one. The contrast at these power levels was 5-8; it could be increased to approximately 10 with larger input powers (45-50 mW).

The differential energy gain varies as a function of the temporal pulse duration, as may be determined from numerical simulations or by experimentation. In simulations the differential energy gain is obtained for pulsed operation by assuming two-pulse excitation. One pulse, the so-called bias pulse, is used to bring the device close to the switching threshold, and the second pulse acts as the switch pulse. The differential energy gain is then obtained as the total transmitted energy (bias and switch pulses present) minus the transmitted energy in the presence of the bias, both divided by the switch-pulse energy. Examples of the results for Gaussian-shaped pulses are plotted in Fig. 16.11. The different curves show that as the lengths of the pulses are shortened, the differential energy gain decreases. Therefore, these studies indicate that passive nonlinear etalon devices cannot be expected to exhibit useful differential energy gain for operating pulses shorter than roughly $10\,\tau$, where τ is the carrier

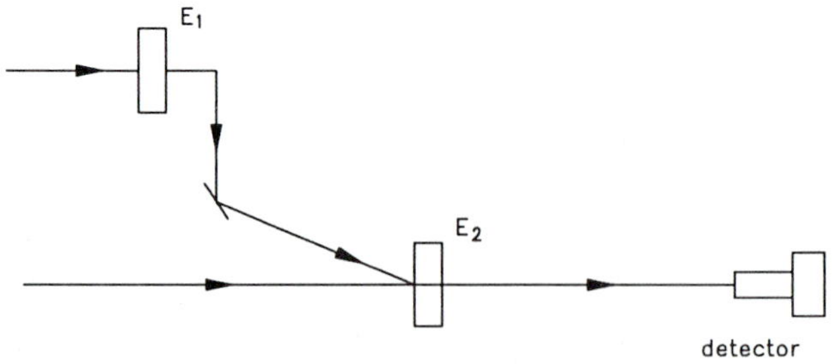

Figure 16.9. The experimental scheme for cascading in transmission. The etalon E_2 is biased below threshold. When the first etalon, E_1, switches to the high state, its transmitted light switches on the second etalon, E_2.

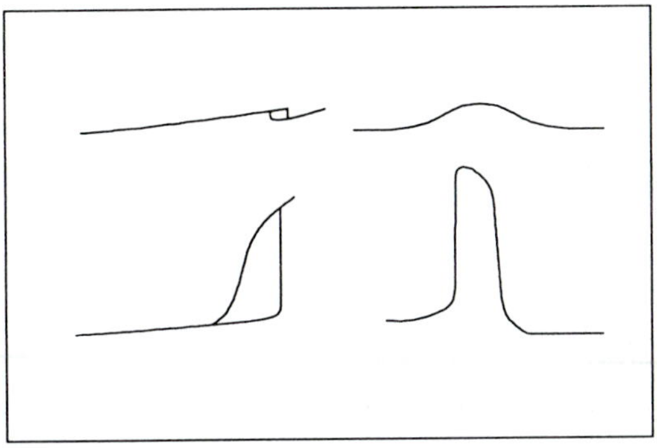

Figure 16.10. Results of the cascading experiment corresponding to Fig. 16.9. The traces on the left are the output versus input plots, and the traces on the right are the outputs from the second stage versus time. The upper traces represent the case when the first stage is "off," and the lower traces represent the case when the first stage is "on." The total input peak power for each device is about 30 mW, and the input pulse length is 1.5 μs (after Ref. 16.10).

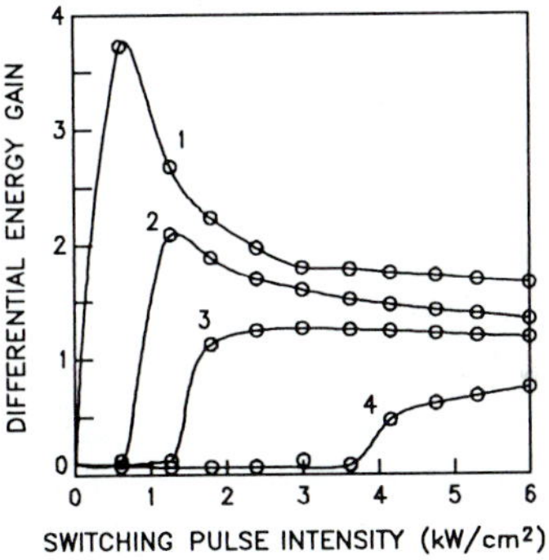

Figure 16.11. Differential energy gain versus switch-pulse intensity for different ratios of temporal pulse width δt to carrier relaxation time τ. The peak intensity of the bias pulse was kept just below the steady-state switch-up value. The parameters are (1) $\delta t/\tau = 100$; (2) $\delta t/\tau = 50$; (3) $\delta t/\tau = 25$; and (4) $\delta t/\tau = 10$ (after Ref. 16.10).

lifetime (Refs. 16.10 and 16.11). In order to achieve cascadable optical logic gates that can operate at a single wavelength with picosecond pulses, it is necessary to reduce the carrier lifetimes in the semiconductor material.

The behavior predicted by the simulation has been observed in semiconductor etalons. A decrease of the gain has been seen as the pulse length was shortened to several hundred nanoseconds. Efforts are pursued to reduce the carrier lifetime in GaAs MQW etalons, such as the use of thin samples without protective AlGaAs "windows" and the increase of surface recombination rates by etching of surface features. The reduction of pixel size also serves this purpose. Another promising approach could be the use of semiconductor microcrystallites.

16-4. Asymmetric Fabry-Perot Modulators*

An asymmetric etalon consists of a nonlinear medium sandwiched between two mirrors of different reflectivities. Such an asymmetric etalon may be used as a high-contrast reflection modulator. We consider an etalon with thickness d and with front and back mirror reflectivities of R_f and R_b, respectively. The etalon reflectance spectrum exhibits minima when the condition

$$m\lambda = 2n(\lambda)d \tag{16.7}$$

is satisfied, where m is the order number, λ is the wavelength, and $n(\lambda)$ is the wavelength-dependent refractive index. At resonance, the etalon normal reflectance is given by

$$R = \left[\frac{\sqrt{R_f} - \sqrt{R_b}\, e^{-\alpha(\lambda)d}}{1 - \sqrt{R_f R_b}\, e^{-\alpha(\lambda)d}} \right]^2 , \tag{16.8}$$

where $\alpha(\lambda)$ is the absorption coefficient of the medium. The reflectance goes through zero when the numerator of Eq. (16.8) is zero, i.e., when

$$R_b = R_f\, e^{2\alpha(\lambda)d} . \tag{16.9}$$

Therefore, the device must be designed with a proper thickness d such that Eq. (16.9) is satisfied and the reflectance reaches zero at the resonance wavelength. When a pump beam is applied to such a device, the condition given by Eq. (16.9) is violated (because the absorption coefficient is intensity dependent) and the reflectivity jumps to a high value. Theoretically, the contrast ratio may in principle become infinite when the original reflectivity is zero. However, the reflectivity does not reach zero, but only a small value, making the contrast ratio a finite but large value.

Figure 16.12 shows the calculated reflectivity for two values of $\alpha(\lambda)$. For curve 1, $\alpha(\lambda)$ is adjusted such that the reflectivity becomes zero at the resonance wavelength. Changing the absorption to a different value increases the reflectivity, making the reflection spectrum similar to that shown by curve 2 in this figure. This clearly demonstrates the high-contrast modulation capability of the device, since the intensity of a probe beam tuned to the reflectivity minimum may be changed from a low to a high value.

As an experimental example, we describe the result obtained for a high-contrast reflection modulator operated at $\simeq 1.3\ \mu m$ using GaAlInAs/AlInAs quantum wells. The etalon front mirror is formed by the air/spacer interface and has a value of $R_f \simeq 0.31$, and the quarter-wave stack back mirror has a reflectance of $R_b \simeq 0.92$. The total spacer thickness is $d = 1.027\ \mu m$. The modulator was designed such that the Fabry-Perot resonance is located on the long wavelength side of the heavy-hole exciton peak to take advantage of large absorptive and refractive nonlinearities induced by the pump beam. In the absence of the pump, the absorption near the heavy-hole exciton is high, thus resulting in a balanced cavity and near-zero reflectance at resonance. When the pump is present, the photogenerated electron and hole populations bleach the absorption near the heavy-hole exciton, resulting in an increase in the reflectance at resonance from unbalancing the Fabry-Perot cavity.

The nonlinear reflectance spectra at pump intensities of 0.0, 6.6, and 41.2 kW/cm^2 are shown in Fig. 16.13. In the linear spectrum (no pump) shown by curve 1, the minimum reflectance occurs at a wavelength of

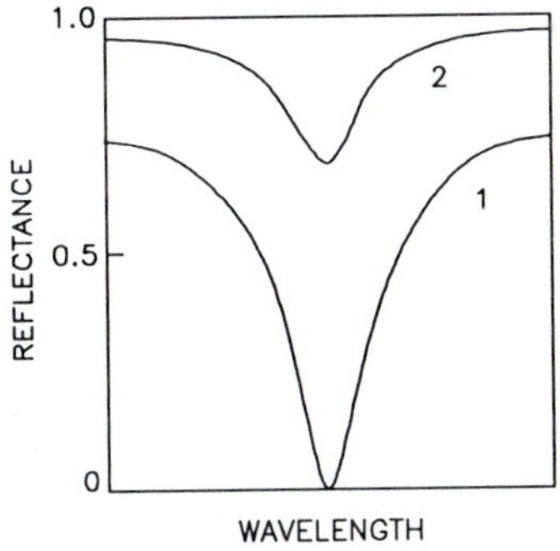

Figure 16.12. Calculated reflectance spectra of an asymmetric reflection modulator for two absorption values. For curve 1, the absorption is adjusted to give a zero reflectance at the Fabry-Perot peak. For curve 2, $\alpha(\lambda)$ is changed to a different value, resulting in larger reflectivity (see Ref. 16.12).

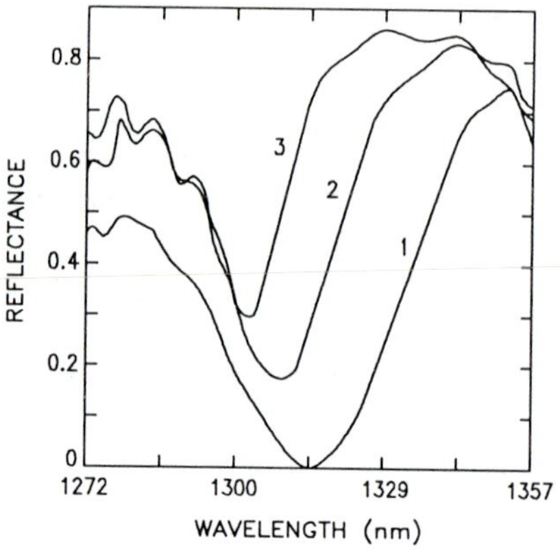

Figure 16.13. Measured reflectance spectra of the GaAlInAs/AlInAs asymmetric reflection modulator for pump intensities of (1) 0.0 (linear), (2) 6.6, and (3) 41.2 kW/cm^2 (after Ref. 16.12).

1314.3 nm, with a value of 0.00055. As the pump intensity increases to 41.2 kW/cm^2, the reflectance at the initial cavity resonance increases to a value of $\simeq 0.7$. As can be seen from curves 2 and 3 of Fig. 16.13, the increased reflectance is the result of the combined absorptive and refractive nonlinearities associated with bleaching the heavy-hole exciton, as indicated by the increase in reflectance and the shift of the resonance to shorter wavelengths. The complicated structure of the reflectance spectra at wavelengths below the resonance are a result of multiple reflections from the unpolished backside of the InP substrate. This device has an on/off contrast ratio of 1060:1, peaking at a wavelength of $\simeq$ 1314 nm.

16-5. All-Optical Waveguide Devices*

In contrast to the case of etalon devices, the light travels parallel to the surface in a waveguide device. There are several possible designs for semiconductor waveguides, such as planar guides, channel structures, ridge waveguides, and strip-loaded structures, to name only a few examples. The merits of the respective designs depend to a large extent on the desired operation and on the ease of fabrication.

In principle, semiconductor waveguide devices can perform useful operations that are not possible in etalon structures. One example is the so-called nonlinear directional coupler consisting of two closely spaced parallel guides [see Fig. 16.14(a)]. The light is inserted into only one of the guides, but it tunnels into the neighboring guide after a characteristic

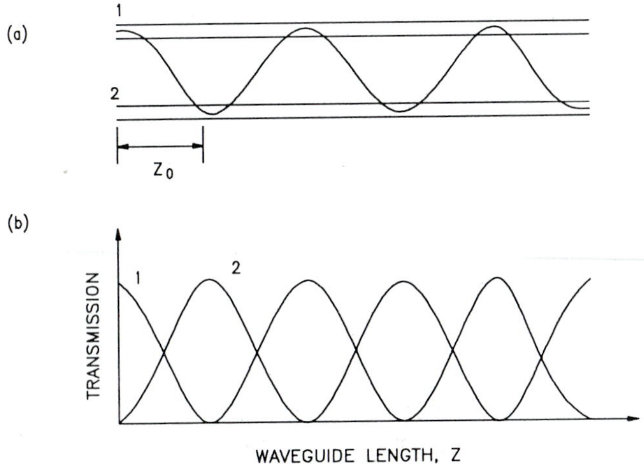

Figure 16.14. (a) Schematic of a nonlinear directional coupler. (b) Transmission from guides 1 and 2 as the length of the coupler is varied.

length [a length of Z_0 in Fig. 16.14(a)], which is the so-called beat length. The coupling of light from one guide to the other through the evanescent field is analogous to the operation of two coupled pendulums where the energy is switched back and forth between the two oscillators. Figure 16.14(b) shows the transmitted light from the two guides with the light transferring back and forth between them as the waveguide length is increased. For a directional coupler of length Z_0, a laser pulse with low intensity coupled into guide 1 completely switches to guide 2, as shown in Fig. 16.15. In this regime the device is a linear directional coupler. Higher intensities induce changes in the refractive index, destroying the balance of the coupler and resulting in the laser pulse staying in guide 1 [see Fig. 16.15(a)]. The beat length is a very sensitive function, not only of the design parameters such as separation between the guides, but also of the refractive indices of the guides. Since the refractive index changes with changing carrier density, the beat length changes too. Hence, for a directional coupler that has just the linear beat length, the light injected into one guide is totally coupled into the other guide for low light intensity, whereas it might stay in the original guide for sufficiently high intensities. An alternative description is shown in Fig. 16.15(b) for a laser pulse coupled initially to channel 1. As the input intensity is increased, the transmission of guide 1 decreases, while that of guide 2 increases until a complete energy transfer occurs. Figure 16.16 shows a scanning electron microscope picture of a GaAs-AlGaAs MQW directional coupler. The ridges are 2-μm wide and are separated from each other by 2 μm. The structure of the nonlinear directional coupler is shown in Fig. 16.17. The

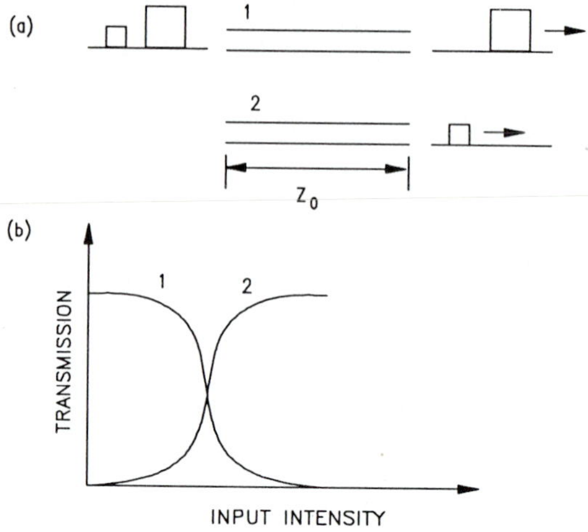

Figure 16.15. (a) Schematic of a nonlinear directional coupler with a length of Z_0 corresponding to one beat length. The low-intensity pulse couples from guide 1 to 2 with the high-intensity pulse staying in guide 1. (b) Transmission from the two guides as the input intensity is varied.

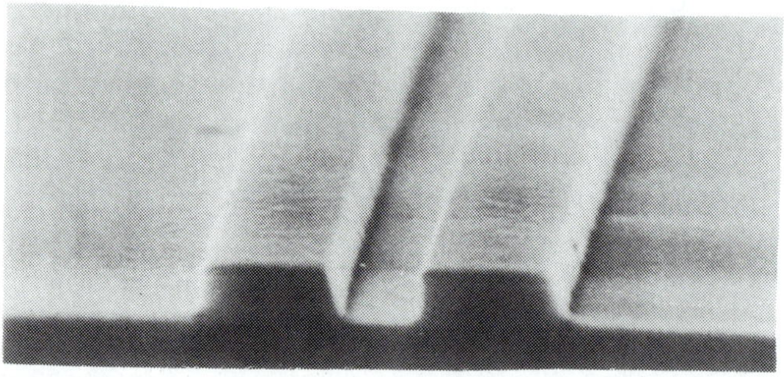

Figure 16.16. A scanning electron microscope graph of a cleaved end of a GaAs MQW nonlinear directional coupler with 1-μm spacing between the two channels (after Ref. 16.13).

guiding in the vertical direction is achieved by two AlGaAs layers with lower effective indices above and beneath the MQW region. Pairs of closely spaced ridges on the top AlGaAs layer result in channel definition of the nonlinear directional coupler. The samples are then cleaved to lengths of the order of 1 to 3 mm.

A typical switching curve showing relative intensity distribution at the output of the two channels as a function of the input intensity is shown in

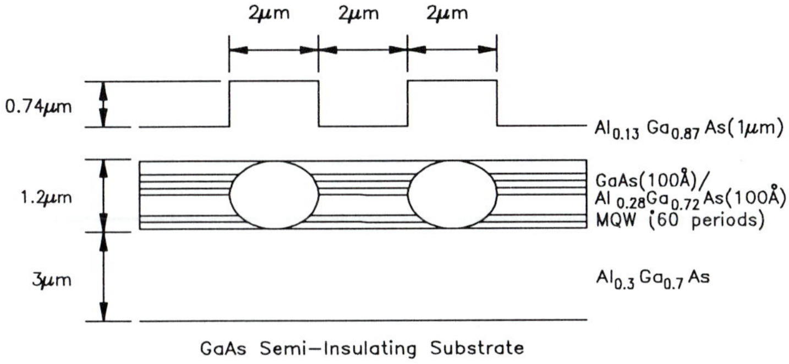

Figure 16.17. Structure of the molecular-beam-epitaxy-grown nonlinear directional coupler. The total thickness of the top AlGaAs layer is 1 μm (after Ref. 16.14).

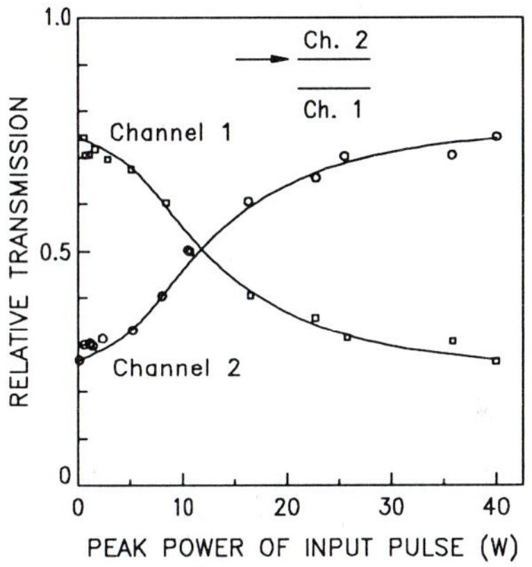

Figure 16.18. Relative transmission of the two channels as a function of pump intensity at a fixed time delay. The intensities are normalized by the transmitted total intensity. The inset shows the geometry of the beam propagation direction (after Ref. 16.14).

Fig. 16.18. The inset of this figure schematically shows the geometry for the operation of the nonlinear directional coupler. As expected at low pump intensities, the probe light that was originally coupled into channel 2 is cross coupled and exits from channel 1. Due to the use of pulsed excitation, and since the length of the waveguide is not completely optimized, not all the probe light is coupled to channel 1. As the pump intensity is increased, the coupling back from channel 1 to channel 2 takes place.

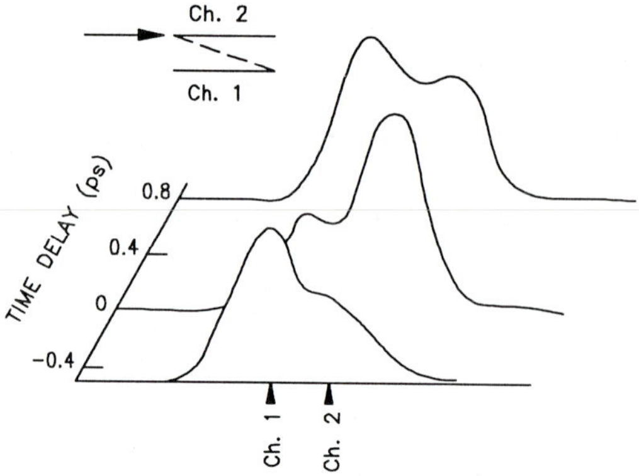

Figure 16.19. Nonlinear directional coupler output intensity profiles for different time delays. The pump intensity was about 2 GW/cm^2 (after Ref. 14.4).

The dynamics of this nonlinear directional coupling operation may be measured in a time-resolved pump-probe experiment. Figure 16.19 displays such an operation, where the output intensity profiles are plotted for different time delays between the pump and the probe pulses. The optical nonlinearity associated with the optical Stark effect was employed for this measurement. Thus, the laser frequency was tuned to the transparency region below the exciton resonance. As shown in Fig. 16.19, at large negative time delays for which the probe pulse precedes the pump pulse in time, most of the probe exits from the device through channel 1 after cross coupling. At zero time delay it is switched back to channel 2 as a result of pump action. For large positive delays it recovers its initial path. Thus, it is possible to obtain a *transient* (i.e., < 1-ps recovery) crossover in the output intensity ratio of the two channels at a pump intensity of approximately 2 GW/cm^2.

16-6. Trade-Offs of All-Optical Devices*

All of the optical nonlinear devices and logic gates that have been described in this chapter follow some common rules. For example, there is a trade-off between speed and the operating power of the device and between the device transmission and the nonlinear index change, which in turn controls the operating power (the larger the index change, the smaller the required power). These fundamental trade-offs may be understood simply from Eq. (16.2). In steady state, this equation reads

$$n = \alpha(\omega, n)\, I_0 \frac{\tau}{\hbar\omega} . \qquad (16.10)$$

Here n is the carrier density, τ is the lifetime of the excited carriers, α is the absorption coefficient, which is carrier-density dependent, and $\hbar\omega$ is the photon energy of the incident light of intensity I_0. (We assume that the intensity inside the sample is the same as the incident intensity.) For switching applications a certain value of $\Delta n(\omega, n)$ is required, which can be obtained with a given value of carrier density n. Equation (16.10) shows that if one reduces τ (increasing the speed of the device), then I_0 needs to be increased in order to keep n constant (power-speed trade-off). Similarly, higher device transmission is possible by moving the operating wavelength away from the exciton peak so that the tail absorption decreases. However, detuning farther from the exciton also reduces $\alpha(\omega, n)$ and, consequently, a smaller value of n is reached, which leads to a smaller nonlinearity and higher required power.

The advantage of nonlinear etalons are their small size, availability of large two-dimensional arrays of etalons, and their high speed. Their disadvantages are the absence of picosecond gain and large fan-out at the present stage of development.

Different applications of nonlinear devices place quite different requirements on the bistable etalons. For example, some applications require that large arrays of pixels (probably > 100 × 100) be operated simultaneously, which means that good uniformity in a relatively large etalon area (on the order of 1 cm^2) is needed. Sometimes one may be able to reduce the speed requirement for a single gate in exchange for a lower operating power. High throughput may be achieved using simple interconnection patterns. Applications along this line also include spatial-light modulators with sub-microsecond (single wavelength, cascadable) or even picosecond (two wavelength) addressing times. Another class of application is high-speed sequential processing, which takes advantage of the fast switching speed of a single device. With GaAs optical logic gates, the data rate of a single channel can be 1 to 10 Gbit/sec (assuming 1000-100 ps pulse spacing). Although the problem of picosecond cascading of passive nonlinear devices is not yet solved, it is conceivable that their fast switching speeds may be used to make logic decisions (such as gating and wavelength conversion), provided the output signals are amplified at the same rate so that the overall bandwidth of the system can still be maintained. The development of high-speed and high-gain laser amplifiers seems promising in providing a possible tool to overcome the pulse energy loss. Thus, optical switches may find some applications in fast sequential processing.

16-7. Problems

16.1. Determine the detuning of the probe laser from the Fabry-Perot peak for the operation of an etalon as AND, OR, XOR, and NAND gates. Refer to Fig. 16.3.

16.2. For an asymmetric etalon with front and back mirror reflectivities of R_f and R_b and mirror spacing of d, calculate the reflectance for a beam impinging on the material at an angle θ. Show that for normal angle of incidence, the reflectance is given by Eq. (16.8). Assume the absorption coefficient of the material is $\alpha(\lambda)$.

16-8. References

1. H. M. Gibbs, *Optical Bistability: Controlling Light with Light* (Academic Press, New York, 1985).

2. E. Abraham and S. D. Smith, Rep. Prog. Phys. **45**, 815 (1982).

3. P. W. Smith and W. J. Tomlinson, IEEE Spect. **18**, 16 (1981).

4. D. A. B. Miller, Laser Focus **18** (4), 79 (1982).

5. N. Peyghambarian and H. M. Gibbs, Phys. Today **39**, 5 (1986).

6. M. Warren, S. W. Koch, and H. M. Gibbs, IEEE Computer **20**, 68 (1987).

7. N. Peyghambarian and H. M. Gibbs, Opt. Engin. **24**, 68 (1985).

8. J. L. Jewell, Y. H. Lee, M. Warren, H. M. Gibbs, N. Peyghambarian, A. C. Gossard, and W. Wiegmann, Appl. Phys. Lett. **46**, 918 (1985).

9. D. Hulin, A. Mysyrowicz, A. Antonetti, A. Migus, W. T. Masselink, H. Morkoc, H. M. Gibbs, and N. Peyghambarian, Appl. Phys. Lett. **49**, 749 (1986).

10. R. Jin, C. Hanson, M. Warren, D. Richardson, H. M. Gibbs, N. Peyghambarian, G. Khitrova, and S. W. Koch, Appl. Phys. B **45**, 1 (1988).

11. R. Jin, D. Richardson, S. W. Koch, and H. M. Gibbs, Opt. Engin. **28**, 344 (1989).

12. M. F. Krol, S. T. Johns, R. K. Boncek, T. Ohtsuki, B. P. McGinnis, C. C. Hsu, G. Khitrova, H. M. Gibbs, and N. Peyghambarian, Proc. SPIE Conference, Boston, MA, Sept. 1992.

13. R. Jin, C. L. Chuang, H. M. Gibbs, S. W. Koch, J. N. Polky, and G. A. Pubanz, Appl. Phys. Lett. **53**, 1791 (1988).

14. C. L. Chuang, R. Jin, J. Xu, P. A. Harten, G. Khitrova, H. M. Gibbs, S. G. Lee, J. P. Sokoloff, N. Peyghambarian, R. Fu, and C. S. Hong, Int. J. Nonlin. Opt. Phys. **1**, (1992).

Chapter XVII
SEMICONDUCTOR LASER

In Chap. XIII we discussed bandfilling effects in semiconductors. Among other phenomena, we showed that negative absorption (i.e., optical gain) may occur in the spectral region between the renormalized bandgap and the chemical potential if the carrier density in the material is sufficiently high. If this optical gain is strong enough to overcompensate the losses, and if one provides feedback for the light in the form of an optical resonator, the semiconductor material becomes a laser. In such a semiconductor laser, the carriers are usually not generated via optical excitation, but rather through injection pumping in a p-n transition.

This semiconductor laser is probably the technologically most important link between optics and electronics. The capability to convert electrical impulses into optical signals is especially useful for combining electronic data manipulation and optical data transmission. The application potential for highly integrated laser diodes is almost unlimited. Present examples range from commercial compact disc drives to optical data storage devices and optical data processing.

From the physics point of view, the laser diode is an example of an active nonlinear semiconductor device. The basic features are determined by the laser resonator characteristics, the carrier/light interactions, and the interaction processes in the system of electrons and holes inside the active semiconductor material. In a passive semiconductor system used for nonlinear guided waves or optical bistability, the excitation density is usually kept below the threshold for population inversion in the frequency region of interest. Moreover, optical pumping can only give rise to gain if two beams are used with one (the pump beam) creating the population inversion that amplifies the second (probe) beam. For one-beam excitation, which is most often realized in passive semiconductor devices, optical gain is not possible. The maximum achievable nonlinearity at any particular frequency is limited by the bleaching of the absorption at the excitation frequency.

In this chapter we discuss some of the basic aspects of semiconductor lasers. First, we briefly discuss p and n doping, and then we describe the operational principles of a laser diode.

17-1. Doping

As mentioned in Chap. VI, the number of conduction-band electrons always equals the number of valence-band holes in intrinsic semiconductors, where no impurities exist. This charge balance can be deliberately disturbed if the crystal is doped, i.e., if different atoms (impurities) are introduced that replace some of the original atoms forming the semiconductor lattice. These impurities generally have a different number of electrons in comparison to the atom they are replacing. For example, let us assume that in GaAs, a column VI atom such as Se replaces an As atom (column V); see Fig. 17.1.

The Se atom has one more valence electron than the As atom. This additional electron is not needed for bond formation in the semiconductor crystal. Its binding to the positively charged Se atom (note that the Se nucleus also has one additional proton) is similar to the binding of an electron to the nucleus of a hydrogen atom with a binding energy of [see Eq. (6.18)]

$$E_d = \frac{e^4 m_e}{2\hbar^2 \epsilon_0^2}\,, \tag{17.1}$$

where m_e is the electron effective mass of GaAs. This binding energy, which is typically a few tens of meV, is the energy required to excite the excess electron from the Se atom to the conduction band of GaAs. Therefore, the Se atom may "donate" an electron to the crystal, which is no longer an insulator but a (bad) "conductor" with one electron in the conduction band; hence, the name semiconductor, which was originally

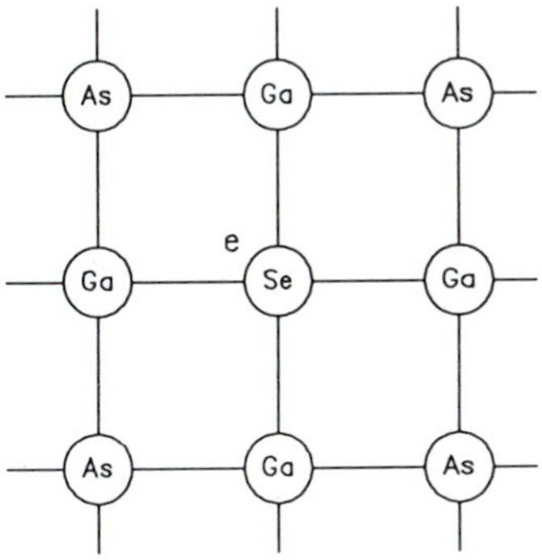

Figure 17.1. A GaAs lattice plane with Se replacing one As atom. The additional valence electron of Se goes into the conduction band of the crystal.

given to doped insulators. Impurities such as Se in GaAs are called donors. Once the Se has donated its electron, it becomes an ion with a net positive charge. A material doped with donors is often called "n-type." The energetic position of the electron at the donor site, with respect to the valence and conduction bands, is schematically shown in Fig. 17.2.

Similarly, if the Ga (group III) in GaAs is replaced with Zn (group II), the Zn acts as an "acceptor," as shown in Fig. 17.3. Zn has one less valence electron than Ga. To satisfy the crystal bonds, one electron out of the originally full GaAs valence band is used to make up for the valence-electron deficiency of Zn. The Zn "accepts" this electron, becoming an ion with a net negative charge. The missing valence electron in the Zn-doped GaAs acts as a hole, like the hole in an optically generated electron-hole pair. A material doped with acceptors is called "p-type." Figure 17.4 schematically displays the energetic position of the acceptor level relative to the band energy levels. The energy E_a is the binding energy of the hole to the Zn atom, and is given by

$$E_a = \frac{e^4 m_h}{2\hbar^2 \epsilon_0^2}, \tag{17.2}$$

where m_h is the hole effective mass of GaAs. The donor and acceptor states in the energy band diagram are schematically shown in Fig. 17.5. These states are localized states, unlike the valence- and conduction-band states, which are delocalized states extending through the whole crystal and obeying the Bloch theorem with wave functions of the form $u_{\mathbf{k}}(\mathbf{r})$ $\exp(i\mathbf{k}\cdot\mathbf{r})$.

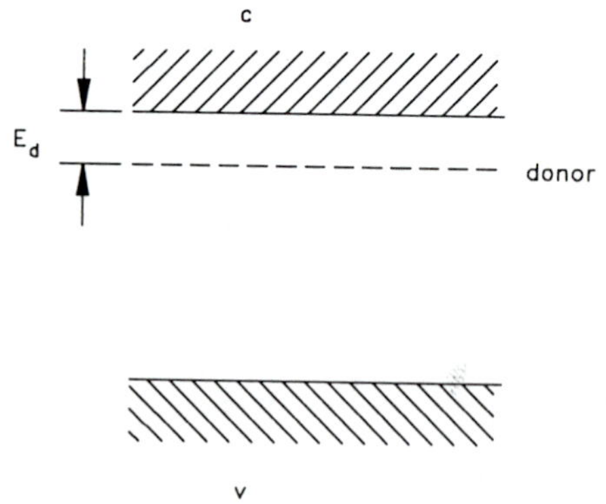

Figure 17.2. The energetic position of the donor level relative to the band edges.

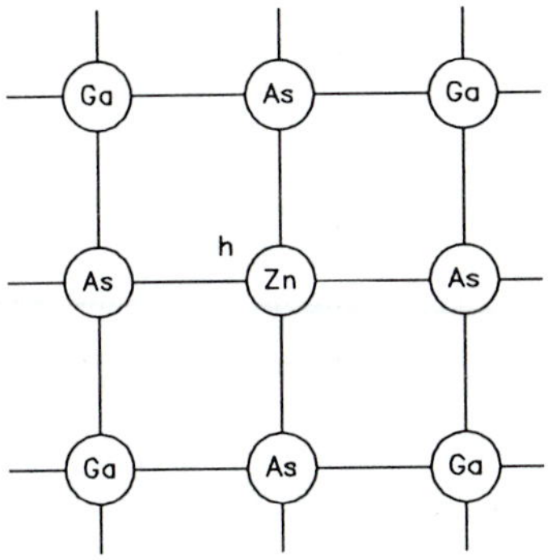

Figure 17.3. A GaAs lattice plane with a Zn atom replacing one Ga atom. The missing electron of Zn is a hole that can be excited to the valence band of GaAs.

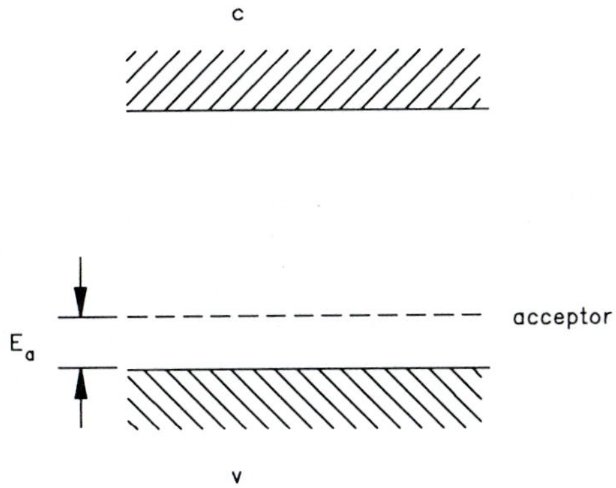

Figure 17.4. The energetic position of the acceptor level relative to the band edges.

Generally, doping can be much more complicated than outlined above. In particular, it is not an easy task to obtain n and p doping in the same material. For the purpose of our discussion here, we only wish to point out that p and n doping of a semiconductor like GaAs is possible and can be done routinely.

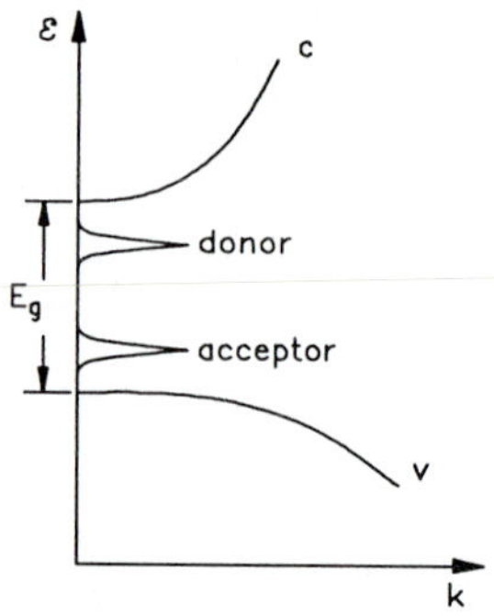

Figure 17.5. The donor and acceptor states in the forbidden gap.

17-2. Chemical Potential for Doped Semiconductors

The electrons in any crystal in thermal equilibrium at a temperature T are described by the Fermi-Dirac distribution function discussed in Chap. II. In total equilibrium, the chemical potential in intrinsic semiconductors is between the valence and conduction bands, whereas in metals, the chemical potential at low temperatures is inside the conduction band. In the following, we first determine the value of the chemical potential for an intrinsic semiconductor in total equilibrium, and then continue the discussion of the case of doped systems.

Consider an intrinsic semiconductor at an equilibrium temperature T. The total electron density in the conduction band may be calculated using Eq. (2.85) as [see Fig. 17.6(a)]

$$n_e = \int d\mathcal{E}\, f_e(\mathcal{E}) g_e(\mathcal{E}) \,, \tag{17.3}$$

where $f_e(\mathcal{E})$ and $g_e(\mathcal{E})$ are the Fermi-Dirac distribution function and the density-of-states function, respectively, given by

$$f_e(\mathcal{E}) = \frac{1}{\exp[(\mathcal{E} - \mu)/k_B T] + 1} \tag{17.4}$$

and

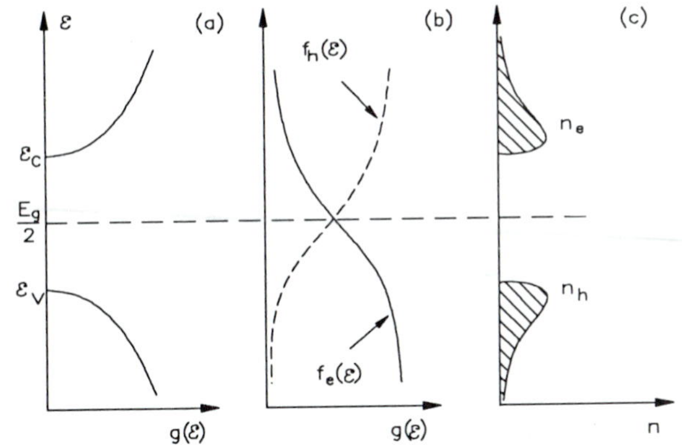

Figure 17.6. Plot of (a) density-of-states functions, (b) probability distribution functions, and (c) carrier densities in an intrinsic semiconductor.

$$g_e(\mathcal{E}) = \frac{1}{2\pi^2}\left(\frac{2m_e}{\hbar^2}\right)^{3/2} (\mathcal{E} - E_g)^{1/2} , \tag{17.5}$$

where the electron energy is written with respect to the top of the valence band, chosen as the reference for zero energy. Here μ is the total chemical potential of the system. For simplicity of argument, we approximate the carrier-distribution function by a Maxwell-Boltzmann function. Then Eq. (17.4) may be written as

$$f_e(\mathcal{E}) \simeq \exp[-(\mathcal{E} - \mu)/k_B T] . \tag{17.6}$$

The integration of Eq. (17.3) should be taken over the conduction band. Thus,

$$n_e = \frac{1}{2\pi^2}\left(\frac{2m_e}{\hbar^2}\right)^{3/2} e^{\mu/k_B T} \int_{E_g}^{\infty} d\mathcal{E}\, (\mathcal{E} - E_g)^{1/2} \exp(-\mathcal{E}/k_B T)$$

$$= \frac{1}{2\pi^2}\left(\frac{2m_e}{\hbar^2}\right)^{3/2} e^{\mu/k_B T} (k_B T)^{3/2} e^{-E_g/k_B T} \int_0^{\infty} dx\, e^{-x} x^{1/2}$$

$$= 2\left(\frac{2\pi m_e k_B T}{h^2}\right)^{3/2} e^{(\mu - E_g)/k_B T} , \tag{17.7}$$

where the value of the integral of $\sqrt{\pi}/2$ was substituted. The hole density may be calculated similarly by noting that the hole density of states is

$$g_h(\mathcal{E}) = \frac{1}{2\pi^2}\left(\frac{2m_h}{\hbar^2}\right)^{3/2} (-\mathcal{E})^{1/2} , \tag{17.8}$$

with $\mathcal{E} < 0$ for holes, since the zero of energy is taken at the top of the valence band (see Fig. 17.6). Also, the hole probability distribution function is given by

$$\begin{aligned} f_h(\mathcal{E}) &= 1 - f_e(\mathcal{E}) \\ &= 1 - \frac{1}{\exp[(\mathcal{E} - \mu)/k_B T] + 1} = \frac{1}{\exp[(\mu - \mathcal{E})/k_B T] + 1} \\ &\simeq \exp[(\mathcal{E} - \mu)/k_B T] . \end{aligned} \tag{17.9}$$

The probability for a state with energy $\mathcal{E}$ to be empty is $f_h(\mathcal{E})$. Equation (17.9) indicates that the state of energy $\mathcal{E}$ is either full or empty; i.e., $f_e(\mathcal{E}) + f_h(\mathcal{E}) = 1$. We also assumed the classical regime with $\exp[(\mu - \mathcal{E})/k_B T] >> 1$. The hole density is then

$$\begin{aligned} n_h &= \int_{\substack{\text{valence} \\ \text{band}}} f_h(\mathcal{E}) g_h(\mathcal{E}) d\mathcal{E} \\ &= \frac{1}{2\pi^2}\left(\frac{2m_h}{\hbar^2}\right)^{3/2} \int_0^{-\infty} d\mathcal{E}\, (-\mathcal{E})^{1/2} \exp[(\mathcal{E} - \mu)/k_B T] \\ &= \frac{1}{2\pi^2}\left(\frac{2m_h}{\hbar^2}\right)^{3/2} (k_B T)^{3/2}\, e^{-\mu/k_B T} \int_0^{\infty} dx\, x^{1/2}\, e^{-x} \\ &= 2\left(\frac{2\pi m_h k_B T}{h^2}\right)^{3/2} e^{-\mu/k_B T} . \end{aligned} \tag{17.10}$$

For intrinsic semiconductors, $n_e = n_h$, and using Eqs. (17.7) and (17.10), one gets

$$2\left(\frac{2\pi m_e k_B T}{h^2}\right)^{3/2} e^{(\mu - E_g)/k_B T} = 2\left(\frac{2\pi m_h k_B T}{h^2}\right)^{3/2} e^{-\mu/k_B T} \quad (17.11)$$

or

$$\boxed{\mu = \frac{1}{2} E_g + \frac{3}{4} k_B T \ln \frac{m_h}{m_e} \, .}$$

chemical potential for an intrinsic semiconductor (17.12)

At room temperature, $k_B T \simeq 0.25$ meV; thus, $E_g >> k_B T$ and Eq. (17.12) gives $\mu \simeq E_g/2$, indicating that the total chemical potential for an intrinsic semiconductor at equilibrium is at midgap. Figure 17.6 schematically shows the density of states, probability distribution functions, and densities of electrons and holes for such an intrinsic compound.

For an extrinsic semiconductor, on the other hand, the chemical potential is no longer at midgap. As the number of donor impurities is increased, the chemical potential moves toward the conduction band. For a sample with a large number of acceptors, the chemical potential is moved toward the valence band. The exact position of the chemical potential may be obtained from the knowledge of the total number of charges, $n = \Sigma\, n_i + n_d + n_a$, where n_i is the intrinsic charge density and n_d and n_a are the density of donor and acceptor states, respectively. For a heavily doped n-type (p-type) semiconductor, the chemical potential may reside in the conduction band (valence band). Such semiconductors are called *degenerate* semiconductors.

So far we have introduced a single total chemical potential under thermal equilibrium condition. This single chemical potential is sufficient in describing both the electron and hole densities. When the thermal equilibrium is lost, either by optical pumping or by electrical injection of carriers, the concept of the chemical potential or Fermi level is no longer relevant. However, we may still assume a quasi-equilibrium condition and employ the concept of quasi-Fermi levels, as we did in Chap. XIII. In this case a single total chemical potential is not enough to describe both electrons and holes. We assume that electrons and holes reach quasi-thermal equilibrium separately in their bands and introduce a quasi-chemical potential for electrons in the conduction band and one for holes in the valence band. The total density of electrons determines the quasi-chemical potential for electrons and, similarly, the hole density gives the hole quasi-chemical potential in the valence band.

17-3. p-n Junctions and Biasing

When a p-type semiconductor is contacted with an n-type semiconductor, a *p-n junction* is formed, as depicted schematically in Fig. 17.7. There are more holes in the p-type semiconductor, in comparison to the n-type material. Therefore, the holes diffuse from the p-type regions close to the junction to the n-type side due to this concentration gradient. This carrier diffusion is similar to heat diffusion in a metallic rod from a hot end to a cold end. The electrons in the n side and holes in the p side are referred to as the respective majority carriers, while the electrons in the p side and holes in the n side are called minority carriers, respectively. The distance the carriers diffuse before recombining is typically a few micrometers and is given by

$$L_D = \sqrt{D\tau}\ , \tag{17.13}$$

where L_D is diffusion length, D is the diffusion coefficient, and τ is the carrier lifetime. The diffusion gives rise to a generation of a *diffusion current* of the majority carriers across the junction.

The departure of holes from the p side leaves behind negatively charged acceptors, generating a negative space-charge layer in the p side close to the junction, as shown in Fig. 17.7. Similarly, the electron diffusion from the n side to the p side leaves behind a positive space-charge layer in the n side. These space-charge layers generate an internal electric field from the n to the p side (see Fig. 17.7). The built-in internal field E_{int} sets up a *drift current* in opposite direction with respect to the

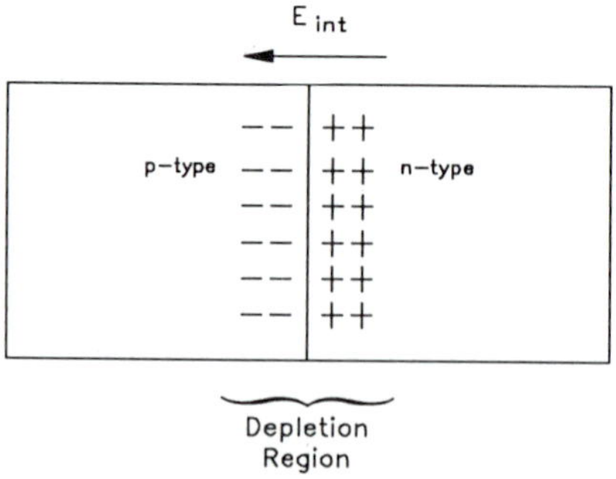

Figure 17.7. Schematics of a p-n junction with the built-in internal field and the depletion layer.

diffusion current. The generated built-in field prevents further diffusion of the majority carriers; thus, in steady state, the total current, which is the sum of the diffusion and drift currents, is zero, as is expected for a p-n junction in thermal equilibrium. The built-in field establishes a built-in potential V_{int} between the two regions, which forces a separation of the energy bands between the p side and the n side, as shown schematically in Fig. 17.8. The larger the doping concentrations and the temperature, the larger are the V_{int} and the band misalignment (see Prob. 17.2). The displacement of the bands is such that the chemical potential is constant across the junction, as seen in Fig. 17.9. This results from the fact that in thermal equilibrium the thermodynamic quantities, such as temperature, pressure, and chemical potential, have to be constant throughout the entire structure. The potential V_{int} behaves like a barrier, preventing the majority carriers, electrons in the n side and holes in the p side, of the junction to cross the junction. The junction region where the space charges build up is referred to as the *depletion region* (because it is almost depleted from majority carriers).

If an external voltage is applied to the p-n junction, the equilibrium is lost and a current can flow. The junction is said to be forward biased if the p region is connected to the positive terminal and the n region is connected to the negative terminal of the external source, such as a battery, as shown in Fig. 17.9. A reverse-biased p-n junction has its p side (n side) connected to the negative terminal (positive terminal), as displayed in Fig. 17.10. For the case of the forward-biased junction, the height of the built-in potential is reduced because the external field E_{ext} is in the opposite direction of the internal built-in field E_{int} (see Fig. 17.9). Since the equilibrium is lost, we no longer have a single chemical potential for

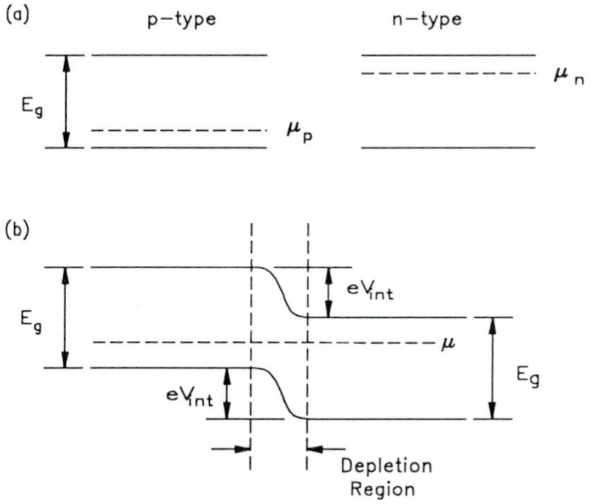

Figure 17.8. (a) The band picture for the n and p sides before contact. (b) The band picture for a p-n junction after contact.

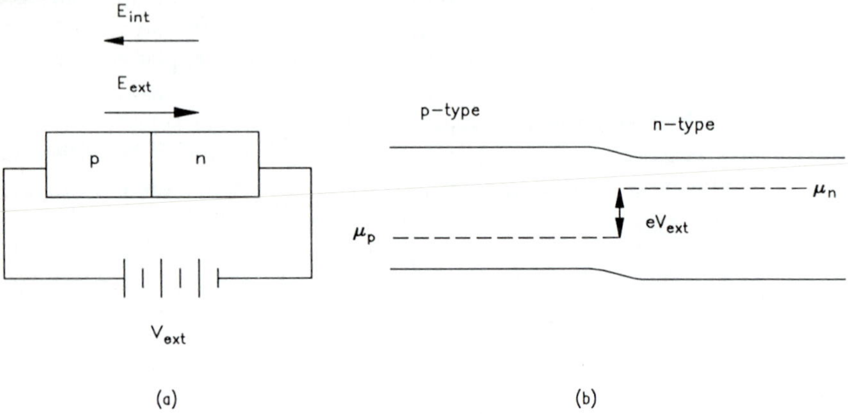

Figure 17.9. (a) A forward-biased p-n junction. (b) The band diagram for the junction. The chemical potentials of the n and p sides are displaced by the external source energy eV_{ext}.

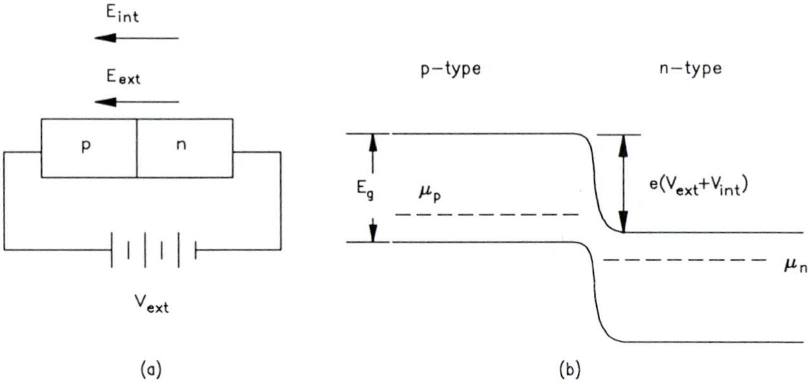

Figure 17.10. (a) A reverse-biased p-n junction. (b) The displacement of the bands in a reverse-biased p-n junction.

the junction. However, we may assume that the electrons and holes reach quasi-equilibrium states through the mechanisms discussed before. The chemical potentials of the n and p sides are, therefore, shifted with respect to each other (see Fig. 17.9), with the separation being eV_{ext}. The reduced built-in potential barrier with the forward bias can no longer prevent the majority carriers from crossing the junction; the majority carriers are *injected* into the junction and a current is established. Some of the electrons may recombine with holes in the depletion region. However, additional carriers from the external source are supplied to keep the current flowing. For the reverse-biased case, the increased built-in potential barrier prevents current flow. Thus, the p-n junction acts as a current rectifier or diode.

17-4. Semiconductor Laser

The doping concentrations may be increased to the point where the chemical potential in the n region is above the conduction band, and the chemical potential in the p region is below the valence band (see Fig. 17.11). Forward biasing of such a heavily doped p-n junction results in electrons and holes being injected into the depletion region, giving rise to a population inversion. In this depletion region, which is also referred to as the "active layer," electrons and holes recombine radiatively, emitting their excess energy as light. Such a device is called a light-emitting diode (LED). If this spontaneous emission encounters positive feedback by being at least partially reflected at the end facets of the device, and if the light is amplified enough to overcome the losses, the p-n junction acts as a semiconductor laser diode.

The amplification of the light is a consequence of the bandfilling mechanism (see Chap. XIII) in the active region. The band occupancy of the active layer is shown in Fig. 17.12. The states in the conduction band are filled by electrons, and the states in the valence band are filled by holes (inverted population). The spectral range between the (renormalized) bandgap and the quasi-equilibrium chemical potential, $E_g' < \hbar\omega < \mu_e + \mu_h + E_g'$, can experience gain.

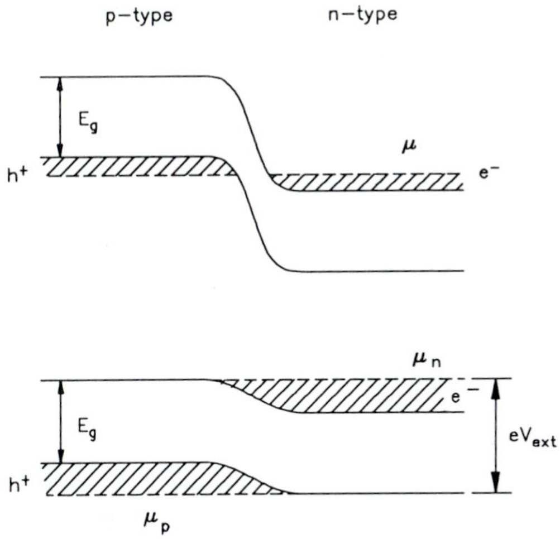

Figure 17.11. (a) The band diagram for a heavily doped p-n junction at equilibrium before applying the voltage and (b) after the voltage is applied.

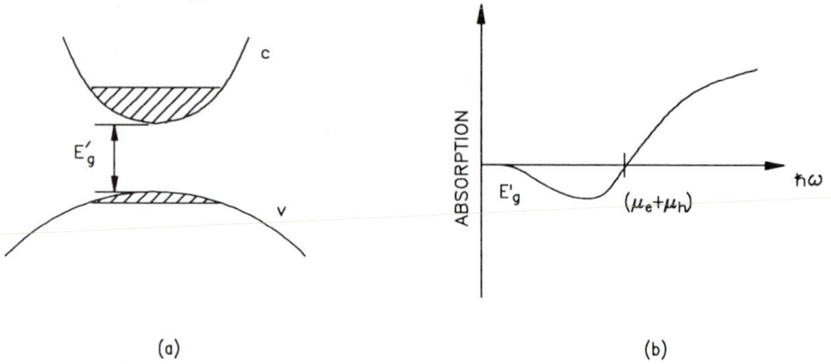

Figure 17.12. (a) The band occupancy of the active region of the p-n junction of Fig. 17.11. (b) The absorption/gain spectrum of the active region of part (a).

The semiconductor laser just described is referred to as a homojunction laser when both the n and p regions are the same material, like GaAs. The active layer in homojunction lasers is the depletion region with a thickness of ~ 1 μm. Figure 17.13 schematically shows such a laser structure. The lasing emission takes place from both of the end facets of the laser. The end facets are usually not coated, and the natural reflectivity of the semiconductor (R ~ 0.3) provides the feedback needed for the lasing action. The emission occurs from the entire junction area. However, the homojunction lasers are not very efficient light sources as a result of the leakage of carriers and light from the active region.

The more commonly used diode lasers are the double heterojunction (DH) lasers. These lasers take advantage of the confinement of carriers and optical waveguiding in heterostructures. Figure 17.13(b) displays a GaAs/AlGaAs DH structure. In contrast to the homojunction structure, the active layer in the DH laser is a thin layer of GaAs of less than 0.1-μm thickness between a p-type and an n-type $Al_xGa_{1-x}As$ layer. The larger bandgap of AlGaAs with respect to GaAs results in confinement of the carriers in the active layer. Furthermore, the index of refraction of GaAs is larger than that of AlGaAs, leading to the confinement of the emitted radiation in the GaAs region. Thus, the emission is not leaking much into the surrounding media, but is waveguided in the active layer as a result of this index difference, $\Delta n(\omega)$, of a few percent. The schematics of the DH structure and the emitted radiation are displayed in Fig. 17.14.

A strip-geometry DH laser is shown in Fig. 17.13(c). The external voltage is applied to only a narrow strip and, therefore, the population inversion and lasing occur only in a narrow region in the active layer, in contrast to a regular DH structure where the entire active layer area lases. For a typical length of the laser of 200-300 μm, a gain of $\simeq$ 200 cm^{-1}, and

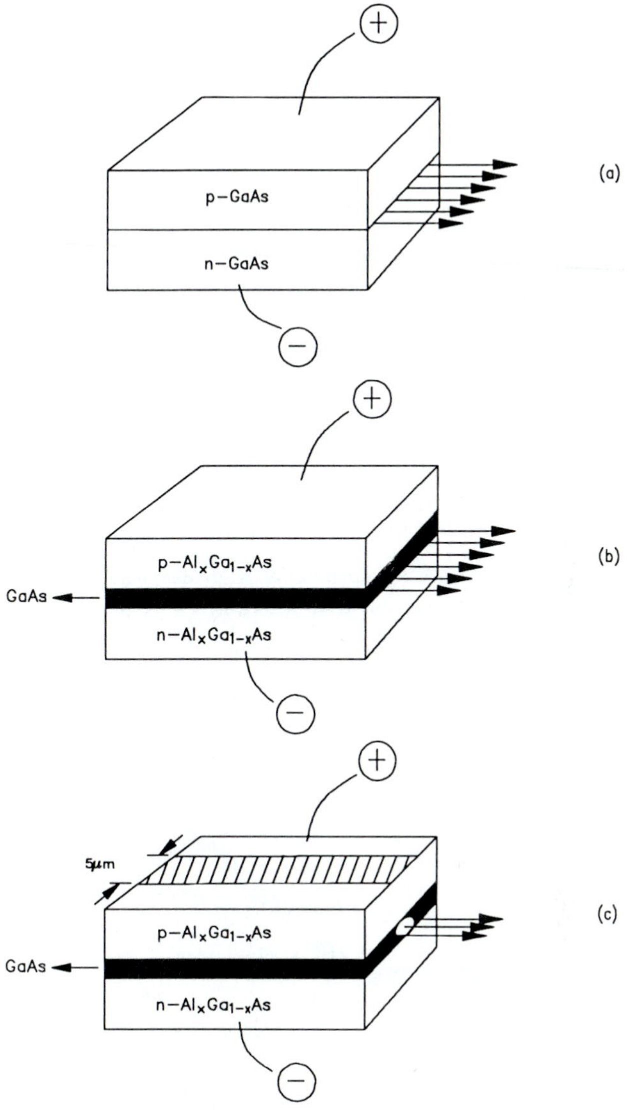

Figure 17.13. The structure of (a) a homojunction laser, (b) a double heterojunction laser, and (c) a strip-geometry double heterojunction laser.

stripe width of 5 μm, the laser threshold current is $\simeq$ 50 mA. The strip geometry of Fig. 17.13(c) is said to be *gain-guided* because it contains no lateral structure, and the optical gain is restricted laterally by the stripe width, making the laser mode laterally restricted also. In contrast, there are structures that are said to be *index-guided*, where the active region is a

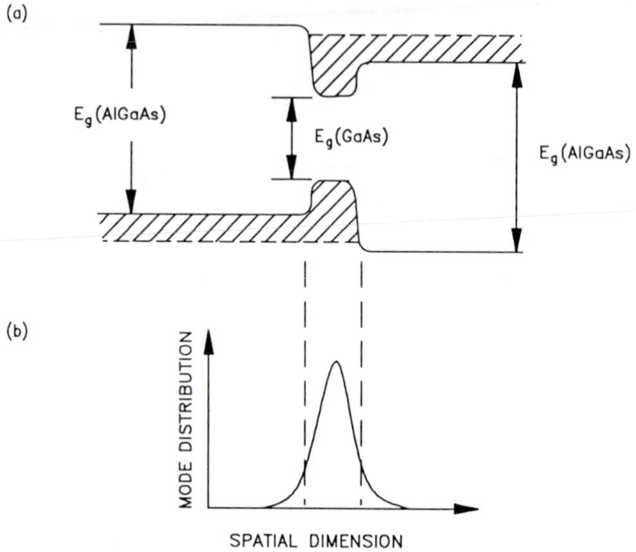

Figure 17.14. (a) The band diagram of a double heterojunction laser. (b) The optical field distribution in the active layer.

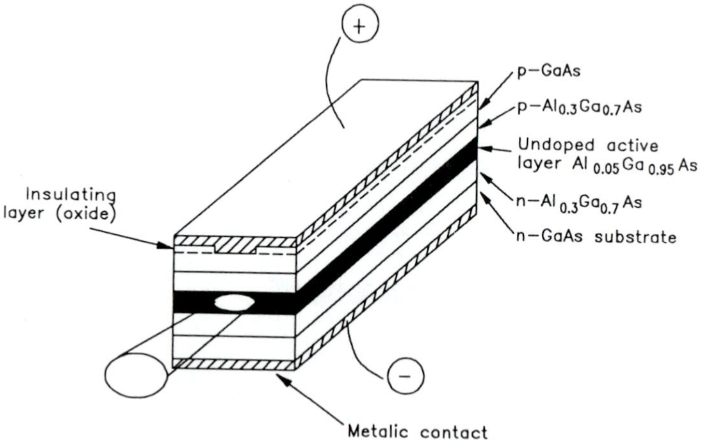

Figure 17.15. A typical structure for a double heterojunction laser. The p-GaAs layer on the top and the n-GaAs layer on the bottom are for contacting purposes (after Ref. 17.1).

narrow channel waveguide surrounded (both laterally and vertically) by a compositionally different material with a lower index of refraction. The difference in the index of refraction of the active region and the surrounding media gives rise to confinement of the radiation mode to the active region. Figure 17.15 shows the schematic of a typical gain-guided semiconductor diode laser with the active layer of $Al_{0.05}Ga_{0.95}As$ and doping layers of p-$Al_{0.3}Ga_{0.7}As$ and n-$Al_{0.3}Ga_{0.7}As$. The 5% Al concentration in the active region leads to light generation at a wavelength of $\lambda \simeq 0.83\ \mu m$. Such solid-state gain-guided lasers can achieve remarkably high overall efficiencies, up to 50%, which is orders of magnitude higher than in most other laser systems.

17-5. Rate Equations

In this section we analyze the basic operation principles of a semiconductor laser. We concentrate on the active region of the laser diode. For that purpose, all the details of the carrier injection can be ignored, and we simply assume that electron-hole pairs are "pumped" into the active region with a constant pump rate p_r. This pump rate is assumed to be high enough that the semiconductor exhibits a spectral region of optical gain, as shown schematically in Fig. 17.16. Furthermore, the optical resonator is designed so there is an eigenmode near the gain maximum. The entire gain grows with increasing pump rate. At the *threshold-pump rate*, the total gain (gain plus spontaneous emission into that mode) for the mode near the gain maximum exactly equals the losses, and this mode begins to lase.

For simplicity, we concentrate only on the mode that reaches threshold first. The number of photons N in this laser mode is given by

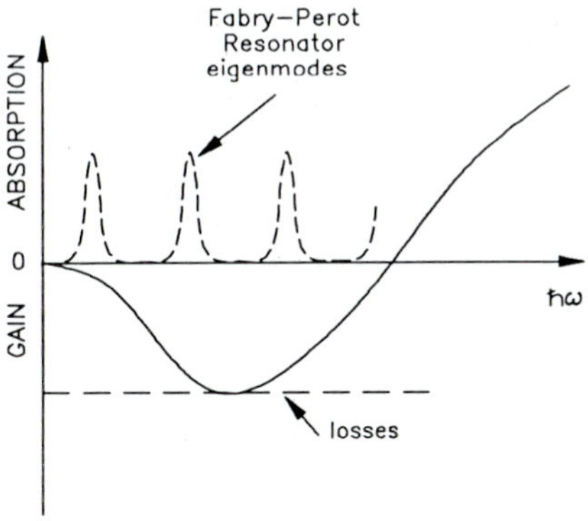

Figure 17.16. Gain spectrum of a semiconductor together with the loss line $2\kappa_\ell$ and the resonator eigenmodes. For the conditions chosen in this figure, the mode closest to the gain maximum is at threshold.

$$\frac{d}{dt} N = -2\kappa_\ell N + NG(n) + R_{cv} , \tag{17.14}$$

where κ_ℓ describes the losses of the mode, R_{cv} is the rate of light spontaneously emitted into the mode, and G(n) is the gain rate, which depends on the carrier density n.

The second relevant equation determines the density of electron-hole pairs in the active region,

$$\frac{d}{dt} n = p_r - N\, G(n) - R(n) , \tag{17.15}$$

where p_r denotes the pump rate (rate of carrier injection) and R(n) is the total carrier-loss rate, which includes the radiative carrier recombination, i.e., the spontaneous emission into the lasing mode R_{cv} and nonradiative recombination of carriers.

Equations (17.14) and (17.15) can be used to compute laser action in semiconductors for several situations of general and applied interest. In this book we only illustrate the simplest case, showing the onset of laser action. For this purpose, we consider only the equations in steady state,

$$0 = -2\kappa_\ell N + NG(n) + R_{cv} \tag{17.16}$$

and

$$0 = p_r - N\, G(n) - R(n) . \tag{17.17}$$

Equation (17.17) yields

$$p_r - R(n) = NG(n) , \tag{17.18}$$

and Eq. (17.16), for $G < 2\kappa_\ell$, gives

$$N = \frac{R_{cv}}{2\kappa_\ell - G(n)} . \tag{17.19}$$

Equation (17.18) shows that pump rate minus the loss rate is equal to the gain times the number of photons in the lasing mode. Equation (17.19) indicates that the mean number of photons in the laser mode is proportional to the rate of spontaneous emission into that mode. Since this is a very small number, there are basically no photons in the laser mode for $G < 2\kappa_\ell$, and Eq. (17.18) yields $p_r \simeq R(n)$. If we use $R(n) = r_1 n + \mathcal{O}(n^2)$ for illustration, where r_1 is a proportionality constant, and keep only the leading term, we obtain

$$n \simeq \frac{p_r}{r_1} \text{ for } G < 2\kappa_\ell \,. \tag{17.20}$$

Hence, the carrier density increases linearly with the pump rate, as shown in Fig. 17.17(a).

With increasing carrier density, the gain G(n) increases until it reaches $2\kappa_\ell$; i.e., the gain starts to compensate the losses of the lasing mode. Equation (17.19) shows that the photon number in the laser mode begins to increase dramatically for $G(n) \to 2\kappa_\ell$. The gain value

$$\boxed{G(n)\Big|_{n = n_{th}} = 2\kappa_\ell}$$

laser threshold (17.21)

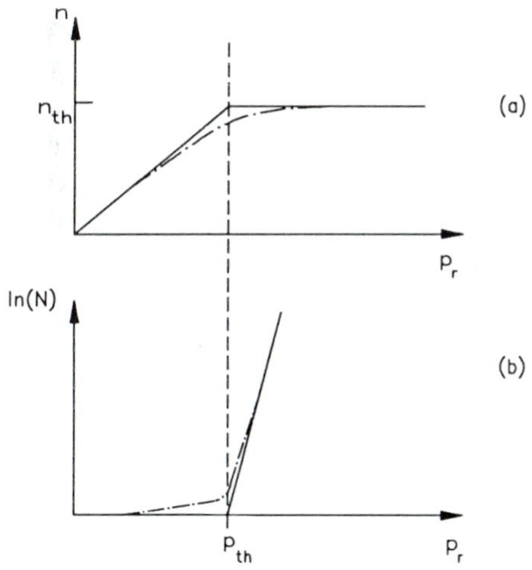

Figure 17.17. (a) Carrier density in a semiconductor laser as a function of pump rate p_r. (b) Logarithm of the photon number in the laser mode as a function of pump rate. The solid lines are the approximate analytical results discussed in this section, and the dashed lines are the numerical solutions of the coupled equations.

defines the laser threshold and, hence, the threshold carrier density n_{th}. If one numerically solves the coupled equations (17.16) and (17.17) using the gain computed, as in Chap. XIII, one of course finds that N does not really diverge, but increases by many orders of magnitude at the laser threshold.

Equation (17.19) shows that G(n) can never exceed $2\kappa_\ell$ because this would imply negative photon numbers, a clearly unphysical solution. Consequently, above the laser threshold, the gain remains at its threshold value,

$$G(n)\Big|_{n > n_{th}} = G(n_{th}) \ . \tag{17.22}$$

It is said that the gain is "clamped." Equation (17.18) then becomes

$$p_r - R(n) = NG(n_{th}) \ . \tag{17.23}$$

Above threshold, the number of photons in the lasing mode becomes very large, so the product on the right-hand side is very large, $G(n_{th})N >> R(n)$. For this situation, Eq. (17.23) yields

$$p_r \simeq N\, G(n_{th}) \ ; \tag{17.24}$$

i.e., any increase in pumping directly leads to an increase in the number of photons in the laser mode, as shown schematically in Fig. 17.17(b).

Unfortunately, many semiconductor lasers do not always operate in a single mode. Multimode emission occurs, which is usually quite undesirable. A discussion of multimode operation and of the design features employed to make semiconductor lasers efficient and as successful a device as they are in practice goes beyond the scope of this book. We refer to Refs. 17.2 and 17.3 for such information.

17-6. Problems

17.1. Estimate the binding energies of a donor like Se and an acceptor like Zn in GaAs. Use Table 1 in Chap. V for the material parameters.

17.2. Show that the built-in potential in the depletion region of a p-n junction can be written as

$$V_{int} = \frac{K_B T}{e} \ln \left(\frac{n_a n_d}{n_i^2} \right), \tag{17.25}$$

where n_a and n_d are the acceptor and donor densities, respectively, and n_i is the intrinsic carrier density.

17.3. Calculate the reflectivity of a bare GaAs end facet [$n(\omega)_{GaAs} \simeq 3.5$ for ω in the range of interest].

17.4. Calculate the resonator mode spacing and the width of each mode for a GaAs semiconductor laser with a length of $d = 250\ \mu m$, and with uncoated facets.

17-7. References

1. W. Streifer, R. D. Burnham, T. L. Paoli, and D. R. Scifres, Laser Focus/Electro-Optics, June 1984.

2. G. P. Agrawal and N. K. Dutta, *Long-Wavelength Semiconductor Lasers* (Van Nostrand Reinhold Company, New York, 1986).

3. G. H. B. Thompson, *Physics of Semiconductor Laser Devices* (Wiley, Chichester, 1980).

4. J. Wilson and J. F. B. Hawkes, *Optoelectronics: An Introduction* (Prentice Hall, Englewood Cliffs, New Jersey, 1983).

5. P. L. Pulfrey and N. G. Tarr, *Introduction to Microelectronic Devices* (Prentice Hall, Englewood Cliffs, New Jersey, 1989).

6. A. Yariv, *Quantum Electronics*, Second Edition (John-Wiley, New York, 1975).

Chapter XVIII
OPTOELECTRONIC DEVICES

In contrast to the all-optical devices discussed in Chap. XVI, the semiconductor laser (Chap. XVII) is usually a hybrid electro-optic device powered through electrical pumping. In this chapter we discuss additional hybrid devices, such as optoelectronic detectors and light modulators. As a specific example, we describe infrared detectors based on intersubband transitions in quantum wells. Quantum well photomultipliers are another subject of discussion in this chapter.

18-1. Bistable Self-Electro-Optic Devices

A self-electro-optic-effect device (SEED) is a cavity-less device that operates on the basis of the quantum-confined Stark effect. Light absorption causes an electro-optic shift of the band edge through electrical feedback. The medium typically consists of a multiple quantum well (MQW) structure incorporated in the device as the active part of a p-i-n diode (see Fig. 18.1). The role of the diode is twofold. First, it acts as a photodetector for the light pulse to be processed. Second, it provides electric feedback to the MQW through an external circuit. The electric field is applied perpendicularly to the MQW layers. For bistability operation, the incident laser frequency is tuned to the vicinity of the exciton resonance in the absence of an external field, i.e., $\hbar\omega_\ell$ in Fig. 18.2. If the diode is negatively biased, its absorption spectrum is shifted to low energies, as shown by the dashed curve (iii), and there is little absorption of the laser. As the light intensity increases, the flow of current through the device, as a result of the generated electron-hole pairs, causes a drop in the applied electric field across the external resistance and, therefore, across the diode. The exciton shifts toward its unbiased position and the device transmission decreases. The larger the laser intensity, the less the field across the MQW structure and the higher the absorption at the laser frequency. The larger absorption creates more electron-hole pairs, which reduces the field further, and so on. This positive feedback continues until the device switches to its off position, characterized by a low transmission.

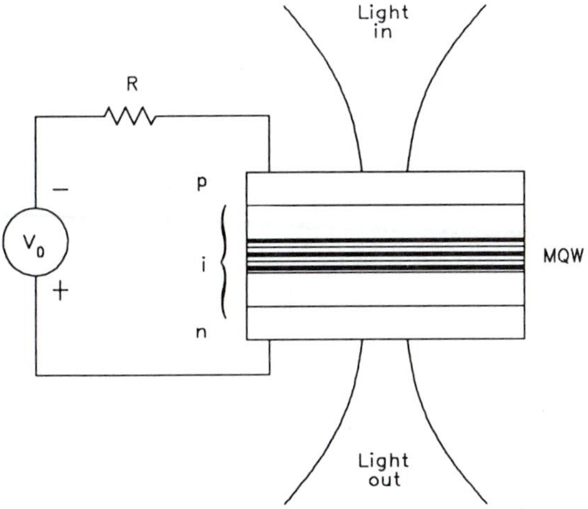

Figure 18.1. Schematic of a MQW SEED.

The absorption of the system in the "off" state is then governed by the exciton peak [see curve (i) in Fig. 18.2]. Figure 18.3 shows the observed optical bistability in a GaAs-AlGaAs MQW p-i-n diode structure.

Typical SEEDs do not work at room temperature with bulk GaAs as the active element since it is essential to have a well-defined exciton resonance, which can be displaced by an applied electric field. Therefore we discuss only quantum-well SEED devices.

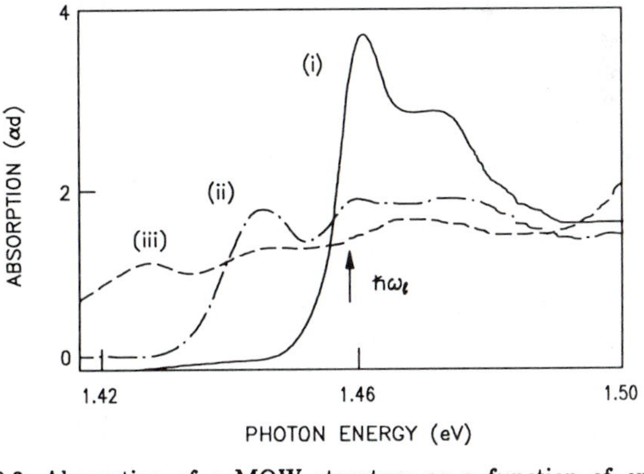

Figure 18.2. Absorption of a MQW structure as a function of energy for various external electric fields applied perpendicular to the quantum well layers. The magnitude of field increases for curves (i) to (iii) from 1.6×10^4 V/cm in (i) to 1.3×10^5 V/cm in (ii) to 1.8×10^5 V/cm in (iii) (after Ref. 18.1).

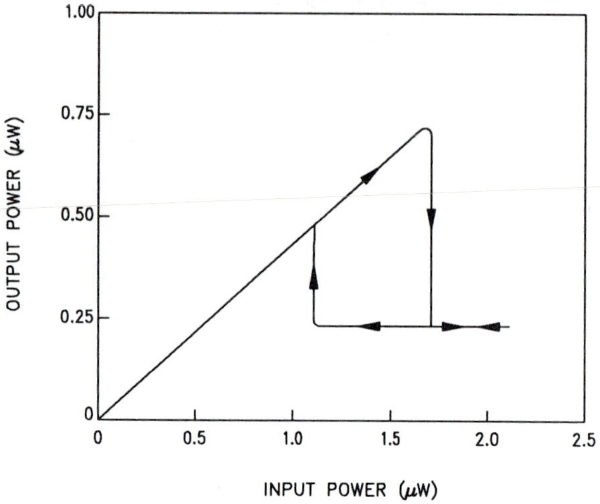

Figure 18.3. Increasing absorption optical bistability observed in a GaAs MQW SEED (after Ref. 18.1).

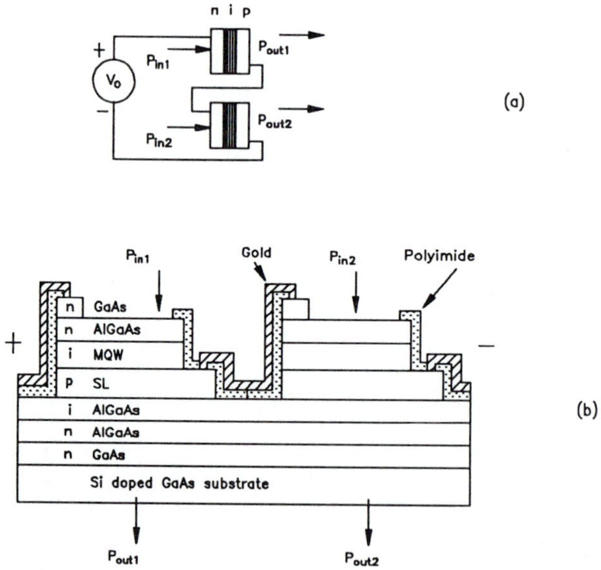

Figure 18.4. (a) Schematic representation of a symmetric SEED. (b) The design of the symmetric SEED (after Ref. 18.1).

SEED devices are hybrid devices, so their capacitance can be increased to trade-away speed for reducing the power. The contrast of a SEED depends upon the absorption difference between the field-on and field-off states and can be enhanced by using devices with thicker active media, whereas its transmission depends upon the lower of the two absorptions. Thus, there is also a tradeoff between bistability contrast and insertion loss. The advantages of the SEED devices are their interface capability to electronic systems, the absence of an external feedback, and their speed-power tunability. Large device size and high absorption (even in the "on" state) are the disadvantages of this type of device.

A variety of SEEDs have been developed. One of the most common structures is called the symmetric SEED, which consists of two quantum well p-i-n diodes in series with reverse biasing, as shown in Fig. 18.4. Illuminating one of the diodes with a constant power results in optical bistability in the other. The symmetric SEEDs show time-sequential gain, an effective input/output isolation, and require no critical biasing.

18-2. Quantum Well Modulators*

The same operation principle of the quantum-confined Stark effect can be exploited in absorption modulators. The basic idea is again to obtain a change in the absorption of the multilayer structure by applying an external electric field, as may be seen from Fig. 18.2. For example, a light beam with a photon having an energy tuned below the exciton resonance when no electric field is applied experiences little absorption. The application of a bias field shifts the exciton to low energies, resulting in an increase of absorption, and thus corresponds to a modulation of the transmitted light-beam intensity. The external field is typically applied through a p-i-n geometry with the absorption changes occurring in the intrinsic layer. Both reflection modulators and waveguide absorption modulators have been demonstrated. As an example, Fig. 18.5 shows the structures of a GaAs MQW reflection modulator and an InGaAlAs/InAlAs MQW waveguide modulator for operation in the 0.8-μm and 1.5-μm regions, respectively. In both structures the MQW constitutes the intrinsic region between n- and p-type materials in the p-i-n junction. The reverse-biased p-i-n structure ensures a large electric field in the intrinsic region. Modulators with high-speed (over 40 GHz) and on-off contrast ratios, over 20 dB, have been demonstrated (see Ref. 18.2) with low voltages of a few volts. Such a large contrast ratio is possible because of the large interaction length, which can be obtained in a waveguide. The disadvantages of these modulators are their narrow operational wavelength range, temperature instability, and their background loss.

Reverse-biased p-i-n waveguide structures can also be employed as directional coupler switches (see, e.g., Ref. 18.4). In these electro-optical directional couplers an external voltage is applied to one of the guides to control the transfer of the passage of the light between the guides. Figure 18.6 shows an example of such a device structure using

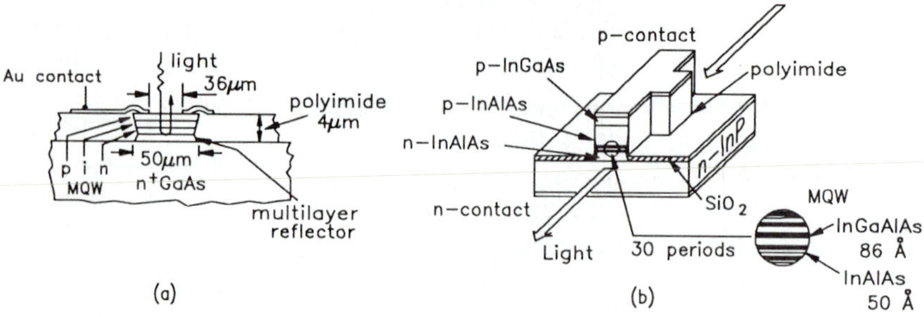

Figure 18.5. (a) Structure of a p-i-n reflection modulator (after Ref. 18.3). (b) Structure of a waveguide absorption modulator using an InGaAlAs/InAlAs MQW heterostructure (after Ref. 18.2).

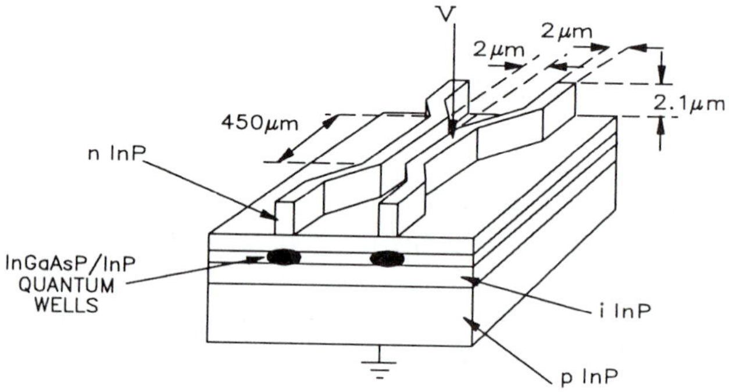

Figure 18.6. Schematic diagram of an electro-optic directional coupler fabricated from p-i-n waveguide InGaAsP/InP heterostructures (after Ref. 18.4).

InGaAsP/InP MQWs suitable for operation at 1.3 μm and 1.55 μm, which corresponds to a useful wavelength for optical fiber communications. Light is originally introduced at zero applied bias into the input guide on the right coupler. Coupling between the guides over distance L leads to the transfer of the light to the left guide. Switching of the light back to the input guide occurs when the input guide is reverse-biased, causing an index change on that guide and destroying the phase match for the coupling. Figure 18.7 displays the switching behavior of the directional coupler. The near-field output intensity pattern at 1.3 μm is shown. The two guides in the switch are laterally separated by 10 μm. Figure 18.7(a) shows that for zero applied bias, 86% of the light that was originally coupled into the input guide appears in the neighboring guide. With -18 V bias voltage applied to the input guide, switching occurs and 86% of the

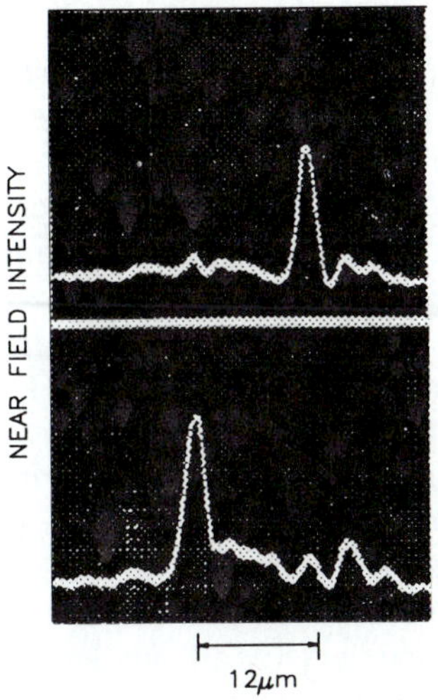

Figure 18.7. Near-field intensity pattern of the directional coupler of Fig. 18.6 (a) under no bias and (b) with a bias of -18 V (after Ref. 18.4).

light emerges from the input channel. The switching response as a function of voltage is shown in Fig. 18.8. The relative output power of the two channels is plotted, showing a cross over at an applied voltage of $\simeq$ 12 V. The losses of the device are due to coupling losses (~ 10 dB), propagation losses (~ 11.8 dB), and radiation losses at the transition between the wide-gap and narrow-gap sections of the device.

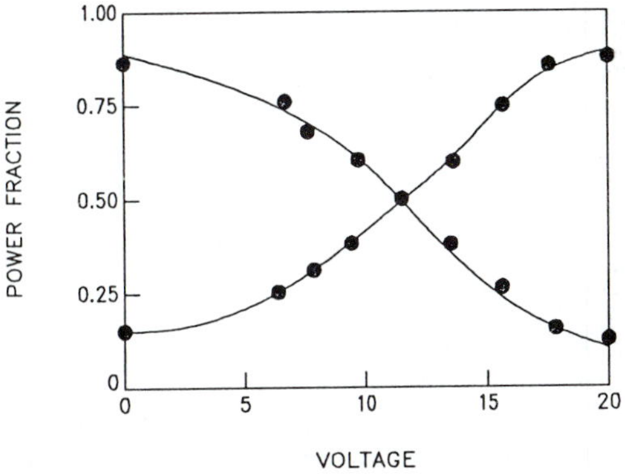

Figure 18.8. Relative power output for the two channels of the directional coupler as a function of applied voltage (after Ref. 18.4).

18-3. Detectors

Semiconductors are not only suitable materials as light emitters, but also can be operated as high-speed detectors. Development in this area is still in its infancy. Here we only focus on some examples of semiconductor-based detectors. We start by describing different types of detection schemes and then give some specific examples.

Semiconductor detection schemes typically involve absorption of light across the bandgap or between quantized states in one band, transport of excited charge carriers, and their collection at electrodes (see Ref. 18.5). Some common detection schemes are as follows.

A *photoconductive detector* is based on the principle that light-induced carrier generation results in a change in resistance (and conductivity) of the material. Photons with energies above the gap are absorbed, generating electrons and holes in the conduction and valence bands, respectively. The density of photogenerated carriers by an incident beam of intensity I_0 and photon energy $\hbar\omega$ is given by Eq. (14.5) as

$$n = \frac{\alpha I_0 \tau}{\hbar\omega} , \tag{18.1}$$

where α is the absorption coefficient, and τ is the carrier lifetime. The change in conductivity resulting from photogenerated carriers is then

$$\Delta\sigma = ne(\mu_e' + \mu_h') , \tag{18.2}$$

with μ_e' and μ_h' being the electron and hole mobilities. The densities of the photogenerated electrons and holes are the same, but their mobilities are different. This photoinduced conductivity change results in a resistance change that can be detected in a circuit such as that shown in Fig. 18.9. A load resistor, R_L, is connected in series with the photoconductor device which is dc biased at V_0. The voltage drop across the load resistor is

$$V_L = \frac{V_0 R_L}{R_L + R_{pc}} , \tag{18.3}$$

with R_{pc} being the photoconductive device impedance. The incident photon beam is detected by measuring the change in this voltage, dV_L, due to the change in photoconductor impedance dR_{pc} as

$$dV_L = \frac{-V_0 R_L \, dR_{pc}}{(R_L + R_{pc})^2} . \tag{18.4}$$

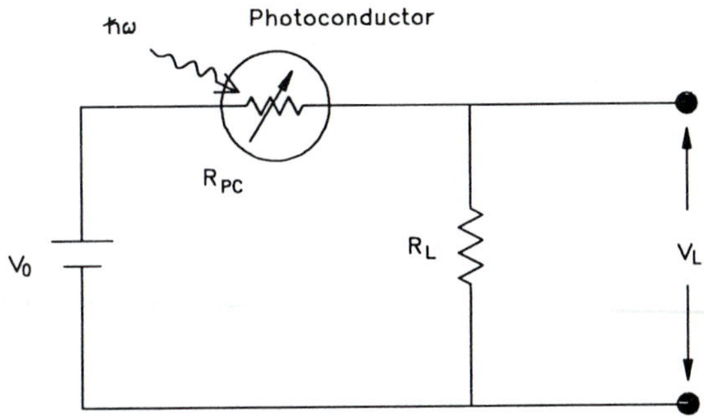

Figure 18.9. A typical circuit for a photoconductive mode detection scheme.

Therefore, the photodetector is operated in a mode in which photoexcitation modulates the current through the circuit.

The spectral responsivity of the detector is determined by the absorption spectrum of the semiconductor. The spectral responsivity of a specific semiconductor may be changed by doping the material and thus allowing absorption involving the impurity levels. The detector response is extended to longer wavelengths using this scheme. For example, for a p-doped semiconductor, electrons may be excited from the valence band to the acceptor level. The electrons become trapped in the acceptor level and the holes are the only mobile carriers. For an n-type material, electrons are the mobile carriers and holes are trapped at the donor levels.

The *photovoltaic* detection mode is based on the generation of a voltage across an *unbiased* p-n junction as a result of photo-excitation. Consider the p-n junction photodiode of Fig. 18.10. Electrons and holes are generated when the junction is illuminated with photons of energy equal to or greater than the bandgap. The contribution to the current from electron and hole transport adds up because of their opposite charges (the motion of electrons and holes is in opposite direction). The equivalent circuit for the device is shown in Fig. 18.11. When the system is in the dark with no incident radiation, the voltage current (I-V) curve is given by (see Prob. 18.1)

$$I = I_s(e^{eV/k_B T} - 1) , \tag{18.5}$$

where I_s is the diode saturation current or the reverse bias leakage current, and V is the voltage drop across the diode. This equation is plotted in Fig. 18.12 as a solid curve. When the junction is illuminated, the photogenerated carriers lead to a generated current I_L, modifying Eq. (18.5) to

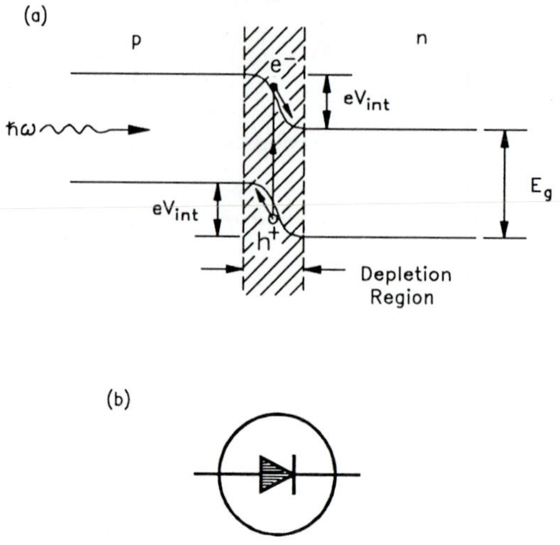

Figure 18.10. (a) Energy diagram of the p-n junction and (b) the symbol typically used for this p-n junction device.

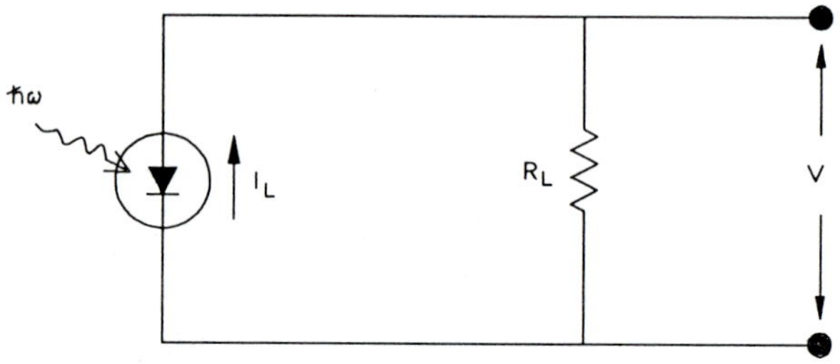

Figure 18.11. Equivalent circuit for the photovoltaic device. A load resistor R_L is added.

$$I = I_s(e^{eV/k_BT} - 1) - I_L \ . \tag{18.6}$$

The I-V curve for the device in the presence of light is shown by the dashed curve in Fig. 18.12. In the short circuit mode, V = 0, the photocurrent varies linearly with the irradiance so the incident light results in constant detector impedance with respect to photon flux. This is in contrast to the photoconductive detectors where the impedance varies with light irradiance. For the open circuit I = 0, the open circuit voltage V_{oc} is obtained from Eq. (18.6) as

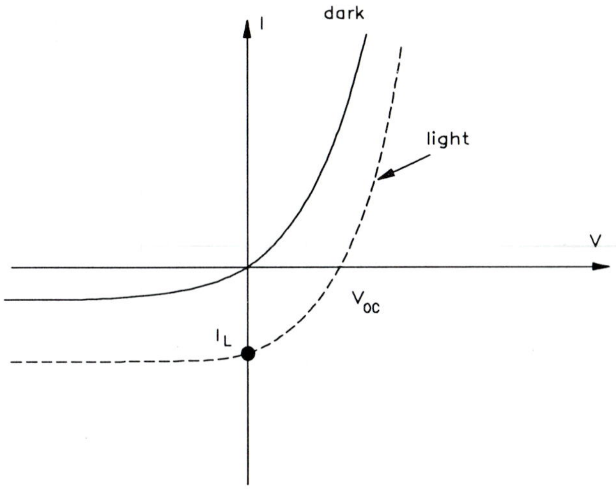

Figure 18.12. The voltage-current characteristic of the photovoltaic device with no incident light present (solid curve) and with photons incident on the device (dashed curve).

$$V_{oc} = \frac{k_B T}{e} \ell n \left[\frac{I_L}{I_s} + 1 \right] . \tag{18.7}$$

A solar cell is a photovoltaic device connected to a load so that the photonic energy is converted to electronic energy. The voltage-current characteristics of the solar cell are given by Eq. (18.6) and are plotted in Fig. 18.12. The cell operates in the quadrant where current is negative and voltage is positive. The largest photonic-to-electric power conversion takes place when the current-voltage product is maximized. A large number of solar cells are made from silicon.

A *photodiode* detector is a p-n junction operated under reverse-bias condition. Illumination of the photodiode with photons generates electrons and holes, resulting in an electric current in the external circuit. The width of the depletion region of the junction determines the response time of the detector. The thinner the depletion region, the shorter the transit time of the carriers across the region and, thus, the faster the device (since it takes less time for the carriers to reach the electrodes). However, the quantum efficiency of the detector decreases with a reduction of the depletion layer width, since a smaller number of photons are absorbed in a thinner depletion layer. To enhance the quantum efficiency, an intrinsic layer is added to the junction, making the device a p-i-n diode, as shown in Fig. 18.13. The depletion region thickness is the thickness of the

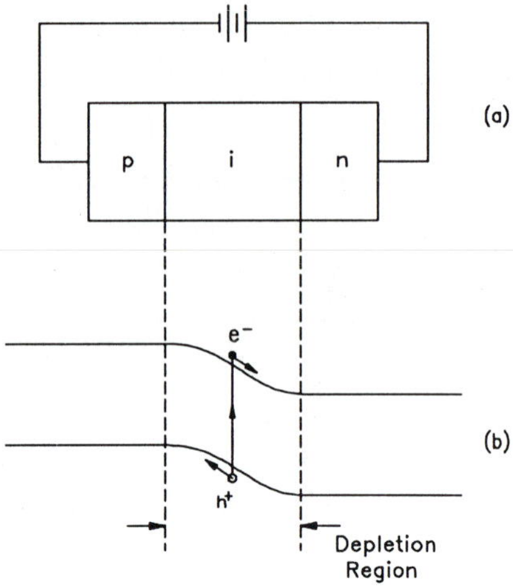

Figure 18.13. (a) Schematic representation of a p-i-n diode under reverse-bias condition. (b) Energy band diagram of the p-i-n diode.

intrinsic material which can be optimized for quantum efficiency and frequency response.

Avalanche photodiodes are p-n junctions operating under a large reverse-bias voltage. The large bias field accelerates the photogenerated charge carriers to the point at which their acquired kinetic energy is sufficient to promote other electrons from the valence band to the conduction band via collision. The process of excitation of additional electron-hole pairs by the photogenerated carriers is referred to as the impact ionization. This process is schematically shown in Fig. 18.14. The incident photon generates the first electron-hole pair (labeled 1). The electron accelerates down the depletion region. After impact ionization, a second electron-hole pair (labeled 2) is excited with the first electron losing its additional energy and falling to the bottom of the conduction band. This process continues during the transit of the carriers in the depletion region. The average number of carriers produced for each incident carrier is referred to as the gain of the avalanche process. A typical avalanche photodiode structure is shown in Fig. 18.15. It is essential to obtain very uniform junctions because of the existence of the large bias field.

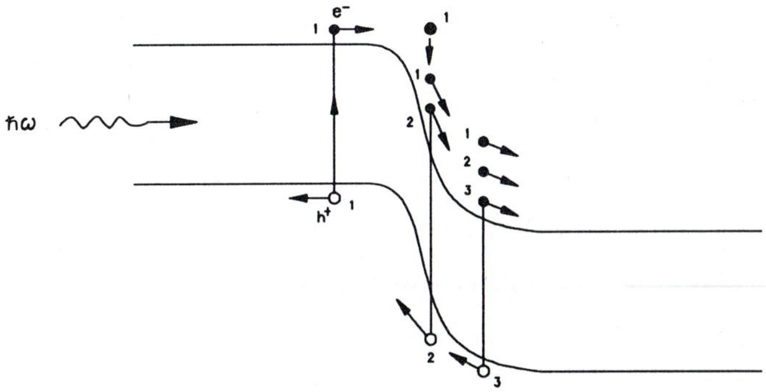

Figure 18.14. Avalanche multiplication under a large reverse bias.

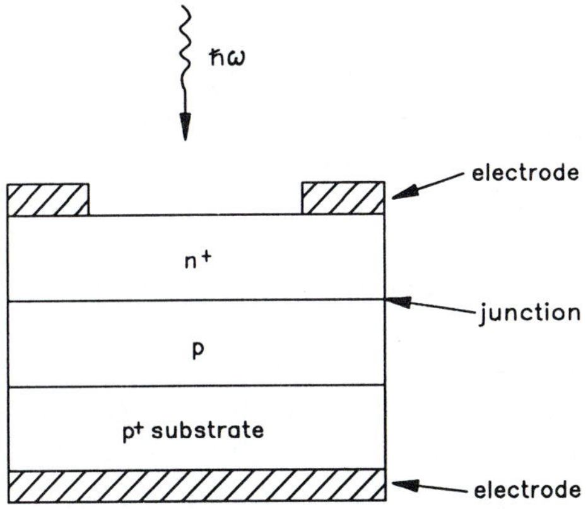

Figure 18.15. Schematic structure of a p-i-n diode.

18-4. Inter-Subband Absorption Tunneling Detectors*

Semiconductor superlattices may be used for high-speed detectors in the infrared. Consider the superlattice of Fig. 18.16 with doped GaAs wells and AlGaAs barriers. The wells are designed to have two confined states. If the lower level, $\mathcal{E}_1$, is populated by appropriate doping of the material, then an incident photon with energy $\hbar\omega$ resonant with the level separation $\mathcal{E}_2 - \mathcal{E}_1 = \hbar\omega$ excites an electron from the $\mathcal{E}_1$ to the $\mathcal{E}_2$ level. The electron may tunnel out of the well through the barrier if the barrier width is thin enough. This photogenerated carrier, with mean free path longer than the superlattice width, generates a photocurrent that can be measured at the other end of the structure. Detectors have been fabricated (see Refs. 18.8 and 18.9) using 50 periods of 70-Å-GaAs wells (doped

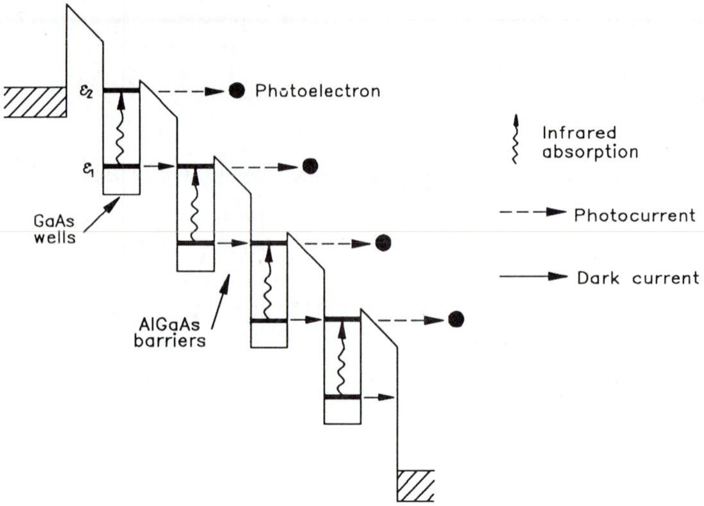

Figure 18.16. A GaAs-AlGaAs superlattice exhibiting photoconductivity, resulting from absorption of intersubband radiation followed by tunneling through the barriers (after Ref. 18.8).

with $n = 1.4 \times 10^{18}\ cm^{-3}$) and 140-Å $Al_{0.36}Ga_{0.64}As$ undoped barriers. With voltages applied to the structure, the device operated at 15 K for 10.3 μm, giving a responsivity R of R = 2 A/W at 9 V bias. The responsivity as a function of bias voltage is shown in Fig. 18.17. As shown in the figure, when the bias voltage is increased beyond 9 V, the responsivity increases, reaching 7A/W at 12 V bias because of avalanche ionization.

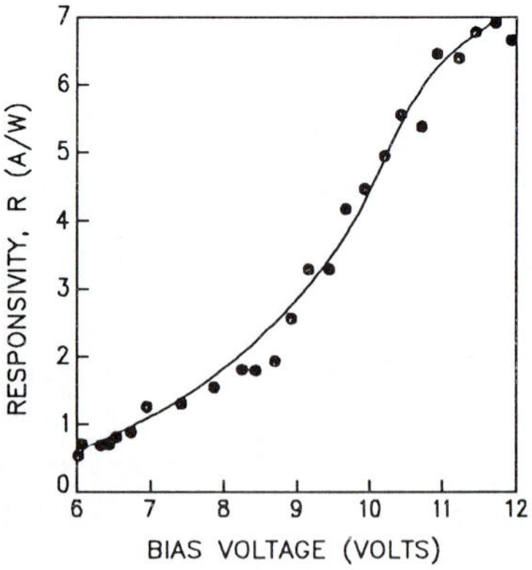

Figure 18.17. Responsivity R versus applied bias voltage. The circles are the experimental values measured at a sample temperature of 15 K. The solid line is the theory including avalanche gain (after Ref. 18.8).

The detector shown in Fig. 18.16 suffers from typically high dark currents. The narrow barriers allow the doped carriers in the $\mathcal{E}_1$ level to tunnel through in the absence of any light. This process is shown by the solid arrows in Fig. 18.16. In order to reduce the dark current, detectors from double barrier structures have been fabricated. Figure 18.18 displays the energy band diagram of such a structure. One period consists of two wells with different widths of W_1 and W_2 and two barriers of widths B_1 and B_2, with only the wider wells being doped. Under a small forward bias, the upper level, $\mathcal{E}_2$, of the wider well comes in resonance with state $\mathcal{E}'_1$ in the thinner well. The photogenerated electrons in state $\mathcal{E}_2$ of the wider well tunnel out through resonant state $\mathcal{E}'_1$ of the thinner well, generating a photocurrent at a very low bias voltage. This structure under forward bias has been shown to exhibit high quantum efficiency at low bias. The device in reverse bias would show a similar quantum efficiency, but at a much larger bias voltage. Therefore, this device has two orders of magnitude lower dark current for the same quantum efficiency compared with the device of Fig. 18.16.

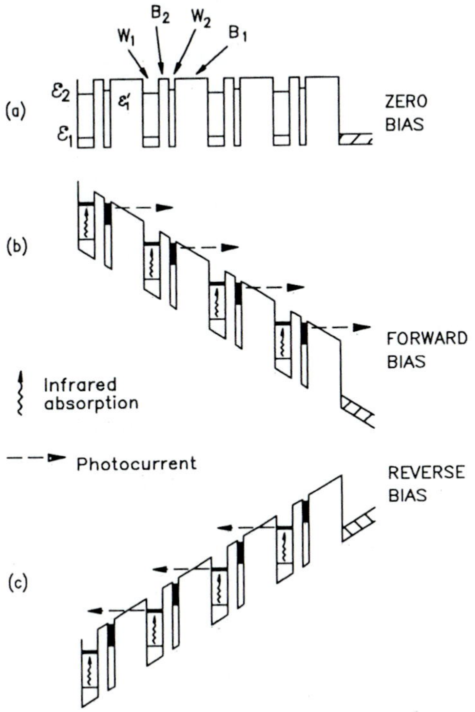

Figure 18.18. (a) The energy-band diagram of the double barrier structure at zero bias. (b) The band diagram under forward bias. (c) The band diagram under reverse bias (after Ref. 18.8).

18-5. Superlattice Photomultipliers*

Low-noise avalanche photodiodes are obtained when the ionization rate of electrons and holes are largely different. It has been demonstrated that this condition can be realized in specially designed GaAs-AlGaAs superlattices (see Refs. 18.10-18.12). An example of such a detector in the form of a p-i-n device is shown in Fig. 18.19. It consists of an intrinsic layer with 50 periods of 45-nm GaAs wells and 55-nm AlGaAs barriers between p^+ and n^+ materials. Under bias, the conduction band discontinuity ΔE_c is enhanced to $\Delta E_c = 0.48$ eV, while valence band discontinuity ΔE_v is lowered to only $\Delta E_v = 0.08$ eV. Therefore, electrons passing across the barrier and reaching the well gain 0.48 eV of kinetic energy and experience a larger ionization probability than holes. The sum of the band discontinuity and the acceleration by the bias field must be larger than the bandgap for impact ionization to take place; the larger the band discontinuity, the more probable the impact ionization.

The hot electrons may also emit phonons and lose the phonon energy of $\simeq$ 30 meV in the process. Electrons need to gain enough average energy between collisions from the bias field to compensate for the energy lost by phonon scattering. Typically, electrons need to cross several barriers and wells before acquiring sufficient energy. Thus, the multiplication process

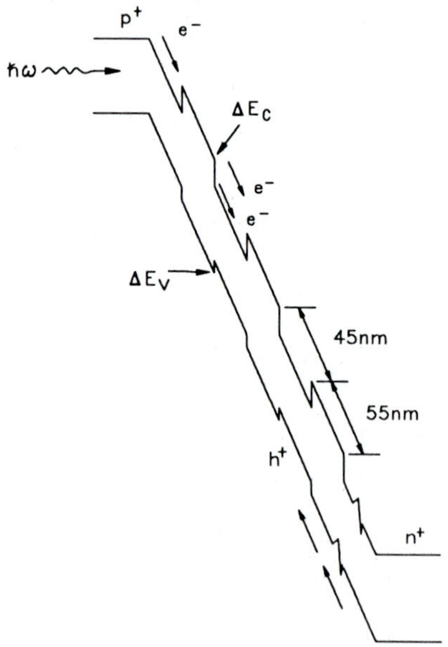

Figure 18.19. Energy band diagram of a superlattice avalanche photodiode under bias. The impact ionization of electrons and holes are different (after Ref. 18.10).

can take place everywhere in the avalanche region and becomes a random statistical process in the sense that only a few ballistic electrons gain the required energy over one superlattice period. Furthermore, once the ionization has taken place, electrons inside the well need to gain enough energy from the bias field to cross the barrier; i.e., their kinetic energy must be larger than the band discontinuity. Some electrons may get trapped inside the wells and, consequently, those electrons do not contribute to conductivity. In order to alleviate the problems associated with trapping and the randomness of the ionization process, the structure of Fig. 18.20 has been proposed (see Ref. 18.12). Here the composition of the multilayer material is linearly graded at each stage; i.e., the value x in $Al_xGa_{1-x}As$ is linearly changed, as shown in Fig. 18.20(a), to gradually increase the barrier bandgap. Under reverse bias, this structure assumes a staircase shape [see Fig. 18.20(b)]. The idea is to make the conduction band discontinuity equal to the electron ionization energy. Electrons gain all the kinetic energy for the ionization at each step, while holes do not ionize, since the bias field is too low for them to cause ionization. In the graded region following the conduction band discontinuity, the electrons drift to the next step under the influence of the applied bias field.

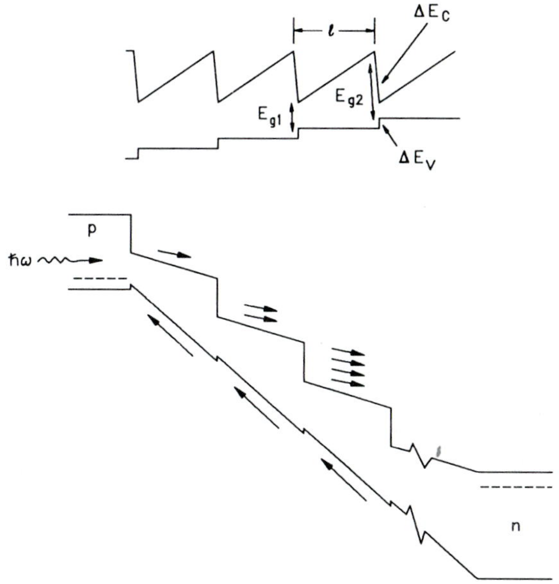

Figure 18.20. (a) Energy band diagram of an unbiased graded multilayer superlattice. (b) The energy band picture for a staircase avalanche photodiode under reverse-bias condition. The electrons impact ionize while the holes do not (after Ref. 18.12).

This structure should eliminate the problem associated with the randomness of the ionization process, as is the case for the device of Fig. 18.19, since ionization takes place at each step with a very high probability. It should also eliminate the problem associated with trapping of electrons in the wells, because after ionization the electrons do not need to cross a potential barrier (the conduction band discontinuity), drifting without ionization through the graded portion to the next step. For a device consisting of m steps, the total gain is $(2 - \delta g)^m$, where δg is the fraction of electrons not experiencing ionization. With δg being small, the gain per stage is nearly 2 and the device behaves like a photomultiplier tube where the series of dynodes cause electron multiplication. This device is the solid state analog of the photomultiplier tube and is referred to as the staircase avalanche photodiode.

The bias voltage of the staircase avalanche photodiode is about 5 volts because all the energy required for ionization is gained at the band discontinuity at each stage. The device has a theoretical bandwidth of $\simeq$ 600 GHz. The noise of the staircase avalanche photodiode is also low because the avalanche process is not random, as discussed.

18-6. Problems

18.1. A p-n junction is assumed to be biased at a voltage V_{ext} with a built-in potential V_{int}. V_{ext} is positive for forward bias and negative for reverse bias. Derive the voltage-current curve for the junction using the following procedure.

(a) Show that the built-in potential can be written as

$$V_{int} = \frac{k_B T}{e} \ell n \frac{n_{ne}}{n_{pe}}, \tag{18.8}$$

where n_{ne} and n_{pe} are the densities of the electrons in the n and p regions, respectively, at thermal equilibrium.

(b) In the presence of the bias voltage show

$$n_n = n_p \, e^{e(V_{int} - V_{ext})/k_B T}, \tag{18.9}$$

where n_n and n_p are the nonequilibrium electron densities in the presence of the bias field at the boundaries of the depletion region in n and p sides, respectively.

(c) For low-injection conditions, where $n_n \simeq n_{ne}$, show that

$$n_p - n_{pe} = n_{pe}\,(e^{eV_{ext}/k_B T} - 1) \tag{18.10}$$

and

$$p_n - p_{ne} = p_{ne}\,(e^{eV_{ext}/k_B T} - 1)\,, \tag{18.11}$$

with p_n and p_{ne} being the density of holes in the n region under nonequilibrium and thermal equilibrium conditions, respectively.

(d) Solve the steady-state continuity equation

$$\frac{d^2 p_n}{dx^2} - \frac{p_n - p_{ne}}{D_p\,\tau_p} = 0\,, \tag{18.12}$$

with appropriate boundary conditions to show

$$p_n - p_{ne} = p_{ne}\,(e^{eV_{ext}/k_B T} - 1)\,e^{-(x - x_n)/L_p}\,, \tag{18.13}$$

where $L_p = \sqrt{D_p\,\tau_p}$ is the diffusion length of holes in the n region. The boundary of the depletion layer in the n region is x_n, D_p is the diffusion constant for holes, and τ_p is the hole lifetime.

(e) At $x = x_n$ show that the hole current is

$$J_p(x_n) = -eD_p \left.\frac{dp_n}{dx}\right|_{x = x_n} = \frac{eD_p\,p_{ne}}{L_p}\,(e^{eV_{ext}/k_B T} - 1)\,. \tag{18.14}$$

(f) Show that the total current through the device is

$$J = J_p(x_n) + J_n(-x_p) = J_s\,(e^{eV_{ext}/k_B T} - 1)\,, \tag{18.15}$$

with

$$J_s = \frac{e\,D_p\,p_{ne}}{L_p} + \frac{e\,D_n\,n_{pe}}{L_n}\,, \tag{18.16}$$

where x_p is the boundary of the depletion layer in the p side, $L_n = \sqrt{D_n \tau_n}$ is the diffusion length of electrons, D_n is the diffusion constant for electrons, and τ_n is the electron lifetime.

18.2. Show that the impedance of the photovoltaic detectors ($R_{pv} = \partial V/\partial I$) is independent of photon flux for the detector being in the short circuit mode.

18.3. Calculate the maximum power conversion for a photovoltaic detector. Note that the conversion power, given by $P = VI$, is maximized when $dP/dV = 0$.

18-7. References

1. D. A. B. Miller, in *Optical Switching in Low-Dimensional Systems*, edited by H. Haug and L. Banyai (Plenum Press, New York, 1989) pp. 1-8; see also, S. Schmitt-Rink, D. S. Chemla, and D. A. B. Miller, Adv. Phys. **38**, 89 (1989); see also, D. A. B. Miller and D. S. Chemla, in *Optical Nonlinearities and Instabilities in Semiconductors*, edited by H. Haug (Academic Press, Inc., Boston, 1988) pp. 325-359.

2. K. Wakita, I. Kotaka, O. Mitomi, H. Asai, Y. Kawamura, and M. Naganuma, J. Lightwave Technol. **8**, 1027 (1990).

3. G. D. Boyed et al., Electron. Lett. **25**, 558 (1989).

4. J. E. Zucker, K. L. Jones, M. G. Young, B. I. Miller, and U. Koren, Appl. Phys. Lett. **55**, 2280 (1989).

5. E. L. Dereniak and D. G. Crowe, *Optical Radiation Detectors* (John Wiley & Sons, New York, 1984).

6. D. Hartman and C. C. Shen, in *Gallium Arsenide Technology*, edited by D. K. Ferry (Howard W. Sams & Co., Inc., Indianapolis, 1985) pp. 343-442.

7. A. Yariv, *Optical Electronics* (Saunders College Publishing, Philadelphia, 1991).

8. B. F. Levine, K. K. Choi, C. G. Bethea, J. Walker, and R. J. Malik, in *Optical Computing and Nonlinear Materials*, edited by N. Peyghambarian, Proc. SPIE Conference, Los Angeles, CA, Vol. 881 (SPIE, Bellingham, Washington, 1988) pp. 74-79.

9. K. K. Choi, B. F. Levine, C. G. Bethea, J. Walker, and R. J. Malik, Phys. Rev. Lett. **59**, 1459 (1987).

10. F. Capasso, W. T. Tsang, A. L. Hutchinson, and G. F. Williams, Appl. Phys. Lett. **40**, 38 (1982).

11. G. F. Williams, F. Capasso, and W. T. Tsang, IEEE Trans. Electron. Devices **3**, 71 (1982).

12. F. Capasso, in *Gallium Arsenide Technology*, edited by D. K. Ferry (Howard W. Sams & Co., Inc., Indianapolis, 1985) pp. 303-330.

Appendix
Conversion of Units

In this book, we have used the CGS-Gaussian system. Distances are expressed in centimeters, mass in grams, time in seconds, and the magnetic field is expressed in gauss. In the legal MKS system (meter, kilogram, second), Maxwell equations are written as follows:

	CGS	MKS
1)	$\nabla \cdot E = 4\pi e(n_i - n_e)$	$\nabla \cdot D = e(n_i - n_e)$
2)	$\nabla \cdot B = 0$	$\nabla \cdot B = 0$
3)	$\nabla \times E = -\frac{1}{c}\frac{\partial B}{\partial t}$	$\nabla \times E = -\frac{\partial B}{\partial t}$
4)	$\nabla \times B = \frac{4\pi}{c}j + \frac{1}{c}\frac{\partial E}{\partial t}$	$\nabla \times H = j + \frac{\partial D}{\partial t}$
	$\epsilon_0 = \mu_0 = 1$	$D = \epsilon_0 E$
		$B = \mu_0 H$

Note that E/B is without dimension in CGS, but has dimension of velocity in MKS. Further, the following conversions are useful to know:

$$\frac{1}{\epsilon_0} = 36\pi \cdot 10^9 \qquad C = (\epsilon_0\mu_0)^{-1/2} = 3 \cdot 10^8 \text{ m/s}$$

Table of Conversion between CGS and MKS units (after L. Cooper, *An Introduction to the Meaning and Structure of Physics*, Harper and Row)

	CGS	MKS	
length	[cm]	m	1 m = 10^2 cm
mass	[g]	kg	1 kg = 10^3 g
time	[s]	s	
force	dyne $\left[g \cdot \frac{cm}{s^2}\right]$	Newton	1 Newton = 10^5 dynes
energy	erg $\left[g \cdot \frac{cm^2}{sec^2}\right]$	joule	1 joule = 10^7 ergs
power	erg/s	watt	1 watt = 10^7 ergs/s

Table of Conversion (continued)

	CGS	MKS	
charge	electrostatic unit (esu)	Coulomb	1 Coulomb = $3 \cdot 10^9$ esu
current	statampere $\left[\frac{\text{esu}}{\text{s}}\right]$	ampere	1 ampere = $3 \cdot 10^9$ esu/s
electric potential	statvolt $\left[\frac{\text{erg}}{\text{esu}}\right]$	volt	1 volt = $\frac{1}{300}$ statvolt
electric field	$\frac{\text{statvolt}}{\text{cm}}$ $\left[\frac{\text{dyne}}{\text{esu}}\right]$	$\frac{\text{volt}}{\text{m}}$	$\frac{1\ \text{volt}}{\text{m}} = \frac{1}{3} \cdot 10^{-4}\ \frac{\text{statvolt}}{\text{cm}}$
resistance	[s/cm]	ohm	1 ohm = $1.11 \cdot 10^{-12}$ s/cm
magnetic field	gauss $\left[\frac{\text{dyne}}{\text{esu}}\right]$	tesla	1 tesla = 10^4 gauss

Glossary of Symbols

This table contains most of the symbols used in this book. We tried to avoid duplication of symbols as much as possible. Symbols with obvious meanings are not listed.

$\mathbf{A}$	vector potential
a	lattice constant
a_B	Bohr radius
a	creation operator
a^+	annihilation operator
C_d	contrast of an optical logic gate
c	speed of light
$\mathbf{D}$	displacement vector
D	diffusion coefficient
$\mathbf{d}$	dipole moment
d	sample thickness
d_{cv}	dipole matrix element between valence and conduction bands
$d_{m\ell}$	dipole matrix element for transition between states ℓ and m
$d\mathbf{k}$	volume element in momentum space
$d\mathbf{r}$	volume element in real space
$\mathbf{E}$	electric field
$\mathbf{E}_0$	constant electric field

E_a	acceptor binding energy
E_B	exciton binding energy
E_{Bxx}	biexciton binding energy
E_d	donor binding energy
E_g	direct bandgap energy
E'_g	renormalized bandgap energy
E_g^{ind}	indirect bandgap energy
E_{XL}	longitudinal exciton energy
E_{XT}	transverse exciton energy
$\hat{e}$	polarization unit
$\mathcal{E}$	energy
$\mathcal{E}_{bx}$	bound exciton energy
$\mathcal{E}_c$	conduction band energy
$\mathcal{E}_D$	binding energy of exciton to an impurity
$\mathcal{E}_e$	electron energy
$\mathcal{E}_F$	Fermi energy
$\mathcal{E}_h$	hole energy
$\mathcal{E}_{kin}$	kinetic energy
$\mathcal{E}_L$	binding energy of an electron-hole pair in a liquid drop
$\mathcal{E}_n$	energy for various orders or modes
$\mathcal{E}_v$	valence band energy
$\mathcal{E}_{pot}$	potential energy
$\mathcal{E}_{xc}$	sum of the exchange and correlation energies in a liquid drop
$\mathcal{E}_{xx}$	biexciton energy
$\mathcal{E}_x^{(0)}$	exciton energy, E_g - E_B
$\mathcal{E}(n_L)$	ground state energy of an electron-hole pair inside a droplet
$\mathbf{F}$	force
$F(\omega)$	photon flux

f	spring force constant
$f(\mathcal{E})$	electron distribution function
G	reciprocal lattice vector
G_D	gain of an optical logic gate
G(n)	gain rate
$g(\mathcal{E})$	density of states
g(k)	density of states in momentum space
H	Hamiltonian
H_{int}	interaction Hamiltonian
$(h\ k\ \ell)$	Miller indices
$\hbar$	Planck constant
I	current
I_0	constant incident intensity
I_{in}	light intensity inside Fabry-Perot resonator
I_t	transmitted intensity
I(x)	space-dependent intensity
i	complex unit ($i^2 = -1$)
J	current density
K	phonon wave vector
K	thermal conductivity
$\mathbf{K}_c$	center of mass wave vector
k	carrier wave vector
k_B	Boltzmann constant
$\mathbf{k}_e$	electron wave vector

$\mathbf{k}_h$	hole wave vector
$\mathbf{k}_F$	Fermi wave vector
$\mathbf{k}_x$	exciton wave vector
$\mathbf{k}_{xx}$	biexciton wave vector
L_D	diffusion length
M	total mass of electron-hole pair, exciton
M′	atomic weight of elements
M_1, M_2	atomic masses
M_{xx}	biexciton mass
m_0	gravitational mass of a free electron
m_e	electron effective mass
m_h	hole effective mass
m_{hh}	heavy-hole mass
m_{ij}	components of the effective mass tensor
m_ℓ	longitudinal effective mass
$m_{\ell h}$	light-hole mass
m_r	electron-hole reduced mass
m_s	spin quantum number of electrons
m_t	transverse effective mass
N	number of electrons, atoms, ions, unit cells, photons, etc.
N_L	total number of electron-hole pairs inside a liquid drop
N_{xx}	biexciton energy distribution
n	electron density
$n(\omega)$	index of refraction
$n'(\omega)$	real part of refractive index
$n''(\omega)$	imaginary part of refractive index

n_0	mean electron density
$n(\mathbf{r})$	real space dependence of electron density
$n(\mathbf{k})$	momentum space dependence of electron density
n_a	acceptor density
n_b	background index of refraction
n_d	donor density
n_i	intrinsic carrier density
n_K	phonon population
n_L	equilibrium density of electron-hole pairs in a liquid drop
O_h	point group notation (octahedral symmetry)
$\mathbf{P}$	polarization
$P(\mathbf{k})$	interband polarization
P_b	polarization due to bound electrons
P_c	polarization due to conduction electrons
P_e	electron contribution to polarization
P_i	ionic contribution to polarization
P_r	principal part of a complex integral
$\mathbf{p}$	momentum
p_r	pump rate
$\mathbf{q}$	photon wave vector in a medium
$\mathbf{q}_0$	photon wave vector in vacuum
$\mathbf{R}$	direct lattice vector
R_{cv}	rate of light spontaneously emitted into a laser mode
R_L	load resistor
R_{pc}	photoconductive impedance

$R(n)$	total carrier loss rate
$\mathbf{r}$	arbitrary space point in a crystal
r_s	a dimensionless parameter
S	Poynting vector
s, p, d, f	notation symbols for electronic configuration
T	temperature
T_2	relaxation time
T_c	critical temperature
T_d	point group notation (tetrahedral symmetry)
T_F	Fermi temperature
t	time
t_p	time delay between two pulses
$\mathcal{T}$	Fabry-Perot transmission
U_j	displacement of atom j from its equilibrium position
$u(\mathbf{r})$	the periodic part of the electron Bloch wave function
V	crystal volume in direct space
$V(\mathbf{k})$	Coulomb potential energy in momentum space
$V(\mathbf{r})$	Coulomb potential energy in real space
V_{conf}	confinement potential
V_k	crystal volume in momentum space
V_L	volume of the electron-hole liquid drop
V_{oc}	open circuit voltage of a photovoltaic device
$V_s(\mathbf{k})$	screened Coulomb potential in momentum space
$V_s(\mathbf{r})$	screened Coulomb potential in real space
$\boldsymbol{v}$	electron velocity

$\bar{v}$	average velocity
$\mathbf{v_g}$	group velocity
v_s	speed of sound
W	periodic crystal potential
W_m	expansion coefficient for lattice potential
W_0	average potential
$W_{kk'}$	interaction matrix element
w	transition probability
$w_{m\ell}$	transition probability from state ℓ to state m
y	Young's modulus
Z	number of valence electrons in an atom
$\alpha(\omega)$	absorption coefficient
β	two-photon absorption coefficient
Γ	point group symmetry notation
γ	damping constant
δ'	two-photon cross section per unit cell
$\delta(x-y)$	delta function
δ_{ij}	Kronecker symbol
$\epsilon(\omega)$	dielectric function
$\epsilon'(\omega)$	real part of the dielectric function
$\epsilon''(\omega)$	imaginary part of the dielectric function
ϵ_{ij}	elements of the dielectric tensor
ϵ_0	dielectric constant at low frequencies

ϵ_∞	dielectric constant at high frequencies
$\theta(x)$	Heavyside function
κ	screening length
κ_ℓ	losses of a laser mode
λ	wavelength in a medium
λ_0	wavelength in vacuum
λ_p	plasma wavelength
μ	chemical potential
μ'	carrier mobility
$\nu_{e,h}$	number of equivalent valleys in conduction and valence bands
ξ	surface plasmon frequency
ρ_m	mass density
ρ_c	charge density
σ	ac conductivity
σ_0	dc conductivity
σ'	real part of ac conductivity
σ''	imaginary part of ac conductivity
τ	carrier lifetime
ϕ	phase shift
$\phi(\mathbf{r})$	wave function
$\Phi(\mathbf{r})$	electrostatic potential
$\Phi(\mathbf{r}_e,\mathbf{r}_h)$	exciton envelope function
χ	optical susceptibility
$\psi(\mathbf{r})$	electronic Bloch wave function
Ω	phonon frequency
Ω_c	volume of unit cell
Ω_L	longitudinal phonon frequency
Ω_T	transverse phonon frequency
ω	angular frequency

ω_{mn}	difference frequency, $\omega_m - \omega_n$
ω_p	plasma frequency
ω_R	Rabi frequency

INDEX